T0228484

continua

with the Houston problem book

PURE AND APPLIED MATHEMATICS

A Program of Monographs, Textbooks, and Lecture Notes

EXECUTIVE EDITORS

Earl J. Taft
Rutgers University
New Brunswick, New Jersey

Zuhair Nashed
University of Delaware
Newark, Delaware

EDITORIAL BOARD

M. S. Baouendi
University of California,
San Diego

Jane Cronin
Rutgers University

Jack K. Hale
Georgia Institute of Technology

S. Kobayashi
University of California,
Berkeley

Marvin Marcus
University of California,
Santa Barbara

W. S. Massey
Yale University

Anil Nerode
Cornell University

Donald Passman
University of Wisconsin,
Madison

Fred S. Roberts
Rutgers University

Gian-Carlo Rota
Massachusetts Institute of
Technology

David L. Russell
Virginia Polytechnic Institute
and State University

Walter Schempp
Universität Siegen

Mark Teply
University of Wisconsin,
Milwaukee

LECTURE NOTES IN PURE AND APPLIED MATHEMATICS

1. *N. Jacobson*, Exceptional Lie Algebras
2. *L.-Å. Lindahl and F. Poulsen*, Thin Sets in Harmonic Analysis
3. *I. Satake*, Classification Theory of Semi-Simple Algebraic Groups
4. *F. Hirzebruch, W. D. Newmann, and S. S. Koh*, Differentiable Manifolds and Quadratic Forms
5. *I. Chavel*, Riemannian Symmetric Spaces of Rank One
6. *R. B. Burckel*, Characterization of C(X) Among Its Subalgebras
7. *B. R. McDonald, A. R. Magid, and K. C. Smith*, Ring Theory: Proceedings of the Oklahoma Conference
8. *Y.-T. Siu*, Techniques of Extension on Analytic Objects
9. *S. R. Caradus, W. E. Pfaffenberger, and B. Yood*, Calkin Algebras and Algebras of Operators on Banach Spaces
10. *E. O. Roxin, P.-T. Liu, and R. L. Sternberg*, Differential Games and Control Theory
11. *M. Orzech and C. Small*, The Brauer Group of Commutative Rings
12. *S. Thomier*, Topology and Its Applications
13. *J. M. Lopez and K. A. Ross*, Sidon Sets
14. *W. W. Comfort and S. Negrepontis*, Continuous Pseudometrics
15. *K. McKennon and J. M. Robertson*, Locally Convex Spaces
16. *M. Carmeli and S. Malin*, Representations of the Rotation and Lorentz Groups: An Introduction
17. *G. B. Seligman*, Rational Methods in Lie Algebras
18. *D. G. de Figueiredo*, Functional Analysis: Proceedings of the Brazilian Mathematical Society Symposium
19. *L. Cesari, R. Kannan, and J. D. Schuur*, Nonlinear Functional Analysis and Differential Equations: Proceedings of the Michigan State University Conference
20. *J. J. Schäffer*, Geometry of Spheres in Normed Spaces
21. *K. Yano and M. Kon*, Anti-Invariant Submanifolds
22. *W. V. Vasconcelos*, The Rings of Dimension Two
23. *R. E. Chandler*, Hausdorff Compactifications
24. *S. P. Franklin and B. V. S. Thomas*, Topology: Proceedings of the Memphis State University Conference
25. *S. K. Jain*, Ring Theory: Proceedings of the Ohio University Conference
26. *B. R. McDonald and R. A. Morris*, Ring Theory II: Proceedings of the Second Oklahoma Conference
27. *R. B. Mura and A. Rhemtulla*, Orderable Groups
28. *J. R. Graef*, Stability of Dynamical Systems: Theory and Applications
29. *H.-C. Wang*, Homogeneous Branch Algebras
30. *E. O. Roxin, P.-T. Liu, and R. L. Sternberg*, Differential Games and Control Theory II
31. *R. D. Porter*, Introduction to Fibre Bundles
32. *M. Altman*, Contractors and Contractor Directions Theory and Applications
33. *J. S. Golan*, Decomposition and Dimension in Module Categories
34. *G. Fairweather*, Finite Element Galerkin Methods for Differential Equations
35. *J. D. Sally*, Numbers of Generators of Ideals in Local Rings
36. *S. S. Miller*, Complex Analysis: Proceedings of the S.U.N.Y. Brockport Conference
37. *R. Gordon*, Representation Theory of Algebras: Proceedings of the Philadelphia Conference
38. *M. Goto and F. D. Grosshans*, Semisimple Lie Algebras
39. *A. I. Arruda, N. C. A. da Costa, and R. Chuaqui*, Mathematical Logic: Proceedings of the First Brazilian Conference
40. *F. Van Oystaeyen*, Ring Theory: Proceedings of the 1977 Antwerp Conference
41. *F. Van Oystaeyan and A. Verschoren*, Reflectors and Localization: Application to Sheaf Theory
42. *M. Satyanarayana*, Positively Ordered Semigroups
43. *D. L Russell*, Mathematics of Finite-Dimensional Control Systems
44. *P.-T. Liu and E. Roxin*, Differential Games and Control Theory III: Proceedings of the Third Kingston Conference, Part A
45. *A. Geramita and J. Seberry*, Orthogonal Designs: Quadratic Forms and Hadamard Matrices
46. *J. Cigler, V. Losert, and P. Michor*, Banach Modules and Functors on Categories of Banach Spaces

Additional Volumes in Preparation

continua

with the Houston problem book

edited by

Howard Cook
University of Houston
Houston, Texas

W.T. Ingram
University of Missouri–Rolla
Rolla, Missouri

K.T. Kuperberg
Auburn University
Auburn, Alabama

Andrew Lelek
University of Houston
Houston, Texas

Piotr Minc
Auburn University
Auburn, Alabama

CRC Press
Taylor & Francis Group
Boca Raton London New York

CRC Press is an imprint of the
Taylor & Francis Group, an **informa** business

Library of Congress Cataloging-in-Publication Data

Continua : with the Houston problem book / [edited by] Howard Cook...
   [et al.].
       p.   cm. — (Lecture notes in pure and applied mathematics ;
   170)
     Includes bibliographical references and index.
     ISBN 0-8247-9650-0 (pbk. : acid-free)
     1. Continuum mechanics.   2. Continuum mechanics—Problems,
   exercises, etc.   I. Cook, Howard.   II. Series: Lecture
   notes in pure and applied mathematics ; v. 170.
   QA808.2.C687   1995
   514'.3—dc20   95-4002
                                                                                        CIP

CRC Press
Taylor & Francis Group
6000 Broken Sound Parkway NW, Suite 300
Boca Raton, FL 33487-2742

© 1995 by Taylor & Francis Group, LLC

CRC Press is an imprint of Taylor & Francis Group, an Informa business

First issued in hardback 2017

No claim to original U.S. Government works

ISBN-13: 978-0-8247-9650-1 (pbk)
ISBN-13: 978-1-138-43028-0 (hbk)

This book contains information obtained from authentic and highly regarded sources.
Reasonable efforts have been made to publish reliable data and information, but the author and
publisher cannot assume responsibility for the validity of all materials or the consequences of
their use. The authors and publishers have attempted to trace the copyright holders of all
material reproduced in this publication and apologize to copyright holders if permission to
publish in this form has not been obtained. If any copyright material has not been acknowledged
please write and let us know so we may rectify in any future reprint.

Except as permitted under U.S. Copyright Law, no part of this book may be reprinted,
reproduced, transmitted, or utilized in any form by any electronic, mechanical, or other means,
now known or hereafter invented, including photocopying, microfilming, and recording, or in
any information storage or retrieval system, without written permission from the publishers.

For permission to photocopy or use material electronically from this work, please access www.
copyright.com (http://www.copyright.com/) or contact the Copyright Clearance Center, Inc.
(CCC), 222 Rosewood Drive, Danvers, MA 01923, 978-750-8400. CCC is a not-for-profit
organization that provides licenses and registration for a variety of users. For organizations that
have been granted a photocopy license by the CCC, a separate system of payment has been arranged.

Trademark Notice: Product or corporate names may be trademarks or registered trademarks,
and are used only for identification and explanation without intent to infringe.

Visit the Taylor & Francis Web site at
http://www.taylorandfrancis.com

and the CRC Press Web site at
http://www.crcpress.com

PREFACE

This book came about as a result of the Special Session "Modern Methods in Continuum Theory" organized by K. T. Kuperberg and Piotr Minc, and held within the 100th Annual Joint Mathematics Meetings in Cincinnati, Ohio, January 12-15, 1994. It is a collection of papers whose authors were invited to participate in the Session and submit their work to the proceedings. Some of the participants were also asked to write expository and survey papers included in **Part I**, whereas **Part II** contains research papers. We thank all of the contributors for their input.

A special feature of the Session and this volume is the "Houston Problem Book," a recently updated set of problems accumulated over several years at the University of Houston by its editors, Howard Cook, W. T. Ingram, and Andrew Lelek. A brief history precedes the collection of over 200 problems.

We are extending our gratitude to George Kozlowski, Head of the Department of Mathematics at Auburn University, for arranging the necessary resources. We are thankful to Robert Henderson for producing the first TEX version of the "Houston Problem Book," and to Lynn Morey for her help with mathematical proofreading and formating the papers. The volume was typeset using the $\mathcal{AMS}$-TEX and the $\mathcal{AMS}$-LATEX macros. The Special Session was supported in part by an NSF Alabama EPSCoR travel grant awarded to the organizers.

Howard Cook
W. T. Ingram
K. T. Kuperberg
Andrew Lelek
Piotr Minc

CONTENTS

CONTRIBUTORS

JAN M. AARTS Technical University Delft, Faculty of Mathematics and Informatics, 2600 GA Delft, The Netherlands

KATHLEEN T. ALLIGOOD Department of Mathematical Sciences, George Mason University, Fairfax, Virginia 22030, USA

MARCY BARGE Department of Mathematics, Montana State University, Bozeman, Montana 59717, USA

DAVID P. BELLAMY Department of Mathematical Sciences, University of Delaware, Newark, Delaware 19716, USA

LOUIS BLOCK Department of Mathematics, University of Florida, Gainesville, Florida 32611, USA

KAREN M. BRUCKS Department of Mathematical Sciences, University of Wisconsin at Milwaukee, Milwaukee, Wisconsin 53201, USA

ALEX CHIGOGIDZE Department of Mathematics and Statistics, University of Saskatchewan, Saskatoon, Saskatchewan, S7N 0W0, Canada

HOWARD COOK Department of Mathematics, University of Houston, Houston, Texas 77204, USA

BEVERLY DIAMOND Department of Mathematics, University of Charleston, Charleston, South Carolina 29424, USA

BERND GÜNTHER Fachbereich Mathematik, Johann Wolfgang Goethe-Universität, 60054 Frankfurt, Germany

CHARLES L. HAGOPIAN California State University at Sacramento, Sacramento, California 95819, USA

JO W. HEATH Department of Mathematics, Auburn University, Auburn, Alabama 36849, USA

W. T. INGRAM Department of Mathematics and Statistics, University of Missouri-Rolla, Rolla, Missouri 65401, USA

ZBIGNIEW KARNO Institute of Mathematics, Warsaw University at Białystok, 15-267 Białystok 1, Poland, and Institute of Mathematics, Polish Academy of Sciences, 00-950 Warsaw, Poland

HISAO KATO Institute of Mathematics, University of Tsukuba, Tsukuba-City, Ibaraki 305, Japan

KAZUHIRO KAWAMURA Institute of Mathematics, University of Tsukuba, Tsukuba-City, Ibaraki 305, Japan

JUDY A. KENNEDY Department of Mathematical Sciences, University of Delaware, Newark, Delaware 19716, USA

KRYSTYNA KUPERBERG Department of Mathematics, Auburn University, Auburn, Alabama 36849, USA

WŁODZIMIERZ KUPERBERG Department of Mathematics, Auburn University, Auburn, Alabama 36849, USA

ANDREW LELEK Department of Mathematics, University of Houston, Houston, Texas 77204, USA

JOE MARTIN Department of Mathematics (Emeritus), University of Wyoming, Laramie, Wyoming 82071 , USA

JOHN C. MAYER Department of Mathematics, University of Alabama at Birmingham, Birmingham, Alabama 35294, USA

JERZY MIODUSZEWSKI Instytut Matematyki, Uniwersytet Śląski, 40-007 Katowice, Poland

PIOTR MINC Department of Mathematics, Auburn University, Auburn, Alabama 36849, USA

LEX G. OVERSTEEGEN Department of Mathematics, University of Alabama at Birmingham, Birmingham, Alabama 35294, USA

JANUSZ R. PRAJS Institute of Mathematics, Opole University, 45-951 Opole, Poland

SHANNON SCHUMANN Department of Mathematics, University of Kentucky, Lexington, Kentucky 40506, USA

CARL R. SEAQUIST Department of Mathematics, Auburn University, Auburn, Alabama 36849, USA

RICHARD B. SHER Department of Mathematics, University of North Carolina at Greensboro, Greensboro, North Carolina 27412, USA

PAUL R. STALLINGS Electronic Data Systems, Maryland Heights, Missouri 63043, USA

RICHARD SWANSON Department of Mathematics, Montana State University, Bozeman, Montana 59717, USA

WILLIAM R. R. TRANSUE Department of Mathematics, Auburn University, Auburn, Alabama 36849, USA

E. D. TYMCHATYN Department of Mathematics and Statistics, University of Saskatchewan, Saskatoon, Saskatchewan, S7N 0W0, Canada

THELMA WEST Department of Mathematics, University of Southwestern Louisiana, Lafayette, Louisiana 70504, USA

Part I

Expository and Survey Papers

1

MATCHBOX MANIFOLDS

JAN M. AARTS Technical University Delft, Faculty Mathematics and Informatics,
P.O.Box 5031, 2600 GA Delft, The Netherlands
LEX G. OVERSTEEGEN Department of Mathematics, University of Alabama at
Birmingham, Birmingham, Alabama 35294, USA

ABSTRACT. A separable metric space X is called a *matchbox manifold* if each point x
of X has an open neighborhood which is homeomorphic to the product $S_x \times \mathbb{R}$ for
some zero-dimensional space S_x. This survey of the theory of matchbox manifolds
starts with the background of the theory. After a discussion of the structure of
matchbox manifolds, an exposition of the most important examples, namely one-
dimensional phase spaces of flows without rest points, is presented. This is followed
by a structure theorem for these spaces. A discussion of the geometry of laminations
fits into the theory in a natural way. Various applications and open problems are
presented.

All spaces under discussion are separable and metrizable.

1. INTRODUCTION

Let G denote either the reals $\mathbb{R}$ or the integers $\mathbb{Z}$. A *dynamical system* on a
space X is a continuous mapping $\pi\colon X \times G \to X$ such that for all x in X and all s,
t in G

(1) $\pi(x,0) = x$,

(2) $\pi\big(\pi(x,s),t\big) = \pi(x,s+t)$.

For each x in X the mapping $\pi_x\colon G \to X$, defined by $\pi_x(t) = \pi(x,t)$, is called
the *motion through* x. It is a continuous mapping of G into X and $\mathcal{O}(x) = \pi_x[G]$

1991 *Mathematics Subject Classification.* primary 54H20, secondary 54F15, 54F50.
Key words and phrases. matchbox manifold, orientable, suspension, lamination.

3

is called the *orbit of x*. For each t in G the mapping $\pi^t\colon X \to X$, defined by $\pi^t(x) = \pi(x,t)$, is called a *transition*. The transition π^t is a homeomorphism with inverse π^{-t}. Thus a dynamical system is the action of a special transformation group, namely $\mathbb{R}$ or $\mathbb{Z}$, on X. In the first case the system is called a *flow*, in the latter a *cascade*. Note that a cascade is fully determined by the transition π^1. Thus a cascade on X may be viewed as a pair (X, f) in which f is a homeomorphism. Dynamical systems, the study of which was initiated by POINCARÉ, are the natural setting for the qualitative study of the long term behavior of solutions of differential equations. In this setting the solutions of the differential equation correspond to the motions of the dynamical system.

We say that the cascades (X, f) and (Y, g) are *conjugate* whenever there exists a homeomorphism $h\colon X \to Y$ such that $h \circ f = g \circ h$. The flows π and ρ on X and Y respectively are called *topologically equivalent* if there exists a homeomorphism h of X to Y which maps each orbit of X onto an orbit of Y and preserves the orientation of orbits (later on we shall explain what orientation of orbits means).

As was already pointed out by POINCARÉ, the interplay between flows and cascades is an important tool for the study of dynamical systems. There is a standard way of making a flow out of a cascade. Let (X, f) be a cascade. On $X \times \mathbb{R}$ we define the flow τ by $\tau\big((x,s),t\big) = (x, s+t)$. On $X \times \mathbb{R}$ we have the equivalence relation $\sim$ defined by

$$(x,s) \sim (y,t) \quad \text{iff} \quad m = s - t \in \mathbb{Z} \text{ and } y = f^m(x).$$

In particular $(x,1) \sim \big(f(x),0\big)$. The equivalence class of (x,s) is denoted by $[x,s]$. The resulting quotient space $X \times \mathbb{R}/\sim$ is called the *mapping torus* and denoted by $\Sigma(X, f)$. For x in X, s, t in $\mathbb{R}$ let

$$\tau^*\big([x,s],t\big) = [x, s+t].$$

This defines the flow on $\Sigma(X, f)$; it is called the *suspension* of the cascade f. An important question is: what flows can be represented as a suspension? And assuming that a space Z can be represented as $\Sigma(X, f)$ and $\Sigma(Y, g)$ how are X, Y, f and g related?

The first step in investigating the representation of flows by suspensions is the study of the local product structure of flows. Suppose π is a flow on X. A closed subset S of X is called a *section* if there exists a real number $\eta > 0$ such that

$$\pi\colon S \times [-\eta, \eta] \to \pi\big[S \times [-\eta, \eta]\big]$$

is a homeomorphism. If $x \in S$ and $U = \pi\big[S \times [-\eta, \eta]\big]$ is a neighborhood of some point x, then we shall also say that S is a *local section at* x and that U is a *flowbox neighborhood of* x. The existence of local sections was first proved by Whitney [36] (see also [26]): every point of the flow that is not a rest point has a flowbox neighborhood. Once we know that flows have a local product structure the natural question is: can the flow be recovered from its local structure? A more general question is: suppose that a space (not necessarily the phase space of a flow) has a local product structure, i.e. each point has arbitrarily small neighborhoods with the topology of a topological product with $\mathbb{R}$ as a factor. Can one reconstruct the space from the local structure? What topological properties of the space follow from this structure? The answer to these questions can hardly be expected to be given without extra conditions. But with a consistency condition on the flowboxes or with a dimension restriction on the space the structure of the space can be recovered. In order to address the above questions we will restrict ourselves mainly to one-dimensional spaces. This gives rise to the definition of a matchbox manifold.

2. Matchbox Manifolds

Definition 1. *A separable metric space X is called a matchbox manifold if each point x of X has an open neighborhood which is homeomorphic to the product $S_x \times \mathbb{R}$ for some zero-dimensional space S_x.*

Suppose that the one-dimensional space X has a local product structure with the reals as a factor, then X is a matchbox manifold. This follows immediately from the following result ([19, 1.9.E)]): for every space Y,

$$\dim(Y \times \mathbb{R}) = \dim(Y) + 1.$$

The following theorem provides an important class of matchbox manifolds.

Theorem 2. *Let $\pi \colon X \times \mathbb{R} \to X$ be a flow on a one-dimensional space X. If π has no rest points, then X is a matchbox manifold.*

Proof. Because there are no rest points, each point x in X has a flowbox neighborhood [36, 26].

Examples 3. Here are some examples of matchbox manifolds.

1. In [1] it was shown that every orbit of a moving point in a flow is a matchbox manifold. Every such orbit is a topological copy of the circle, the real line or the suspension $\Sigma(\mathbb{Q}, f)$ where f is a homeomorphism of $\mathbb{Q}$ onto itself such that $\mathbb{Q} = \mathcal{O}(x)$ for each point x of $\mathbb{Q}$. These matchbox manifolds were called P-manifolds in [1]. The P-manifolds have been employed in [20] to show that there are continuum many topologically distinct orbits. We shall discuss this further in Section 4.

2. A compact and connected space is called a *Cantor bundle* if each point has a neighborhood that is homeomorphic to the product of the Cantor set and $\mathbb{R}$. These matchbox manifolds were investigated in [23] and [24].

3. Many topological structures in dynamics are related to laminations [35, 14]. A *lamination* is a Cantor bundle that can be embedded in a surface and that does not contain a closed curve. Laminations will be further discussed in Section 5.

For a further discussion of the local structure of matchbox manifolds we fix some notation.

Standing notation 4. Throughout the paper the following notation is used. Let S be any zero-dimensional subset of $\mathbb{R}$. We define

$$
\begin{aligned}
F_S &= \{(x,y) \in \mathbb{R}^2 : x \in S, -1 \le y \le 1\} \\
E_S &= \{(x,y) \in \mathbb{R}^2 : x \in S, -1 < y < 1\}
\end{aligned}
$$

The set F_S is called a *standard matchbox*. For each x in S the set $\{x\} \times [-1,1]$ is called a *match in F*. If no confusion is likely to arise, we simply write F and E instead of F_S and E_S respectively. The natural projections of F_S onto S and $[-1,1]$ are denoted by pr_1 and pr_2 respectively. Both projections are open. By compactness of $[-1,1]$, the map pr_2 is closed as well.

The building blocks of a matchbox manifold are the matchboxes.

Definition 5. *If $h\colon F_S \to X$ is a topological embedding such that $h[F_S]$ is closed and $h[E_S]$ is open in X, then $V = h[F_S]$ is called a matchbox in X. We also say that V is a matchbox neighborhood of $h(x,0)$, $x \in S$. The induced map $h\colon F_S \to V$ is called a parameterization of V.*

In a matchbox manifold every point has arbitrarily small matchbox neighborhoods, [1, 7]. Now we know that matchboxes are the building blocks, we must study how to put them together.

Definition 6. *Let X be a matchbox manifold. Suppose that J is an arc in X with parameterization $g\colon [0,1] \to J$ and that $V = h[F_S]$ is a matchbox in X. Then V is called a matchbox along J if for some x in S*

(1) $J \cap V = h(\{x\} \times [-1,1])$, *and*
(2) *the map $t \to g^{-1}\big(h(x,t)\big)$ is increasing.*

If J is a parameterized arc in a matchbox manifold and x is not an endpoint of J, then x has arbitrarily small neighborhoods V such that V is a matchbox along J. The Pasting Theorem [1, 7] tells how to paste the matchboxes: if V_1 and V_2 are matchboxes along an arc J and if $J \cap V_1 \cap V_2$ is an arc, then there is a matchbox V along J such that $V \subset V_1 \cup V_2$ and $J \cap V = J \cap (V_1 \cup V_2)$. By a compactness argument with the Pasting Theorem we can produce long matchboxes[1, 7].

Theorem 7 (Long Box). *Let J be a parameterized arc in a matchbox manifold X with first point x_1 and last point x_2. Suppose that $V_1 = h_1[F_{S_1}]$ and $V_2 = h_2[F_{S_2}]$ are disjoint neighborhoods of x_1 and x_2 respectively. Then there is a matchbox $V = h[F_S]$ along J and s in S such that*

$x_1 = h(s,-1)$ *and* $h[S \times \{-1\}]$ *is a closed and open subset of* $h_1[S_1 \times \{0\}]$, *and*

$x_2 = h(s,1)$ *and* $h[S \times \{1\}]$ *is a closed and open subset of* $h_2[S_2 \times \{0\}]$.

A part of the theory of matchbox manifolds can be generalized to higher dimensional objects [8, 9]. The Pasting Lemma however is typical for the one-dimensional theory. This explains the presence of the consistency condition in the following definition.

Definition 8. *A space X is called a flowbox manifold if it has the following two properties.*

(1) *Local product structure: there exists a base $\mathcal{U} = \{ U_\beta : \beta \in B \}$ for the open sets such that for each β in B there exists a space S_β and a homeomorphism $h_\beta : S_\beta \times \mathbb{R} \to U_\beta$.*

(2) *Consistency: suppose that $U_\alpha = h_\alpha(S_\alpha \times \mathbb{R})$ and $U_\beta = h_\beta(S_\beta \times \mathbb{R})$ are elements of $\mathcal{U}$. If $U_\alpha \subset U_\beta$, then for each s in S_α there exists t in S_β such that $h_\alpha[\{ s \} \times \mathbb{R}] \subset h_\beta[\{ t \} \times \mathbb{R}]$.*

It is easily seen that each matchbox manifold is a flowbox manifold; the consistency property follows from the fact that matches and components of the (standard) matchbox coincide. Other examples of flowbox manifolds are dimension-one C^0-foliations [15] and flows without rest points.

3. Orientability

The definition of orientability of a matchbox manifold involves the notion of parameterization of its arc components. If the arc component C is compact, it is a topological circle and any covering map of $\mathbb{R}$ to C is called a *parameterization*. If the arc component C is not compact, then any continuous bijection $p \colon \mathbb{R} \to C$ is called a *parameterization*. The existence of such bijections was proved in [1]. The fundamental property of parameterizations is the arc lifting property which is formulated in the following lemma. See [1, 2] for details.

Lemma 9. *Suppose that p_1 and p_2 are parameterizations of an arc component of a matchbox manifold X. Then for all a_1, a_2 in $\mathbb{R}$ such that $p_1(a_1) = p_2(a_2)$ there is a unique homeomorphism $h \colon \mathbb{R} \to \mathbb{R}$ such that $h(a_2) = a_1$ and $p_2 = p_1 \circ h$.*

As any homeomorphism of $\mathbb{R}$ onto itself is either increasing or decreasing, the parameterizations of a given arc component fall into two classes. Now we come to the definition of orientability of a matchbox manifold. Every match in a matchbox has the order given by the second coordinate. Every parameterized arc component inherits the order of $\mathbb{R}$. If for some simultaneous parameterization of the arc components of the matchbox manifold X each point of X has a matchbox neighborhood such that the order of each of its matches agrees with the order of the arc components that hit this matchbox, we say that X is orientable. The following definition makes this precise.

Definition 10. *Let X be a matchbox manifold and let $\{\, C_\alpha : \alpha \in A \,\}$ denote the collection of its arc components. Then X is called orientable if there exist parameterizations $p_\alpha \colon \mathbb{R} \to C_\alpha$ for all α in A such that each point x of X has a matchbox neighborhood $V = h[F_S]$ with the following property: for each $\alpha \in A$ and each $t \in \mathbb{R}$ with $p_\alpha(t) \in h[E_S]$ there exists an interval J around t such that $\mathrm{pr}_2 \circ h^{-1} \circ p_\alpha$ is increasing.*

If the phase space X of a flow without rest points is one-dimensional, the arc components of X coincide with the orbits. As the orbits are parameterized by the motions in a natural way, it is easily seen that X is orientable. The converse of this statement is also true [7]:

Theorem 11. *A one-dimensional space X is the phase space of a flow without rest points if and only if X is an orientable matchbox manifold.*

Examples 12. Here are examples of matchbox manifolds that are not orientable.

1. Let X be the subset of the plane defined by

$$X = \big\{\, (x,y) : y = \sin \frac{1}{x}, x > 0 \,\big\} \cup \{(0,y) : -1 < y < 1\}.$$

It is clear that X is a matchbox manifold. It should be noticed that the homeomorphism type of the matchbox neighborhoods may vary with the points. If V is a matchbox neighborhood of a point $(0,y)$ then the $\sin \frac{1}{x}$-curve keeps crossing V in opposite directions as x converges to 0. This shows that X is not orientable.

2. Let g be a strictly increasing function of $\mathbb{R}$ onto $(1,2)$. Using polar coordinates we define

$$Y = \big\{\, (1,\varphi) : 0 \le \varphi < 2\pi \,\big\} \cup \big\{\, (2,\varphi) : 0 < \varphi \le 2\pi \,\big\} \cup \big\{\, (r,\varphi) : \varphi \in \mathbb{R}, r = g(\varphi) \,\big\}.$$

The space X obtained by identifying $(1,\varphi)$ and $(2, 2\pi - \varphi)$ for each φ in $[0,2\pi)$ is a compact matchbox manifold that is not orientable.

3. Let K be the Knaster bucket handle without the endpoint. The space K is a matchbox manifold that is not orientable. That K is not orientable may be seen as follows. If K were orientable, by Theorem 11, one could define a flow on K. By the Poincaré-Bendixson theorem (see also [13]) it would follow that K contains a circle, a contradiction. Note that each arc component of K is a homogeneous matchbox manifold that is not orientable. The arc components of the bucket handle are studied in [11] and [3]. In the latter paper matchbox manifold techniques are employed.

In view of Theorem 11 it is of interest to determine sufficient conditions for a matchbox manifold to be orientable.

Definition 13 ([21, 4]). *We will say that a matchbox manifold X is minimal provided that X does not contain a simple closed curve and each arc component of X is dense in X.*

Now we make the following observation. Suppose that X is a minimal matchbox manifold such that there exist parameterizations $p_\alpha \colon \mathbb{R} \to C_\alpha$ for all α in A and a matchbox $V = h[F_S]$ satisfying the property stated in Definition 10, then X is orientable.

Using this observation and a theorem by Effros [18, 10] one can prove the following theorem [6].

Theorem 14. *If a matchbox manifold is compact and homogeneous, then it is orientable.*

It follows from Examples 12, 2 and 3, that both conditions are required.

4. SUSPENSIONS

In Theorem 11 it is stated that every orientable matchbox manifold admits a flow without rest points. The proof in [7] reveals that such a space admits a rich structure.

Theorem 15 (Structure Theorem). *Let X be a space. Then X is an orientable matchbox manifold if and only if there exists a zero-dimensional space S and a homeomorphism $f \colon S \to S$ such that X is homeomorphic with the mapping torus $\Sigma(S, f)$. In particular, the one-dimensional space X admits a flow without rest points if and only if this flow is the suspension of a cascade f on some zero-dimensional set S.*

We shall develop the topological counterpart of the theory of Kakutani [27, 33] who classified measure preserving transformations through suspensions. Recall that a point x of a cascade is *positively [negatively] recurrent* if for each neighborhood U of x there exists $n > 0$ $[n < 0]$ such that $f^n(x) \in U$.

Definition 16. *Let (X, f) be a cascade such that each point x is positively and negatively recurrent. Let A be a clopen subset of X. For each x in A let $n(x)$ be the least integer $m \geq 1$ such that $f^m(x) \in A$. Then define the first return homeomorphism $r(A, f) \colon A \to A$ by*

$$r(A, f)(x) = f^{n(x)}(x), \quad x \in A.$$

That $r(A, f)$ is a well defined homeomorphism follows from the condition that each point is two-sided recurrent. In order to guarantee that this condition is satisfied some extra conditions are inserted in the following theorems. One such condition is the minimality of the matchbox manifold. This condition is used in the formulation of the following result [21].

Theorem 17. *Suppose that $\Sigma(X, f)$ and $\Sigma(Y, g)$ are compact minimal matchbox manifolds. Then $\Sigma(X, f)$ and $\Sigma(Y, g)$ are homeomorphic if and only if there exist closed-and-open subsets A and B of X and Y respectively such that $r(A, f)$ and $r(B, g)$ are conjugate.*

It is not difficult to see that if $\Sigma(X, f)$ is a compact minimal matchbox manifold, then X is homeomorphic to the Cantor set C. So in Theorem 17 the spaces X and Y are actually copies of C. We remark that in Theorem 17 it is not necessarily true that (X, f) and (Y, g) are conjugate, as is illustrated by the following example.

Example 18. Let $P = (p_1, p_2, \ldots)$ be a sequence of prime numbers. By C_P we denote the Cantor set represented as topological product $\prod_{i=0}^{\infty} \{ 0, \ldots, p_{i-1} \}$. The *adding machine* (C_P, σ) is defined by

$$\sigma(x_1, x_2, \ldots) = (x_1 + 1, x_2, \ldots) \quad \text{if } x_1 < p_1 - 1,$$
$$\sigma(p_1 - 1, \ldots, p_{k-1} - 1, x_k, \ldots) = (0, \ldots, 0, x_k + 1) \quad \text{if } x_k < p_k - 1,$$
$$\sigma(p_1 - 1, p_2 - 1, \ldots) = (0, 0, \ldots).$$

The *P-adic addition* σ is a homeomorphism. We note that $\Sigma(C_P, \sigma)$ is the *P-adic solenoid* S_P. Let $X = C_P \times \{ 0, 1 \}$ and define $f : X \to X$ by

$$\begin{aligned} f(x, 0) &= (x, 1) &, x \in C_P, \\ f(x, 1) &= (\sigma(x), 0) &, x \in C_P. \end{aligned}$$

Let $Y = C_P \times \{ 0, 1, 2 \}$ and define $g : Y \to Y$ by

$$\begin{aligned} g(y, i) &= (y, i + 1) &, y \in C_P, i = 0 \text{ or } i = 1, \\ g(y, 2) &= (\sigma(y), 0) &, y \in C_P. \end{aligned}$$

Both $\Sigma(X, f)$ and $\Sigma(Y, g)$ are homeomorphic to S_P, which obviously is compact and minimal. The set $C_P \times \{ 0 \}$ is a closed-and-open subset of both X and Y and the first return maps $r(C_P \times \{ 0 \}, f)$ and $r(C_P \times \{ 0 \}, g)$ are both "equal" to σ. However the cascades (X, f) and (Y, g) are not conjugate, as follows from the distinct behavior of f^2 and g^2.

In Example 3,1 it was mentioned that each non-compact orbit in a flow can be represented as a suspension $\Sigma(\mathbb{Q}, f)$ where f is a homeomorphism of $\mathbb{Q}$ onto itself such that $\mathbb{Q} = \mathcal{O}(x)$ for each point of $\mathbb{Q}$.

Theorem 19 ([1, 5]). *Suppose that $\Sigma(\mathbb{Q}, f)$ and $\Sigma(\mathbb{Q}, g)$ are orbits of flows. Assume that each point of $\mathbb{Q}$ is two-sided recurrent for both f and g. Then $\Sigma(\mathbb{Q}, f)$ and $\Sigma(\mathbb{Q}, g)$ are homeomorphic if and only if there exist closed-and-open subsets A and B of $\mathbb{Q}$ such that $r(A, f)$ and $r(B, g)$ are conjugate.*

This result has been employed in [20, 21] to show that there exist continuum many homeomorphism types of orbits in flows. In [29] it is used to prove that any two arc components of any two solenoids are homeomorphic. Related to these results there is the following open question.

Question 20. Let f and g denote rotations of the circle S through irrational multiples of 2π. Suppose that Γ_1 and Γ_2 are orbits of the irrational flows on the torus $\Sigma(S, f)$ and $\Sigma(S, g)$ respectively. Are Γ_1 and Γ_2 homeomorphic?

The result of Theorem 15 is typical for matchbox manifolds and cannot be generalized to higher dimensional flows.

Example 21. Here is an example of a flow that cannot be represented as a suspension. In $\mathbb{R}^2$ we consider the differential equation (cf. [31, page 30])

$$\frac{dx}{dt} = \sin y, \quad \frac{dy}{dt} = \cos^2 y.$$

The solutions are the curves $x + c = 1/\cos y$ and the straight lines $y = k\pi + \pi/2$, $k = 0, \pm 1, \ldots$. If we assume that this flow is a suspension over some space Y, then it can be shown that Y is connected and locally homeomorphic to an arc [26, Chapter VI]. It can be seen that this is impossible.

5. GEOMETRY OF LAMINATIONS

In this section we will focus on the topological structure of matchbox manifolds which are subsets of the plane. In order to avoid cumbersome statements we will focus mainly on a special type of matchbox manifolds, namely, minimal laminations (Example 3, 3, Definition 13). Minimal laminations are indecomposable continua. It is customary to call the arc components of laminations *leaves*. The Plykin example [17] of a planar continuum which admits an expansive homeomorphism is a prototype of a minimal lamination. We will establish that this continuum is the simplest possible minimal planar lamination. This follows from the following theorems which were established in [22].

Theorem 22. *Let X be a minimal planar lamination, the X has at least four complementary domains.*

This theorem follows from the fact that it is possible to extend the lamination to a foliation of the 2-sphere with finitely many singularities. These singularities correspond to complementary domains of X. By using the index theorem and analyzing the singularities which occur, one can use the Euler characteristic of the sphere to obtain at least four singularities which correspond to at least four complementary domains.

Historically, indecomposable continua in the plane have been constructed by a Lakes of Wada type construction. By this we mean a construction which removes from a disc with finitely many holes (i.e. the "lakes") infinitely long channels, with asymptotic boundaries, starting from each of these lakes. By lifting the lamination to the Poincaré disk, one can use hyperbolic geometry to obtain the following results [22].

Theorem 23. *A minimal planar lamination can be obtained by a lakes of Wada construction.*

In attempting to use a lakes of Wada type construction to construct distinct laminations, one might be tempted to "dig" different number of channels from different numbers of lakes. Theorem 22 shows that one must have at least three lakes if one requires the resulting continuum to be a lamination (one channel can be started from the unbounded complementary domain). The following result puts additional restrictions on such a construction.

Theorem 24. *Every complementary domain of a minimal planar lamination is bounded by finitely many leaves. Moreover, all but finitely many have two boundary leaves.*

The following results alludes to the fact that laminations are closely related to Denjoy modifications of the irrational flow on the torus. Like those examples, planar laminations are far from homogeneous.

Theorem 25. *All continuous functions of a minimal planar lamination to itself are either homotopic to a homeomorphism or a constant map. Furthermore, continuous surjections must permute boundary leaves.*

Theorem 22 can be generalized to arbitrary compact matchbox manifolds in the following sense.

Theorem 26. *Let X be a compact matchbox manifold, then X is not tree-like.*

6. APPLICATIONS

We have already mentioned various applications of the theory of matchbox manifolds. In this section we discuss some more.

In [25] Hagopian gave the following characterization of the solenoids.

Theorem 27. *If X is a homogeneous continuum such that every proper nondegenerate subcontinuum is an arc, then X is a solenoid.*

In [6] this result is proved by employing matchbox manifold techniques. Let X be as mentioned in the theorem. First by using Effros' theorem [18, 10] it is shown that the family of arc components of X is a so-called *regular family of curves*. Then, by applying techniques of Whitney [36], it is proved that X has a local product structure with the reals as a factor (see also [8]). By Theorem 14, X is an orientable matchbox manifold. Finally by adapting an argument of Thomas [34] it is shown that X is a solenoid.

Related to the last proof and Example 12, 3 there is the following open question.

Question 28. Does there exist a homogeneous curve (i.e. one-to-one continuous image of the real line) that is not a matchbox manifold?

In Example 18 we have defined the P-adic solenoid for a sequence P of prime numbers. Sequences of prime numbers P and Q are said to be *equivalent* if a finite number of primes can be deleted from each sequence so that every prime number occurs the same number of times in the deleted sequence. There is the following classification theorem.

Theorem 29. *The solenoids S_P and S_Q are homeomorphic if and only if the sequences P and Q are equivalent.*

This theorem was conjectured by Bing [12], who proved the correctness of the "if"-part. The "only if"-part was proved by McCord [30]. The proof makes use of the fact that solenoids can be endowed with the structure of a topological group and invokes Čech cohomology and Pontryagin duality. An elementary proof of the theorem employing matchbox manifold techniques is presented in [3].

Finally we mention that matchbox manifold techniques are employed in [16] in the study of exactly two-to-one maps from continua onto arc-continua.

REFERENCES

1. J.M. Aarts, *The structure of orbits in dynamical systems*, Fundamenta Mathematicae **129** (1988), 39–58.

2. J.M. Aarts, *Orientation of orbits in flows*, Papers on General Topology and Related Category Theory and Topological Algebra, Annals of the New York Academy of Sciences **552** (1989), 1–7.

3. J.M. Aarts and R.J. Fokkink, *The classification of solenoids*, Proceedings American Mathematical Society **111** (1991), 1161–1163.

4. J.M. Aarts and R.J. Fokkink, *On composants of the bucket handle*, Fundamenta Mathematicae **139** (1991), 193–208.

5. J.M. Aarts and Z. Frolík, *Homeomorphisms of $\mathbb{Q}$ and the orbit classification of flows*, Supplemento ai Rendiconti del Circolo Matematico di Palermo, Serie II, **18** (1988), 15–25.

6. J.M. Aarts, C.L. Hagopian and L.G. Oversteegen, *The orientability of matchbox manifolds*, Pacific Journal of Mathematics **150** (1991), 1–12.

7. J.M. Aarts and M. Martens, *Flows on one-dimensional spaces*, Fundamenta Mathematicae **131** (1988), 53–67.

8. J.M. Aarts and L.G. Oversteegen, *Whitney's regular families of curves revisited*, Proceedings of the 1989 Joint Summer Conference on Continuum Theory and Dynamical systems, M. Brown (ed.), Contemporary Mathematics **117** (1991), 1–7.

9. J.M. Aarts and L.G. Oversteegen, *Flowbox manifolds*, Transactions American Mathematical Society **327** (1991), 449–463.

10. F.D. Ancel, *An alternative proof and applications of a theorem of E.G. Effros*, Michigan Mathematical Journal **34** (1987), 39–55.

11. C. Bandt, *On the composants of the bucket handle*, to appear in Fundamenta Mathematicae.

12. R.H. Bing, *A simple closed curve is the only homogeneous bounded plane continuum that contains an arc*, Canadian Journal of Mathematics **12** (1960), 209–230.

13. H. Bohr and W. Fenchel, Ein Satz über stabile Bewegungen in der Ebene, Dansk Mat. Fys. Medd. **14** (1936), 3–15.

14. A.J. Casson and S.A. Bleiler, *Automorphisms of surfaces after Nielsen and Thurston*, Cambridge 1988.

15. C. Camacho and A.L. Neto, *Geometric theory of foliations*, Birkhäuser, Basel, 1985.

16. W. Dębski, J. Heath and J. Mioduszewski, *Exactly two-to-one maps from continua onto arc-continua*, preprint.

17. R.L. Devaney, *An introduction to chaotic dynamical systems*, Addison-Wesley, Reading (Mass.), 1989.

18. E.G. Effros, *Transformation groups and C*-algebras*, Annals of Mathematics **81** (1965), 38–65.

19. R. Engelking, *Dimension theory* PWN, Warszawa 1978

20. R.J. Fokkink, *There are uncountably many homeomorphism types of orbits in flows*, Fundamenta Mathematicae **136** (1990), 147–156.

21. R.J. Fokkink, *The structure of trajectories*, Thesis, Delft University of Technology, 1991.

22. R.J. Fokkink and L.G. Oversteegen, *The geometry of laminations*, preprint.

23. A. Gutek, *On compact spaces which are locally Cantor bundles*, Fundamenta Mathematicae **108** (1980), 27–31.

24. A. Gutek and J. van Mill, *Continua that are locally a bundle of arcs*, Topology Proceedings **7** (1982), 63–69.

25. C.L. Hagopian, *A characterization of solenoids*, Pacific Journal of Mathematics **68** (1977), 425–435.

26. O. Hájek, *Dynamical systems in the plane*, London 1968.

27. S. Kakutani, *Induced measure preserving transformations*, Proceedings Imperial Academy Tokyo **19** (1943), 635–641

28. H.B. Keynes and M. Sears, *Modelling expansion in real flows*, Pacific Journal of Mathematics **85** (1979), 111–124.

29. A.J. de Man, *On composants of solenoids*, preprint, 1994

30. M.C. McCord, *Inverse limit sequences with covering maps*, Transactions American Mathematical Society **114** (1965), 197–209.

31. V.V. Nemytskii and V.V. Stepanov, *Qualitative theory of differential equations*, Princeton University Press, Princeton 1960.

32. B. Parry and D. Sullivan, *A topological invariant of flows on 1-dimensional spaces*, Topology **14** (1975), 297–299.

33. K. Petersen, *Ergodic Theory*, Cambridge University Press, Cambridge 1983.

34. E.S. Thomas, *One-dimensional minimal sets*, Topology **12** (1973), 233–242.

35. W.P. Thurston, *On the geometry and dynamics of diffeomorphisms of surfaces*, Bulletin American Mathematical Society **19** (1988), 418–431.

36. H. Whitney, *Regular families of curves*, Annals of Mathematics **34** (1933), 244–270.

E-mail address: aarts@twi.tudelft.nl "Jan Aarts"
E-mail address: overstee@math.uab.edu "Lex Oversteegen"

2
ROTATION SETS FOR INVARIANT CONTINUA

KATHLEEN T. ALLIGOOD Department of Mathematical Sciences, George Mason University, Fairfax, Virginia 22030, USA

ABSTRACT. Birkhoff attractors for area–contracting twist maps of the annulus contain orbits of all rotation numbers in a closed interval. We describe results which lead to an analogous theorem for invariant continua (generalized attractors). Mild nondegeneracy and smoothness assumptions replace the strict twist condition in order to assign rotation intervals to attractors for a large class of maps.

An orientation–preserving homeomorphism g of the circle S^1 has a well–defined rotation number. This number measures the average rate of rotation of a point under iteration by g—averaged, in the limit, over the entire orbit. This limit exists for every point in the circle and is independent of the point chosen. (A complete treatment of rotation numbers for circle homeomorphisms is given, for example, in [11].) For a homeomorphism f of the annulus $S^1 \times I$ (where $I \subset R$ is a compact interval) rotation numbers are measured with respect to the angular component. In the two–dimensional case, different points can have different rotation numbers, and a point may have no rotation number at all (i.e., the limit may not exist). In this paper, we focus on non–degenerate continua which are invariant under orientation–preserving, area–contracting diffeomorphisms of the plane or annulus. In particular, we describe results which characterize their rotation sets, i.e., the set of all rotation numbers which occur on the continuum. For a planar continuum Θ, rotations are measured with respect to a fixed point in the continuum.

1991 *Mathematics Subject Classification.* primary 58F12, secondary 58F13, 54F15, 54H25.

Key words and phrases. rotation number, accessible point, prime end, heteroclinic pair, attractor.

In Section 1 we describe rotations of points on Θ which are "accessible" from the complement $R^2 \backslash \Theta$. A point p on Θ is called *accessible* if there is a path s, $s : [0,1) \to R^2 \backslash \Theta$, such that $\lim_{t \to 1^-} s(t) = p$, (i.e., p is the only limit point of the path). The accessible points on Θ have a well–defined rotation number. When this rotation number is rational, under some mild nondegeneracy and smoothness assumptions, there is a complete characterization of the dynamics of the accessible points. In Sections 2 and 3, we use this characterization to describe invariant continua whose rotation sets contain nontrivial intervals of rotation numbers. My collaborators in various aspects of this work have been James Yorke and Marcy Barge.

1. ACCESSIBLE ORBITS AND THE PRIME END ROTATION NUMBER

Let f be an area–contracting, orientation–preserving diffeomorphism of the plane, and let Θ be a non–degenerate continuum which is invariant under f. Since f is area–contracting and Θ is closed and connected, $R^2 \backslash \Theta$ is open and connected. Let $Z = R^2 \cup \{\infty\}$, and let $W = Z \backslash \Theta$. Then W is simply–connected and thus is homeomorphic to an open disk. Notice that Θ is the boundary of W in Z. Since f is an area–contracting map of the plane, Θ cannot be homeomorphic to a circle. Using the theory of "prime ends" due to Carathéodory [7], however, we can study the dynamics on the accessible points of Θ through the dynamics on the circle of prime ends. (The "prime end" compactification of W is a non–standard compactification of the open disk. Background on prime ends appears in [2]. A more complete treatment of the subject can be found in [8] and [10].)

An orientation–preserving homeomorphism g of the circle S^1 has a unique rotation number, defined as follows:

Let $\tilde{g}$ be a lift of g, and let

$$\rho_{\tilde{g}}(x) = \lim_{n \to \infty} (\tilde{g}^n(\tilde{x}) - \tilde{x})/n,$$

for $x \in S^1$ and $\tilde{x} \in R$ lying over x. The *rotation number* $\rho(g)$ is the fractional part of $\rho_{\tilde{g}}(x)$. This number is independent of both x and the particular lift $\tilde{g}$ of g.

A rational rotation number determines the dynamics of individual orbits on the circle. In particular (see, e.g., [11]), the rotation number is rational if and only if g has periodic orbits. If $\rho(g) = a/b$, then the orbit of every point converges to a periodic point of (minimum) period b. If $\rho(g) = 0$, then the orbit of every point converges to a fixed point.

If the map induced by f on the circle of prime ends has a rational rotation number, then there exist periodic prime ends. Without any additional hypotheses, however, we can draw few conclusions about periodic behavior for the underlying map f. Figure 1(a) indicates the action of a planar map on an invariant continuum; Figure 1(b) shows the induced map on the associated circle of prime ends. The rotation number on the prime end circle is 0, and the induced map has two fixed points.

These fixed points represent the two limit circles in (a). If points on each limit circle in (a) rotate counterclockwise through an irrational angle (i.e., irrationally related to the circumference of the circle), then Θ has no periodic points. Thus we have a rational prime end rotation number and no periodic points on Θ. Notice, however, that this continuum cannot be an invariant set for an area–contracting map of the plane.

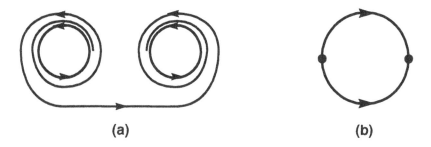

(a)   (b)

FIGURE 1. An invariant continuum and the associated prime end circle

The topic of accessible points on basin boundaries, when the basins of attraction are homeomorphic to a disk, is dealt with in [2]. This topic is analogous to the present one in the following sense: If we invert the area–contracting map f and view it as a homeomorphism of the one–point compactification Z of R^2, then Θ is the boundary of the basin $W = Z \backslash \Theta$. This approach is reminiscent of Birkhoff [6] and also of Cartwright and Littlewood [8] who studied attractors in forced Van der Pol type equations. The property of basin boundaries which we need in order to satisfy the hypotheses of the theorems in [2] is that the boundary be "unstable" in the basin; that is, there must exist an epsilon neighborhood of the boundary and a dense set of points within this neighborhood such that forward iterates of points within this dense set eventually leave the neighborhood. In our case, the inverse of f will be area–expanding, and thus any epsilon neighborhood of the compact set Θ will have the required dense set of points. The following theorem appears in [2]:

Theorem 1A. *Let f be an area–contracting, orientation–preserving homeomorphism of the plane, and let Θ be a nondegenerate continuum, invariant under f. Assume that the rotation number on the circle of prime ends of $W = Z \backslash \Theta$ induced by f is the reduced fraction a/b. If the fixed points of f^b on Θ are isolated, then there is an accessible fixed point of f^b on Θ.*

When the accessible periodic points guaranteed by Theorem 1A are "hyperbolic", then further conclusions may be obtained. A periodic point p of period k is called *hyperbolic* if none of the eigenvalues of the Jacobian matrix $Df^k(p)$ has magnitude one. For planar homeomorphisms, there are three types of hyperbolic periodic points: sinks (all eigenvalues inside the unit circle), sources (all eigenvalues outside the unit circle), and saddles (two real eigenvalues—one of absolute value greater than one and the other of absolute value smaller than one). Planar saddles have a one–dimensional stable manifold and a one–dimensional unstable manifold. The *stable manifold* of p is the set

$$W^s(p) = \{x : |f^n(x) - f^n(p)| \to 0\}.$$

The *unstable manifold* of p is the set

$$W^u(p) = \{x : |f^{-n}(x) - f^{-n}(p)| \to 0\}.$$

For ease of exposition, assume that the prime end rotation number is 0. (Otherwise, replace f by f^b, where b is the denominator of the rotation number.) Start with an accessible point $x \in \Theta$ which is not fixed, and iterate x under f^{-1}. Accessible points map to accessible points, each representing a distinct prime end. Since the prime end rotation number is 0, these prime ends must converge to a fixed prime end. Under the hypotheses that fixed points of f are isolated and that all accessible periodic points are hyperbolic, it can be shown (see [2]) that the impression of the fixed limit prime end contains an accessible fixed point p of f^{-1}, and that the forward orbit of x under f^{-1} converges to this fixed point. Now p cannot be a source, since the orbit of x converges to it; nor can it be a sink, since the "boundary" Θ is unstable (under f^{-1}) in the domain Z. Therefore, p must be a saddle. Since the orbit of x converges to p under f^{-1}, x must be on the unstable manifold of p under f. The following theorem can be obtained directly from the corresponding theorem about basin boundaries in [2]:

Theorem 1B. *In addition to the hypotheses of Theorem 1A, assume that all accessible periodic points are hyperbolic and that the prime end rotation number is not $1/2$. Then every accessible point on Θ is either a periodic point of minimum period b or is in the unstable manifold of such a point.*

When the prime end rotation number is $1/2$, there is a simple example in which Θ is the unstable manifold of a "flipped" fixed point saddle (i.e., one in which the eigenvalues of Df are negative), and each branch of the unstable manifold ends at a period-two point (one period–two orbit). In this example, there is one accessible periodic orbit of period two and one accessible fixed point.

2. ROTATION INTERVALS FOR HETEROCLINIC PAIRS

Understanding the dynamical behavior of accessible orbits is useful in determining the rotation set of an invariant continuum. The analysis of accessible orbits, however, depends on rotations for one–dimensional (i.e., circle) maps. Calculating rotation numbers for two–dimensional maps—say for points under the action of an annulus map—is a little trickier than for circle maps since the limit which defines the rotation number may not exist and there may not be a unique rotation number which characterizes the action on the entire annulus.

Formally, let $U = R \times I$ be the universal cover of the annulus A. Let $\tilde{f} : U \to U$ be a lift of f, and let $\pi_x : U \to R$ be the projection onto the first coordinate. Then the forward and backward rotation numbers, ρ^+ and ρ^-, respectively, are defined for a point (x, y) of A as

$$\rho^+(x, y) = \lim_{n \to \infty} \frac{1}{n}[\pi_x \tilde{f}^n(\tilde{x}, \tilde{y}) - \tilde{x}],$$

where $(\tilde{x}, \tilde{y})$ in U lies over (x, y), if this limit exists, and

$$\rho^-(x, y) = -\lim_{n \to \infty} \frac{1}{n}[\pi_x \tilde{f}^{-n}(\tilde{x}, \tilde{y}) - \tilde{x}],$$

where $(\tilde{x}, \tilde{y})$ in U lies over (x, y), if this limit exists.

Notice that for a periodic orbit $\rho^+ = \rho^-$. For a circle homeomorphism, ρ^+ always equals ρ^-. For annulus maps, ρ^+ need not equal ρ^- and may not even exist for any given point, and different points can have different rotation numbers. We define the *rotation set* J_Θ of an invariant set Θ, $\Theta \subset A$, under f as

$$J_\Theta = \{\rho \mid \rho = \rho^+(x) = \rho^-(x), \text{ for some } x \in \Theta\}.$$

While one often studies $\rho \mod(1)$ in dealing with maps on the circle, for the results described here, it is important to view J_Θ as a subset of all real numbers, not just those between 0 and 1. A different choice of the lift will translate J_Θ by an integer.

There are several now classic situations in which J_Θ has been shown to equal or to contain a nontrivial interval. For an area–preserving, invertible, monotone twist map of the annulus, (see [17] for the definition and a discussion of such maps), the rotation set is an interval whose endpoints are the rotation numbers of the boundary circles [16, 14]. The rotation set of an area–contracting monotone twist map is not necessarily an interval; however, the rotation set of the Birkhoff attractor contained within the annulus is an interval [15, 9].

We can similarly describe the rotation of points around a fixed point p of an orientation–preserving planar homeomorphism f. In this case, $R^2 - \{p\} \approx S^1 \times (0, 1)$ is invariant under f. Angles are measured in polar coordinates (centered at p). A lift $\tilde{f}$ of f is defined on the universal cover $R \times (0, 1)$, analogous to the definition on $U = R \times [0, 1]$.

Two theorems are particularly relevant to the result presented here. First is a theorem of M. Barge and R. Gillette [5]: Let f be an orientation–preserving homeomorphism of the plane that leaves invariant a continuum Λ which irreducibly separates the plane into exactly two domains. Then the rotation set of Λ contains all rational numbers in its convex hull, and each such rational number is the rotation number of a periodic orbit in Λ. This theorem, together with a result of M. Handel

[12] that the rotation set of an annulus homeomorphism is closed, shows that the rotation set of an invariant continuum which irreducibly separates the plane (or annulus) is a closed interval.

The second theorem is originally due to Aronson, Chory, Hall, and McGehee [4] (with a different proof given by Hockett and Holmes [13]). In their theorem, f is an orientation preserving diffeomorphism of the annulus A isotopic to the identify with a fixed point saddle p. They assume that branches of the stable and unstable manifolds of p cross in such a way so as to encircle the inner boundary A. Then, by choosing $N > 0$ large enough so that f^N forms a hyperbolic horseshoe containing p, it is shown that the rotation set of f contains the interval $[0, 1/N]$. In the result described in this section, by comparison, we require a second point with larger (i.e., non–zero) rotation number to provide a minimum upper bound for the rotation interval.

We assume that f is a diffeomorphism of the annulus or the plane. Let p and q be periodic saddles. In [3] p and q are called a *heteroclinic pair* if one branch of $W^u(p)$ crosses a branch of $W^s(q)$ and one branch of $W^u(q)$ crosses a branch of $W^s(p)$ so as to form a topological rectangle D. See Figure 2. This geometric condition is a natural one for chaotic sets which typically contain many periodic orbits whose stable and unstable manifold cross in grid–like fashion. We label the four sides of D as L (left), R (right), T (top), and B (bottom).

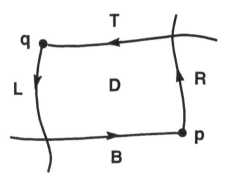

FIGURE 2. A heteroclinic pair of saddles

Notice that the rotation number of one point of the pair is defined only up to an integer (which depends on the particular lift $\tilde{f}$). Then, however, the rotation number of the other periodic orbit in the pair is determined. Changing the lift will simply shift the rotation set by an integer. The following theorem is proved in [3]:

Theorem 2. *Let f be an orientation–preserving diffeomorphism of the plane or annulus. Assume that p and q form a heteroclinic pair with $\rho(p) = a/b$, and $\rho(q) = c/d, (c/d \geq a/b)$. Then for each $\alpha, \alpha \in [a/b, c/d]$, there is a point in D with rotation number α.*

Remark. Specifically, for each $\alpha, \alpha \in [a/b, c/d]$, there is a continuum Γ in D (extending from L to R) such that $\rho^+(x) = \alpha$ for each $x \in \Gamma$. Similarly, there is a continuum Λ in D (extending from B to T) such that $\rho^-(y) = \alpha$ for each $y \in \Lambda$. Points in $\Gamma \cap \Lambda$ have $\rho = \rho^- = \rho^+ = \alpha$.

Notice that Theorem 2 does not require that the diffeomorphism f be area–contracting. The ideas behind the proof of this theorem are outlined below.

Let r be a rational number between a/b and c/d. Then r can be expressed as the Farey sum of k copies of a/b and j copies of c/d (in any order); i.e., $r = \dfrac{ka}{kb} \oplus \dfrac{jc}{jd} = \dfrac{ka+jc}{kb+jd}$. (See, for example, [13].) Let $\alpha \in [a/b, c/d]$, rational or irrational, be given. Choose a sequence $\{a_n\}$ such that each a_i is a/b or c/d and the sequence S_n of partial Farey sums $\{a_1 \oplus a_2 \oplus \cdots \oplus a_n\}$ converges to α. Such a sequence may be constructed by choosing $a_{n+1} = a/b$, if $S_n \geq \alpha$, and $a_{n+1} = c/d$, if $S_n < \alpha$.

Figure 3(a) shows forward iterates of the rectangle D under f^b and f^d. After b iterates, the original rectangle will be contracted along the bottom (along $W^s(p)$) toward p and stretched out along the right side (along $W^u(p)$). Similarly, $f^d(D)$ is contracted along the top (along $W^s(q)$) toward q and stretched down (along $W^u(q)$). (Of course, an arbitrary map f may not produce images as straight as those shown here. We address the general case later.) It follows that points along a subrectangle D_p located at the bottom of D map back to D after b iterates (see Figure 3(b)); points along a subrectangle D_q located at the top of D map back to D after d iterates.

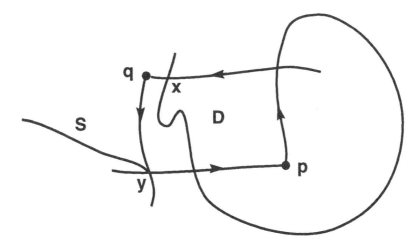

FIGURE 3. (a) Forward iterates of the rectangle D under f^b and f^d. (b) Subrectangles D_p and D_q, containing points which map back to D after b iterates and d iterates, respectively.

We define inductively a sequence $\{D_n\}$ of nested topological rectangles corresponding to the sequence $\{S_n\}$. Let h_n (resp., l_n) be defined as the number of times a/b (resp., c/d) occurs in the set $\{a_1, \ldots, a_n\}$. (In particular, $h_n + l_n = n$.)

$$\text{Let } D_1 = \left\{ \begin{array}{ll} D_p & \text{if } a_1 = a/b \\ D_q & \text{otherwise.} \end{array} \right.$$

$$\text{Let } D_{n+1} = \left\{ \begin{array}{ll} f^{-i}(D_p \cap f^i(D_n)) & \text{if } a_{n+1} = a/b \\ f^{-i}(D_q \cap f^i(D_n)) & \text{otherwise,} \end{array} \right. \quad \text{where } i = h_n b + l_n d.$$

Since D_n is a compact, connected subset of D and extends from L to R for each $n, n \geq 1$, and the sequence $\{D_n\}$ is nested, $\Gamma = \bigcap_{n \geq 1} D_n$ is connected, containing points in L and R. In the lift, points in D_n move through $h_n a + l_n c$ fundamental domains after $h_n b + l_n d$ iterates. Since the sequence of partial Farey sums

$$\{a_1 \oplus \cdots \oplus a_n = \frac{h_n a + l_n c}{h_n b + l_n d}\}$$

converges to α, the rotation number ρ^+ of each x in Γ is α.

The proof for the existence of the continuum Λ all of whose points have backward rotation number ρ^- equal to α is the same as above if we apply it to f^{-1} instead of f. In this case, however, the subrectangle D_{n+1} is defined as the appropriate forward iterate of $D_p \cap D_n$ or $D_q \cap D_n$.

Details of this proof (and special cases, such as negative eigenvalues of $Df(p)$ or $Df(q)$) appear in [3]. One particular complication which arises in the proof is that the bth or dth image of the rectangle D may not extend through D as they are shown in Figure 3(a). In particular, the image $f^d(R)$ of the right side R, for example, may not intersect the bottom side B. It must, however, intersect the top T, since it contains a heteroclinic crossing of $W^u(p)$ and $W^s(q)$—the dth image of the upper right "corner" of D. Therefore, it will be the case that for some higher iterate, say the kth iterate, $f^{kd}(R)$ will extend through D, intersecting B. This property follows from the λ–lemma (see, for example, [18]) and will also hold for any iterate of f^d greater than k.

By having to take higher iterates of f to accomplish the partitioning described above, it may be the case that periodic orbits of rational rotation number r/s have minimum period a multiple of s. In particular, if $f^{kd}(D)$ and $f^{kb}(D)$ both pass through D, as shown in Figure 3(a), then there will a periodic orbit of rotation number $\frac{kr}{ks}$.

3. Rotation Intervals for Attractors

In this section we put the theory of accessible points described in Section 1 together with the "partitioned rectangle" approach of Section 2. In particular, we assume that f is an area–contracting diffeomorphism *into* the annulus, isotopic to the identity, and that Θ is an invariant continuum which separates the annulus into two complementary domains, U_i and U_e—one of which contains the interior boundary circle and one of which contains the exterior boundary circle. In this case, there are two prime end rotation numbers: ρ_e, the prime end rotation number from the exterior domain U_e, and ρ_i, the prime end rotation number from the interior domain U_i. We assume throughout that these prime end rotation numbers are both rational. Our aim then is to describe a general class of maps for which the rotation set of Θ contains all real numbers between ρ_e and ρ_i, *and* for which rational rotations are represented by periodic orbits with the smallest possible period.

When the periodic points of a heteroclinic pair, p and q, are accessible points on Θ (one accessible from the exterior boundary circle and one from the interior boundary circle), then the necessity of having to take higher iterates to have $W^u(p)$ extend through the rectangle D is eliminated. Figure 4 shows such a heteroclinic pair. Since q and all points on $W^u(q)$ are accessible, there is a path s contained in the exterior complementary domain which limits on y, the point where $W^u(q)$ and $W^s(p)$ cross. Let x be the first crossing of $W^u(p)$ with $W^s(q)$ from inside R to outside. Since $W^u(p)$ is contained in Θ, it cannot cross s, and thus must cross $W^s(p)$ between y and p before it crosses $W^s(q)$ at the point x.

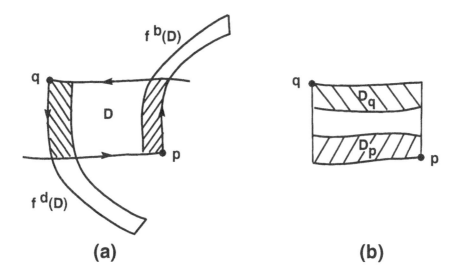

FIGURE 4. A heteroclinic pair of accessible saddles

Let $\rho_i = a/b$ and $\rho_e = c/d$. Assuming the hypotheses of Theorem 1B for both internal and external accessible orbits, we are guaranteed that all points on Θ accessible from the external boundary circle are on unstable manifolds of saddles of rotation number a/b; all points on Θ accessible from the interior boundary circle are on unstable manifolds of saddles of rotation number c/d. (In the annulus setting, we do not need to exclude cases where either ρ_i or ρ_e is $1/2$.) In the analysis of [1], a condition similar to that of the existence of a heteroclinic pair takes the place of a strict "twist" condition to obtain an entire interval of rotation numbers.

Specifically, we require that a branch of the stable manifold of one of the saddles accessible from the interior domain U_i intersects the exterior domain U_e, and that a branch of the stable manifold of one of the saddles accessible from the exterior domain U_e intersects the interior domain U_i. This condition is illustrated in Figure 5 for accessible saddles p and q. We say that such saddles form an *accessible twist pair*. Notice that if accessible orbits p and q form a heteroclinic pair, then they form an accessible twist pair. The converse does not necessarily hold, however, since $W^s(p)$ and $W^s(q)$ may not cross $W^u(q)$ or $W^u(p)$ at all, but rather cross the unstable manifolds of other accessible saddles. The following theorem appears in [1]. We assume that the eigenvalues of Df evaluated at each saddle of the accessible twist pair are positive.

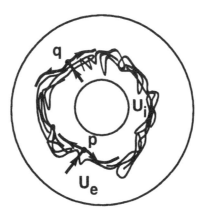

FIGURE 5. An accessible twist pair of saddles

Theorem 3. *Let f be an area–contracting diffeomorphism into the annulus, isotopic to the identity. Assume ρ_i and ρ_e are rational with $\rho_i < \rho_e$, and assume the hypotheses of Theorem 1B on Θ and on both internal and external accessible periodic orbits. If there exists an accessible twist pair for Θ, then, for each $\alpha \in [\rho_i, \rho_e]$, there is a point x in Θ such that $\rho(x) = \alpha$. If $\alpha = r/s$, then there is a periodic orbit x in Θ with rotation number r/s and minimum period s.*

The proof of Theorem 3 relies on the technique of partitioning a rectangle, as described in Section 2. In this case, however, branches of the stable and unstable manifolds of the accessible twist pair of orbits do not necessarily form a topological rectangle. Although pieces of stable manifolds of p and q again form two opposite sides of the rectangle, segments of unstable manifolds and accessible periodic saddles may not be path connected. In such cases, the other two sides will be made up of pieces of unstable manifolds together with cross–cuts. An analysis of the dynamics near Θ associated with different types of prime ends shows that cross-cuts (as needed in U_i or U_e) can be chosen so that pieces of accessible unstable manifolds together with these cross–cuts form the boundary of a topological annulus $\hat{A}$ whose forward iterates converge to Θ. See Figure 6. The analysis leading to the choice of crosscuts is the content of the proof of the following lemma in [1]:

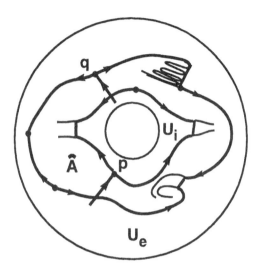

FIGURE 6. A closed annulus is constructed from cross–cuts and segments of unstable manifolds.

Lemma. *Assume the hypotheses of Theorem 1B on Θ and on both internal and external accessible periodic orbits. Then there exists a closed (topological) annulus $\hat{A} \subset A$ such that $f(\hat{A}) \subset \hat{A}$ and $\bigcap f^n(\hat{A}) = \Theta$.*

Adding the hypothesis that there exists an accessible twist pair now gives the rectangle necessary for the proof of Theorem 3. The fact that forward interates of the annulus $\hat{A}$ converge to Θ guarantees that a point x of specified rotation number α, as in Theorem 3, is actually in Θ.

Results related to Theorem 3 and the associated Lemma in the case where f is an area–contracting map of the plane and Θ is the closure of the unstable manifold of a planar saddle are given in [1].

REFERENCES

1. K. Alligood and M. Barge, Rotation intervals for attractors, in preparation.
2. K. Alligood and J. Yorke, Accessible saddles on fractal basin boundaries, Ergod. Th. and Dynam. Sys. 12 (1992), 377–400.
3. K. Alligood and J. Yorke, Rotation intervals for chaotic sets, preprint.
4. D. Aronson, M. Chory, G. Hall, R. McGehee, Bifurcations from an invariant circle for two–parameter families of maps of the plane: a computer assisted study, Commun. Math. Phys. 83 (1982), 303–354.
5. M. Barge and R. Gillette, Rotation and periodicity in plane separating continua, Ergod. Th. and Dynam. Sys. 11 (1991), 619–631.

6. G. Birkhoff, Sur quelques courbes fermées remarquables, Bull. Soc. Math. France 60 (1932), 1–26.

7. C. Carathéodory, Uber die Begrenzung einfach zusammenhangender Gebiete, Math. Ann. 73 (1913), 323–370.

8. M. Cartwright and J. Littlewood, Some fixed point theorems, Ann. Math. 54 (1951), 1–37.

9. M. Casdagli, Periodic orbits for dissipative twist maps, Ergod. Th. Dynam. Syst. 7 (1987), 165–173.

10. E. Collingwood and A. Lohwater. 1966. *Theory of Cluster Sets. Cambridge Tracts in Mathematics and Mathematical Physics, No. 56*, Cambridge University Press, Cambridge.

11. R. Devaney. 1986. *An Introduction to Chaotic Dynamical Systems*, Benjamin/Cummings, Menlo Park, CA.

12. M. Handel, The rotation set of a homeomorphism of the annulus is closed, Commun. Math. Phys. 127, 339–349.

13. K. Hockett and P. Holmes, Josephson's junction, annulus maps, Birkhoff attractors, horseshoes and rotation sets, Ergod. Th. Dynam. Syst. 6 (1986), 205–239.

14. A. Katok, Some remarks on Birkhoff and Mather twist theorems, Ergod. Th. Dynam. Syst. 2 (1982), 183–194.

15. P. LeCalvez, Properties des attracteurs de Birkhoff, Ergod. Th. Dynam. Syst. 8 (1988), 241–310.

16. J. Mather, Existence of quasi–periodic orbits for twist homeomorphisms of the annulus, Topology 21 (1982), 457–467.

17. K. Meyer and G. Hall. 1992. *Introduction to Hamiltonian Dynamical Systems and the N–Body Problem*, Springer–Verlag, New York.

18. J. Palis and W. deMelo. 1982 *Geometric Theory of Dynamical Systems: An Introduction*, Springer–Verlag, New York.

E-mail address: alligood@gmu.edu "Kathleen T. Alligood"

3
THE FIXED POINT PROPERTY IN DIMENSION ONE

DAVID P. BELLAMY Department of Mathematical Sciences, University of Delaware, Newark, Delaware 19716, USA

ABSTRACT. In 1969, R. H. Bing [3] gave an overview of the fixed point property for compact metric spaces. Much has been learned since the publication of Bing's article. This article reviews a few of the results since that time and poses some new questions. The emphasis is on results where "one–dimensional methods" can be used.

Let X be a topological space and let $f : X \to X$ denote a continuous function. A *fixed point* of f is a point $x \in X$ such that $f(x) = x$. The space X has the fixed *fixed point property* provided every continuous $f : X \to X$ has a fixed point; at times it is also useful to refer to the fixed point property for certain types of maps; such as surjections, homeomorphisms, etc.

It is easy to prove that every space with the fixed point property is connected; but many connected spaces fail to have the fixed point property, notably spheres. A *continuum* here means a compact connected metric space. The primary concern here will be the fixed point property for continua, though similar results are also known in the compact Hausdorff case in many cases.

The celebrated result of L. E. J. Brouwer that all closed n-cells have the fixed point property set the ball rolling and led to the belief, or hope, that the fixed point property was related to the property of being acyclic in one sense or another, especially since the failure of the property for all spheres was so elegantly apparent (using the map $f(x) = -x$ on $\{x \in \mathbb{R}^n \mid ||x|| = 1\}$). There have been both deep

1991 *Mathematics Subject Classification.* primary 54H25, secondary 54C05, 54F15.
Key words and phrases. fixed point property, continua.

positive results and surprising counterexamples in this area since Brouwer. One manifestation of this hope is Problem 107 in the Scottish Book [13, p. 190], which asks whether a subcontinuum of the plane with connected complement always has the fixed point property.

It became clear in the 1950's and 60's that indecomposability was intimately connected with the fixed point property, or more precisely, with the failure of the fixed point property. In fact, the first indication of this connection had come much earlier in [9]. A continuum X is *indecomposable* if it is not decomposable. It is *decomposable* if it has a pair of proper subcontinua H and K with $X = H \cup K$.

Another property, called *pathwise* indecomposability may also be of interest here. A pathwise connected continuum X is *pathwise indecomposable* if and only if it is not the union of two proper subcontinua both of which are path connected. Although formally similar to indecomposability, this property is possessed by much simpler continua than those which are actually indecomposable. The connection is probably more semantic than real, except for the rather curious question on the subject below.

In the 1960's H. Bell [2] and K. Sieklucki [19] proved that a plane continuum with connected complement which admits a fixed point free map f must contain an indecomposable continuum M in its boundary such that $f(M) \subseteq M$. This guaranteed, for example, that hereditarily decomposable, non–separating plane continua have the fixed point property. Borsuk [4] had established this fact earlier for locally connected non-separating continua.

In [22], G. S. Young proved that a continuum in which every increasing union of arcs is contained in an arc had the fixed point property. L. E. Ward, Jr., [20] proved an analogous theorem in the non-metrizable case, and in [21] Young considered whether this result could be generalized to continua which are arcwise connected but contain no simple closed curves, called uniquely arcwise connected continua. He provided a counterexample which was not homeomorphic with a subset of the plane, and had a fixed point free map which was not one-to-one.

This example consists of two $\sin(\frac{1}{x})$ circles attached along a common arc in such a way that a 180° rotation about the point p is a homeomorphism, together with a ray which winds counterclockwise around the outside of the figure so that the entire symmetric difference of the two $\sin(\frac{1}{x})$ circles is in the closure of the ray. The two points labeled b then go to the point p, and p is pushed forward to a (along an arc, not pictured, which curves out of the plane) the points of the ray are then all pushed ahead 180° to make the map continuous, and fixed point free.

This example leads to two questions.

1. Does every uniquely arcwise connected continuum have the fixed point property for homeomorphisms?
2. Does every uniquely arcwise connected continuum in the plane have the fixed point property?

Both these questions have now led to theorems, and there are some further questions in this area as well.

I will use [p, q] to denote the unique arc from p to q in any uniquely arcwise connected continuum. The notations [p, q), (p, q], and (p, q) are then self-explanatory. In [16], L. E. Mohler uses some techniques from measure theory and descriptive set theory to prove the following result.

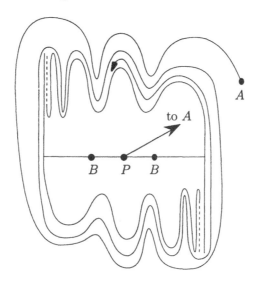

Figure 1.

Theorem. *Every homeomorphism of a uniquely arcwise connected continuum has a fixed point.*

Proof. This proof itself is quite elegant and will be included here. It depends on the following two results which are stated without proof.

Theorem. *Let X be a compact metric space, and let $h : X \to X$ be a homeomorphism. Then there is a Borel measure, μ, invariant under h; that is, $\mu(h(A)) = \mu(A)$ for every Borel set A.*

Remark on proof. : This follows from a generalization of the Brouwer fixed point theorem to compact convex subsets of Banach spaces, specifically, the set of Borel probability measures on X.

Theorem. *Every analytic set is universally measureable.*

Explanation. An *analytic set* in a complete separable metric space is a set which is a continuous image of some Borel set in such a space. A set $A \subseteq X$ is *universally measurable* if and only if it is measurable in the completion of every Borel measure on X. The *completion* of a Borel measure μ is the measure obtained by declaring all subsets of sets of measure zero to be measurable, and to have measure zero. It is defined on the σ - algebra generated by the Borel subsets of X and the newly defined sets of measure zero. This σ - algebra then depends on the measure μ; however the theorem states that it always includes the analytic sets.

Mohler's theorem can then be established as follows:

Proof of Mohler's Theorem. Let X be a uniquely arcwise connected continuum and let $h : X \to X$ be a homeomorphism. By way of contradiction, assume h is a fixed point free. Let $a \in X$ be arbitrary. The union of the arcs $\left[h^n(a), h^{n+1}(a)\right] n \in \mathbb{Z}$, is a one-to-one continuous image of the real line $\mathbb{R}$. In particular, let $w : \mathbb{R} \to X$ be a one-to-one map such that $w(n) = h^n(a)$ and $h^n(w(t)) = w(t + n)$ for each $n \in \mathbb{Z}$ and each $t \in \mathbb{R}$.

(This map may be constructed as follows: Let $w_0 : [0, 1] \to [a, h(a)]$ be a homeomorphism with $w_0(0) = a$ and $w_0(1) = h(a)$. Define $w : \mathbb{R} \to X$ by $w(t) = h^n(w_0(t - n))$ where n is any integer (unique unless $t \in \mathbb{Z}$) such that $t - n \in [0, 1]$. The pasting theorem guarantees that w is continuous, and the facts that X contains no simple closed curve and h is one-to-one and fixed point free guarantee that w is one-to-one.)

Thus, $w(\mathbb{R}) = \bigcup\limits_{n=-\infty}^{+\infty} w[n, n + 1)$, and these sets are pairwise disjoint.

Now, for every $x \in X$, the unique arc $[a, x]$ is the union of an arc $[a, t(x)] \subseteq w(\mathbb{R})$ and an arc $[t(x), x]$ disjoint from $w(\mathbb{R})$ except for the point $t(x)$. Each $t(x)$ is then in exactly one set $w[n, n + 1)$. Let $X_n = \{x \in X \mid t(x) \in w[n, n + 1)\}$ (Of course, $t(x) = x$ if $x \in w(\mathbb{R})$) Let $A \subseteq X - \{a\}$ and let $C(X, A)$ denote the set of continuous functions $\alpha : [0, 1] \to X - A$, such that $\alpha(0) = a$. $C(X, A)$ is a complete separable metric space if A is an F_σ subset of X. In particular, let $A = \left\{w(-\frac{1}{n}) \mid n \text{ a positive integer }\right\} \cup \{w(1)\}$. The evaluation-at-1 map $e : C(X, A) \to X$ defined by $e(\alpha) = \alpha(1)$ is continuous and maps $C(X, A)$ onto X_0. Thus (since $C(X, A)$ is Borel in itself), X_0 is analytic, and by applying positive and negative powers of h, so is every X_n. Further, $\{X_n \mid n \in \mathbb{Z}\}$ is a pairwise disjoint collection, and $\bigcup\limits_{n=-\infty}^{+\infty} X_n = X$.

Now, let μ_0 be a probability measure invariant under h and let μ denote its completion. Then every X_n, being analytic, is μ–measurable, and by invariance under h (easily shown to apply to μ as well as μ_0), all X_n's have the same measure. This is impossible since

$$\mu(X) = \sum_{n=-\infty}^{+\infty} \mu(X_n) = 1.$$

This contradiction completes the proof.

The next theorem proven by C. L. Hagopian [8] answers Young's other question.

Theorem. *(Hagopian) Uniquely arcwise connected plane continua have the fixed point property.*

The proof of this is far more intricate. It involves deep arguments in the topology of the plane. The arcwise connectedness prevents too much folding, since there are strong restrictions on how an indecomposable continuum can lie in the given continuum. A major refinement of Bing's dog and rabbit argument comes from constructing a ray one gets by always moving on an arc from x toward $f(x)$, finally settling into a special subcontinuum and then, with some more plane topology, to a fixed point.

I am not going to try to summarize Hagopian's argument better than this. Reading it and studying it will certainly improve almost any mathematician's understanding of the plane. He has recently generalized it to homotopically simply connected plane continua. In his 1969 paper mentioned earlier, Bing asked whether tree-like continua have the fixed point property. A continuum X is *tree-like* if and only if it has arbitrarily fine open covers with nerve an acyclic one-complex; that is, a tree. (The *nerve* of a finite covering ζ is the abstract simplicial complex whose vertices are the members of the covering ζ and a nonempty set $M \subseteq \zeta$ is a simplex if and only if $\bigcap M \neq \phi$.) A *dendroid* is a continuum X which is arcwise connected and hereditarily unicoherent (which means that the intersection of every two subcontinua H and K of X is a continuum unless it is empty). A λ-*dendroid* is a continuum which is both hereditarily decomposable (which means that every subcontinuum with more than one point is decomposable) and hereditarily unicoherent. Howard Cook [6] proved that dendroids and λ-dendroids are tree like. Every dendroid is a λ-dendroid, but not conversely.

Roman Mańka [13] proved that λ-dendroids have the fixed point property, settling an important special case of the tree-like question. Mańka's theorem, like Hagopian's, is very deep and intricate and makes use of a major generalization of Bing's dog and rabbit argument. In this case, however, there is a different proof of the theorem of Young [22] mentioned earlier which gives the flavor of Mańka's proof and should serve as a springboard for understanding it.

Theorem. *(Young) Let X be an arcwise connected continuum in which every union of a collection of arcs linearly ordered by the subset relation is contained in an arc. Then X has the fixed point property.*

Proof. Suppose X is such a continuum. Notice that X contains no simple closed curves and so is uniquely arcwise connected. Then suppose $f : X \to X$ is a fixed point free continuous map. Let $a_0 \in X$. Then $f\left([a_0, f(a_0)]\right) \cap [a_0, f(a_0)] \neq \phi$ since $f(a_0)$, at least, belongs to the intersection. Applying Zorn's lemma to the collection $\{[p, q] \subseteq [a_0, f(a_0)] \mid [p, q] \cap f[p, q] \neq \phi\}$ we obtain an arc $[a, b]$ minimal with respect to the property that $[a, b] \cap f[a, b] \neq \phi$. Necessarily, either $f(a) = b$ or $f(b) = a$. We label the points so that $f(a) = b$. Then, it follows that $[a, b] \cap [b, f(b)] = \{b\}$. Define $\mathcal{L} = \{[a, x] \mid x \in X \text{ and } [a, x] \cap [x, f(x)] = \{x\}\}$. $\mathcal{L}$ is a family of arcs partially ordered by $\subseteq$, and $[a, b] \in \mathcal{L}$. By Hausdorff maximality there is a maximal totally ordered subset ζ of $\mathcal{L}$, and by hypothesis, $\bigcup \zeta$ is contained in an arc $[a, q]$. By choosing q appropriately (making $[a, q]$ as short as possible) we can guarantee that $\bigcup \zeta$ is not a subset of $[a,p]$ for any $p \in [a,q)$. There are two cases to consider: either $[a,q] \in \zeta$ or $[a,q] \notin \zeta$. In the first case, $[a,q] \cap [q,f(q)] = \{q\}$. Let U and V denote open sets such that $q \in U, f(q) \in V, U \cap V = \emptyset$, and $U \subseteq f^{-1}(V)$. Choose $r \in (q, f(q)]$ such that $[q,r] \subseteq U$. Then $f[q,r] \subseteq V$, and $[f(q),f(r)] \subseteq f[q,r]$, since $f[q,r]$ is necessarily arcwise connected. Thus, $[f(q),f(r)] \subseteq V$. Also, $[r,f(q)] \cup [f(q),f(r)]$ is arcwise connected, so $[r,f(r)] \subseteq [r,f(q)] \cup [f(q),f(r)]$. Now $[f(q),f(r)] \cap U = \emptyset$, so $([f(q),f(r)] \cup [r,f(q)]) \cap [a,r]$ is a connected set which meets U only in the point r. Consequently, $([f(q),f(r)] \cup [r,f(q)]) \cap [a,r] = \{r\}$. But then, $[a,r] \in \mathcal{L}$, and every $[a,b] \in \zeta$ is a subset of $[a,r]$. By maximality, $[a,r] \in \zeta$, and therefore $[a,r] \subseteq [a,q]$. But $[a,q]$ is a proper subset of $[a,r]$, so this is a contradiction.

In the alternate case, if $[a,q] \notin \zeta$, we also have $[a,q] \notin \mathcal{L}$ by maximality of ζ. There are two subcases. Either $f(q) \in [a,q]$ or $f(q) \notin [a,q]$. In the latter case, $[q,f(q)] \cap [a,q] = [p,q]$ for some $p \neq q$. Let U, V, and W be disjoint open sets with $p \in W, q \in U, f(q) \in V$, and $f(U) \subseteq V$. Now, by choice of q, there exists $r \in [p,q]$ such that $[a,r] \in \zeta$, and $[r,q] \subseteq U$. Then $f[r,q]$ is an arcwise connected continuum containing $[f(r),f(q)]$ and missing p. Since $[r,f(r)]$ misses p also, $[r,f(q)]$ does not contain p. But, $[r,q] \cup [r,f(q)]$ contains $[q,f(q)]$ and hence must contain p. This is a contradiction.

In the final case, $[a,q] \notin \mathcal{L}$ and $f(q) \in [a,q]$. By assumption $f(q) \neq q$, so $[a,q] = [a,f(q)] \cup [f(q),q]$, and $[f(q),q]$ is a nondegenerate arc. Let U and V be disjoint open sets with $f(q) \in V$, $q \in U$, and $f(U) \subseteq V$. Again, there exists $x \in [f(q),q]$ such that $[a,x] \in \zeta$, and $[x,q] \subseteq U$. Thus, $f([x,q]) \subseteq V$, so that $[f(x),f(q)] \subseteq V$. Now $[x,f(x)] \cap [a,x] = \{x\}$ and $[x,f(x)] \subseteq [x,f(q)] \cup [f(x),f(q)]$ and since $[x,f(q)] \cap [a,x] = [x,f(q)]$, no point of $[x,f(q)]$ except x can belong to $[x,f(x)]$. Thus, $[x,f(x)] \subseteq \{x\} \cup [f(x),f(q)]$. This is impossible since $x \in U$ and $[f(x),f(q)] \subseteq V$ and $U \cap V = \emptyset$. This contradiction completes the proof.

The preceeding argument is basically a version of Mańka's argument in [13] with the technical details suppressed by considering a simplified form of the hypotheses. With this argument in mind, it should be fairly easy for anyone to understand Mańka's proof for λ-dendroids. (Maćkowiak gave a similar, but shorter proof in [12], but Mańka's is easier to follow in my opinion.) This argument can also be easily adapted to prove Ward's theorem for the non-metric case [20].

A continuum is *weakly chainable* [7] and [11] if and only if it is a continuous image of a pseudo-arc. I will take this as the definition, for our purposes, though it is actually a characterization theorem.

In [15], Piotr Minc proved that weakly chainable, non-separating plane continua have the fixed point property; actually he proved more.

Theorem. *(Minc) Any non-separating plane continuum X with the property that every indecomposable continuum in the boundary of X is contained in a weakly chainable subcontinuum of X must have the fixed point property.*

The next question is then especially interesting. Does every weakly chainable tree-like continuum have the fixed point property?

It is not true that every weakly chainable acyclic continuum has the fixed point property; as mentioned in Bing's article, Knill [9] showed that the cone over the plane continuum $\{(r,\theta) \mid r \le 1, \text{ or } \theta \ge 1 \text{ with } r = 1 + \frac{1}{\theta}\}$ (expressed in polar coordinates) is a counterexample. Every contractible continuum is weakly chainable, since it must be a continuous image of the cone over the Cantor set which is in turn a continuous image of a pseudo-arc. Hence, tree-likeness is an essential part of the preceeding question; no assumptions about mere acyclicity can suffice.

These four theorems, by Mańka, Minc, Mohler, and Hagopian, represent the major achievements in terms of positive results in the one-dimensional case since 1969. In terms of negative results, that is to say counterexamples, the situation is as follows:

The author [1] gave the first example of a tree-like continuum which admits a fixed point free mapping.

Essentially the only way to get around the dog and rabbit argument in a tree-like continuum seems to be the "split, switch, & push" method. This is illustrated conceptually as follows: The half-open interval $[0,\infty)$ has the fixed point property for onto maps. To sidestep this define

$$H = \left\{(x,y) \in \mathbb{R}^2 \mid y = 0 \ \& \ x \ge 2\right\} \cup \left\{(x,y) \in \mathbb{R}^2 \mid x = 2 - |y| \ \& \ -2 \le y \le 2\right\}.$$

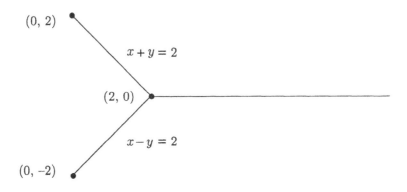

Figure 2.

There is a unique map $h : H \to H$ which fixes the points $(0,2)$ and $(0,-2)$ and doubles the x coordinate of every point of H; h is two-to-one for those points (x,y) with $1 \le x < 2$. Otherwise, it is one-to-one. If r is the reflection, $r(x,y) = (x,-y)$, then $r \circ h : H \to H$ is a fixed point free map.

This easy example is not compact, and finding ways to make the same thing happen in a compact space has been the theme of much of the search for counterexamples in fixed point theory in the one-dimensional case. The standard example of a uniquely arcwise connected continuum which admits a fixed point free map (see figure 1, above) uses this, as do the author's [1] and others [17] and [18] tree-like examples. These tree like examples are indecomposable continua. Apart from questions in the plane, probably the most interesting open question concerning tree like continua and fixed points is:

Does every hereditarily indecomposable tree-like continuum have the fixed point property?

It seems entirely possible that hereditary indecomposability, like its "opposite," hereditary decomposability, could defeat the split, switch, and push strategy.

Another question which is intriguing is:

Does every uniquely arcwise connected continuum X which admits a fixed point free map f contain a pathwise indecomposable subcontinuum M invariant under f? The example in Figure 1 is pathwise indecomposable, and is also not weakly chainable. This suggests another question:

Does every weakly chainable, uniquely arcwise connected continuum have the fixed point property?

Finally, returning to the question of whether non-separating plane continua have the fixed point property, it has long been my suspicion that the answer may turn out to be "yes" for tree-like continua and "no" for continua which contain disks. Thus, I am going to formally state them as two separate questions:

Does every non-separating plane continuum with nonempty interior in $\mathbb{R}^2$ have the fixed point property? This question is of interest also for homeomorphisms, open mappings, and possibly for some other classes of mappings.

Does every tree-like plane continuum have the fixed point property? The same special cases may be of interest.

References

1. D. P. Bellamy, *A tree-like continuum with the fixed point property*, Houston J. Math. **6** (1980), 1–13.
2. H. Bell, *On fixed point properties of plane continua*, Trans. Amer. Math. Soc. **128** (1967), 539–548.
3. R. H. Bing, *The elusive fixed point property*, Amer. Math. Monthly **76** (1969), 119–132.
4. K. Borsuk, *Einige Satz über stetige Streckenbilder*, Fund. Math. **18** (1932), 198–213.
5. _____, *A theorem on fixed points*, Bull Acad. Polon. Sci. Cl. III **2** (1954), 17–20.
6. H. Cook, *Tree-likeness of dendroids and λ-dendroids*, Fund. Math. **68** (1970), 119–22.
7. L. Fearnley, *Characterization of the continuous images of the pseudo-arc*, Trans. Amer. Math. Soc. **111** (1964), 380–399.
8. C. L. Hagopian, *Uniquely arcwise connected plane continua have the fixed point property*, Trans. Amer. Math. Soc. **248** (1979), 85–104.

9. O. H. Hamilton, *Fixed points under transformations of continua which are not connected in Kleiner*, Trans. Amer. Math. Soc. **44** (1938), 18–24.

10. R. J. Knill, *Cones, products, and fixed points*, Fund. Math. **60** (1967), 35–46.

11. A. Lelek, *On weakly chainable continua*, Fund. Math. **51** (1962/3), 271–282.

12. T. Maćkowiak, *Fixed point property for λ-dendroids*, Bull. Acad. Polon. Sci. Sér. Sci. Math. Astronom. Phys. **26** (1978), 61–64.

13. R. Mańka, *Association and fixed points*, Fund. Math. **91** (1976), 105–121.

14. R. D. Mauldin (ed.), *The Scottish Book - Mathematics from the Scottish Café*, Birkhauser, Boston, 1981.

15. P. Minc, *A fixed point theorem for weakly chainable plane continua*, Trans. Amer. Math. Soc. **317** (1990), 303–312.

16. L. Mohler, *The fixed point property for homeomorphisms of 1-arcwise connected continua*, Proc. Amer. Math. Soc. **52** (1975), 451–456.

17. L. G. Oversteegen, and J. T. Rogers, Jr., *Fixed point free maps on tree-like continua*, General Topology and its applications **13** (1982), 85–95.

18. _____ , *An inverse limit description of an atriodic tree-like continuum and an induced map without a fixed point*, Houston J. Math. **6** (1980), 549–564.

19. K. Sieklucki, *On a class of plane acyclic continua with the fixed point property*, Fund. Math. **63** (1968), 257–278.

20. L. E. Ward, Jr., *A fixed point theorem for chained spaces*, Pacific J. Math. **9** (1959), 1273–1278.

21. G. S. Young, *Fixed point theorems for arcwise connected continua*, Proc. Amer. Math. Soc. **11** (1960), 880–884.

22. G. S. Young, *The introduction of local connectivity by change of topology*, Amer. J. Math **68** (1946), 479–494.

E-mail address: bellamy@math.udel.edu "David P. Bellamy"

4

MENGER MANIFOLDS

ALEX CHIGOGIDZE Department of Mathematics and Statistics, University of Saskatchewan, Saskatoon, Saskatchewan, S7N 0W0, Canada
KAZUHIRO KAWAMURA Institute of Mathematics, University of Tsukuba, Tsukuba-City, Ibaraki 305, Japan
E. D. TYMCHATYN Department of Mathematics and Statistics, University of Saskatchewan, Saskatoon, Saskatchewan, S7N 0W0, Canada

ABSTRACT. We present a survey of Menger manifold theory and applications of this theory to several areas of modern geometric topology.

CONTENTS

1991 *Mathematics Subject Classification*. Primary 57Q12.
Key words and phrases. Menger compacta, n-homotopy, autohomeomorphism, continuum.
* the first named author was partially supported by NSERC grant No. OGP0155552
** the second named author was supported by an NSERC International Fellowship
*** the third named author was partially supported by NSERC grant No. A5616
The paper was written while the second author was visiting the University of Saskatchewan.

1. INTRODUCTION

The main purpose of the present paper is to give a survey of the theory of Menger Manifolds and to outline some possible directions for applications of the ideas, techniques and philosophy of the field to other branches of modern geometric topology. Of course, the presented directions reflect to some extent our own prejudices. However, we have tried to represent as broad a perspective of research as possible. Particularly, application areas include continuum theory, dimension theory, homotopy theory, shape theory, homogeneity, fixed points, infinite-dimensional topology, topological dynamics, homeomorphism groups and group actions. On the other hand, we would like to emphasize that because of the significant body of related work (especially over the past decade, since Bestvina's fundamental thesis) we have missed with probability one some of the interesting results.

Before we start a detailed discussion, let us outline the construction of the k-dim-ensional universal Menger compactum $\mu^k, k \geq 0$. We partition the standard unit cube I^{2k+1}, lying in $(2k + 1)$-dimensional Euclidean space R^{2k+1}, into 3^{2k+1} congruent cubes of the "first rank" by hyperplanes drawn perpendicular to the edges of the cube I^{2k+1} at points dividing the edges into three equal parts, and we choose from these 3^{2k+1} cubes those which intersect the k-dimensional skeleton of the cube I^{2k+1}. The union of the selected cubes is denoted by $I(k, 1)$. In an analogous way, we divide every cube entering as a term in $I(k, 1)$ into 3^{2k+1} congruent cubes of the "second rank" and the union of all analogously selected cubes of the second rank is denoted by $I(k, 2)$. If we continue the process we get a decreasing sequence of compacta

$$I(k,1) \supset I(k,2) \supset \cdots .$$

The compactum $\mu^k = \cap \{I(k,i) : i \in N\}$ is called the *k-dimensional universal Menger compactum*. Note that $\mu^k = M_k^{2k+1}$, where $M_k^n, 0 \le k < n$, denotes the "k-dimensional Menger compactum constructed in the n-dimensional cube I^n " (precise definitions of M_k^n are given in section 2.1).

It is clear that μ^0 coincides with the Cantor discontinuum 2^ω and, consequently, is the only zero-dimensional compactum with no isolated points [Br]. It was shown by Sierpiński [Sie$_2$] that μ^0 is a universal space for the class of all zero-dimensional compacta.

Positive dimensional Menger compacta M_k^n were originally defined within the classical dimension theory by Menger in [Me$_1$]. They are generalizations of the Cantor discontinuum and of Sierpiński's universal curve M_1^2 [Sie$_1$]. Let us recall that the space M_1^2 is universal for the class of all at most 1-dimensional planar compacta [Sie$_1$]. Further, it was shown by Menger [Me$_1$] that the 1-dimensional Menger compactum $\mu^1 = M_1^3$ is universal for the class of all at most 1-dimensional compacta. Generally, it was conjectured in [Me$_1$] that M_k^n is a universal space for the class of all at most k-dimensional compacta embeddable in R^n (Menger's problem). This well-known problem has an interesting and long history. As was already mentioned, this problem was known to have a positive solution for $(k,n) = (1,2)$ and $(k,n) = (1,3)$. A positive solution in the case $(k,n) = (n-1,n)$ was also given by Menger [Me$_1$]. Results of Lefschetz [Le] and Bothe [Bot$_1$] produced a positive solution in the case $(k,n) = (k, 2k+1)$. The ultimate result solving Menger's problem affirmatively was given in 1973 by Štanko [Š$_2$].

If a complete verification of Menger's conjecture took half a century, one might imagine what kinds of difficulties confronted researchers who were trying to characterize topologically the Menger compacta. The first result in this direction (dealing with μ^0) has already been mentioned. Note also that μ^0 being a topological group is homogeneous. Two 1-dimensional Menger compacta M_1^2 and $\mu^1 = M_1^3$ were also characterized topologically by Whyburn [Why] and Anderson [An$_1$] (see also [An$_2$]), respectively. Anderson's theorem characterizes μ^1 as a 1-dimensional locally connected continuum with no local separating points and with no non-empty open subspaces embeddable in the plane.

Comparing the characterizations of μ^0 and μ^1, it is very hard to see what are the roots of these results, what is common between them and, finally, how they can be generalized to higher dimensions in order to (at least) make a reasonable conjecture concerning the characteristics of M_k^n. Unfortunately, in full generality some of these questions still remain open. To the best of our knowledge, there are not even conjectures concerning the characteristic properties of the compacta M_k^n when $1 < k < n \le 2k$. Even in the most important cases, i.e. for $n = 2k+1$ ($k > 1$), until very recently no solutions had been given to the fundamental problems of topological characterization of the μ^k and their topological homogeneity. To illustrate, it was not even known whether we obtain a homeomorphic copy of μ^k if we carry out the Lefschetz construction [Le], i.e. if we follow the procedure described above beginning, not with the $(2k+1)$-dimensional cube but, let us say, with the $(2k+1)$-dimensional unit simplex. In 1-dimensional cases this follows from the above mentioned characterization theorems for M_1^2 and μ^1. If $k = n - 1$, this fact was also known [C].

It was only in 1984 that answers were obtained to all the questions listed above. In a remarkable work [Be], Bestvina laid down the basis of the whole theory of k-dimensional Menger manifolds (briefly, μ^k-manifolds) and gave solutions to almost all the fundamental problems.

In order to present the main properties of μ^k, it is useful to go back to another classical result of dimension theory (see [Hur],[Le],[Nö], [PT]) which states that $(2k + 1)$-dimensional Euclidean space R^{2k+1} has a much stronger property, than simple universality with respect to the class of all at most k-dimensional compacta. Namely, R^{2k+1} is strongly k-universal, which means that any (continuous) map $f : X \longrightarrow R^{2k+1}$ of any at most k-dimensional compactum X into R^{2k+1} can be approximated by embeddings. It can be easily shown that an LC^{k-1}-compactum is strongly k-universal if and only if it has DD^kP (Disjoint k-Disks Property), the property crucial in finite and infinite dimensional manifold theories. The fundamental theorem of Bestvina can be stated now as follows:

μ^k **is the only** k**-dimensional, strongly** k**-universal** AE(k)**-compactum.**

A simple comparison of the above result with the fundamental theorem of Toruń-czyk [Tor$_2$] (stating that the Hilbert cube is the only strongly ∞-universal, i.e. having DD^kP for each k, AE(∞)-compactum) shows that, from a certain point of view, the k-dimensional analogue of the Hilbert cube Q should be considered to be, not the usual k-dimensional cube I^k, but the k-dimensional Menger compactum μ^k (and, conversely, the Hilbert cube may be viewed as the "infinite-dimensional Menger compactum"). In addition, one can observe a fairly deep analogy between the theories of μ^k-manifolds and Q-manifolds themselves. We would like to emphasize that philosophically the first part of the present paper is built so that each more or less significant result of μ^k-manifold theory is coupled with its infinite-dimensional prototype. We hope that this approach has certain advantages and simplifies the understanding of the subject.

The reader can also find examples of a situation when a purely intuitive expectation - that the finite-dimensional theory should be, at least in some way, simpler than the infinite-dimensional one - is indeed the case. In all such examples we indicate differences and, when possible, outline their roots.

Finally, we present several results, which, strictly speaking, are not formally tied with μ^k-manifold theory, but which use ideas and tools of the field.

The authors would like to express their thanks to Professors R. J. Daverman, J.

Keesling, R.B.Sher and R. Pol for helpful information and suggestions.

BASIC DEFINITIONS AND NOTATIONS

In most cases all necessary notions are defined in the text. All spaces under consideration are assumed to be *separable and metrizable and maps are continuous*. N, I and R stand for the natural numbers, the closed unit segment and the real line respectively.

Throughout the paper, the following notations will also be used:

(a) Let X be a metric space and $\mathcal{B}$ be a collection of subsets of X.

    (1) $\operatorname{ord}\mathcal{B} = \max\{k$ members of $\mathcal{B}$ with nonempty intersection $\}$

    (2) To simplify notation, we sometimes denote $\cup\mathcal{B}$ by $|\mathcal{B}|$

    (3) For a subset S of $X, ST(S,\mathcal{B})$ denotes the collection of members of $\mathcal{B}$ which intersect S and $st(S,\mathcal{B}) = |ST(S,\mathcal{B})|$

    (4) $\operatorname{mesh}\mathcal{B} = \sup\{\operatorname{diam} B : B \in \mathcal{B}\}$

    (5) For a subset S of X, $\operatorname{int}(S)$ and $\operatorname{Fr}(S)$ denote the (topological) interior and the boundary of S respectively.

(b) Let M be a locally compact polyhedron with a triangulation L. We use the notation $M = |L|$.

    (1) The barycentric subdivision of L is denoted by βL and inductively, $\beta^i L = \beta(\beta^{i-1}L)$. For a simplex σ of L, b_σ denotes the barycenter of σ.

    (2) The k-skeleton of L is denoted by $L^{(k)}$ and $M^{(k)} = |L^{(k)}|$. We often identify $M^{(k)}$ with $L^{(k)}$.

    (3) As in a), $\operatorname{mesh} L = \sup\{\operatorname{diam} \sigma : \sigma \in L\}$. For a subset S of $M, ST(S,L)$ denotes the collection of simplexes of L which meet S and $st(S,L) = |ST(S,L)|$.

    The same notations apply to cell complexes.

(c) When M is a PL manifold, its triangulation is always assumed to be combinatorial. The (manifold) boundary is denoted by ∂M and $\operatorname{int} M = M - \partial M$.

(d) A space X is *k-connected* $(X \in C^k)$ if $\pi_i(X) = 0$ for each $i \le k$. A space X is *locally k-connected* $(X \in LC^k)$ if for every $x \in X$ and every neighborhood U of x there exists a smaller neighborhood V of x with the property that every map $\alpha : \partial B^{i+1} \longrightarrow V(i = 0, 1, \ldots, k)$ extends to a map $\beta : B^{i+1} \longrightarrow U(B^i$ denotes the i-cell).

(e) A compactum X is an *UV^k-compactum* if there is an embedding of X into the Hilbert cube Q such that for every neighborhood U of X in Q there exists a smaller neighborhood V of X in Q such that every map $\alpha : \partial B^i \longrightarrow V(i = 0, 1, \ldots, k)$ extends to a map $\beta : B^i \longrightarrow U$. Obviously, the class of UV^0-compacta coincides with the class of continua. A map with UV^k fibers is called a UV^k-map.

(f) A map $f : X \longrightarrow Y$ is *locally k-soft* [Šče] if for every at most k-dimensional space B, its closed subspace A and two maps $g : A \longrightarrow X$ and $h : B \longrightarrow Y$ with $fg = h|A$, there exist a neighborhood U of A in B and a map $s : U \longrightarrow X$ such that $s|_A = g$ and $h|_U = fs$. If, in this definition, $U = B$, then we say that f is *k-soft*. If in the definition of k-soft maps the spaces A and B are assumed to be polyhedra, then we get the definition of *polyhedrally k-soft* maps.

(g) A space X is an *absolute (neighborhood) extensor* in dimension k (notation: $X \in A(N)E(k)$) if its constant map is (locally) k-soft. It is well-known that within the class of Polish spaces (i.e. completely metrizable and separable) $\mathrm{ANE}(k) = LC^{k-1}$ and $\mathrm{AE}(k) = LC^{k-1} \cap C^{k-1}$. Note also that we sometimes use the abbreviation $X \in \mathrm{AE}(\infty)$ to denote the fact that X is an absolute extensor.

(h) $C(X,Y)$ denotes the set of all continuous functions from X into Y endowed with the limitation topology. If X is compact this topology coincides with the compact-open topology.

Partitions of Menger compacta. The reader who is already familiar with Bestvina's thesis [Be] may skip this part and proceed to §2 , after noting the definitions. As an introduction to the area, [Be] may be somewhat demanding. Anderson's characterization [An₁] of μ^1 is also a difficult entry. [MOT] is a gentle introduction to μ^1. The following is a very brief outline of these results.

The proof of the characterization theorem of μ^k, due to Bestvina, consists of two parts.

(A) every k-dimensional $\mathrm{AE}(k)$-compactum with DD^kP can be obtained from a $(k-1)$-connected, $(2k+1)$-dimensional, compact, PL manifold by the construction described in §2 below.

(B) compacta obtained in (A) are mutually homeomorphic.

The main technique of the proof of (B) in [Be] is a powerful generalization of the partition argument originally due to Anderson [An₁] and Bing [Bi₂]. Instead of working directly with partitions of Menger manifolds, Bestvina consistently considered partitions of their manifold neighborhoods (see the construction in §2). This approach made it possible to achieve the necessary connectivity degrees of members of partitions as well as correct intersection properties of the partitions. On the other hand, if $k = 1$, the connectivity requirement is rather mild ($= LC^0 \cap C^0$), so, in this case, we may work with partitions of μ^1 directly (in this way we sometimes even obtain slightly stronger results than in the general case). Here we review partition properties of μ^1 from the above point of view. It is not obvious whether this approach can be generalized to higher dimensions, and even if it can be, it is not clear at all whether it would have an advantage over the approach of Bestvina.

We decided to outline it here, having in mind that the 1-dimensional case is much easier and the 1-dimensional arguments sometimes serve as models of proofs in the general case.

Recall that a partition (weak partition) $\mathcal{U}$ of a compactum X is a finite cover by regularly closed subcontinua (subcompacta) of X such that $\{\operatorname{int} U : U \in \mathcal{U}\}$ forms a mutually disjoint collection (see also section 2.1).

Definition 1.1. A partition $\mathcal{U}$ of X is called a *brick partition* if for each pair $\{U_1, U_2\}$ of elements of $\mathcal{U}$ (possibly $U_1 = U_2$), $\operatorname{int}(U_1 \cup U_2)$ is uniformly LC^0 and connected.

Each locally connected continuum admits a brick partition of arbitrarily small mesh [Bi₂]. For 1-dimensional, locally connected continua this can be strengthened with a much simpler proof [MOT] (see also [An₂, Theorem 1]. Observe the similarity between the partitions in the next theorem and the partitions obtained from Bestvina's construction in section 2.1.

Theorem 1.2. [MOT, Theorem 2.9]. *Let X be an 1-dimensional locally connected continuum. Then there is a sequence $\mathcal{U}_i$ of brick partitions of X so that for each i:*

    (1) $\mathcal{U}_{i+1} < \mathcal{U}_i$ and $\operatorname{mesh} \mathcal{U}_i \to 0$,

    (2) $\dim Bd(U) = 0$, whenever $U \in \mathcal{U}_i$,

    (3) $\operatorname{ord} \mathcal{U}_i \leq 2$.

Theorem 1.4 below provides useful machinery for constructing homeomorphisms of Menger compacta.

Definition 1.3. Let $\mathcal{U}_1$ and $\mathcal{V}_1$ be two partitions of continua X and Y respectively. A bijection $\phi_1 : \mathcal{U}_1 \longrightarrow \mathcal{V}_1$ is called an isomorphism if, for each $U_1, \ldots, U_n \in \mathcal{U}_1$, $\cap\{U_i : i = 1, \ldots, n\} \neq \emptyset$ if and only if $\cap\{\phi_1(U_i) : i = 1, \ldots, n\} \neq \emptyset$ (isomorphisms in this sense are called one-to-one correspondences in [Be]). Suppose that $\mathcal{U}_2$ and $\mathcal{V}_2$ are refinements of $\mathcal{U}_1$ and $\mathcal{V}_1$ respectively and let $\phi_2 : \mathcal{U}_2 \longrightarrow \mathcal{V}_2$ be an isomorphism. If, for each pair $(U_1, U_2) \in \mathcal{U}_1 \times \mathcal{U}_2$, $U_2 \subseteq U_1$ if and only if $\phi_2(U_2) \subseteq \phi_1(U_1)$, then ϕ_2 is called an isomorphism with respect to ϕ_1.

Theorem 1.4. ([An₁], [Bi₂]). *Suppose that $\mathcal{U}_i$ and $\mathcal{V}_i$ are sequences of partitions of locally connected continua X and Y respectively such that*

    (1) $\mathcal{U}_{i+1} < U_i$ and $\mathcal{V}_{i+1} < \mathcal{V}_i$ for each i,

    (2) $\operatorname{mesh} \mathcal{U}_i \to 0$ and $\operatorname{mesh} \mathcal{V}_i \to 0$, and

    (3) *there is a sequence $\{\phi_i : \mathcal{U}_i \longrightarrow \mathcal{V}_i\}$ of bijections so that for each i, ϕ_{i+1} is an isomorphism with respect to ϕ_i.*

Suppose further that C and D are closed sets of X and Y respectively and $h : C \longrightarrow D$ is a homeomorphism such that

    (4) $\phi_i(st(x, \mathcal{U}_i)) \subseteq st(h(x), \mathcal{V}_i)$ *for each $x \in C$ and each i.*

Then there is an homeomorphism $\phi : X \longrightarrow Y$ which is an extension of h.

Suppose that X and Y are 1-dimensional locally connected continua with DD^1P (= with no local separating points and with no open subspaces embeddable in the plane). In [An$_2$] and [MOT] it is proved that we may define sequences $\{\mathcal{U}_i\}$ and $\{\mathcal{V}_i\}$ of brick partitions of X and Y respectively such that they are all of order 2 with 0-dimensional boundaries and satisfy the conditions of Theorem 1.4 ($C = D = \emptyset$). Thus X and Y are homeomorphic.

The sequences $\{\mathcal{U}_i\}$ and $\{\mathcal{V}_i\}$ are constructed inductively. Starting with $\mathcal{U}_1 = \{X\}$, $\mathcal{V}_1 = \{Y\}$ and a trivial function $\phi_1 : \mathcal{U}_1 \longrightarrow \mathcal{V}_1$, we assume that $\mathcal{U}_i$, $\mathcal{V}_i$ and $\phi_i : \mathcal{U}_i \longrightarrow \mathcal{V}_i$ are already defined. Let $\mathrm{Fr}(\mathcal{V}_i) = \cup\{\ \mathrm{Bd}\ V : V \in \mathcal{V}_i\}$. Take a "fine" refinement $\mathcal{V}_{i+1}$ of $\mathcal{V}_i$. Then $\mathcal{V}_{i+1}$ determines a decomposition $\mathcal{V}_{i+1}|_{\mathrm{Fr}(\mathcal{V}_i)}$ of $\mathrm{Fr}(\mathcal{V}_i)$. Using the 0-dimensionality of $\mathrm{Fr}(\mathcal{U}_i)$, it is easy to "imitate" the pattern of $\mathcal{V}_{i+1}|_{\mathrm{Fr}(\mathcal{V}_i)}$ by a decomposition of $\mathrm{Fr}(\mathcal{U}_i)$, i.e. we may define a closed decomposition $\mathcal{D}_{i+1}$ of $\mathrm{Fr}(\mathcal{U}_i)$ and an isomorphism $\delta : \mathcal{D}_{i+1} \longrightarrow \mathcal{V}_{i+1}|_{\mathrm{Fr}(V_i)}$ with respect to ϕ_i. Using DD^1P, we may "extend" $\mathcal{D}_{i+1}$ to a partition $\mathcal{U}_{i+1}$ of X which admits an isomorphism $\phi_{i+1} : \mathcal{U}_{i+1} \longrightarrow \mathcal{V}_{i+1}$ with respect to ϕ_i such that $\phi_{i+1}|\mathcal{D}_{i+1} = \delta$. Note that the mesh of $\mathcal{U}_{i+1}$ may be close to the mesh of $\mathcal{U}_i$. Reversing the roles of $\mathcal{U}_{i+1}$ and $\mathcal{V}_{i+1}$ and repeating the argument we may suppose that $\mathcal{U}_{i+1}$ also has small mesh.

To prove the Z-set unknotting theorem for μ^1 we need the following lemma.

Definition 1.5. [MOT]. Let S be a finite set and let A be a subset of $S \times S$ satisfying the following properties:

 (1) A is symmetric (i.e. $(x, y) \in A$ if and only if $(y, x) \in A$) and contains the diagonal of $S \times S$,
 (2) for each pair $x, y \in S$ there is a sequence $x = s_1, \ldots, s_m = y$ such that $(s_i, s_{i+1}) \in A$ for each $i = 1, \ldots, m-1$,

Then A is called a coherent subset of $S \times S$.

Theorem 1.6. [MOT, Theorem 5.7]. *Let m be a positive integer and suppose that a coherent subset A of $\{1, \ldots, m\} \times \{1, \ldots, m\}$ is given. Let K be a Z-set of μ^1 and let $\mathcal{U} = \{U_1, \ldots, U_m\}$ be an order ≤ 2, weak partition of K such that $\dim(U_i \cap U_j) \leq 0$, whenever $i \neq j$, and*

 (1) $(i, j) \in A$ *whenever* $U_i \cap U_j \neq \emptyset$.

Then there is an order 2 brick partition $\mathcal{V} = \{V_1, \ldots, V_m\}$ of μ^1 such that

 (2) $\mathcal{V}|_K = \mathcal{U}$ *and*
 (3) $V_i \cap V_j \neq \emptyset$ *if and only if* $(i, j) \in A$.

In order to prove the Z-set unknotting theorem, we proceed as in the proof of the characterization theorem described above. Let K and L be Z-sets in μ^1 and suppose that a homeomorphism $h : K \longrightarrow L$ is given. The homeomorphism h determines a sequence of at most order 2, weak partitions $\{\mathcal{K}_i\}$ and $\{\mathcal{L}_i\}$ of K and

L which are isomorphic in the sense of Definition 1.3. Theorem 1.6 asserts that we may extend $\mathcal{K}_i$ and $\mathcal{L}_i$ to brick partitions $\mathcal{U}_i$ and $\mathcal{V}_i$ of μ^1 which are also isomorphic. Then we apply Theorem 1.4 and get the desired extension of h.

The underlying idea of the proof of the statement (B) in the higher dimensional cases is essentially the same as above. We need to construct sequences of partitions (of manifold neighborhoods) as in Theorem 1.4 and with the necessary connectivity of elements of these partitions (and their intersections as well). If we ignore the last requirement, the construction of sequences of isomorphic partitions is not terribly difficult [Be, 2.4]. To achieve the necessary connectivity properties of partition elements, we need to attach cells of various dimensions to these partition elements. However, this operation destroys the correct intersection property of partitions. This difficulty is overcome in [Be,. sections 2.4 - 2.7].

2. THEORY OF MENGER MANIFOLDS

2.1. Construction. In this section we describe the constructions of Menger compacta given by Menger, Lefschetz and Bestvina. Although the Menger construction has already been described in the Introduction, we restate it in a slightly different (but equivalent) form for later use. Throughout this section we fix integers $0 \leq k \leq n$.

I. Menger's construction [Me$_1$]

For a metric on R^n, we use the maximum metric, i.e. $d(\{x_i\}, \{y_i\}) = \max\{|x_i - y_i| : 1 \leq i \leq n\}$ for each $\{x_i\}, \{y_i\} \in R^n$. If A is a subset of R^n, then $N(A, \varepsilon)$ denotes the ε-neighborhood of A with respect to the above metric.

Let I^n be the n-cell in R^n with the standard linear structure $L_0 = \{I^n\}$. For each integer $i \geq 0$, L_i denotes the cell complex structure of I^n whose n-cells are of the form

$$\Pi\{[m_t/3^i, (m_t + 1)/3^i] : m_t = 0, 1, \ldots, 3^i - 1\}.$$

We define the Menger compactum M_k^n as follows: Let $M_0 = I^n$ and (by induction) for each integer $i \geq 1$ let

$$M_{i+1} = st(|L_i^{(k)}|, L_{i+1}) \cap M_i = st(|L_i^{(k)}|, L_{i+1}|M_i) = st(|L_i^{(k)}|M_i|, L_{i+1}|M_i).$$

Clearly, $\{M_i\}$ is a decreasing sequence of compacta and $M_k^n = \cap M_i$ is called the *Menger compactum of type* (k, n). When $n = 2k + 1$, we use the symbol $\mu^k = M_k^{2k+1}$ to denote the k-*dimensional universal Menger compactum*.

We will also use another description of M_k^n([Di$_2$, Chap.2]).

Let $V_i = \{(2t + 1)/2 \cdot 3^i : t = 0, 1, \ldots, 3^i - 1\}$ and $V = \cup V_i$. Note that $B_i = V_i \times \cdots \times V_i$ (n factors) is the set of centers of n-cells of L_i. Let $\mathcal{P}$ be the finite collection of homeomorphisms of R^n defined by permutations of coordinates of

R^n. For each i we define $D_i = \cap\{\alpha(\{c\} \times I^{n-k-1}) : \alpha \in \mathcal{P}$ and $c \in V_i^{k+1}$ and $N_i = N(D_i, 1(2 \cdot 3^i))$. Then D_i can be regarded as the "dual $(n - k - 1)$-skeleton" of L_i and N_i as the regular neighborhood of D_i. It is easy to see that $M_k^n = I^n - \cup\{\text{int}(N_i) : i = 1, \ldots, n\}$, $I^n = M_i \cup N_i$ and $\partial M_i = M_i \cap (N_i \cup \partial I^n)$.

If we perform the above construction starting with R^n (instead of I^n) we get a closed subspace U_k^n of R^n which is a countable union of copies of M_k^n (in this case L_0 is the partition of R^n into unit cubes).

II. Lefschetz's construction [Le]

Replacing cell complexes in (I) by simplicial complexes, we obtain Lefschetz's construction. We describe it in slightly general form.

Let M be a PL n-manifold with a (combinatorial) triangulation L. Inductively, we define a sequence $\{M_i\}$ of PL n-manifolds and their triangulations L_i as follows.

Let $M_0 = M$ and $L_0 = L$. Let $M_1 = st(L^{(k)}, \beta^2 L_0)$, $L_1 = \beta^2 L_0|M_1$ and suppose that M_i and L_i have already been defined. Consider $\beta^2 L_i$ and let $M_{i+1} = st(L_i^{(k)}, \beta^2 L_i)$ and $L_{i+1} = \beta^2 L_i|M_{i+1}$. Then $\{M_i\}$ is a decreasing sequence and $\cap M_i \neq \emptyset$.

If M is the n-simplex with the standard simplicial complex structure, then the resulting compactum $\cap M_i$ is denoted by L_k^n. In particular, $L_k^{2k+1} = \mu^k$. (We use the same symbol as in (I). This notation is justified by the Characterization Theorem 2.4.1 (due to Bestvina). Notice that M_{i+1} may be regarded as a regular neighborhood of the k-skeleton of M_i (with respect to L_i).

III. Bestvina's construction [Be]

In Bestvina's construction, the k-skeleta in (II) are replaced by the dual k-skeleta. Suppose that M is a PL n-manifold with a (combinatorial) triangulation L. As in (II), we define a sequence $\{M_i\}$ of PL n-manifolds and their triangulations $\{L_i\}$ as follows:

Let $M_0 = M$, $L_0 = L$ and suppose that we already have defined M_i and L_i. Then $M_{i+1} = \cup\{st(b_\sigma, \beta^2 L_i) : b_\sigma$ is the barycenter of $\sigma \in L_i$ with $\dim \sigma \geq n - k\}$ and $L_{i+1} = \beta^2 L_i|M_{i+1}$.

If M is the n-simplex with the standard simplicial complex structure, the resulting compactum $\cap M_i$ is denoted by T_k^n. In particular, T_k^{2k+1} is denoted by μ^k. (Again this is justified by the Characterization Theorem). Observe that M_{i+1} is regarded as a regular neighborhood of the dual k-skeleton of M_i (with respect to L_i).

It might be worth noting the differences among these constructions. Consider the properties of partitions which are naturally induced by each of the above constructions. For simplicity, we formulate these properties only for M_k^n, L_k^n and T_k^n.

Proposition 2.1.1. *There are sequences* $\{P_i\}, \{Q_i\}$ *and* $\{R_i\}$ *of partitions of* M_k^n, L_k^n *and* T_k^n *respectively satisfying the following conditions:*

(a) P_{i+1}, Q_{i+1} *and* R_{i+1} *are refinements of* P_i, Q_i *and* R_i *respectively.*

(b) $\lim \operatorname{mesh} P_i = \lim \operatorname{mesh} Q_i = \lim \operatorname{mesh} R_i = 0$.

(c) $\operatorname{ord} P_i = n + 1, \lim \operatorname{ord} Q_i = \infty$ *and* $\operatorname{ord} R_i = k + 1$.

(d) *for each* $p_1, p_2, \ldots, p_t \in P_i$, $\cap\{p_j : j = 1, \ldots, t\}$ *is an at most is* k-*dimensional* $LC^{k-1} \cap C^{k-1}$-*compactum* **or** *an at most* k-*dimensional cell.*

(e) *for each* $q_1, q_2, \ldots, q_t \in Q_i$, $\cap\{q_j : j = 1, \ldots, t\}$ *is an at most* k-*dimensional* $LC^{k-1} \cap C^{k-1}$-*compactum* **or** *an at most* k-*dimensional simplex.*

(f) *for each* $r_1, r_2, \ldots, r_t \in R_i$, $\cap\{r_j : j = 1, \ldots, t\}$ *is a* $(k - t + 1)$-*dimensional* $LC^{k-t} \cap C^{k-t}$ *-compactum.*

Proof. The partitions defined below satisfy the desired conditions:

$$P_i = \{e \cap M_k^n : e \in L_i\}, Q_i = \{s \cap L_k^n : s \in L_i\} \text{ and } R_i = \{s \cap T_k^n : s \in L_i\}.$$

Remark 2.1.2. (1) In the last case, if $(k, n) = (1, 3)$, we have a partition of μ^1 with 0-dimensional intersections of all adjacent elements. In this sense, the partition determined by Bestvina's construction can be regarded as a generalization of the partition of the Menger curve considered in [An$_{1-2}$] and [MOT].

(2) We may obtain characterizations of (compact) Menger manifolds as well as (compact) Q-manifolds in terms of the existence of certain types of partitions [Ka$_2$] (The proof is based on the characterization theorems of Bestvina (see 2.4.1) and Toruńczyk).

2.2. n-homotopy. In this section we describe an adequate homotopy language for μ^{n+1}-manifold theory. This is the so called n-homotopy. The related notion of μ^{n+1}-homotopy was first exploited in [Be].

Definition 2.2.1. [Chi$_2$]. Two maps $f, g : X \longrightarrow Y$ are said to be n-homotopic (written $f \overset{n}{\simeq} g$) if the compositions $f\alpha$ and $g\alpha$ are homotopic in the usual sense for any map $\alpha : Z \longrightarrow X$ of an arbitrary at most n-dimensional space Z.

It can be easily seen [Chi$_2$, Proposition 2.3] that if $\dim X \leq n + 1$ and $Y \in LC^n$ then maps $f, g : X \longrightarrow Y$ are μ^{n+1}-homotopic in the sense of Bestvina [Be, Definition 2.1. 7] if and only if they are n-homotopic. Note also that if, in the above definition, we consider, instead of compact, only polyhedral Z, then we get Fox's definition of n-homotopy [Fo].

In practice it is convenient to use the following statement [Chi$_2$, Proposition 2.4].

Proposition 2.2.2. *Maps* $f, g : X \longrightarrow Y$ *are* n-*homotopic if and only if for some (or, equivalently, any)* n-*invertible map* $\alpha : Z \longrightarrow X$ *with* $\dim Z \leq n$ *the compositions* $f\alpha$ *and* $g\alpha$ *are homotopic.*

Definition 2.2.3. ([Ho], [Chi$_1$]). *A map $\alpha : A \longrightarrow X$ is said to be n-invertible if for any map $\beta : B \longrightarrow X$ with $\dim B \leq n$ there is a map $\gamma : B \longrightarrow A$ such that $\gamma\alpha = \beta$.*

Note that 0-invertible maps between metrizable compacta are precisely the so-called Milutin maps, i.e. surjections having a regular averaging operator (see [Pe], [Ho]). Note also that each compactum is an n-invertible image of an n-dimensional compactum.

Of course, homotopic maps are n-homotopic for each $n \geq 0$ but not conversely. Indeed, consider the identity map and the constant map of an arbitrary non-contractible $LC^\infty \cap C^\infty$-compactum. Nevertheless, n-homotopic maps have several useful properties. Here we mention only three of them [Be, Section 2.1], [Chi$_2$] (compare with [Hu]).

Proposition 2.2.4. *For each $Y \in LC^n$, there exists an open cover $\mathcal{U} \in \operatorname{cov}(Y)$ such that any two $\mathcal{U}$-close maps of any space into Y are n-homotopic.*

Proposition 2.2.5. *(n-Homotopy Extension Theorem). Let $Y \in LC^n$. Then for each $\mathcal{U} \in \operatorname{cov}(Y)$, there exists $\mathcal{V} \in \operatorname{cov}(Y)$ refining $\mathcal{U}$ such that the following condition holds:*

$(*)_n$ For any at most $(n+1)$-dimensional space B, any closed subspace A and any two $\mathcal{V}$-close maps $f, g : A \longrightarrow Y$ such that f has an extension $F : B \longrightarrow Y$ it follows that g also has an extension $G : B \longrightarrow Y$ which is $\mathcal{U}$-close to F.

Proposition 2.2.6. *Let $Y \in LC^n$. Suppose that A is closed in B and $\dim B \leq n + 1$. If maps $f, g : A \longrightarrow Y$ are n-homotopic and f admits an extension $F : B \longrightarrow Y$, then g also admits an extension $G : B \longrightarrow Y$, and it may be assumed that $F \stackrel{n}{\simeq} G$.*

A map $f : X \longrightarrow Y$ is an n-homotopy equivalence if there is a map $g : Y \longrightarrow X$ such that $gf \stackrel{n}{\simeq} \operatorname{id}_X$ and $fg \stackrel{n}{\simeq} \operatorname{id}_Y$ [Chi$_2$]. The spaces X and Y in this case are said to be n-homotopy equivalent. For example, any map between arcwise connected spaces is a 0-homotopy equivalence. Note also that the $(n+1)$-dimensional sphere S^{n+1} is n-homotopy equivalent to the one-point space.

In general we have the following algebraic characterization of n-homotopy equivalences [Whi, Theorem 2].

Proposition 2.2.7. *A map $f : X \longrightarrow Y$ between at most $(n+1)$-dimensional locally finite polyhedra is an n-homotopy equivalence if and only if it induces isomorphisms of homotopy groups of dimension $\leq n$, i.e., f induces a bijection between the components of X and Y and the homomorphism $\pi_k(f') : \pi_k(C_X) \longrightarrow \pi_k(C_Y)$ is an isomorphism for each $k \leq n$ and each pair of components $C_X \subseteq X$ and $C_Y \subseteq Y$ with $f(C_X) \subseteq C_Y$, where $f' : C_X \longrightarrow C_Y$ denotes the restriction of f.*

It is well-known that each ANR-compactum is homotopy equivalent to a finite polyhedron [We$_2$]. The following statement [Chi$_5$, Proposition 1.5] is an "n-homotopy version" of West's result.

Proposition 2.2.8. *Every at most $(n+1)$-dimensional locally compact LC^n-space is properly n-homotopy equivalent to an at most $(n+1)$-dimensional locally finite polyhedron.*

Therefore, 2.2.7 holds even for maps between at most $(n+1)$-dimensional locally compact LC^n-spaces.

Proper n-homotopies and all associated notions are defined in the natural way and we do not repeat them here. In order to state an algebraic characterization of proper n-homotopy equivalences similar to 2.2.7 we need some preliminary definitions. We say that a proper map $f : X \longrightarrow Y$ between locally compact spaces induces an epimorphism of i-th homotopy groups of ends ($i \geq 0$) if for every compactum $C \subseteq Y$ there exists a compactum $K \subseteq Y$ such that for each point $x \in X - f^{-1}(K)$ and every map $\alpha : (S^i, *) \longrightarrow (Y - K, f(x))$ there exists a map $\beta : (S^i, *) \longrightarrow (X - f^{-1}(C), x)$ and a homotopy $f\beta \simeq \alpha(rel *)$ in $Y - C$. We say that $f : X \longrightarrow Y$ induces a monomorphism of i-th homotopy groups of ends if for every compactum $C \subseteq Y$ there exists a compactum $K \subseteq Y$ such that for every map $\alpha : S^i \longrightarrow X - f^{-1}(K)$ with the property that $f\alpha$ is null-homotopic in $Y - K$ it follows that α is null-homotopic in $X - f^{-1}(C)$. As usual, f is said to induce an isomorphism of i-th homotopy groups of ends if it simultaneously induces an epimorphism and a monomorphism.

Proposition 2.2.9. *A proper map $f : X \longrightarrow Y$ between at most $(n+1)$-dimensional locally compact LC^n-spaces is a proper n-homotopy equivalence if and only if it induces isomorphisms of homotopy groups of dimension $\leq n$ and isomorphisms of homotopy groups of ends of dimension $\leq n$.*

Note that proper n-homotopies have been studied from the categorical point of view [HP].

The following proposition [Chi9], will be used below and indicates a major difference between n-homotopy and usual homotopy theories (compare with [W]).

Proposition 2.2.10. *Let M be an at most $(n+1)$-dimensional locally finite polyhedron. Suppose that there exists an at most $(n+1)$-dimensional finite polyhedron K and two maps $f : M \longrightarrow K$ and $g : K \longrightarrow M$ such that $gf \overset{n}{\simeq} \mathrm{id}_M$ (i.e. g is an n-homotopy domination). Then there exists an at most $(n+1)$-dimensional finite polyhedron T, containing K as a subpolyhedron, and an n-homotopy equivalence $h : T \longrightarrow M$, extending g such that f is a n-homotopy inverse of h.*

The analogous statement for proper n-homotopy dominations (near ∞) will be discussed in section 2.9.

2.3. Z-set unknotting and topological homogeneity. The notion of Z-set is very important in infinite-dimensional topology, especially in Q-manifold and l_2-manifold theories. We have the same situation in Menger manifold theory. Let us recall the definition.

Let A be a closed subset of a compactum X. It is easy to see that if I^0 denotes a one point space then the set $\{f \in C(I^0, X) : f(I^0) \cap A = \emptyset\}$ is dense in $C(I^0, X)$ if and only if A is nowhere dense. This simple observation allows one to define higher degrees of "nowhere denseness " as follows.

Definition 2.3.1. A closed subset A of a space X is a Z_n-set if the set $\{f \in C(I^n, X) : f(I^n) \cap A = \emptyset\}$ is dense in $C(I^n, X)$.

Definition 2.3.2. A closed subset A of a space X is a Z-set if the set $\{f \in C(Q, X) : f(Q) \cap A = \emptyset$ is dense in $C(Q, X)$.

Originally the concept of a Z-set was introduced in $[\text{An}_7]$ (see also, [Cha, Chapter 2] and $[\text{Tor}_1]$).

Proposition 2.3.3. *Let A be a closed subset of a Polish $\text{ANE}(n)$-space X . Then the following conditions are equivalent:*

  (1) *A is a Z_n-set .*
  (2) *for each at most n-dimensional locally finite polyhedron P the set $\{f \in C(P, X) : f(P) \cap A = \emptyset\}$ is dense in $C(P, X)$.*
  (3) *for each at most n-dimensional Polish space Y the set $\{f \in C(Y, X) : f(Y) \cap A = \emptyset\}$ is dense in $C(Y, X)$*

Note that each Z_n-set in any at most n-dimensional LC^{n-1}-space is a Z-set. If, in 2.3.3, X is locally compact, then the listed conditions are equivalent to the following:

  (4) *for each at most n-dimensional Polish space Y the set $\{f \in C(Y, X) : \text{cl}(f(Y)) \cap A = \emptyset\}$ is dense in $C(Y, X)$.*

Closed subsets satisfying the property (iv) are called *strong Z_n-sets*. These sets are especially important in the non-locally compact setting. Note that even in topologically complete ANR-spaces there are examples of Z_n-sets which are not strong Z-sets [BBMW].

Proposition 2.3.4. *One-point subsets of Menger manifolds are Z-sets.*

The following statements are versions of the powerful Z-set unknotting theorem.

Theorem 2.3.5. [Be, Theorem 3.1.4]. *Let Z_1 and Z_2 be two Z-sets in a μ^{n+1}-manifold M, and let $h : Z_1 \longrightarrow Z_2$ be a homeomorphism. Denote by $i_j : Z_j \longrightarrow M$ the inclusion map ($j = 1, 2$). If i_1 and $i_2 \circ h$ are n-homotopic, then h extends to a homeomorphism $H : M \longrightarrow M$ which is n-homotopic to id_M.*

Corollary 2.3.6. *Every homeomorphism between Z-sets of μ^n can be extended to an autohomeomorphism of μ^n.*

For $n = 0$ this result is well-known. A closed subset of μ^1 is a Z-set if and only if it does not locally separate μ^1. The result for $h = 1$ originally appears in [MOT].

A compactum X is called *strongly locally homogeneous* if for each point $x \in X$ and for each neighborhood U of x , there is a neighborhood V of x contained in U such that the following condition holds: for each point $y \in V$ there is a homeomorphism $h : X \longrightarrow X$ such that $h(x) = y$ and $h|_{(X-U)} = \mathrm{id}$.

Corollary 2.3.7. [Be, Theorems 3.2.1-2]. *μ^n is topologically homogeneous. Moreover, it is strongly locally homogeneous.*

Theorem 2.3.8. [Be, Theorem 3.1.3]. *Let M be a μ^n-manifold. For each open cover $\mathcal{U} \in \mathrm{cov}(M)$ there is an open cover $\mathcal{V} \in \mathrm{cov}(M)$ with the following property:*

(*) *if a homeomorphism $h : Z_1 \longrightarrow Z_2$ between two Z-sets of M is $\mathcal{V}$-close to i_1 (see notations in 2.3.5), then h can be extended to a homeomorphism $H : M \longrightarrow M$ which is $\mathcal{U}$-close to id_M.*

2.4. Topological characterization. The following characterization theorem [Be, Theorem 5.2.1 and Chapter 6] is central to the whole theory. We recall that a space X has DD^nP (Disjoint n-Disks Property) if for each open cover $\mathcal{U} \in \mathrm{cov}(X)$ and any two maps $\alpha : I^n \longrightarrow X$ and $\beta : I^n \longrightarrow X$ there are maps $\alpha_1 : I^n \longrightarrow X$ and $\beta_1 : I^n \longrightarrow X$ such that α_1 is $\mathcal{U}$-close to α, β_1 is $\mathcal{U}$-close to β and $\alpha_1(I^n) \cap \beta_1(I^n) = \emptyset$.

An embedding $f : Y \longrightarrow X$ is closed if $f(Y)$ is a closed subset of X. If additionally $f(Y)$ is a Z-set in X then f is called a *Z-embedding*.

Theorem 2.4.1. *The following conditions are equivalent for any n-dimensional locally compact $\mathrm{ANE}(n)$-space X :*

(1) *X is a μ^n-manifold.*

(2) *X has DD^nP.*

(3) *Each map of the discrete union $I^n \oplus I^n$ into X can be approximated arbitrarily closely by embeddings.*

(4) *Each proper map of any at most n-dimensional locally compact space into X can be approximated arbitrarily closely by closed embeddings.*

(5) *Each proper map of any at most n-dimensional locally compact space into X can be approximated arbitrarily closely by Z-embeddings.*

(6) *Each proper map $f : Y \longrightarrow X$ of any at most n-dimensional locally compact space Y into X such that the restriction $f|_{Y_0}$ onto a closed subset Y_0 is a Z-embedding can be arbitrarily closely approximated by Z-embeddings coinciding with f on Y_0.*

If additionally, X is compact and $(n-1)$-connected (i.e. $X \in \mathrm{AE}(n)$), then the conditions (ii)-(vi) give a topological characterization of the compactum μ^n.

Note that the theorem remains true even in the case $n = \infty$ (see Introduction).

If $n = 0$ and X is compact, the condition (ii) trivially implies that X has no isolated points. Therefore, in this case, X is homeomorphic to the Cantor discontinuum as has already been noted in the introduction.

Applying the above characterization in the case $k = 1$, we see that a compactum is homeomorphic to μ^1 if and only if it is a 1-dimensional, locally connected continuum with DD^1P. It is known (see [An$_2$] or [MOT]) that a locally connected continuum has DD^1P if and only if it has no local separating points and has no open subspaces embeddable in the plane. In this sense, Bestvina's characterization of μ^1 reduces to Anderson's.

In 2.1 three major constructions of the universal Menger compactum have been presented. Let us indicate another, spectral construction, given in [Pa, 14]. Let $\{G_i\}$ be a basis of open sets of S^1, such that G_i is an open cell with the property that diam $G_i \to 0$. Let us construct an inverse sequence $\{X_i, p_i^{i+1}\}$ as follows. We set $X_0 = S^1$ and get X_{i+1} from X_i by "bubbling over G_i", i.e. X_{i+1} is the quotient space obtained from the disjoint union $X_i \oplus X_i$ identifying the two copies of $x \in X_i$ precisely when $p_0^i(x) \notin G_i$ (here $p_0^i : X_i \longrightarrow X_0$ denotes the corresponding projection). The projection $p_i^{i+1} : X_{i+1} \longrightarrow X_i$ is defined as shown in Figure 1.

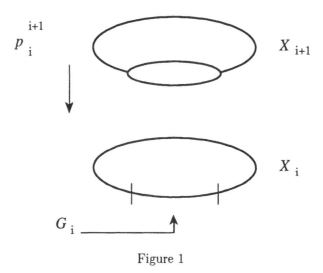

Figure 1

Note that p_i^{i+1} is a retraction. One can check directly that $X = \lim\{X_i, p_i^{i+1}\}$ is an 1-dimensional AE(1)-compactum with DD^1P and hence, by 2.4.1, X is a copy of μ^1. More careful consideration shows [Be] that we get μ^n if we start with $S^n = X_0$ (and proceed as above).

There are several other constructions of μ^n. To the best of our knowledge, all of them are defined as the limit spaces of inverse sequences(see, for example, [Dr$_2$], [Chi$_6$], [GHW], [Ka$_1$]).

2.5. Resolution, triangulation and stability theorems. It is a well-known result of Q-manifold theory (see, for example, [Cha]) that each locally compact ANR space X is a proper CE-image of a Q-manifold Q_X. Moreover, as the desired CE-map one can take the projection $\pi_X : Q_X = X \times Q \longrightarrow X$. In the n-dimensional case one should observe that CE-maps can be successfully replaced by UV^{n-1}-maps. On the other hand it is not clear what is the analogue of the operation of "taking the product by Q" in μ^n-manifold theory. We will discuss the corresponding results below. Let us start with the following resolution theorem [Be, Theorem 5.1.8].

Theorem 2.5.1. *Every locally compact LC^{n-1}-space is an image of a μ^n-manifold under a proper UV^{n-1}-map.*

A decisive step in finding a "full" analogue of the above mentioned resolution theorem for Q-manifolds, i.e. justified interpretation of the operation of "taking the product by Q" in μ^n-manifold theory, was made in [Dr₂]. A crucial point in Dranishnikov's argument is the following fundamental result.

Theorem 2.5.2. [Dr₂, Theorems 1-3]. *There are maps $f_n : \mu^n \longrightarrow Q$ and $g_n : \mu^n \longrightarrow \mu^n$ having the following properties:*

  (1) *The maps f_n and g_n are n-invertible, $(n-1)$-soft and polyhedrally n-soft.*
  (2) *All the fibers of the maps f_n and g_n are homeomorphic to μ^n.*
  (3) *The inverse images of the LC^{n-1}-compacta under the maps f_n and g_n are μ^n-manifolds.*
  (4) *The maps f_n and g_n have the parametric version of DD^nP, that is, any two maps $\alpha, \beta : I^n \longrightarrow \mu^n$ can be arbitrarily closely approximated by maps $\alpha_1, \beta_1 : I^n \longrightarrow \mu^n$ such that $f_n\alpha_1 = f_n\alpha$, $f_n\beta_1 = f_n\beta$, $g_n\alpha_1 = g_n\alpha$, $g_n\beta_1 = \beta_n\beta$) and $\alpha_1(I^n) \cap \beta_1(I^n) = \emptyset$.*

Note that polyhedral n-softness of f_n and g_n implies that both maps are UV^{n-1}-maps. The 0-soft maps between metrizable compacta are precisely the open surjections. Therefore 2.5.2 contains several earlier results in this direction (see, for example, [Wil, Theorem 1]).

Note that the multiplication operation by the Hilbert cube featured in the statements on Q-manifold theory at the beginning of this section may be interpreted as taking inverse images (of the space under consideration) under the projection $\pi_1 : Q \times Q \longrightarrow Q$. Consider the map $g_n : \mu^n \longrightarrow \mu^n$ in 2.5.2 to be the analogue of the projection π_1 in the theory of μ^n-manifolds. Now theorems 2.5.3 and 2.5.4 below may be considered as Triangulation and Stability results respectively. In the compact case they were proved in [Dr₂, Theorems 6 - 7] (for the non-compact case see [Chi₃]).

Theorem 2.5.3. *For any μ^n-manifold M there is an n-dimensional polyhedron P such that for any embedding of P in μ^n the inverse image $g_n^{-1}(P)$ is homeomorphic to M.*

Theorem 2.5.4. *For any μ^n-manifold M in μ^n the inverse image $g_n^{-1}(M)$ is homeomorphic to M.*

Using the map f_n as a guide it is possible to prove the following two results [Dr₃].

Theorem 2.5.5. *Any metrizable LC^n-compactum is an UV^{n-1}-image of:*

    (1) *an $(n+1)$-dimensional ANR-compactum.*

    (2) *a Q-manifold.*

    (3) *a $(2n+1)$-dimensional (topological) manifold $(n \geq 3)$.*

Theorem 2.5.6. *Any metrizable $LC^n \cap C^n$-compactum is an UV^{n-1}-image of:*

    (1) *an $(n+1)$-dimensional AR-compactum.*

    (2) *the Hilbert cube Q.*

    (3) *the $(2n+1)$-cell $(n \geq 3)$.*

Chernavskii's result [Che] on the existence of an UV^{n-1}-map of a $(2n+1)$-cell onto a $(2n+2)$-cell is a direct corollary of 2.5.6 (iii).

Note that the maps f_n and g_n can not be made n-soft. Also both of them fail to satisfy the property of preservation of Z-sets in the inverse direction. The map in the following resolution theorem [Chi$_6$] has this property and is "almost" n-soft. First, we need some definitions. If a map $f : X \longrightarrow Y$ is given, then a closed subset Z of X is said to be a *fibered Z-set* (with respect to f) if the identity map id_X can be arbitrarily closely approximated by maps $g : X \longrightarrow X$ such that $Im(g) \cap Z = \emptyset$ and $fg = f$. An embedding of some space into X is said to be a *fibered Z-embedding* if its image is a fibered Z-set.

Theorem 2.5.7. *There exists an n-invertible (see 2.4.4 for the definition) UV^{n-1}-map $h_n : \mu^n \longrightarrow Q$ satisfying the following conditions:*

    (1) *for each at most n-dimensional compactum Z, each open cover $\mathcal{U}$ of μ^n and each map $f : Z \longrightarrow \mu^n$ there exists a fibered Z-embedding (with respect to h_n) $g : Z \longrightarrow \mu^n$ which is $\mathcal{U}$-close to f and such that $h_n g = h_n f$.*

    (2) *there exists a subspace A of μ^n such that the restriction $h_n|_A : A \longrightarrow Q$ is an n-soft map and the complement $\mu^n - A$ is a σZ-set in μ^n.*

    (3) *if Z is a Z_n-set in Q, then $h_n^{-1}(Z)$ is a Z-set in μ^n.*

We conclude this section with the following statement [Chi$_8$].

Theorem 2.5.8. *For each locally finite polyhedron K there exists an n-invertible, proper, UV^{n-1}-surjection $h_K^n : M_K^n \longrightarrow K$ of some μ^n-manifold M_K^n onto K satisfying the following conditions:*

    (1) *if L is a closed subpolyhedron of K then its inverse image $(h_{K^n})^{-1}(L)$ is a μ^n-manifold.*

    (2) *if L is a closed subpolyhedron of K and Z is a Z-set in L, then the inverse image $(h_K^n)^{-1}(Z)$ is a Z-set in $(h_K^n)^{-1}(L)$.*

In [IS$_1$], it is proved that the Freudenthal compactification of any connected μ^n-manifold can be obtained from μ^n by a suitable identification of a Z-set. If a μ^n-manifold is compact, then the result can be slightly improved as follows [Ka$_4$].

Theorem 2.5.9. *Let M be a compact connected μ^n-manifold. Then there exist a surjection $f : \mu^n \longrightarrow M$ and a Z-set Z in μ^n such that*

(1) $f^{-1}f(Z) = Z$ *and $M - f(Z)$ is dense in M.*

(2) $f|_Z : Z \longrightarrow f(Z)$ *is a finite to one map.*

(3) $f|_{(\mu^n - Z)} : \mu^n - Z \longrightarrow M - f(Z)$ *is a homeomorphism.*

A Q-manifold version of this result has earlier been obtained in [B], [Br-M].

2.6. Approximation by homeomorphisms. It is a well-known result of Q-manifold theory [Cha, Section 43] that proper CE-maps between Q-manifolds are near-homeomorphisms (i.e. can be approximated arbitrarily closely by homeomorphisms). The μ^n-manifold version of this result was discovered in [Be, Theorem 4.3.1 and Chapter 6].

Theorem 2.6.1. *Proper UV^n-maps between μ^{n+1}-manifolds are near-homeomorphisms.*

Another important statement of Q-manifold theory [Cha, Sections 38-39] says that an infinite simple homotopy equivalence between Q-manifolds is homotopic to a homeomorphism. Let us note that this is not correct for homotopy equivalences (i.e. there exist non-homeomorphic but homotopy equivalent compact Q-manifolds). In μ^n-manifold theory we do not have simple homotopy obstructions and this significantly simplifies the corresponding result [Be, Theorem 2.8.6].

Theorem 2.6.2. *Each proper n-homotopy equivalence between μ^{n+1}-manifolds is proper n-homotopic to a homeomorphism.*

The following result, due to Ferry [F$_3$, Proposition 1.7] and improved slightly in [Ka$_6$], also illustrates this situation.

Theorem V 2.6.3. *Let $f : P \longrightarrow L$ be a map between compact polyhedra which induces an isomorphism between i-th homotopy groups for each $i \leq n$. Then there is a compact polyhedron Z and UV^n-maps $\alpha : Z \longrightarrow P$ and $\beta : Z \longrightarrow L$ such that $f\alpha \overset{n}{\simeq} \beta$.*

2.7. n-homotopy kernel: Open embedding and classification theorems.
The Open Embedding theorem for Q-manifolds states that for each Q-manifold X the product $X \times [0, 1)$ can be embedded into Q as an open subspace [Cha, Section 13]. Note that identifying X with $X \times [0, 1]$ (stability of Q-manifolds) the product $X \times [0, 1)$ is simply the complement of the image of an appropriately chosen Z-embedding of X into itself. Using this observation as a guide, we introduce the following notion [Chi$_8$].

Definition 2.7.1. An n-homotopy kernel $Ker_n(M)$ of a μ^{n+1}-manifold M is defined to be the complement $M - f(M)$, where $f : M \longrightarrow M$ is an arbitrary Z-embedding properly n-homotopic to id_M.

2.3.5. shows that the definition does not depend on the choice of Z-embedding.

The Open Embedding theorem for μ^{n+1}-manifolds can now be formulated in full analogy with its infinite-dimensional version [Chi$_8$, Theorem 2.1].

Theorem 2.7.2. *The n-homotopy kernel of each μ^{n+1}-manifold admits an open embedding into μ^{n+1}.*

In fact, one can show [I] that each map from an n-homotopy kernel of a μ^{n+1}-manifold into a μ^{n+1}-manifold is n-homotopic to an open embedding.

A Q-manifold X is $[0,1)$-stable if it is homeomorphic to $X \times [0,1)$. In the light of the above discussion, it is natural to call a μ^{n+1}-manifold *stable* if it is homeomorphic to its n-homotopy kernel. The following statement [Chi$_8$], [I] (compare with [Chi$_3$]) shows the connection between stable μ^{n+1}-manifolds and $[0,1)$-stable Q-manifolds.

Proposition 2.7.3. *A μ^{n+1}-manifold M is stable if and only if M admits a proper UV^n-surjection onto a $[0,1)$-stable Q-manifold.*

Another illustration of "stability" in the above sense is the fact [Chi$_8$, Proposition 2.2] that n-homotopy kernels of μ^{n+1}-manifolds are homeomorphic to their own n-homotopy kernels.

Theorem 2.6.2 completely solves the proper n-homotopy classification problem for μ^{n+1}-manifolds. The following statement [Chi$_8$] corresponds to the homotopy classification theorem for Q-manifolds, which states [Cha] that $X \times [0,1)$ and $Y \times [0,1)$ are homeomorphic whenever X and Y are homotopy equivalent Q-manifolds.

Theorem 2.7.4. *n-homotopy kernels of n-homotopy equivalent μ^{n+1}-manifolds are homeomorphic.*

In particular, n-homotopy equivalent stable μ^{n+1}-manifolds are homeomorphic. The homeomorphism guaranteed in 2.7.4 can be chosen to be n-homotopic to the given n-homotopy equivalence [I].

2.8. n-shape and the complement theorem. The famous Complement Theorem of Chapman [Cha, Section 25] states that if X and Y are Z-sets in Q then their complements $Q - X$ and $Q - Y$ are homeomorphic if and only if the shapes of X and Y coincide, i.e. $Sh(X) = Sh(Y)$. The obvious form of The Complement Theorem fails for μ^{n+1}. The equality of shapes of two Z-sets X and Y in μ^{n+1} is sufficient but far from necessary for the complements $\mu^{n+1} - X$ and $\mu^{n+1} - Y$ to be homeomorphic. Indeed, it can be easily seen that if the $(n+1)$-dimensional sphere S^{n+1} is embedded into μ^{n+1} as a Z-set, then $\mu^{n+1} - S^{n+1}$ is homeomorphic to $\mu^{n+1} - \{pt\}$. At the same time $Sh(S^{n+1}) \neq Sh(pt)$.

The problem was solved in [Chi$_2$] (see also [Chi$_3$],[Chi$_5$]) where the notion of n-shape was introduced. Relations between n-SHAPE and n-HOMOTOPY category are of the same nature as those between the categories of SHAPE and HOMOTOPY. Roughly, n-SHAPE is a "spectral completion" of n-HOMOTOPY. The main result in this direction is the following.

Theorem 2.8.1. *Let X and Y be Z-sets in μ^{n+1}. The complements $\mu^{n+1} - X$ and $\mu^{n+1} - Y$ are homeomorphic if and only if $n - Sh(X) = n - Sh(Y)$.*

We would like to mention some corollaries of this theorem and the definition of n-shape itself.

Corollary 2.8.2. *If $Sh(X) = Sh(Y)$, then $n - Sh(X) = n - Sh(Y)$.*

Corollary 2.8.3. *If X and Y are at most n-dimensional, then $Sh(X) = Sh(Y)$ if and only if $n - Sh(X) = n - Sh(Y)$.*

Corollary 2.8.4. *If Z-sets X and Y in μ^{n+1} are μ^{n+1}-manifolds, then the complements $\mu^{n+1} - X$ and $\mu^{n+1} - Y$ are homeomorphic if and only if the compacta X and Y are homeomorphic.*

Let us emphasize that the notion of n-equivalence introduced by Ferry [F$_3$, §2] as a generalization of Whitehead's notion of n-type coincides in several important cases with the notion of n-shape. Relations between these two concepts have been studied in [Chi$_6$]. We conclude this section by noting that 2.8.1 was extended [Sh] to a larger class of subspaces than Z-sets. These are the so-called weak Z-sets.

2.9. Menger manifolds with boundaries. The problem of putting a boundary on various types of manifolds has been considered in [BLL] (PL manifolds), [S] (smooth manifolds) and [CS] (Q-manifolds). It was proved in [CS] that if a Q-manifold M satisfies certain minimal necessary homotopy-theoretical conditions (finite type and tameness at ∞), then there are two obstructions $\sigma_\infty(M)$ and $\tau_\infty(M)$ to M having a boundary. These obstructions are elements of the groups defined in terms of projective class group and Whitehead group of ends of M respectively. It was shown in [CS] that the different boundaries that can be put on M constitute a whole shape class and a classification of all possible ways of putting boundaries on M can be done in terms of an appropriately defined group. It should be emphasized that the above mentioned obstructions essentially involve the Wall's finiteness obstruction [W].

The natural analogue of Wall's obstruction vanishes in the n-homotopy category. This is exactly what was stated in 2.2.10. We will see that this observation significantly simplifies the situation for μ^{n+1}-manifolds.

We say that a μ^{n+1}-manifold M admits a boundary if there exists a compact μ^{n+1}- manifold N such that $M = N - Z$, where Z is a Z-set in N. In this case we shall say that N is a compactification of M corresponding to the boundary Z, and conversely, Z is a boundary of M corresponding to the compactification N. We also need the following definition [Chi$_9$].

Definition 2.9.1. A space X is said to be n-tame at ∞ if for each compactum $A \subseteq X$ there exists a larger compactum $B \subseteq X$ such that the inclusion $X - B \longrightarrow X - A$ factors up to n-homotopy through an at most $(n + 1)$-dimensional finite polyhedron.

The compactification of a given μ^{n+1}-manifold which is n-tame at ∞ can be obtained as the limit space of some carefully chosen inverse sequence consisting of n-clean submanifolds and corresponding retractions, where n-clean submanifolds are defined as follows [Chi$_9$].

Definition 2.9.2. A μ^{n+1}-manifold M lying in a μ^{n+1}-manifold N is said to be n-clean in N provided that M is closed in N and there exists a closed subspace $\delta(M)$ of M such that the following conditions are satisfied:

(1) $\delta(M)$ is a μ^{n+1}-manifold.
(2) $(N - M) \cup \delta(M)$ is a μ^{n+1}-manifold.
(3) $\delta(M)$ is a Z-set in M.
(4) $\delta(M)$ is a Z-set in $(N - M) \cup \delta(M)$.
(5) $M - \delta(M)$ is open in N.

Sometimes we say that M is n-clean with respect to $\delta(M)$. Here are the corresponding results [Chi$_9$, Theorems 2.7,3.1, 3.3].

Theorem 2.9.3. A μ^{n+1}-manifold admits a boundary if and only if it is n-tame at ∞.

Proposition 2.9.4. If a compactum X is a boundary for a μ^{n+1}-manifold M, then a compactum Y is also a boundary for M if and only if $\dim Y \leq n + 1$ and $n - Sh(Y) = n - Sh(X)$.

Proposition 2.9.5. Every two μ^{n+1}-manifold compactifications of a μ^{n+1}-manifold are equivalent in the sense of Chapman-Siebenmann.

Recall that compactifications P and T of M are equivalent in the sense of Chapman-Siebenmann if for every compactum $A \subseteq M$ there is a homeomorphism of P onto T fixing A pointwise.

A classical example, due to Whitehead, of a 3-manifold W without boundary is defined as the complement of a continuum Wh in S^3 (called the *Whitehead continuum*), where Wh is obtained as the intersection of a nested sequence (T_i) of tori in S. The manifold W has an infinitely generated fundamental group at ∞. Let $n \geq 2$ and consider a μ^n-manifold M and a proper UV^{n-1}-map $f : M \longrightarrow W$. Since $n \geq 2$, f induces an isomorphism of fundamental groups of ends. Then it is easy to see that M is not 1-tame at ∞ and, consequently, cannot have a boundary.

The idea of using the Whitehead manifold first appeared in [Chi$_9$].

On the other hand, it is not difficult to show that the Freudenthal compactification of any connected μ^1-manifold contains its end as a Z-set. This implies that the Freudenthal compactification of every connected μ^1-manifold is homeomorphic to μ^1. In other words, every connected μ^1-manifold has a boundary.

The problem of whether a μ^{n+1}-manifold has a boundary which is itself a μ^{n+1}-manifold was also considered [CKS].

Definition 2.9.6. A proper map $f : Y \longrightarrow X$ between at most $(n+1)$-dimensional locally compact spaces is an *n-domination near* ∞ provided that there exists a cofinite subspace X_1 of X (i.e. X_1 is closed and $X - X_1$ has compact closure) and a proper map $g : X_1 \longrightarrow Y$ such that fg is properly n-homotopic to the inclusion map $X_1 \longrightarrow X$. If, in addition, for some cofinite subspace Y_1 of Y the composition $gf|_{Y_1}$ is properly n-homotopic to the inclusion map $Y_1 \longrightarrow Y$, then we say that f is an equivalence near ∞.

Definition 2.9.7. A space X is *finitely n-dominated near* ∞ if there exists a finite polyhedron P and an n-domination near $\infty, f : (P \times [0,1))^{(n+1)} \longrightarrow X$. If, in addition, f is an equivalence near ∞, then we say that X has finite n-type near ∞.

The following statement corresponds to 2.2.10.

Proposition 2.9.8. *Each finitely n-dominated near* ∞*, at most* $(n+1)$*-dimensional, locally compact* LC^h*-space has finite n-type near* ∞*.*

Using 2.9.8, it is possible to prove [CKS] the following theorem.

Theorem 2.9.9. *A* μ^{n+1}*-manifold M has a boundary which itself is a* μ^{n+1}*-manifold if and only if M is finitely n-dominated near* ∞*.*

It follows from 2.9.4 and 2.6.2 that if such a boundary exists, then it is uniquely determined.

The corresponding problem for Q-manifolds was considered in [CF].

3. HOMEOMORPHISM GROUPS

Let M be a Menger manifold and Auth(M) its group of autohomeomorphisms. The study of Auth(M) is well-developed. It is in their spaces of autohomeomorphisms that we see one of the major differences between Q-manifolds on the one hand and Menger manifolds on the other. Homeomorphism groups of Q-manifolds are l_2-manifolds [F_1] while homeomorphism groups of μ^n-manifolds M are totally disconnected. If $n = 0$ then Auth(M) is 0-dimensional and if $n \geq 1$ then Auth(M) is 1-dimensional. Hence, the only compact Lie groups to act effectively on a μ^n-manifold are finite groups. On the other hand, all compact 0-dimensional metric groups act effectively on all Menger manifolds. The algebraic properties of homeomorphism groups of Menger manifolds are rather similar to the properties of homeomorphism groups of n- and Q-manifolds. For an autohomeomorphism $H : X \longrightarrow X$, suppH $=$ cl $\{x \in X : H(x) \neq x\}$. If suppH is contained in a subset A, we say that H is supported on A.

3.1. Dimension of Auth(μ^n). Anderson proved that Auth(μ^1) is totally disconnected. The proof first appeared in [Bre] and extends to all Menger manifolds (see 3.1.4).

Definition 3.1.1. [Bre]. A continuum X is *locally setwise homogeneous* if there exists a basis $\mathcal{U}$ of connected open subsets of X and a dense subset $B \subseteq X$ such that for each $E \in \mathcal{U}$ and $a, b \in B \cap E$ there exists $h \in \mathrm{Auth}(X)$ supported on E such that $h(a) = b$.

Clearly, a locally setwise homogeneous continuum is locally connected. The Sierpiński curve M_1^2 is locally setwise homogeneous but not homogeneous. The solenoids are homogeneous but not locally setwise homogeneous.

Theorem 3.1.2. [Bre, Theorem 2.1]. Let X be a locally setwise homogeneous continuum. Then $\mathrm{Auth}(X)$ is at least 1-dimensional.

We outline a proof of 3.1.2. Let $\varepsilon > 0$ be a sufficiently small number and let U be a neighborhood of id_X in $\mathrm{Auth}(X)$ with $\mathrm{diam}\, U < \varepsilon$. Let $x, y \in X$ with $d(x, y) = \varepsilon$ and let A be a small diameter arc from x to y in X. One can define using local setwise homogeneity a convergent sequence $\{h_i\}$ in $\mathrm{Auth}(X)$ "sliding x towards y along A" and such that $h = \lim h_i \in \mathrm{Auth}(X)$ and $h \in Bd(U)$. Thus, each ε-neighborhood of id_X has non-empty boundary and so $\dim \mathrm{Auth}(X) \geq 1$.

Let M be a μ^n-manifold. Since M is strongly locally homogeneous (2.3.7), M is locally setwise homogeneous.

A space X is *almost 0-dimensional* if it has a basis $\mathcal{B}$ of open sets such that for each $B \in \mathcal{B}$, $X - \mathrm{cl}\, B = \cup\{U_i : i \in N\}$ where each U_i is both open and closed. Clearly, each 0-dimensional space is almost 0-dimensional. Each almost 0-dimensional space is totally disconnected. Complete Erdős space $\mathcal{E} = \{x \in l_2 : x_i$ is irrational for each $i\}$ is a 1-dimensional space which is almost 0-dimensional. By [OT], every almost 0-dimensional space X imbeds in the set of endpoints of an R-tree. It is easy to reimbed X into the endpoints $\mathrm{End}(Y)$ of a universal separable R-tree Y. By [KOT], $\mathrm{End}(Y)$ is homeomorphic to $\mathcal{E}$.

Theorem 3.1.3. [OT]. *Each almost 0-dimensional space is at most 1-dimensional.*

We outline a proof that was communicated to us by R.Pol.

Let X be an almost 0-dimensional space and let $\mathcal{B}$ be a countable basis witnessing this fact. Let $\mathcal{F} = \{f_i : i \in N\}$ be a collection of continuous functions $f_i : X \longrightarrow \{0, 1\}$ such that if $B, B' \in \mathcal{B}$ with $\mathrm{cl}(B) \cap \mathrm{cl}(B') = \emptyset$ there is a $f_i \in \mathcal{F}$ with $f_i(B) = 0$ and $f_i(B') = 1$. Let e be the metric on X given by $e(x, y) = \sum_i 2^{-i}|f_i(x) - f_i(y)|$.

Let d be a totally bounded metric on X. By [Na, 32.5], it suffices to show that the metric dimension $\mu \dim(X, d) \leq 1$. Now, $d' = d + e$ is also a totally bounded metric on X. Let Y be the completion of X with respect to d'. It suffices to show, by an observation of M. Levin, that for each $t > 0$ there is an open set U of Y containing X such that each continuum in U has diameter less than t.

Let $\mathcal{U} = \{U \text{ open in } Y : \mathrm{diam}\, U < t/3 \text{ and } U \cap X \in \mathcal{B}\}$. Let C be any continuum in $W = \cup\mathcal{U}$. If $B, B' \in \mathcal{B}$ with $B \cap C \neq \emptyset \neq B' \cap C$ then $\mathrm{cl}(B) \cap \mathrm{cl}(B') \neq \emptyset$, otherwise, there is $f_i \in \mathcal{F}$ with $f_i(B) = 0$ and $f_i(B') = 1$. But $f_i(X) = \{0, 1\}$ and so, by the definition of d', $Y = \mathrm{cl}_Y(f_i^{-1}(0)) \cup \mathrm{cl}_Y(f_i^{-1}(1))$, where $\mathrm{cl}_Y(f_i^{-1}(0))$ and $\mathrm{cl}_Y(f_i^{-1}(1))$ are disjoint closed sets. Hence, C being a continuum, can not meet both $\mathrm{cl}_Y(f_i^{-1}(0))$ and $\mathrm{cl}_Y(f_i^{-1}(1))$ which is a contradiction. It follows that diam $C < t$.

Theorem 3.1.4. *If M is a M_n^k-manifold with $0 \leq n < k < \infty$, then* $\mathrm{Auth}(M)$ *is almost 0-dimensional. Hence,* dim $\mathrm{Auth}(M) \leq 1$.

Outline of Proof. It suffices, by the last result, to show that $\mathrm{Auth}(M)$ is almost 0-dimensional. Let $g \in \mathrm{Auth}(M)$ and $\varepsilon > 0$. Let $h \in \mathrm{Auth}(M)$ with $d(g,h) > \varepsilon$. We show that there is an open and closed set U containing h such that $d(g,j) > \varepsilon$ for each $j \in U$. Now, $d(g,h) = \varepsilon + 4\delta$ for some $\delta > 0$, and there is $x \in M$ so that $d(g,h) = d(g(x), h(x))$. Choose an n-sphere S in M such that $g(S) \subseteq N(g(x), \delta)$ and $h(S) \subseteq N(h(x), \delta)$. Since dim $M = n$ there is a retraction $r : M \longrightarrow h(S)$ such that $r(M - N(h(x), 2\delta))$ is constant. Let $U = \{f \in \mathrm{Auth}(M) : r \cdot f|S \not\simeq *\}$. Then U is both open and closed in $\mathrm{Auth}(M)$ because close maps into S are homotopic. Also $h \in U$. Let $f \in N(g, \varepsilon)$. Then $f(S) \subseteq N(g(x), \varepsilon + \delta)$. Hence, $r \cdot f(S)$ is a point and $f \notin U$.

Corollary 3.1.5. *If M is a μ^n-manifold, where $1 \leq n < \infty$, then* dim $\mathrm{Auth}(M) = 1$.

Problem 3.1.6. If M is a μ^n-manifold, is $\mathrm{Auth}(M)$ homeomorphic to complete Erdős space $\mathcal{E}$? Is $\mathrm{Auth}(M)$ a dispersed set ($\mathrm{Auth}(M)$ embeds in $\mathcal{E}$ and $\mathcal{E}$ is a dispersed set)?

Dobrowolski and Grabowski [DG] give an uncountable family of non-isomorphic weakly closed and line-free subgroups of l_2. It can be shown that each of these is homeomorphic to $\mathcal{E}$. Thus, $\mathcal{E}$ admits uncountably many algebraically distinct topological group structures.

It can be proved using [Whit$_{1-2}$] that Menger manifolds are distinguished by the algebraic structure of their homeomorphism groups.

3.2. Algebraic properties of $\mathrm{Auth}(\mu^n)$.
Simplicity

Anderson [An$_{3-4}$] originated a technique for identifying minimal, non-trivial, normal subgroups of $\mathrm{Auth}(X)$ for spaces X with certain dilation and homogeneity properties.

Definition 3.2.1. Let X be a space. A subset A of X is deformable if for every non-empty open set U in X there is $h \in \mathrm{Auth}(X)$ with $h(A) \subseteq U$. Let U be an open set. A collection $(\{B_i : i \in N\}, h)$ is called a *dilation system* in U if $\{B_i\}$ is a sequence of disjoint non-empty open sets in U with $\lim B_i = \{p\}$ for some $p \in U$ and $h \in \mathrm{Auth}(X)$ supported on U such that $h(B_{i+1}) = B_i$ for each i.

Theorem 3.2.2. [Fi]. *Let X be a metrizable space in which each non-empty open set contains a dilation system. Let G be a subgroup of $\mathrm{Auth}(X)$ generated by all homeomorphisms which are supported by deformable subsets of X. If $G \neq \{e\}$, then G is the smallest non-trivial normal subgroup of $\mathrm{Auth}(X)$.*

In case X is a finite-dimensional manifold without boundary then $\mathrm{Auth}_0(X)$, the subgroup of homeomorphisms isotopic to the identity, is simple using [Fi] and [EK]. It is known [Mc] and [Wo$_2$] that $\mathrm{Auth}(l_2)$ and $\mathrm{Auth}(Q)$ are simple.

Definition 3.2.3. Let M be a μ^{n+1}-manifold. A pair $(W, \delta(W))$ is an n-clean pair if W is n-clean with respect to $\delta(W)$ in the sense of 2.9.2 and if, in addition, both W and $\delta(W)$ are homeomorphic to μ^{n+1}.

By the Z-set unknotting theorem (section 2.3) and the existence of n-clean pairs in μ^{n+1}[IS$_1$], it follows that every open set in μ^{n+1} has a dilation system. Also, every proper closed set in μ^{n+1} is deformable. Since every element of $\mathrm{Auth}(\mu^{n+1})$ is stable (see 3.2.6 below) we have:

Theorem 3.2.4. ([IS$_2$], [An$_4$]). $\mathrm{Auth}(\mu^{n+1})$ *is simple.*

Iwamoto and Sakai [IS$_2$] show that if M is the μ^{n+1}-manifold which admits a UV^{n-}-surjection onto the wedge of two circles, then $\mathrm{Auth}(M)$ is not simple.

Problem 3.2.5. [IS$_2$]. For a compact μ^{n+1}-manifold M, let $\mathrm{Auth}_0(M)$ be the set of all autohomeomorphisms of M which are n-homotopic to the identity. Is $\mathrm{Auth}_0(M)$ the smallest non-trivial normal subgroup of $\mathrm{Auth}(M)$? Is it simple? Is every non-trivial normal subgroup G of $\mathrm{Auth}(M)$ open in $\mathrm{Auth}(M)$?

Stable homeomorphisms

An autohomeomorphism of a space X is said to be *stable* [BG] if it can be expressed as the composition of finitely many autohomeomorphisms each of which is the identity on some non-empty open subspace of X.

It is well-known that all autohomeomorphisms of the Hilbert cube Q and the Hilbert space l_2 are stable (see, for example, [BP, chapter 5]). Every orientation-preserving homeomorphism of R^n is stable [Ki].

Theorem 3.2.6. [Sa$_2$]. *Let M be a connected μ^{n+1}-manifold. If $h \in \mathrm{Auth}(M)$ then h is a composition of two homeomorphisms each of which is the identity on some non-empty open set.*

A special case ($M = \mu^{n+1}$) was proved earlier by Chigogidze [Chi$_7$]. Sakai's techniques can be used to prove the known corresponding results for manifolds modelled on Q and on l_2 (see [Wo$_2$], [Mc]).

Factorwise rigidity

K.Kuperberg, W.Kuperberg and W.R.R.Transue [KKT$_1$] proved that the product $\mu^1 \times \mu^1$ is not 2-homogeneous by proving that $\mu^1 \times \mu^1$ is factorwise rigid, i.e.

each homeomorphism of $\mu^1 \times \mu^1$ is either (1) a product of homeomorphisms of μ^1 onto itself or (2) a product of such homeomorphisms composed with the homeomorphism of $\mu^1 \times \mu^1$ which switches coordinates (see [KKT$_2$] for a generalization). J. Kennedy-Phelps [Ken] proved an extension of these theorems for arbitrary products of μ^1. Her proof can be extended to products of μ^n-manifolds for arbitrary n.

Definition 3.2.7. Let $Y = \Pi\{X_\alpha : \alpha \in A\}$ where each X_a is homeomorphic to a fixed space X. We say that Y is factorwise rigid if for each $f \in \text{Auth}(Y)$ there exist a bijection $\phi : A \longrightarrow A$ and $f_\alpha \in \text{Auth}(X_\alpha), \alpha \in A$, such that $\pi_{\phi(\alpha)}f = f_\alpha \pi_\alpha$ for each $\alpha \in A$.

Theorem 3.2.8. *If X_α is a μ^n-manifold $X(n > 0)$, for each a in an indexed set A, then $\Pi\{X_\alpha : \alpha \in A\}$ is factorwise rigid. In particular, $\Pi\{X_\alpha : \alpha \in A\}$ is not 2-homogeneous.*

Hint of proof. Let $\alpha \in A$ and let $f, g : S^n \longrightarrow X_\alpha$ be maps such that $f(S^n)$ and $g(S^n)$ lie in disjoint copies of μ^n in X. By the proof of 3.1.3, f and g are not homotopic if f is essential. With this observation the proof in [Ken] extends to prove 3.2.8. The validity of 3.2.8 was known to T.Yagasaki [Y].

We remark that 3.2.8 is also true if X is taken to be the pseudo- arc [BK].

Acyclicity

Mather [M] proved that if G_n is the group of all homeomorphisms of R^n with compact support, then G_n is acyclic in the sense that homology of G_n, considered as a discrete group, vanishes in all positive dimensions, i.e. $H_k(G_n, Z) = 0$ for $k > 0$ (see [Br-K] for the definition).

A group G is *pseudo-mitotic* if for each finitely generated subgroup F of G there is a homomorphism $\phi : F \longrightarrow G$ and $g \in G$ such that for all $x, y \in F$ we have:

$$yxy^{-1} = \phi(y)x\phi(y^{-1}) \text{ and } x = g^{-1}\phi(x^{-1})g\phi(x) .$$

Varadarajan [Var] gave an alternative proof of Mather's theorem based on the following:

Theorem 3.2.9. [Var]. *Each pseudo-mitotic group is acyclic.*

Sankaran and Varadarajan [SV] showed that G_n and the group of homeomorphisms with bounded support of the irrationals are pseudo-mitotic.

Kawamura [Ka$_5$] has recently shown that the group of autohomeomorphisms of $\mu^{n+1} - \{pt\}$ with compact supports is pseudo-mitotic and, hence, acyclic.

3.3. Group actions on Menger manifolds, connections with the Hilbert-Smith conjecture. The Hilbert Smith conjecture asks whether every compact group acting effectively on a manifold is a Lie group. This is equivalent to asking whether the group A_p of p-adic integers acts effectively on a manifold. See [E$_4$] for some related results.

For Menger manifolds the situation is rather different. Anderson [An$_5$] proved that if G is any compact 0-dimensional topological group then G acts freely on μ^1 so that the orbit space μ^1/G is homeomorphic to μ^1. This is in contrast with group actions on manifolds.

Theorem 3.3.1. [Y]. *If the group A_p of p-adic integers acts freely on a manifold M then $\dim_Z M/A_p = \dim M + 2$, where $\dim_Z$ denotes the cohomological dimension with integer coefficients. In particular, either $\dim M/A_p = \dim M + 2$ or $\dim M/A_p = \infty$.*

Dranishnikov [Dr$_5$, Theorem 5] extended Yang's result to n-dimensional ANR's X with rank $H^n(X; \mathbb{Z}) \neq 0$.

There are a number of constructions of interesting group actions on μ^n-manifolds:

Theorem 3.3.2. *Let M be a μ^n-manifold. Then:*

(1) *Every compact 0-dimensional metrizable group G acts effectively on M so that the orbit space M/G is homeomorphic to M. This is done by Dranishnikov [Dr$_5$] and Mayer and Stark [MS] using Pasynkov's partial product description of μ^n. See [Sa$_1$] for another construction.*

(2) *A_p acts freely on M so that $\dim M/A_p = n + 1$. This example depends on work of Bestvina, Edwards, Mayer and Stark (see [MS], Theorem 5.2]).*

(3) *A_p acts on each μ^n-manifold M so that $\dim M/A_p = n + 2$. This is due to Mayer and Stark [MS] and is based on a construction of Raymond and Williams [RW].*

Definition 3.3.3. Let X be a space and G a compact group acting freely on X. We call X a free G-space. We say that X is free n-universal if any equivariant map $A \longrightarrow X$ of a closed invariant subset A of a free G-space Z with $\dim Z/G \leq n$ extends to all of Z.

Theorem 3.3.4. [Ag$_2$]. *For each compact 0-dimensional group G and each n there is a free n-universal free G-space μ^n.*

This has been proved by Dranishnikov [Dr$_5$] for finite groups.

Definition 3.3.5. The orbit space X/G of a free G-space is called n-classifying for free G-actions if:

(1) Any free G-space Z with $\dim Z/G \leq n$ embeds equivariantly in X.
(2) For any at most $(n-1)$-dimensional metric space A two maps $f, g : A \longrightarrow X/G$ are homotopic if and only if $f^*(X) = \{(a, x) \in X : f(a) = \pi(x)\}$ is homeomorphic to $g^*(X)$ over A, where π is the projection of X onto X/G, i.e. there is a homeomorphism $s : f^*(X) \longrightarrow g^*(X)$ that preserves the first coordinate.

Theorem 3.3.6. [Ag$_1$]. *The orbit space X/G of any free G-space X, where $X \in$ AE(n), is n-classifying for free G-actions.*

Conjecture 3.3.7. [Ag$_1$]. Let m and n be positive integers and G a 0-dimensional compact metric group. If μ^{m+n} and μ^n are free G-spaces then there is no equivariant map $\mu^{m+n} \longrightarrow \mu^n$.

A positive solution to this conjecture would prove that there is no free A_p-action on a connected manifold X with $\dim X/A_p < \infty$.

Fathi [Fa] proved that if M is a connected l_2-manifold and if G is a locally compact, non-compact, metrizable topological group, then there is a continuous and minimal action by G (i.e. every orbit $G \cdot x$ is dense in M) on M.

Not much is known on minimal actions on Menger manifolds.

Theorem 3.2.8. [St]. *For each $n \geq 0$ there is a finitely generated free group acting minimally on μ^n.*

Problem 3.2.9. [St]. Are there minimal actions of discrete or profinite groups on all μ^n-manifolds?

4. EMBEDDINGS INTO EUCLIDEAN SPACES

In this section we discuss the embeddings of Menger compacta into Euclidean spaces. This subject is directly connected with the theory of demension, developed by Štanko [Š$_{1-3}$]. Štanko also introduced the notion of "tame" compacta which behave like subpolyhedra from the point of view of general position. Using these concepts, Štanko proved [Š$_2$] the universality of M_k^n with respect to all k-dimensional compacta in R^n (the solution of Menger's problem). Connections with Hausdorff dimension are also discussed.

4.1. Štanko's demension theory. Following the excellent article due to Edwards [E$_1$], we introduce the notion of *demension* of compacta. A homotopy $(f_t) : R^n \longrightarrow R^n$ is called an *isotopy* if each f_t is a homeomorphism. If f_t is a homeomorphism for all t except 1 , we call it a *pseudo-isotopy*. A (pseudo-)isotopy of R^n is called an *ambient* (pseudo-)isotopy if $f_0 = \text{id}$. If $\text{diam} \{f_t(x) : t \in [0,1]\} < \varepsilon$ for each $x \in R^n$, it is called an ε *-(pseudo)-isotopy*. For a homotopy (f_t), the *support* of (f_t) denotes the set $\{x \in R^n : x \to f_t(x) \text{ is not constant}\}$.

Definition 4.1.1. (1) For a map $f : X \longrightarrow Y$, the mapping cylinder $M(f)$ is defined by $M(f) = X \times [0,1] \cup Y/_{(x,1)\sim f(x)}$. As usual, X is identified with $X \times \{0\}$ and the standard CE retraction onto the base is denoted by $c : M(f) \longrightarrow Y$. The open mapping cylinder $M(f)$ is $X \times (0,1] \cup Y/_\sim$. The source of $M(f)$ is $X \times 0$ and the target of $M(f)$ is Y.

(2) An open neighborhood U of a closed set K in R^n (or more generally, in a topological manifold) is called an open mapping cylinder neighborhood with spine

K if there are a space W and a proper surjection $p : W \longrightarrow K$ such that (U, K) is homeomorphic to $(M(p), K)$ as pairs. The restriction $r = c|_U : U \longrightarrow K$ is called the mapping cylinder retraction. If $\varepsilon > 0, U$ is said to be an open ε-mapping cylinder neighborhood if r is an ε-map (i.e. diam $r^{-1}(x) < \varepsilon$ for each $x \in K$).

Observe that:

(*) For any compactum C in U and for each $\delta > 0$, there is an ambient isotopy
    $(h_t : R^n \longrightarrow R^n)$ with compact support in U such that:
(1) (h_t) fixes some neighborhood of K and
(2) $h_1(C) \subseteq N(K, \delta)$.

If K is a subpolyhedron of R^n, the ε-regular neighborhood of K is an example of an open ε-mapping cylinder neighborhood.

Definition 4.1.2. A compactum X in R^n has *demension* $\leq k$ (notation: dem $K \leq k$) if for any $\varepsilon > 0$ there is an open ε-mapping cylinder neighborhood U whose spine K is a closed subpolyhedron in R^n such that dim $K \leq k$. We write dem $X = k$ if dem $X \leq k$ and dem $X \nleq k - 1$.

Remark. The above definition is equivalent to one given in $[E_1]$.

Theorem 4.1.3. ($[\check{S}_1], [E_1,$ Proposition 1.1]$)$. *For compacta* $X, Y \subseteq R^n :$

(1)   dim $X \leq$ dem X.
(2)   *If* (R^n, X) *and* (R^n, Y) *are homeomorphic as pairs, then* dem $X =$ dem Y.
(3)   *If* $Y \subseteq X$, *then* dem $Y \leq$ dem X.
(4)   *If* $X = \cup \{X_i : i \in N\}$, *then* dem $X = \sup\{$dem $X_i : i \in N\}$.

Let N_k^n be the k-dimensional Nöbeling space in R^n. This set is defined as follows: $N_k^n = \{x \in R^n :$ at most k coordinates of x are rational$\}$. Clearly, the complement of N_k^n in R^n can be represented as a countable union of $(n - k - 1)$-dimensional hyperplanes.

The conditions (2) in the following theorem is the original definition of dem due to Štanko.

Theorem 4.1.4. ($[\check{S}_1], [E_1,$ Proposition 1.2]$)$. *The following conditions are equivalent for any compactum* $X \subseteq R^n :$

(1) dem $X \leq k$.

(2) *For any closed subpolyhedron* L *in* R^n *with* dim $L \leq n - k - 1$, *there is an arbitrary small ambient isotopy* (h_t) *of* R^n *whose support is arbitrarily close to* $X \cap L$ *and such that* $h_1(L) \cap X = \emptyset$.

(3) *There is an arbitrarily small ambient isotopy* (h_t) *of* R^n *whose support is arbitrarily close to* X *and such that* $h_1(X) \subseteq N_k^n$.

(4) *There is an ambient isotopy* (g_t) *of* R^n *with compact support such that* $g_1(X) \subseteq M_k^n$.

Remark. In [GS], a compactum X with the property (2) is called a strong Z_{n-k-2}-set (see section 5).

The above equivalences show that a compactum with dem $\leq k$ behaves like a k-dimensional subpolyhedron in R^n. The following theorem also illustrates this similarity.

Theorem 4.1.5. [D, p.164]. *Let X and Y be compacta in R^n (or more generally, in a n-manifold). Then for each $\varepsilon > 0$, there is a homeomorphism $h : R^n \longrightarrow R^n$ such that* dem $(h(X) \cap Y) \leq$ dem $X +$ dem $Y - n$ *and* $d(h, \mathrm{id}) < \varepsilon$.

Definition 4.1.6. A subset X of R^n is called locally homotopically k-co-connected $(k - LCC)$ if, for each $z \in R^n$ and for each neighborhood U of z in R^n, there is a neighborhood V of z in R^n such that $V \subseteq U$ and any map $\alpha : S^k \longrightarrow V - X$ is null-homotopic in $U - X$.

The following result is of fundamental importance in the theory of demension.

Theorem 4.1.7. ([E_1, Theorem 1.4], [Bot$_2$], [McR]). *Let X be a compactum in R^n.*

(1) *For $n \neq 3$.*

    (1.1) *If* dim $X \geq n - 2$, *then* dem $X =$ dim X.

    (1.2) *If* dim $X \leq n - 3$, *then* dem $X =$ dim X *or* dem $X = n - 2$. *Moreover,* dem $X =$ dim X *if and only if X is 1-LCC in R^n.*

(2) *If $n = 3$, then the above holds except that if* dim $X = 1$ *then* dem $X \in \{1, 2\}$.

The authors are grateful to R.J.Daverman for the information on the case $n = 4$.

Definition 4.1.8. An embedding $e : X \longrightarrow R^n$ of a compactum X is said to be tame if dem $e(X) =$ dim X.

By 4.1.7, if dim $X \geq n - 2$ and $(n, \dim X) \neq (3, 1)$ (i.e. e is a codimension ≤ 2 embedding), then e is tame in the above sense. If the codimension is sufficiently big, there are many tame embeddings.

Theorem 4.1.9. ([$\check{S}_1$, 1.10 Proposition 7], [GS$_1$, Theorem 3.3]). *Let X be an n-dimensional compactum. Then the set of all tame embeddings of X into R^{2n+1} is a dense G_δ-subset in $C(X, R^{2n+1})$.*

If the codimension of X in R^n is at least 3 , the following beautiful theorem holds.

Theorem 4.1.10. ([$\check{S}_3$], [E$_3$]). *Let X be a compactum with* dim $X \leq n - 3$. *Then any embedding $f : X \longrightarrow R^n$ of X into R^n can be arbitrarily closely approximated by tame embeddings.*

Suppose that X is a k-dimensional compactum in R^n, where $k < n$. If $k < n-2$, by 4.1.10, we may obtain a compactum X^* which is homeomorphic to X and such that dem $X^* = \dim X$. If $k \geq n - 2$, 4.1.7 (for $n \geq 4$) and 4.1.9 (for $n = 3$) also guarantee the existence of such a compactum. Applying 4.1.4, we can embed X^* into M_k^n or N_k^n. Therefore, we have the solution of Menger's problem.

Theorem 4.1.11. [Š$_2$]. *Let $0 \leq k < n$. Then M_k^n and N_k^n are universal with respect to at most k-dimensional compacta in R^n.*

Next we discuss wild embeddings of μ^k into R^{2k+1}. For $k = 1$, Bothe [Bot$_2$] and McMillan and Row [McR] independently constructed wild embeddings of μ^1 into R^3. The following construction of a wild embedding of μ^k into $R^{2k+1}, k \geq 2$, is based on an observation of R.B.Sher. There is a wild embedding $\alpha : C \longrightarrow R^{2k+1}$ of a Cantor set so that dem $\alpha(C) = 2k - 1$. We can regard C as a subspace of μ^k. Since R^{2k+1} has DD^kP, we can construct an embedding $e : \mu^k \longrightarrow R^{2k+1}$ such that $f|_C = \alpha$. By 4.1.3, dem $e(\mu^k) \geq$ dem $\alpha(C) = 2k - 1 > k$ showing that e is a wild embedding.

The wildness of the above embedding was detected by referring to a wild Cantor set. The existence of a wild surface all subarcs of which are tame suggests the following question:

Problem 4.1.12. Is there a wild embedding $h : \mu^k \longrightarrow R^{2k+1}$ such that each cell-like set in $h(\mu^k)$ is cellular?

On the other hand, if $n \geq 2k+2$, every tame embedding of μ^k into R^n is equivalent to the standard embedding of μ^k as M_k^n. This follows from the following theorem which is a natural generalization of the Bing-Kister theorem to tame compacta.

Theorem 4.1.13. ([Š$_1$, 2.3 Theorem 2], [Bry$_{1-2}$], [GS$_1$, Theorem 2.5]). *Suppose that X and Y are tame k-dimensional compacta in R^n and $n \geq 2k + 2$. Then each homeomorphism $f : X \longrightarrow Y$ can be extended to an autohomeomorphism of R^n.*

The authors do not know whether there are inequivalent tame embeddings of μ^k into R^{2k+1}.

Problem 4.1.14. Are there two inequivalent tame embeddings of μ^k into R^{2k+1}?

Since μ^k is LC^{k-1} and C^{k-1}, we may apply the complement theorem [ISV] to conclude that the complements of all tame embeddings are mutually homeomorphic if $n \geq 2$.

4.2. Hausdorff dimension. Hausdorff dimension is another concept dealing with the complexity of an embedding of a compactum into Euclidean space and there is a close connection between Hausdorff dimension and demension in the sense of section 4.1. Again, very little is known about Hausdorff dimension of Menger compacta in Euclidean spaces. Here we present some known results, their easy consequences and discuss some problems. We introduce the notion of Hausdorff dimension following [Ke].

Definition 4.2.1. Let X be a separable metric space with a metric d and let p be a positive real number. For each $\varepsilon > 0$, let

$$\wedge_{p,\varepsilon}(X) = \inf\{\sum_i (\operatorname{diam} U_i)^p : \{U_i\} \text{ is an open cover of } X\}.$$

Then the Hausdorff p-measure is defined by

$$\wedge_p(X) = \sup\{\wedge_{p,\varepsilon}(X) : \varepsilon > 0\} .$$

Note that $\wedge_p(X)$ might be infinite. If $p < q$ and $\wedge_p(X)$ is finite, then $\wedge_q(X) = 0$ [Ke, Theorem 1.2].

Definition 4.2.2. The Hausdorff dimension of a space X is defined by $\wedge(X) = \inf\{p > 0, \wedge_p(X) = 0\}$.

It follows from the above remark that if $q < \wedge(X)$, then $\wedge_q(X) = \infty$ and if $q > \wedge(X)$, then $\wedge_q(X) = 0$. In the sequel, any metric space is assumed to be endowed a metric d.

Theorem 4.2.3. ([Ke], [HW]). *Let X be a separable metric space.*

  (1) *If $A \subseteq X$, then $\wedge(A) \leq \wedge(X)$, where A is endowed with the restriction of the metric on X.*
  (2) *If $X = \cup\{A_i : i \in N\}$, then $\wedge(X) = \sup\{\wedge(A_i) : i \in N\}$.*
  (3) $\dim X \leq \wedge(X)$.
  (4) $\wedge(I^n) = n$, *where I^n is the n-cube with the standard metric.*

If X is any compactum which contains a Cantor set, then $\wedge(X)$ is infinite for some metric on X by 4.2.4 and 4.2.3.

Theorem 4.2.4. [Ke, Theorem 6.3]. *For each compactum X there is a cantor set C in X such that $\wedge(C) = \wedge(X)$.*

The connection between Hausdorff dimension and dimension for compacta in Euclidean spaces is given in the following theorem.

Theorem 4.2.5. [V] *If $X \subseteq R^n$ (occupied with the standard metric) is a compactum, then:*

  (1) $\dim X \leq \operatorname{dem} X \leq \wedge(X)$.
  (2) *There is a homeomorphism $f : R^n \longrightarrow R^n$ such that $\wedge(f(X)) = \operatorname{dem} X$.*

Proving the above theorem, Väisälä constructed a homeomorphic copy M of M_k^{2k+1} in R^{2k+1} such that $\wedge(M) = k$. The basic idea of his construction is "dig out bigger holes than usual " . Notice that the standard construction of M_k^{2k+1} in section 2.1 gives us a compactum whose Hausdorff dimension is greater than k.

From 4.2.5, it is natural to ask what are the possible values of Hausdorff dimensions of compacta in R^{2k+1} which are homeomorphic to μ^k. Following the suggestion of J.Keesling, we prove the following result.

Theorem 4.2.6. *For each $t \in [k, 2k + 1]$, there is a homeomorphic copy X_t of μ^k in R^{2k+1} such that $\wedge(X_t) = t$.*

Outline of Proof. There is a tame Cantor set C_t in R^{2k+1} such that $\wedge(C_t) = t$ (see, for example, [R, the end of Section 2]). Using the construction in the proof of 4.2.5, we can construct a non-compact, $(k-1)$-connected, μ^k-manifold M_t in R^{2k+1} such that

(1) $\wedge(M_t) = k$ and

(2) $X_t = M_t \cup C_t$ is a k-dimensional compactum containing C_t as a Z-set.

        (2.1) implies that X_t is homeomorphic to μ^k. The condition (1) and 4.2.3(2) ensure that $\wedge(X_t) = t$.

5. PSEUDO-BOUNDARIES AND PSEUDO-INTERIORS

Skeletoids and absorbers play an important role in infinite-dimensional topology. The existence and uniqueness of skeletoids (and absorbers) with respect to certain collections allow one to study completely metrizable infinite-dimensional manifolds by using the technique of compact manifolds, incomplete manifolds by those of complete ones and so on. For instance, several statements of l_2-manifold theory can be easily proved just exploiting the following fundamental fact: the pseudo-interior of the Hilbert cube is homeomorphic to l_2 (see [BP], for further details).

Skeletoids and absorbers in finite dimensional spaces were studied by Geoghegan and Summerhill [GS]. They defined the pseudo-boundary B_k^n, an absorber with respect to the collection of k-dimensional tame compacta in R^n, and the pseudo-interior $s_k^n = R^n - B_{n-k-1}^n$. It was shown in [Di$_2$] that B_k^n is in fact a skeletoid with respect to the same collection.

Based on the properties of μ^n, Chigogidze [Chi$_3$, §1], introduced the pseudo-boundary Σ^n and the pseudo-interior ν^n of μ^n and conjectured the existence of an entire ν^n-manifold theory as the n-dimensional analogue of l_2-manifold theory. In [CKT], it is proved that ν^n is homeomorphic to the universal Nöbeling space N_n^k and that Σ^n is homeomorphic to $B_n^k, k \geq 2n + 1$.

5.1. General information. In this section we introduce the basic concepts of skeletoids and absorbers [BP, Chapter 4], [Di$_2$, Chapter 1]. Let $\mathcal{K}$ be an additive, invariant collection of closed sets of a Polish space X, that is, for each $A, B \in \mathcal{K}$ and for each $F \in \mathrm{Auth}(X)$, we have $F(A \cup B) \in \mathcal{K}$.

Definition 5.1.1. (1) By a $\mathcal{K}$-*skeleton*, we mean any increasing sequence $\{A_i : i \in N\}$ of members of $\mathcal{K}$ such that for every $i \in N$, for every $A, B \in \mathcal{K}$ with $B \subseteq A_i$ and for every open cover $\mathcal{U}$ of X there are an index $j > i$ and a $H \in \mathrm{Auth}(X)$ which is $\mathcal{U}$-close to id_X and such that $H(A) \subseteq A_j$ and $H|_B = \mathrm{id}$. A subset of X which can be expressed as the union of the members of a $\mathcal{K}$-skeleton is called a $\mathcal{K}$-skeletoid.

(2) By a $\mathcal{K}$-*absorber* we mean a countable union $\cup\{K_j : j \in N\}$ of members of $\mathcal{K}$ such that for every $A \in \mathcal{K}$, every closed set E of X and for every open cover $\mathcal{U}$ of

X there is a homeomorphism $H \in \text{Auth}(X)$ which is $\mathcal{U}$-close to id_X and such that $H(A - E) \subseteq \cup\{K_j : j \in N\}$ and $H|_E = \text{id}$.

It is easy to see that any $\mathcal{K}$-skeletoid is a $\mathcal{K}$-absorber. Any two $\mathcal{K}$-skeletoids ($\mathcal{K}$-absorbers) are equivalent in the following sense.

Theorem 5.1.2. (1) [BP, Chap. 4, Theorem 2.1]. *If A and B are $\mathcal{K}$-skeletoids of X, then the set $\{H \in \text{Auth}(X) : H(A) = B\}$ is dense in $\text{Auth}(X)$.*

(2) ([W_1], [Di_2]). *If A and B are $\mathcal{K}$-absorbers, then for every collection $\mathcal{U}$ of open subsets of X, there exists $f \in \text{Auth}(X)$ which is $\mathcal{U}$-close to id such that $f(A \cap \cup\mathcal{U}) = B \cap \cup\mathcal{U}$ and $f|_{(X - \cup\mathcal{U})} = \text{id}$.*

A closed subset A of a Polish space X is called a *thin* set, if for each open neighborhood V of A and for each open cover $\mathcal{U}$ of X there is a $F \in \text{Auth}(X)$ which is $\mathcal{U}$-close to id and satisfies the following two conditions: $F|_{(X-V)} = \text{id}$ and $F(A) \cap A = \emptyset$.

Definition 5.1.3. An additive invariant collection $\mathcal{K}$ of closed subsets of a Polish space X is called a *perfect collection* if it satisfies the following conditions:

(1) Each member of K is compact and thin.
(2) If B is closed in $A \in \mathcal{K}$, then $B \in \mathcal{K}$.
(3) (Estimated Extension Property) For each $A \in \mathcal{K}$, for each neighborhood V of A and for each open cover $\mathcal{U}$ of X there exists an open cover $\mathcal{V}$ of X refining $\mathcal{U}$ such that: for each $B \in \mathcal{K}$ contained in V and each homeomorphism $f : A \longrightarrow B$ which is $\mathcal{V}$-close to id, there is an extension $F \in \text{Auth}(X)$ of f which is U-close to id and $F|_{(X-V)} = \text{id}$.

The following result allows us to recognize skeletoids with respect to perfect collections.

Theorem 5.1.4. [BP, Chap 4, Proposition 4.1]. *Let $\{A_i : i \in N\}$ be an increasing sequence of members of a perfect collection $\mathcal{K}$. Then $\{A_i : i \in N\}$ is a $\mathcal{K}$-skeleton if and only if for every $K \in \mathcal{K}$, every $i > 0$ and every $\varepsilon > 0$, there exist an index $j > i$ and an embedding $f : K \longrightarrow A_j$ such that $d(f, \text{id}_K) < \varepsilon$ and $f|_{(K \cap A_i)} = \text{id}$.*

Theorem 5.1.4 implies the following useful result.

Proposition 5.1.5. *If A is a $\mathcal{K}$-skeletoid in X and B is a countable union of members of $\mathcal{K}$, then $A \cup B$ is also a $\mathcal{K}$-skeletoid.*

Example 5.1.6. The pseudo-boundary $\Sigma = \{(x_n) \in \Pi\{[-1,1]_n : n \in N\} : x_n \in \{-1,1\}$ for some $n \in N\}$ of the Hilbert cube Q is a $\mathcal{Z}$-skeletoid, where $\mathcal{Z}$ denotes the collection of all Z-sets in Q (see [BP], [vM]).

5.2. Pseudo-boundaries and pseudo-interiors of Euclidean spaces and Menger compacta. In this section we introduce pseudo-boundaries and pseudo-interiors of Euclidean spaces which were constructed in [GS]. Below $\mathcal{M}_k^n$ denotes the collection of all at most k-dimensional tame (in the sense of 4.1.8) compacta in R^n. A set is called a $\sigma\mathcal{M}_k^n$-set if it is a countable union of members of $\mathcal{M}_k^n$.

Let us recall (section 2.1) that for each integer $i > 0$, L_i denotes the cell complex structure of R^n whose n-cells are of the form $\Pi\{[m_t/3^i, (m_t + 1)/3^i] : m_t = 0, 1, \ldots, 3^i - 1\}$. For each integer $i \geq 1$, let $K_j = st(|L_j^{(k)}|, L_{j+1})$ and then let $B_k^n(i) = \cap\{K_j : j \geq i\}$. Clearly, $B_k^n(i) \subseteq B_k^n(i+1)$. The union $B_k^n = \cup\{B_k^n(i) : i \in N\}$ is called the *universal k-dimensional pseudo-boundary* (of R^n). The complement $s_k^n = R^n - B_{n-k-1}^n$ is called the *universal k-dimensional pseudo-interior* (of R^n). It is known [GS, Theorem 3.12] that B_k^n is a $\mathcal{M}_k^n$-absorber (compare with [Bot$_1$], [Š$_2$]). This result was improved in [Di$_1$].

Theorem 5.2.1. B_k^n *is a $\mathcal{M}_k^n$-skeletoid.*

As was mentioned in 5.1.6, the Hilbert cube Q also has a pseudoboundary Σ. Moreover, it was shown in [Di$_2$] that for each $k \geq 0$ the Hilbert cube has a k-dimensional pseudo-boundary B_k^∞, i.e. a skeletoid with respect to the collection of all at most k-dimensional Z-sets in Q. The following result illustrates the ties between the finite and infinite-dimensional cases.

Theorem 5.2.2. [DMM]. *Let $0 \leq k \leq n, m \leq \infty$. Then the following conditions are equivalent:*

  (1) B_k^n *and* B_k^m *are homeomorphic.*

  (2) $m = n$ *or* $m, n \geq 2k + 1$.

The following result is an analogue of 5.1.6 for Menger compacta.

Theorem 5.2.3. [Chi$_3$], [CKT]. *Let $0 \leq k \leq n \leq \infty$. There is a $\mathcal{Z}(k,n)$-skeletoid $\Sigma(k,n)$ in μ^n, where $\mathcal{Z}(k,n)$ denotes the collection of all at most k-dimensional Z-sets in μ^n.*

For convenience $\mathcal{Z}(n,n)$ is simply denoted by $\mathcal{Z}_n$ and $\Sigma(n,n)$ by Σ^n. If $k < n = \infty$, then, by 5.1.2, $\Sigma(k,\infty)$ and B_k^∞ are homeomorphic. If $k = n = \infty$, for the same reason we conclude that $\Sigma(\infty,\infty)$ is homeomorphic to Σ. The following result shows that pseudo-boundaries of Euclidean spaces and of Menger compacta are topologically equivalent.

Theorem 5.2.4. [CKT]. Σ^k *and* B_k^{2k+1} *are homeomorphic.*

Note that the space Σ is topologically characterized [Mo] (see also [BM]) as the only σ-compact AR-space X satisfying the following condition for $k = \infty$:

  $(*)_k$ for each map $f : A \longrightarrow X$ of an at most k-dimensional compactum A into X, for each closed subspace B of A such that $f|_B$ is an embedding, and for each open cover $\mathcal{U} \in \mathrm{cov}(X)$ there exists an embedding $g : A \longrightarrow X$ such that g is $\mathcal{U}$-close to f and $g|_B = f|_B$.

It is easy to see that the space B_k^{2k+1} (or Σ^k, see 5.2.4) is a k-dimensional σ-compact $LC^{k-1} \cap C^{k-1}$-space which satisfies the condition $(*)_k$. The question whether the above conditions characterize B_k^{2k+1} remains open.

Problem 5.2.5. ([DM], [DMM], [We$_3$]). Is it true that every k-dimensional σ-compact $LC^{k-1} \cap C^{k-1}$-space satisfying the condition $(*)_k$ is homeomorphic to B_k^{2k+1}?

Let us consider now the universal k-dimensional pseudo-interior of R^{2k+1}, i.e. the space $s_k^{2k+1} = R^{2k+1} - B_k^{2k+1}$. The infinite-dimensional version of this space, i.e. the space $Q - \Sigma$ is topologically equivalent to the Hilbert space l_2. Recall that l_2 was characterized topologically by Toruńczyk [Tor$_3$] as a Polish AR-space X satisfying the following condition for $k = \infty$:

$(**)_k$ *For each at most k-dimensional Polish space Y the set*

$\qquad \{f \in C(Y,X) : f$ is a closed embedding $\}$ is dense in the space $C(Y,X)$.

As in the case of pseudo-boundaries, it can be easily seen that s_k^{2k+1} satisfies the condition $(**)_k$. In addition, s_k^{2k+1} is itself a k-dimensional Polish AE(k)-space. Another well-known space having all of the above mentioned properties of s_k^{2k+1} is the k-dimensional Nöbeling space N_k^{2k+1}. It was shown by Toruńczyk that N_k^{2k+1} can be represented as the complement $R^{2k+1} - D_k^{2k+1}$, where D_k^{2k+1} denotes an absorber with respect to the collection of all at most k-dimensional tame polyhedra in R^{2k+1}. As is indicated by the following result, the space N_k^{2k+1} can be studied using the techniques of Menger manifold theory.

Theorem 5.2.6. [CKT]. ν^k and N_k^{2k+1} are homeomorphic.

Now we can formulate another important unsolved problem, which is, in some sense, dual to 5.2.5.

Problem 5.2.7. ([DM],[DMM],[We$_3$], [CF]). Is it true that every k-dimensional Polish AE(k)-space satisfying the condition $(**)_k$ is homeomorphic to N_k^{2k+1}?

6. OTHER TOPICS

6.1. Menger compacta $M_k^n(1 \leq n \leq 2k)$. Very little is known about Menger compacta of the type M_k^n if $n \leq 2k$. As was mentioned in section 4.1, M_k^n is an universal compactum for the class of all at most k-dimensional compacta embeddable in R^n. Compacta M_k^n are not homogeneous if $n \leq 2k$ [Lew$_2$].

Below we consider only the case $n = k + 1$. By the construction, M_k^{k+1} is obtained from I^{k+1} (or, equivalently, from S^{k+1}) by removing the interiors of tame $(k+1)$-dimensional cells forming a null-sequence. It follows from 6.1.2 that this construction uniquely determines the resulting compactum up to a homeomorphism. 6.1.2 was proved by Whyburn [Why] for $k = 1$ and by Cannon [C] for $k > 1, k \neq 3$. Cannon's proof relies on the Annulus Conjecture which was unsolved for dimension 4 at that time. The solution of the Annulus Conjecture due to Quinn [Qu$_1$] allows us to state the theorem in full generality (Both Whyburn and Cannon deal with the case when the underlying manifold is a sphere. Their approaches can be easily extended to the general case). First we need the following definition.

Definition 6.1.1. A compactum X in a closed topological manifold M^{k+1} is called a $\mathcal{S}(M)$-compactum if the collection $\{U_i\}$ of components of $M^{k+1} - X$ satisfies the following conditions:

(1) The closure $\mathrm{cl}(U_i)$ of each U_i is a $(k+1)$-cell with a (locally) flat k-sphere as boundary.
(2) $\mathrm{cl}(U_i) \cap \mathrm{cl}(U_j) = \emptyset$ whenever $i \neq j$.
(3) $\cup U_i$ is dense in M.
(4) $\mathrm{diam}\, U_i \to 0$ as $i \to \infty$

A compactum X is called a k-dimensional Sierpiński compactum of M^{k+1} if there is an embedding $h : X \longrightarrow M$ such that $h(X)$ is a $\mathcal{S}(M)$-compactum in M^{k+1}.

Theorem 6.1.2. ([Why],[Ca]). *For each $k \geq 1$ and for each $(k+1)$-dimensional closed topological manifold M, the k-dimensional Sierpiński compactum is topologically unique. It is denoted by $S(M)$.*

Any μ^k-manifold X can be obtained as the intersection of a decreasing sequence $\{M_i\}$ of PL manifolds with $\dim M_i \geq 2k + 1$ for each i. However, only the low dimensional homotopies of M_1 are reflected in the topology of X. For example, if there is a map $f : M_1 \longrightarrow M_1'$ between compact PL manifolds of dimension $\geq 2k + 1$ which induces isomorphisms of i-dimensional homotopy groups for each $i \leq k - 1$, then the corresponding μ^k-manifolds are homeomorphic. In contrast, the topological type of the Sierpiński compactum $S(M)$ is completely determined by the topology of M.

Proposition 6.1.3. *Let M and N be closed topological manifolds. Then $S(M)$ and $S(N)$ are homeomorphic if and only if M and N are homeomorphic.*

Proof. Let X and Y be $\mathcal{S}(M)$– and $\mathcal{S}(N)$-compacta and let $\{U_i\}$ and $\{V_i\}$ be the components of $M - X$ and $N - Y$ respectively. If there is a homeomorphism $h : X \longrightarrow Y$, it is easy to see that the boundary ∂U_i of each U_i is mapped onto the boundary ∂V_j of some V_j. Since $\mathrm{cl}(U_i)$ and $\mathrm{cl}(V_j)$ are cells of the same dimension, $h|_{\partial U_i}$ has an extension $H_i : \mathrm{cl}(U_i) \longrightarrow \mathrm{cl}(V_j)$. Taking these extensions for each U_i, we get a homeomorphism $H : M \longrightarrow N$ which extends h.

6.2. n-homotopy idempotents. The authors believe that there are potential applications of Menger manifold theory to homotopy theory. In this section we demonstrate one such application.

A map $f : X \longrightarrow X$ is called a *homotopy idempotent* if $f^2 \simeq f$. f *splits* (through a space Y) if there are maps $g : X \longrightarrow Y$ and $h : Y \longrightarrow X$ such that $hg \simeq f$ and $gh \simeq \mathrm{id}_Y$. Corresponding notions are defined in the pointed homotopy category as well. Clearly, f is a homotopy idempotent provided that f splits. But the converse of this statement is not always true (Dydak-Minc, Freyd-Heller). It is well-known that pointed homotopy idempotents of polyhedra [EG] (see also [Fr]), as well as unpointed homotopy idempotents of finite-dimensional polyhedra [HH] split. However even in the case of finite polyhedra pointed homotopy idempotents do not necessarily split through finite polyhedra [W].

The "n-homotopy" version of the above result was proved in [CKS] (corresponding notions in the (pointed) n-homotopy category are defined in a natural way).

Theorem 6.2.1. *Each pointed n-homotopy idempotent of an at most $(n+1)$-dimensional LC^n-compactum splits through an at most $(n+1)$-dimensional finite polyhedron.*

The question concerning the validity of the unpointed version of 6.2.1 is still open.

The proof of 6.2.1 (see also section 6.3) exploits an "n-homotopy version of (infinite) mapping cylinder"([CKS], implicitly in [Chi$_8$]). We briefly describe the construction.

Let $f : X \longrightarrow Y$ be a map between at most $(n+1)$-dimensional LC^n-compacta. Consider the usual mapping cylinder $M(f)$ of f. Consider the $(n+1)$-invertible UV^n-map $r : M_n(f) \longrightarrow M(f)$, where $M_n(f)$ is an $(n+1)$-dimensional LC^n-compactum given by (2.5.2). By shrinking fibers of points in $X \times \{0\} \cup Y$ into points if necessary, we may assume without loss of generality that the restrictions $r|_{r^{-1}(X \times \{0\})}$ and $r|_{r^{-1}(Y)}$ are one-to-one maps. We identify $r^{-1}(X \times \{0\})$ with $X, r^{-1}(Y)$ with Y and call them the source and the target of $M_n(f)$ respectively. Note that X is a Z-set in $M_n(f)$. The UV^n-retraction $c_n : M_n(f) \longrightarrow Y$ defined as the composition $c_n = r^{-1}cr$ plays a role similar to the collapse onto the target and $c_n|X = f$. Note also that for each point $x \in X$ there is an arc $A(x)$ connecting x with $f(x)$ in $c_n^{-1}(f(x))$. Indeed, consider the arc $I(x) \subseteq c^{-1}(f(x))$ connecting x and $f(x)$ in the usual mapping cylinder $M(f)$. Since r is at least 1-invertible $(n \geq 0)$ and $\dim I(x) = 1$, there is a map $a : I(x) \longrightarrow M_n(f)$ such that $r\alpha = \mathrm{id}_{I(x)}$. Let $A(x) = \alpha(I(x))$.

The proof of 6.2.1 follows the original proof due to Freyd [HH, §1]. Suppose that $f : X \longrightarrow X$ is a pointed n-homotopy idempotent. Sewing together infinitely many copies of the space $M_n(f)$ along their naturally identified sources and targets, we define a space $\mathrm{Map}_n(f)$ (an "infinite mapping cylinder") as shown in Figure 2.

It can be shown that f splits (in the pointed n-homotopy category) through $\mathrm{Map}_n(f)$. In particular, $\mathrm{Map}_n(f)$ is n-homotopy dominated by X. Note also that $\mathrm{Map}_n(f)$ is a $(n+1)$-dimensional, locally compact, LC^n-space. By 2.2.8 and 2.2.10, we conclude that $\mathrm{Map}_n(f)$ is n-homotopy equivalent to an at most $(n+1)$-dimensional finite polyhedron.

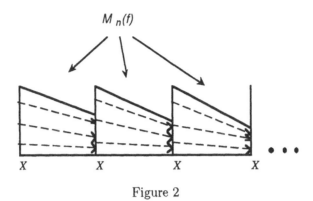

Figure 2

6.3. Homogeneous continua. Recall that a continuum X is said to be *homogeneous* if, for each pair x, y of points of X there is a homeomorphism $h : X \longrightarrow X$ such that $h(x) = y$.

All connected topological manifolds without boundary, connected Menger manifolds, connected Q-manifolds, connected l_2-manifolds as well as all topological groups are homogeneous. The classification of homogeneous 1-dimensional continua is one of the most important problems in continuum theory. If, in addition, the continua are locally connected, then the situation is quite simple.

Theorem 6.3.1. [An$_2$]. *The simple closed curve and the Menger curve are the only locally connected, 1-dimensional, homogeneous continua.*

The general problem is still unsolved. Even planar homogeneous continua have not been classified yet. There are several techniques allowing construction of homogeneous continua from known examples. We briefly discuss some of them.

A continuum X is said to be *strongly locally homogeneous* if, for each point $x \in V$ and for each neighborhood U of x, there is a neighborhood V of x, contained in U, such that for each point $y \in V$ there is a homeomorphism $h : X \longrightarrow X$ with $h(x) = y$ and $h|_{(X-U)} = \mathrm{id}$.

Theorem 6.3.2. [Ro$_1$, Theorem 7]. *Let $G \longrightarrow E \longrightarrow B$ be a principal G-bundle over a strongly locally homogeneous continuum. If G is abelian, then the total space E is homogeneous.*

Corollary 6.3.3. *Let $X = \varprojlim\{X_i, p_i^{i+1}\}$ be a continuum such that each X_i is a strongly locally homogeneous continuum and each $p_i^{i+1} : X_{i+1} \longrightarrow X_i$ is a regular covering. Then X is homogeneous.*

Using this result, Minc and Rogers produced uncountably many, mutually distinct, homogeneous curves which are total spaces of a Cantor set bundle over the Menger curve [MR].

Theorem 6.3.4. [MR]. *Suppose that X is an m-dimensional strongly locally homogeneous continuum such that there is a retraction $r : X \longrightarrow B_n$, where B_n is the wedge of n simple closed curves. Then for each n-tuple $S = \{S_1, \ldots, S_n\}$ of solenoids, there is a homogeneous m-dimensional continuum $X(S)$ such that the following two conditions are satisfied:*

(a)   *$X(S)$ admits a retraction onto each of the S_i.*

(b)   *$X(S)$ is the total space of a Cantor set bundle over X.*

As is shown in [Vi], we can not perform the same construction for infinitely many solenoids.

The simple connectivity of $\mu^n (n \geq 2)$ prevents us from constructing homogeneous continua which are the total spaces of Cantor set bundles over μ^n. In this case, the following theorem may be applied.

Theorem 6.3.5. [Pa, §12]. *For each strongly locally homogeneous continuum X, there is another homogeneous continuum which admits an open, 0-dimensional map onto X.*

Another way of producing homogeneous curves has been developed by Lewis. Here is the result.

Theorem 6.3.6. [Lew$_1$]. *For each homogeneous curve X there is a homogeneous curve X^* which admits an open map onto X whose fibers are all homeomorphic to the pseudo-arc.*

It is not known if X^* in the above theorem is uniquely determined by the above property. Lewis announced [Lew$_3$] that if the open map is atomic the uniqueness holds as well. It is not known whether there are higher dimensional analogues of 6.3.6. Notice that an easy application of [LW] shows that for any n-dimensional compact polyhedron P, there is an n-dimensional continuum P^* which admits an open map onto P such that all fibers are homeomorphic to the pseudo-arc.

A continuum X is said to be *bihomogeneous* if for each pair of points $x, y \equiv X$ there is a homeomorphism $h : X \to X$ such that $h(x) = y$ and $h(y) = x$. The first example of an homogeneous, non-bihomogeneous continuum was constructed by K.Kuperberg [Ku$_1$]. Her example is 7-dimensional and locally connected. Recently she announced that there is even a 4-dimensional one [Ku$_2$]. Later, Minc [Mi] gave a simpler construction of an infinite- dimensional, non-locally connected example. His construction can be modified to produce a 4-dimensional, non-locally connected example (P.Minc, a private communication). Using the "mapping cylinder", described in section 6.2, we may construct a 2-dimensional example. It is not known whether there is a 1-dimensional one.

Theorem 6.3.7. ([Mi], [Ka$_5$]). *For each $n \geq 2$ there is an n-dimensional, homogeneous, non-bihomogeneous continuum.*

There are several versions of homogeneity and the Menger curve has been applied to provide various examples. See, for example, $[K_1]$, $[OP]$ and $[Pat_1]$.

6.4. Fixed points. A space X has the *complete invariance property* (CIP) provided that each of the nonempty closed subsets of X is the fixed point set of some self-map of X [Wa]. Some spaces known to have CIP are n-cells [R], dendrites $[Sc_3]$, convex subsets of Banach spaces [W], compact manifolds without boundary $[Sc_2]$.

On the other hand, it was shown in [MN] that there is a locally connected continuum having CIP whose wedge with itself at a specified point does not have CIP. This shows that the union of two locally connected continua having CIP need not have CIP. Examples of this type can not be 1-dimensional. This follows from the following statement.

Theorem 6.4.1. [MT]. *Every 1-dimensional locally connected continuum has CIP.*

The interesting question [M] of whether every metrizable ANR-compactum has CIP remains open. Surprisingly, there is no non-metrizable ANR-compactum with CIP [CM] (see also [Ko], where the case of uncountable powers of the unit segment was considered).

If each non-empty closed subset of a space X is the fixed point set of some auto-homeomorphism of X, then we say [M] that X has CIP with respect to homeomorphisms (CIPH). Closed surfaces, even-dimensional cells and positive-dimensional spheres are known to have CIPH $[Sch_{1-2}]$.

Theorem 6.4.2. [M]. *The Hilbert cube Q, as well as every $[0,1)$-stable Q-manifold, have CIPH.*

The following questions are inspired by 6.4.2.

Problem 6.4.3. Is it true that μ^n has CIPH (or CIP for $n \geq 2$)? In particular, does μ^1 have CIPH ?

Problem 6.4.4. Is it true that every stable (in the sense of section 2.7) μ^n-manifold has CIPH ?

A recent result of Sakai $[Sa_3]$ shows that every Z-set is the fixed point set of some homeomorphism of μ^n.

6.5. Topological dynamics. Let $(f_t) : X \to X$ be a continuous flow of X and A be a closed subset of X. A is called an attractor of f_t if there is an open neighborhood U of A such that for each $t \geq 0$ $f_t(U) \subseteq U$ and $\cap\{f_t(U) : t \geq 0\} = A$. An attractor of a homeomorphism is defined similarly. The following result shows that Menger manifolds cannot be attractors of continuous flows of manifolds.

Theorem 6.5.1. [GuS]. *If A is an attractor of a continuous flow of a manifold, then A has the same shape type as a compact polyhedron.*

On the other hand, Stark [St] realized the Menger curve as an invariant set of an expanding (differentiable) map on the 3-dimensional torus. For information on the attractors of homeomorphisms on Menger, topological, Hilbert cube manifolds, see [Kenn].

A homeomorphism $f : X \to X$ is said to be *expansive* if there is a positive constant $c > 0$ such that, for each pair x, y of distinct points of X, there is an integer n such that $d(f^n(x), f^n(y)) > c$. A homeomorphism $g : X \to X$ is said to have the *pseudo-orbit tracing (or shadowing) property* if for each $\epsilon > 0$ there is a $\delta > 0$ such that for each sequence $\{x_i : i \in Z\}$ of points of X with $d > (g(x_i), x_{i+1}) < \delta$ for each i, there is a point $x \in X$ such that $d(g^i(x), x_i) < \epsilon$ for each $i \in Z$. Expansive homeomorphisms with the pseudo-orbit tracing property are topological analogues of hyperbolic diffeomorphisms.

Theorem 6.5.2. *There are no expansive homeomorphisms with the pseudo-orbit shadowing property on compact Menger manifolds.*

It is easy to see that, in μ^n, any continuum which is the product of two non-degenerate continua has empty interior. Thus the above result follows immediately from the Local Product Structure Theorem [Ao] , [Hi].

It is not known whether there is an 1-dimensional locally connected continuum which admits an expansive homeomorphism. Anderson's characterization of the Menger curve and the following theorem show that the Menger curve is essentially the only candidate for such a continuum.

Theorem 6.5.3. [KTT]. *Suppose that a 1-dimensional locally connected continuum X has an expansive homeomorphism. Then X is nowhere locally planar. In particular, if a 1-dimensional locally connected continuum X with no local separating points admits an expansive homeomorphism, then it is homeomorphic to the Menger curve.*

One of the constructions of expansive homeomorphisms on 1-dimensional continua consists of defining a map $f : G \to G$ of a graph G and taking the inverse limit $G^* = \varprojlim \{G \xleftarrow{f} G \xleftarrow{f} G \xleftarrow{f} \dots\}$ of the associated inverse sequence. The map f naturally defines a homeomorphism of G^*, called the shift map. However this approach does not work for the Menger curve, because the Menger curve can not be represented as the limit of an inverse sequence consisting of a single graph.

What we might expect next would be to obtain an expansive homeomorphism by constructing an inverse sequence of (different) graphs $(G_1 \xleftarrow{f_1} G_2 \xleftarrow{f_2} G_3 \xleftarrow{f_3} \dots)$ whose limit is the Menger curve, and constructing a sequence of maps $(h_i : G_i \to G_i)$ such that $h_i \circ f_i = f_i \circ h_{i+1}$ and $\lim h_i$ is expansive. However, in [KTT], it is proved that this is impossible.

It is proved by Kato [K_2] that there are no expansive homeomorphisms on planar locally connected continua.

References

[AO]   J. Aarts and L. G. Oversteegen, *The dynamics of the Sierpiński curve*, preprint.

[Ag$_1$]   S. M. Ageev, *Classifying spaces for free actions and the Hilbert-Smith conjecture, Russian*, Acad. Sci. Sbornik Math. **75** (1993), 137–144.

[Ag$_2$]   _____, *Menger compacta with actions of compact groups*, General Topology: Spaces, Mappings and Functors, Moscow Univ. Press, 1991.

[An$_1$]   R. D. Anderson, *A characterization of the universal curve and a proof of its homogeneity*, Ann. of Math. **67** (1958), 313-324.

[An$_2$]   _____, *One-dimensional continuous curves and a homogeneity theorem*, Ann. of Math. **68** (1958), 1–16.

[An$_3$]   _____, *The algebraic simplicity of certain groups of homeomorphisms*, Amer. J. Math. **80** (1958), 955–963.

[An$_4$]   _____, *On homeomorphisms as products of conjugates of a given homeomorphism and its inverse*, Topology of 3-manifolds and related topics (M. K. Fort, ed.), Prentice-Hall, 1961, pp. 231–234.

[An$_5$]   _____, *Zero-dimensional compact groups of homeomorphisms*, Pac. J. Math. **7** (1957), 797–810.

[An$_6$]   _____, *A continuous curve admitting monotone open maps onto all locally connected metric continua*, Bull. Amer. Math. Soc. **62** (1956), 264–265.

[An$_7$]   _____, *On topological infinite deficiency*, Mich. Math. J. **14** (1967), 365–383.

[Ao]   N. Aoki, *Topological dynamics, Chap. 15*, General Topology (K.Morita and J.Nagata, eds.), Elsevier Pub. BV., 1989, pp. 231–234.

[BK]   D. Bellamy and J. Kennedy, *Factorwise rigidity of products of pseudo-arcs*, Topology Appl. **24** (1986), 197–205.

[B]   R. Berlanga, *A mapping theorem for topological sigma-compact manifolds*, Compositio Math. **63** (1987), 209–216.

[BP]   C. Bessaga and A. Pełczynski, *Selected topics in infinite-dimensional topology*, PWN, Warszawa, 1975.

[Be]   M. Bestvina, *Characterizing k-dimensional universal Menger compacta*, Mem. Amer. Math. Soc. #380 **71** (1988).

[BBMW] M. Bestvina, P. Bowers, J. Mogilski and J. Walsh, *Characterizing Hilbert space manifolds revisited*, Topology Appl. **24** (1986), 53–69.

[BM]   M. Bestvina and J. Mogilski, *Characterizing certain incomplete infinite-dimensional absolute retracts*, Michigan Math. J. **33** (1986), 291–313.

[Bi$_1$]   R. H. Bing, *A characterization of three-space by partitionings*, Trans. Amer. Math. Soc. **70** (1951), 15–27.

[Bi$_2$]   _____, *Partitioning continuous curves*, Bull. Amer. Math. Soc. **58** (1952), 536–556.

[Bor$_1$] K. Borsuk,, *Theory of Retracts*, PWN, Warszawa, 1966.

[Bor$_1$] _____, *Theory of Shape*, PWN, Warszawa, 1975.

[Bot$_1$] H. G. Bothe, *Eine Einbettung m-dimensionaler Mengen in einen $(m+1)$-dimensionalen absoluten Retrakt*, Fund. Math. **52** (1963), 209–224.

[Bot$_2$] _____, *Ein eindimensionale Kompaktum in E^3, das sich nicht lagetreu in die Mengersche Universalkurve einbetten labt*, Fund. Math. **54** (1964), 251–258.

[Bow$_1$] P. Bowers, *General position properties satisfied by finite products of dendrites*, Trans. Amer. Math. Soc. **288** (1985), 739–753.

[Bow$_2$] _____, *Dense embeddings of nowhere locally compact separable metric spaces*, Topology Appl. **26** (1987), 1–12.

[Bre] B. Brechner, *On the dimensions of certain spaces of homeomorphisms*, Trans. Amer. Math. Soc. **121** (1966), 516–548.

[Br] L. E. J. Brouwer, *On the structure of perfect sets of points*, Proc. Akad. Amsterdam **12** (1910), 785–794.

[BLL] W. Browder, J. Levine and G. R. Livesay, *Finding a boundary for an open manifold*, Amer. J. Math. **87** (1965), 1017–1028.

[Br-K] K. S. Brown, *Cohomology of groups*, Springer-Verlag, New York, 1982.

[Br-M] M. Brown, *A mapping theorem for untriangulated manifolds*, in Topology of 3–manifolds and related topics (M. K. Fort, ed.), Prentice-Hall, 1961, pp. 92–94.

[BG] M. Brown and H. Gluck, *Stable structures on manifolds I. Homeomorphisms of S^n*, Ann. Math. **79** (1964), 1–17.

[Bry$_1$] J. L. Bryant, *On embeddings of compacta in Euclidean space*, Proc. Amer. Math. Soc. **23** (1969), 46–51.

[Bry$_2$] _____, *On embeddings of 1-dimensional compacta in E^5*, Duke Math. J. **38**, 1971 265–270.

[BFMW] J. Bryant, S. Ferry, W. Mio and S. Weinberger, *Topology of homology Manifolds*, Bull. Amer. Math. Soc. **28** (1993), 324–328.

[C] J. W. Cannon, *A positional characterization of the $(n-1)$-dimensional Sierpiński curve in S^n ($n\Pi 4$)*, Fund. Math. **79** (1973), 107–112.

[Cha] T. A. Chapman, *Lectures on Hilbert Cube Manifolds*, vol. 28, CBMS, Providence, 1976.

[CF] T.A. Chapman and S. Ferry, *Obstructions to Finiteness in the Proper Category*, unpublished manuscript..

[CS] T. A. Chapman and L. Siebenmann, *Finding a boundary for a Hilbert cube manifolds*, Acta Mathematica **137** (1976), 171–208.

[Che] A. V. Chernavskii, *Generalization of L. V.Keldysh's construction of monotone mappings of the cube onto a higher dimensional cube*, Uspekhi Mat. Nauk **40**, #4 (1985), 209–211.

[Chi$_1$] A. Chigogidze, *Noncompact absolute extensors in dimension n, n-soft mapppings and their applications*, Math. USSR Izvestiya **28** (1987), 151–174.

[Chi$_2$] _____, *Compact spaces lying in an n-dimensional universal Menger compact space and having homeomorphic complements in it*, Math. USSR Sbornik **61** (1988), 471–484.

[Chi$_3$] _____, *The theory of n-shapes*, Russian Math. Surveys #5, **44** (1989), 145–174.

[Chi$_4$] _____, *n-soft maps of n-dimensional spaces*, Mat. Zametki **46** (1989), 88–95.

[Chi$_5$] _____, *n-shapes and n-cohomotopy groups of compacta*, Math. USSR Sbornik **66** (1990), 329–342.

[Chi₆]   _____, UV^n-equivalence and n-equivalence, Topology Appl. **45** (1990),
          283–291.

[Chi₇]   _____, Autohomeomorphisms of the universal Menger compacta are sta-
          ble, Bull.Acad. Sci. Georgia **#3, 142** (1991), 477–479.

[Chi₈]   _____, Classification theorem for Menger manifolds, Proc. Amer. Math.
          Soc. **116** (1992), 825–832.

[Chi₉]   _____, Finding a boundary for a Menger manifold, Proc. Amer. Math.
          Soc. **121** (1994), 631–640.

[CF]   A. Chigogidze and V. V. Fedorchuk, Absolute retracts and infinite dimen-
          sional manifolds, Nauka, Moscow, 1992. (in Russian)

[CKS]   A. Chigogidze, K. Kawamura and R. B. Sher, in preparation.

[CKT]   A. Chigogidze, K. Kawamura and E. D. Tymchatyn, Nöbeling spaces and
          the pseudo-interiors of Menger compacta, preprint.

[CM]   A. Chigogidze and J. R. Martin, Fixed point sets of Tychonov cubes,
          Topology Appl. (to appear).

[D]   R. J. Daverman, Decomposition of manifolds, Academic Press, New York,
          1986.

[Di₁]   J. J. Dijkstra, k-dimensional skeletoids in R^n and the Hilbert cube, Topol-
          ogy Appl. **19** (1985), 13–28.

[Di₂]   _____, Fake topological Hilbert spaces and characterization of dimension
          in terms of negligibility, vol. CWI Tract 2, Mathematisch Centrum, Am-
          sterdam, 1984.

[DMM]   J. J. Dijkstra, J. van Mill and J. Mogilski, Classification of finite-dimension-
          al pseudo-boundaries and pseudo-interiors, Trans. Amer. Math. Soc. **332**
          (1992), 693–709.

[DG]   T. Dobrowolski and J. Grabowski, Subgroups of Hilbert spaces, Math. Z.
          211 (1992), 657–669.

[DM]   T. Dobrowolski and J. Mogilski, Problems on topological classification of
          incomplete metric spaces, "Open Problems in Topology" (J.van Mill and
          G.M.Reed, eds.), New York,, pp. 409–429.

[Dr₁]   A. N. Dranishnikov, Absolute extensors in dimension n and dimension
          raising n-soft maps, Math. Surveys **# 5,39** (1984), 63–111. (Russian)

[Dr₂]   _____, Universal Menger compacta and universal mappings, Math. USSR
          Sbornik **57** (1987), 131–149.

[Dr₃]   _____, On resolution of LC^n-compacta, Lecture Notes in Math. **1283**
          (1987), 48–64.

[Dr₄]   _____, Homological dimension theory, Russian Math. Surveys **# 4,43**
          (1988), 11–63.

[Dr₅]   _____, On free actions of zero-dimensional compact groups, Izv. Akad.
          Nauk USSR **32** (1988), 217–232.

[EG]   D. A. Edwards and R. Geoghegan, Shapes of complexes, ends of manifolds,
          homotopy limits, and the Wall obstruction, Ann. of Math. **101** (1975),
          521–535.

[E₁]   R. D. Edwards, Demension Theory, I, Geometric Topology (L. C. Glaser
          and T. B. Rushing, eds.), Lective Notes in Math., vol. 438, Springer, 1975,
          pp. 195–211.

[E₂]   _____, Characterizing infinite-dimensional manifolds topologically (after
          Henryk Toruńczyk), Lecture Notes in Math. **770** (1980), 278–302.

[E₃]   _____, *Approximating codimension ≥ 3 σ-compacta with locally homo-topically unkotted embeddings*, Top. Appl. **24** (1986), 95–122.

[E₄]   _____, *Some remarks on the Hilbert-Smith Conjecture*, Proceedings of the 4th Annual Western Workshop in geometric topology 1987, Oregon State Univ. (1987), 1–8.

[EK]   R. D. Edwards and R. C. Kirby, *Deformations of spaces of embeddings*, Ann. Math. **93** (1971), 63–88.

[En]   R. Engelking, *Dimension Theory*, PWN, Warszawa, 1978.

[Fa]   A. Fathi, *Skew products and minimal dynamical systems on separable Hilbert manifolds*, Ergodic Th. Dynamical Systems **4** (1984), 213–224.

[F₁]   S. Ferry, *The homeomorphism group of a compact Hilbert cube manifold is an ANR*, Ann. of Math. **106** (1977), 101–119.

[F₂]   _____, *A stable converse to the Vietoris-Smale theorem with applications to shape theory*, Trans. Amer. Math. Soc. **261** (1980), 369–380.

[F₃]   _____, *UV^k-equivalent compacta*, Lecture Notes in Math. **1283** (1987), 88–114.

[Fi]   G. Fisher, *On the group of all homeomorphisms of a manifold*, Trans. Amer. Math. Soc. **97** (1960), 193–212.

[Fo]   R. H. Fox, *On the Lusternik-Schnirelman category*, Ann. of Math. **42** (1941), 333–370.

[Fr]   F. Freyd, *Stable homotopy theory*, Proc. Conf. on categorical Algebra, La Jolla, Springer, Berlin-Heidelberg-New York, 1965.

[GHW]   D. Garity, J. P. Henderson and D. G. Wright, *Menger spaces and inverse limits*, Pacific J. Math. **131** (1988), 249–259.

[GS]   R. Geoghagen and R .Summerhill, *Pseudo-boundaries and pseudo-interiors in Euclidean spaces and topological manifolds*, Trans. Amer. Math. Soc. **194** (1974), 141–165.

[GvM]   H. Gladdiness and J. van Mill, *Hyperspaces of Peano continua of Euclidean spaces*, Fund. Math. **142** (1993), 173–188.

[GuS]   B. Gunther and J. Segal, *Every attractor of a flow of a manifold has the shape of finite polyhedron*, Proc. Amer. Math. Soc. **119** (1993), 321–329.

[Ha]   O. G. Harold, Jr., *A characterization of locally Euclidean spaces*, Trans. Amer. Math. Soc. **118** (1965), 1–16.

[HH]   H. M. Hastings and A. Heller, *Homotopy idempotents of finite-dimensional complexes split*, Proc. Amer. Math. Soc. **85** (1982), 619–622.

[HP]   L. J. Hernandez and T. Porter, *An embedding theorem for proper n-types*, Topology Appl. **48** (1992), 215–233.

[Hi]   K. Hiraide, *Expansive homeomorphisms of compact surfaces are pseudo-Anosov*, Osaka J. Math. **27** (1990), 117–162.

[Ho]   B. Hoffman, *A surjective characterization of Dugundji spaces*, Proc. Amer. Math. Soc. **76** (1979), 151–156.

[Hu]   Sze-Tsen Hu, *Theory of Retracts*, Wayne State Univ. Press, Detroit, 1965.

[Hur]   W. Hurewicz, *Dimensionstheorie und Cartesishe Raume*, Proc. Akad. Amsterdam **31** (1931), 916–922.

[HW]   W. Hurewicz and H. Wallman, *Dimension Theory*, Princeton, 1941.

[ISV]   I. Ivansic, R. B. Sher and G. A. Venema, *Complement theorems beyond the trivial range*, Illinois J. Math. **25** (1981), 209–220.

[I]   Y. Iwamoto, *Menger manifolds homeomorphic to their n-homotopy ker-nels*, Proc. Amer. Math. Soc. (to appear).

[IS$_1$]   Y. Iwamoto and K. Sakai, *A mapping theorem for Q-manifolds and m^{n+1}-manifolds*, preprint.

[IS$_2$]   _____, *A private communication..*

[K$_1$]   H. Kato, *Generalized homogeneity of continua and a question of J.J.Chara-tonik*, Houston J. Math. **13** (1987), 51–64.

[K$_2$]   _____, *The nonexistence of expansive homeomorphisms of Peano con-tinua in the plane*, Topology Appl. **34** (1990), 161–165.

[K$_3$]   _____, *Expansive homeomorphisms in continuum theory*, Topology Appl. **45** (1992), 223–243.

[K$_4$]   _____, *Striped structure of stable and unstable sets of expansive home-omorphisms and a theorem of K. Kuratowski on independent sets*, Fund. Math. **143** (1993), 153–165.

[Ka$_1$]   K. Kawamura, *An inverse system approach to Menger manifolds*, Topol-ogy Appl. (to appear).

[Ka$_2$]   _____, *Characterizations of Menger manifolds and Hilbert cube mani-folds in terms of partitions*, this volume.

[Ka$_3$]   _____, *Acyclicity of a homeomorphism group of a Menger manifold*, preprint.

[Ka$_4$]   _____, *A proof of Mapping theorem for compact Menger manifolds*, preprint.

[Ka$_5$]   _____, *A construction of homogeneous, non-bihomogeneous continua of P. Minc*, preprint.

[Ka$_6$]   _____, *On n-homotopy equivalences between compact ANR's*, in prepa-ration.

[KOT]   K. Kawamura, L. G. Oversteegen and E. D. Tymchatyn, *On homogeneous totally disconnected 1-dimensional spaces*, preprint.

[KTT]   K. Kawamura, H. M. Tuncali and E. D. Tymchatyn, *Expansive homeo-morphisms on Peano curves*, preprint.

[Ke]   J. Keesling, *Hausdorff dimension*, Topology Proc. **11** (1986), 349–383.

[Kenn]   J. Kennedy, *The topology of attractors*, preprint..

[Ken]   J. Kennedy-Phelps, *Homeomorphisms of products of universal curves*, Houston J. Math. **6** (1980), 127–134.

[Ki]   R. C. Kirby, *Stable homeomorphisms and the annulus conjecture*, Ann. Math. **89** (1969), 575–582.

[Ko]   P. Koszmider, *On the complete invariance property in some uncountable products*, Canad. Math. Bull. **35** (1992), 221–229.

[Ku$_1$]   K. Kuperberg, *On the bihomogeneity problem of Knaster*, Trans. Amer. Math.Soc. **321** (1990), 128–143.

[Ku$_2$]   _____, *Bihomogeneity and Menger manifolds*, preprint.

[KKT$_1$] K. Kuperberg, W. Kuperberg and W. R. R. Transue, *On the 2-homogeneity of Cartesian products*, Fund. Math. **110** (1980), 131–134.

[KKT$_2$] _____, *Homology separation and 2-homogeneity*, Marcel Dekker, this volume.

[La]   R. C. Lacher, *Cell-like mappings and their generalizations*, Bull. Amer. Math. Soc. **83** (1977), 495–552.

[Le]   S. Lefschetz, *On compact spaces*, Ann. of Math. **32** (1931), 521–538.

[Lew₁]   W. Lewis, *Continuous curves of pseudo-arcs*, Houston J. Math. **11** (1985), 91–99.

[Lew₂]   _____, *Non-homogeneity of intermediate universal continua*, Top. Proc. **12** (1987), 193–196.

[Lew₃]   _____, *Continuum Theory Problems*, Topology Proc. **8, No.2** (1983), 361–394.

[LW]   W. Lewis and J. Walsh, *A decomposition of the plane into pseudo-arcs*, Houston J. Math. **4** (1978), 209–222.

[M]   J. R. Martin, *Fixed point sets of homeomorphisms of metric products*, Proc. Amer. Math. Soc. **103** (1988), 1293–1298.

[MR]   J. R. Martin and S. M. Nadler, *Examples and questions in the theory of fixed point sets*, Canad. J. Math. **31** (1979), 1017–1032.

[MT]   J. R. Martin and E. D. Tymchatyn, *Fixed point sets of 1-dimensional Peano continua*, Pacific J.Math. **89** (1980), 147–149.

[Ma]   J. N. Mather, *The vanishing of the homology of certain groups of homeomorphisms*, Topology **10** (1971), 297–298.

[MOT]   J. C. Mayer, L. G. Oversteegen and E. D. Tymchatyn, *The Menger curve: characterization and extension of homeomorphisms of non-locally-separating closed subsets*, Diss. Math. **CCLII** (1986).

[MS]   J. C. Mayer and C. W. Stark, *Group actions on Menger manifolds*, preprint.

[Mc]   R. A. McCoy, *Groups of homeomorphisms of normed linear spaces*, Pacific J. Math. **39** (1971), 735–743.

[Me₁]   K. Menger, *Allgemeine Raume und Cartesische Raume Zweite Mitteilung: "Uber umfas - sendste n-dimensionale Mengen"*, Proc.Akad.Amsterdam **29** (1926), 1125–1128.

[Me₂]   _____, *Dimensionstheorie*, Leipzig, 1928.

[vM]   J. van Mill, *Infinite-Dimensional Topology, Prerequisites and Introduction*, North-Holland, Amsterdam, 1989.

[Mi]   P. Minc, *Solenoids and bihomogeneity*, Marcel Dekker, this volume.

[MR]   P. Minc and J. T. Rogers, Jr., *Some new examples of homogeneous curves*, Topology Proc. **10** (1985), 347–356.

[McR]   D. R. McMillan and H. Row, *Tangled embeddings of one-dimensional continua*, Proc. Amer. Math. Soc. **22** (1964), 378–385.

[Mo]   J. Mogilski, *Characterizing the topology of infinite-dimensional σ-compact manifolds*, Proc. Amer. Math. Soc. **92** (1984), 111–118.

[NTT]   J. Nikiel, H. M. Tuncali and E. D. Tymchatyn, *Dense embeddings into cubes and manifolds*, Continuum Theory and Dynamical Systems (T.West, ed.), Marcel Dekker, 1994, pp. 243–260.

[N]   G. Nöbeling, *Über eine n-dimensionale Universalmenge im R^{2n+1}*, Math. Ann. **104** (1931), 71–80.

[OP]   K. Omiljanowski and H. Patkowska, *On the continua which are Cantor homogeneous or arcwise homogeneous*, Colloq. Math. **58** (1990), 201–212.

[OT]   L. G. Oversteegen, and E. D. Tymchatyn, *On the dimension of certain totally disconnected spaces*, Proc. Amer. Math. Soc. (to appear).

[Pa]   B. A. Pasynkov, *Partial topological products*, Trans. Moscow Math. Soc. **13** (1965), 153–271.

[Pat₁]   H. Patkowska, *On the homogeneity of Cartesian products of Peano continua*, Bull. Polish Acad. Sci. **32** (1984), 343–350.

[Pat$_2$] ———, *On LC^1-spaces which are Cantor and arcwise homogeneous*, Fund. Math. **142** (1993), 139–146.

[Pe] A. Pełczynski, *Linear extensions, linear averaging and their applications to linear topological classification of space of continuous functions*, Diss. Math. **58** (1968).

[Pol] R. Pol, A private communication.

[PT] L. Pontrjagin and G. Tolstowa, *Beweis des Mengerschen Einbettungssatzes*, Math. Ann. **105** (1931), 734–747.

[Qu$_1$] F. Quinn, *Ends of Maps III*, J. Diff. Geom. **17** (1982), 503–521.

[Qu$_2$] ———, *Resolution of homology manifolds and the topological characterization of manifolds*, Invent. Math. **72** (1983), 267–284.

[RW] F. Raymond, and R. F. Williams, *Examples of p-adic transformation groups*, Ann. Math. **78** (1963), 92–106.

[R] H. Robbins, *Some complements to Brouwer's fixed point theorem*, Israel J. Math. **5** (1967), 225–226.

[Ro$_1$] J. T. Rogers, Jr., *An aposyndetic homogeneous curve that is not locally connected*, Houston J. Math. **9** (1983), 433–440.

[Ro$_2$] ———, *Aposyndetic continua as bundle spaces*, Trans. Amer. Math. Soc. **283** (1984), 49–54.

[Ru] T. B. Rushing, *Hausdorff dimension of wild fractals*, Trans. Amer.Math. Soc. **334** (1992), 597–613.

[Sa$_1$] K. Sakai, *Free actions of zero-dimensional compact groups on Menger manifolds*, prepr..

[Sa$_2$] ———, *All autohomeomorphisms of connected Menger manifolds are stable*, preprint.

[Sa$_3$] ———, *Semi free actions of zero-dimensional compact groups on Menger compacta*.

[SV] P. Sankaran and K. Varadarajan, *Acyclicity of certain homeomorphism groups*, Canad. J. Math. **XLII** (1990), 80–94.

[Šče] E. V. Ščepin, *Soft maps of manifolds*, Russian Math. Surveys #5; **39** (1984), 251–270.

[Sc$_1$] H. Schirmer, *On fixed point sets of homeomorphisms of the n-ball*, Israel J. Math. **7** (1969), 46–50.

[Sc$_2$] ———, *Fixed point sets of homeomorphisms of compact surfaces*, Israel J. Math. **10** (1971), 373–378.

[Sc$_3$] ———, *Properties of fixed point sets on dendrites*, Pacif. J. Math. **36** (1971), 795–810.

[Sh] R. B. Sher, *A complement theorem on Menger manifolds*, Proc. Amer. Math. Soc. **121** (1994), 611–618.

[Si] L. C. Siebenmann, *The obstruction to finding a boundary for an open manifold of dimension greater than five*, Doctoral Dissertation, Princeton Univ, Princeton, 1965.

[Sie$_1$] W. Sierpiński, *Sur une courbe cantorienne qui contient une image biunivoque et continue de toute courbe donnee*, C.R. Acad. Paris **162** (1916), 629–632.

[Sie$_2$] ———, *Sur les ensembles connexes et non connexes*, Fund. Math. **2** (1921), 81–95.

[Š₁]   M. A. Štanko, *The embedding of compacta in Euclidean space*, Math. USSR Sbornik **12** (1970), 234–254.

[Š₂]   ———, *Solution of Menger's problem in the class of compacta*, Soviet Math. Dokl. **12** (1971), 1846–1849.

[Š₃]   ———, *Approximation of compacta in E^n in codimension greater than two*, Math. USSR Sbornik **19** (1973), 615–626.

[St]   C. W. Stark, *Some problems concerning dynamics on Menger manifolds*, Talk given at AMS special session; Continuum Theory and Dynamical Systems, University of Tennessee, March 1993.

[Tor₁] H. Toruńczyk, *Concerning locally homotopy negligible sets and characterization of l_2-manifolds*, Fund. Math. **101** (1978), 98–110.

[Tor₂] ———, *On CE-images of the Hilbert cube and the characterization of Q-manifolds*, Fund. Math. **106** (1980), 31–40.

[Tor₃] ———, *Characterizing Hilbert space topology*, Fund. Math. **111** (1981), 247–262.

[Tor₄] ———, *A correction of two papers concerning Hilbert manifolds*, Fund. Math. **125** (1985), 89–93.

[V]   J. Väisälä, *Dimension and Measure*, Proc. Amer. Math. Soc. **76** (1979), 167–168.

[Var]   K. Varadarajan, *Pseudo-mitotic groups*, J. Pure Applied Alg. **37** (1985), 205–213.

[Vi]   K. Villarreal, *The space obtained by spinning the Menger curve about infinitely many of its holes is not homogeneous*, Topology Proc. **16** (1991), 223–238.

[W]   C. T. C. Wall, *Finiteness conditions for CW-complexes*, Ann. of Math. **81** (1965), 56–69.

[Wa]   L. E. Ward, Jr., *Fixed point sets*, Pacif.J.Math. **47** (1973), 553–565.

[We₁]   J. E. West, *The ambient homeomorphy of an incomplete subspace of infinite-dimensional Hilbert spaces*, Pacific J.Math. **34** (1970), 257–267.

[We₂]   ———, *Mapping Hilbert cube manifolds to ANR's: A solution of a conjecture of Borsuk*, Ann. of Math. **106** (1977), 1–18.

[We₃]   ———, *Open Problems in Infinite Dimensional Topology*, Open Problems in Topology (J. van Mill and G.M.Reed, eds.), New-York, 1990, pp. 523–597.

[Whi]   J. H. C. Whitehead, *Combinatorial homotopy, I*, Bull. Amer. Math. Soc. **55** (1949), 213–245.

[Whit₁] J. E. Whittaker, *On isomorphic groups and homeomorphic spaces*, Ann. of Math. **78** (1963), 74–91.

[Whit₂] ———, *On normal subgroups of homeomorphisms*, Trans. Amer. Math. Soc. **123** (1966), 88–98.

[Why]   G. T. Whyburn, *Topological characterization of the Sierpiński curve*, Fund. Math. **45** (1958), 320–324.

[Wil]   D. C. Wilson, *Open mappings of the universal curve onto continuous curves*, Trans. Amer. Math. Soc. **168** (1972), 497–515.

[Wo₁]   R. Y. T. Wong, *A note on stable homeomorphisms of infinite dimensional manifolds*, Proc. Amer. Math. Soc. **28** (1971), 271–272.

[Wo₂]   ———, *Periodic actions on (I-D) normed linear spaces*, Fund. Math. **80** (1973), 133–139.

[Y]   T. Yagasaki, A private communication.

[Ya]   C. T. Yang, *p-adic transformation groups*, Michigan Math. J. **7** (1960),
         201–218.

E-mail address: chigogid@snoopy.usask.ca

E-mail address: kawamura@math.tsukuba.ac.jp

E-mail address: tymchat@snoopy.usask.ca

5

EXACTLY k-TO-1 MAPS: FROM PATHOLOGICAL FUNCTIONS WITH FINITELY MANY DISCONTINUITIES TO WELL-BEHAVED COVERING MAPS

JO W. HEATH Department of Mathematics, Auburn University, Alabama 36849, USA

ABSTRACT. Many mathematicians encounter k-to-1 maps only in the study of covering maps. But, of course, k-to-1 maps do not have to be open. This paper touches on covering maps, and simple maps, but concentrates on ordinary k-to-1 functions (both continuous and finitely discontinuous) from one metric continuum to another. New results, old results, and ideas for further research are given; and a baker's dozen of questions are raised.

1. INTRODUCTION

Requiring a function from one metric continuum X to another, Y, to be finite-to-one, or even to be light, adds a strong hypothesis. But if the function must be k-to-1, meaning that each inverse has exactly k points, then the collection of available maps shrinks drastically and may even disappear. For instance, if Y is a dendrite, then there is a wealth of finite-to-one maps that map onto Y, but there are no k-to-1 maps, [15]. What is it about the dendrite that repels these maps? Now, consider the domain X. It may be that every metric continuum X admits a k-to-1 map for $k > 2$ (see Question 9 later), but for the special case $k = 2$ many interesting situations arise. For instance, the unit interval does not admit an exactly 2-to-1 map, [16], but some dendrites do. What is the crucial topological difference between an arc and a dendrite? And one of the big questions today in this field is whether or not the pseudo-arc admits such a map. The central purpose of this survey paper

1991 *Mathematics Subject Classification.* primary 54C10, secondary 26A03.
Key words and phrases. 2-to-1 map, k-to-1 map, continua, graphs, dendrites, indecomposable spaces, covering maps, simple maps, finitely discontinuous functions, tree-like continua.

is to describe what is known about the domains and images of exactly k-to-1 maps, with special emphasis on the important $k = 2$ case, and to list some of the many questions that still need to be answered. Secondly we will see what happens when finitely many discontinuities are allowed; the surprising thing is that there is still a lot of control. Thirdly and fourthly we will touch on two related topics: covering maps when the spaces do not have the usual textbook connectedness properties and, in variance with the title of this survey, simple maps, that is, maps such that each point in the image has an inverse of cardinality 1 or 2.

By *continuum* we mean a connected compact metric space. There is a glossary at the end of the paper containing other definitions.

2. DOMAINS AND IMAGES OF K-TO-1 MAPS.

2.1. 2-to-1 images. If one toys with the question of which continua are 2-to-1 images (of continua), it is quickly seen that it is easy to map 2-to-1 onto a circle and onto other continua with subcontinua that are similar (in some sense) to a circle and it is hard otherwise. In fact, in [44], Nadler and Ward show how to construct a straightforward 2-to-1 map (or k-to-1 for any $k > 1$) onto any continuum that contains a non-unicoherent subcontinuum. In this same paper, they ask if any tree-like continuum could be the image of an exactly 2-to-1 map. Their question, still unanswered today, is the basis of the conjecture that a continuum is a 2-to-1 image iff it is not tree-like. Furthermore, each of the examples of non-tree-like continua, that the author has tested, is a 2-to-1 image of some continuum. Hence the following two questions, if answered affirmatively by a helpful reader, would neatly classify continua that are 2-to-1 images.

Question 1. [44] Is it true that no tree-like continuum can be the 2-to-1 image of a continuum?

Question 2. Is it true that every continuum that is not tree-like is the 2-to-1 image of a continuum? Must there always exist either a 2-to-1 retraction or a 2-fold covering map?

Regarding Question 1, we know that the following types of continua, if tree-like, cannot be the 2-to-1 image of any continuum: dendrites ([15]), hereditarily indecomposable continua ([25]), and indecomposable arc-continua ([9]). Furthermore, if a continuum has any of the following properties, then it cannot be the 2-to-1 image of a continuum: (1) every subcontinuum has a cut point, [44] and [15], (2) every subcontinuum has a finite separating set and the continuum is hereditarily unicoherent, [23], or (3) every subcontinuum has an endpoint, [44].

Furthermore, we know that whatever the tree-like continuum Y, there is no confluent or crisp ([25]) 2-to-1 map from any continuum onto Y.

Regarding Question 2, the 2-to-1 maps onto orientable or non-orientable inde-composable arc-continua that are local Cantor bundles (which includes all solenoids for instance) have recently been studied in [10]. (These definitions are in the glossary.) It was proved that in the non-orientable case, every 2-to-1 map onto the continuum is a 2-fold covering map and in either case every 2-to-1 map onto the continuum is either a 2-fold cover or a retraction. Furthermore, every orientable local Cantor bundle is the 2-to-1 image of a continuum.

2.2. k-to-1 images. If we consider integers larger than $k = 2$, the situation is murky. For each of these larger k, there is indeed a tree-like continuum that is the k-to-1 image of a continuum, [23]. On the other hand, no dendrite is the k-to-1 image of a continuum for any $k > 1$, [15]. There isn't a lot of elbow room between dendrites and tree-like continua, and we do not even have a conjecture as to what the classification might be:

Question 3. For integers $k > 2$, which continua are k-to-1 images?

A related question asks which continua are k-to-1 images of dendrites [40]. The topological structure of a dendrite dictates, [22], that any k-to-1 image must be one-dimensional, it must contain a simple closed curve, and it cannot contain un-countably many disjoint arcs. And of course it must be a Peano continuum. Is this sufficient? Yes, *if* the continuum contains only finitely many simple closed curves ([22], [40]); but sometimes the answer is yes when the continuum does contain infinitely many simple closed curves [22].

Question 4. [40] Exactly which continua are k-to-1 images of dendrites?

Question 5. [22] If each of Y_1 and Y_2 is the k-to-1 image of a dendrite, and if $Y_1 \cap Y_2$ is a single point, then must $Y_1 \cup Y_2$ be the k-to-1 image of a dendrite?

Question 6. [22] Might the answer to Question 4 depend on k?

That is, does there exist a continuum Y and integers k and m, both greater than 1, such that Y is the k-to-1 image of a dendrite, but Y is not the m-to-1 image of a dendrite?

2.3. 2-to-1 domains. It has been known for over fifty years that no 2-to-1 map can be defined on an arc [16], or, in fact, any connected graph with odd Euler number [14], but extending these results has been difficult. Many dendrites admit 2-to-1 maps, and Wayne Lewis, [23], has constructed a decomposable arc-like continuum that admits a 2-to-1 map. In [7] W. Dębski uses the time-honored technique (see for instance [42], [4] and [6]) of classifying continuous involutions on a space in order to indirectly study 2-to-1 maps on the space. He applies this to determine that a solenoid admits a 2-to-1 map iff there are at most finitely many even integers in an integer sequence that defines the inverse limit structure of the solenoid.

It would be nice of course to know exactly which continua admit 2- to-1 maps but we don't even know the answers to these more restricted questions:

Question 7 [23] Is there an indecomposable arc-like continuum that admits a 2-to-1 map?

(No use trying the classic Knaster Buckethandle space; J. Mioduszewski proved [42] over thirty years ago that it does not admit a 2-to-1 map. See also [7].)

Question 8 [42] Does the pseudo-arc admit a 2-to-1 map?

It is known [27] that there is no weakly confluent 2-to-1 map defined on the pseudoarc. In fact, if f is a weakly confluent 2- to-1 map defined on any hereditarily indecomposable continuum, then neither the domain or the image can be tree-like. For more information related to Question 8, see [26].

George Henderson [33] has proved that if the domain, X, of a 2-to-1 map is a mod 2 homology sphere and $\phi(X)$ denotes the homology dimension of X, then $H_i(Y, Z_2) = Z_2$ if $0 \le i \le \phi(X)$ and $H_i(Y, Z_2) = 0$ otherwise, where Y is the image. Thus additive mod 2 homology cannot be used to distinguish 2-to-1 images of such a sphere. As a corollary, he has that the circle is the only sphere that maps 2- to-1 onto a sphere.

2.4. k-to-1 domains. As is evident by the question below, very little is known about which continua can be the domain of a k-to-1 map, if $k > 2$.

Question 9. Is there a continuum X and an integer $k > 2$ such that there is no exactly k-to-1 map defined on X?

There is extensive literature concerning which continua can or cannot be covering spaces, i.e. domains of very special k-to-1 maps, namely covering maps. We will not attempt to survey these results, but we will mention two relatively recent papers. R. Myers [43] has constructed contractible open 3-manifolds which cannot cover closed 3-manifolds; and David Wright [47] gives a general method of determining when a contractible manifold cannot be a covering space of a manifold.

2.5. k-to-1 maps between graphs. There should be some way to look at the adjacency matrices of two given graphs and decide if a k-to-1 map exists from one onto the other. Although good progress has been made on this question, a direct answer has not been found (see Question 10 below). In this discussion we assume that the two given graphs have enough vertices to eliminate loops, are non-trivial (consist of more than just one vertex), and are connected, even though many of the known results are true, with little or no modification, for disconnected graphs.

Given a positive integer k, and graphs G and H, there are some preliminary filters to rule out the existence of a k-to-1 map from G onto H. For instance, is k times the Euler number of H at least as large as the Euler number of G (or, if $k = 2$, is the Euler number of G twice that of H)? If not, then there is no finitely discontinuous k-to-1 function from G onto H, much less a continuous one. (See the theorem stated later in the subsection on finitely discontinuous functions.) Since only endpoints (vertices of order one) of G can map to endpoints of H, one can count them and make sure G has at least k times as many as H. A more subtle requirement, true for odd integers k, is that each vertex of H with odd order must have an odd number of vertices in G with odd order mapping to it ([28] or [35]). So, one can make sure that G has at least as many odd-order vertices as H does. But these tests can only give a definite "no". S. Miklos has one of the few definite yes results in [41]; namely, if k is odd, then a graph admits a k-to-1 map onto itself iff it has no endpoints.

The original paper [28] that worked on Question 10 started with graphs G and H and a k-to-1 function f from a vertex set of G onto a vertex set of H, and answered whether or not f extended to a k-to-1 map from all of G onto all of H. Similar questions for $\leq$k-to-1 maps are answered in [30] and [31]. The answers are algebraic in terms of the adjacency matrix for H and the "inverse adjacency" matrix for G and f (defined in the glossary). An example of one of the theorems is as follows:

Theorem. [28] Suppose G and H are graphs, k is an odd integer, and f is a k-to-1 function from a vertex set of G onto a vertex set of H. Then f extends to a k-to-1 map from all of G onto H iff f, the adjacency matrix A, and the inverse adjacency matrix B satisfy:

(1) For each vertex p in H, k times the order of p is at least as large as the sum of the orders of the vertices in $f^{-1}(p)$,

(2) each diagonal element of $k \cdot A - B$ is even and non-negative, and

(3) each entry of $B - A$ is nonnegative.

The shortcoming of this theorem and the other similar results is clear. If a given k-to-1 function from the vertex set of G onto the vertex set of H fails to extend, that does *not* mean that there is no k-to- 1 function from G onto H. Perhaps we just started with the wrong vertex function.

A favorite approach is to change the question. Given two non-trivial, connected graphs, G and H, does there *exist* an integer k (or odd integer k or even integer k) and a k-to-1 map from G onto H [34]? Another variation is: suppose that G and H are compatible enough to admit a k-to-1 map, and suppose m is a larger integer (perhaps with the same parity); must G and H admit a m-to-1 map [32] ? Many cases of these and other similar questions have affirmative answers and the answers depend loosely on how close H is to being a simple closed curve. If $H \neq S^1$, then H is inspected to see if it is at least Eulerian (every vertex has even order). If not, how many odd-order vertices does H have compared to the number of odd-order vertices in G; and, most important, how many endpoints does H have? The graph G seems to have little to do with the answer in many cases. Three nice results by A.J.W. Hilton [34] are: (1) if $H = S^1$ and k is greater than the number of vertices in G, then all that is needed for the existence of a k-to-1 map from G onto H is that G not have more edges than vertices. (2) If $H \neq S^1$ but H has no endpoints, then for all sufficiently large even integers k, there is a k-to-1 map from G onto H, and (3) if $H \neq S^1$, H has no endpoints and there are at least as many odd-order vertices in G as there are in H, then there are k- to-1 maps from G onto H for all sufficiently large odd k. The conditions given for G and H are, in each of the three cases, also necessary. Hilton [35] has also studied the relationships, for each parity, between the *initial k* (the least k such that there is a k-to-1 map) and the *threshold k* (the least k such that k and every integer larger than k admits a k-to-1 map) and found that, for each parity, in many cases they are the same integer. See [36] and [29] for some explicit constructions of k-to-1 maps from graphs onto a simple closed curve.

Question 10. Given an integer $k > 1$ and two graphs, G and H, when does a k-to-1 map exist from G onto H?

3. FINITELY DISCONTINUOUS K-TO-1 FUNCTIONS.

With many studies involving functions, so many theorems go out the window if a discontinuity is allowed that much of the power is lost. But not so with k-to-1 functions, especially when the image is required to be a continuum. (Even with k-to-1 functions, a single discontinuity can easily destroy both connectivity and compactness.) The process is this: Suppose the domain is a graph G. Remove a finite set, N, of points from G. Now reassemble the components of the complement of N, along with the points of N, in a k-to-1 fashion in such a way that the resulting space is a continuum. From this mental picture emerges the fact that the Euler number of G is all-important. (We define the Euler number of a graph to be the number of edges minus the number of vertices.) In fact, the following theorem is a concise characterization of exactly which pairs of graphs have k-to-1 finitely discontinuous functions between them. No such characterization, based entirely on Euler numbers, is possible for continuous k-to-1 functions between graphs; in fact, we have no characterization at all for the continuous case (see earlier subsection). So, in the case of graphs, allowing finitely many discontinuities actually clarifies the picture. For some studies of finitely discontinuous functions from an arc *into* an arc, where the image is not required to be compact, see [38], [39] and [19].

Theorem. [21] If G is a graph with Euler number m and H is a graph with Euler number n, then there is a k-to-1 function from G onto H with finitely many discontinuities:

(1) iff $m \leq kn$, if $k > 2$, and

(2) iff $m = 2n$, if $k = 2$.

Up to now, the study of finitely discontinuous k-to-1 functions has remained mostly in the safe haven of locally connected continua for domains, images, or both. Perhaps a good starting place to branch out would be the Knaster Buckethandle continuum (description in [23]). This indecomposable continuum can be neither the domain [42] nor the image [9] of a 2-to-1 (continuous) map, but in [23] an example is given of an exactly 2-to-1 map from a hereditarily decomposable tree-like continuum onto the Knaster Buckethandle continuum with exactly one discontinuity. However, the following is not known:

Question 11. Is there a 2-to-1 finitely discontinuous function defined on the Knaster Buckethandle continuum?

There are a number of important facts, true about continuous k-to-1 functions, that remain true if finitely many discontinuities are allowed, even if the image is not required to be a continuum. For instance, (i) the dimension of the image is the dimension of the domain for continuous k-to-1 functions [17], and, if the image is compact, for finitely discontinuous k-to-1 functions [20]; (ii) Gottschalk's result [15] that no dendrite can be the continuous k-to-1 image of any continuum is still true if the function is allowed to have finitely many discontinuities [20]; and (iii) Harrold's original theorem in [16] that there is no continuous 2- to-1 map defined on $[0, 1]$ also extends to the finitely discontinuous case, [18]. But, oddly enough, the similar result, that the n-ball does not support a continuous 2-to-1 map (Roberts [45] for $n = 2$, Civin [6] for $n = 3$ and Černavskii [5] for $n > 2$), is not true for finitely discontinuous functions. Krystyna Kuperberg constructed a 2-to-1 function defined on the unit square that has exactly one discontinuity, and her example can be modified to make a 2-to-1 function defined on the n-ball, for any $n > 2$, with only one discontinuity. This example has not appeared in print, and we will describe it here:

Example (K. Kuperberg) A 2-to-1 function defined on the unit disk with exactly one discontinuity.

We will use the following notation:

(1) $D = \{(x,y)|x^2 + y^2 \leq 1\}$, the unit disk,

(2) for each integer $n > 0$, $S_n = \{(x,y)|(x + (n-1)/n)^2 + y^2 = 1/n^2; y \geq 0\}$, and

(3) $S = D \setminus \cup_{n=1}^{\infty} S_n$.

The domain of the function is some unit square. We will first remove a point p from the boundary of this unit square and let h denote a homeomorphism from the square minus p onto S. (Note: we are not suggesting that the boundary of S is homeomorphic to the boundary of the square minus p.) We will, in the next paragraph, construct a continuous map f on S that is 2-to-1 everywhere except for the one point $(0, -1)$ at the bottom of S at which it is 1-to-1. We will then extend the composition $f \circ h$ to all of the square by mapping p to $f((0, -1))$, to complete the construction of the function of the example.

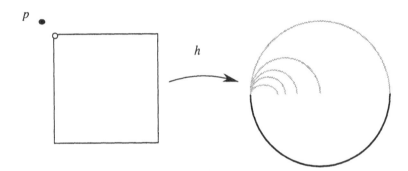

The function f on S will be described as a series of identifications. First, for each point (x,y) in the top half of S (meaning $y > 0$), identify (x,y) and $(x,-y)$. Now the set of points of S that have not been identified is the union of a countable collection $\mathcal{I}$ of disjoint open intervals such that (1) each interval lies either on the x-axis or in the bottom half of S, (2) the sequence of intervals converges to the point $(-1,0)$, and (3) the endpoints of each interval are not in S. Next, locate the interval in $\mathcal{I}$ containing $(0,-1)$ and identify each point (x,y) in this interval with $(-x,y)$; thus the point $(0,-1)$ itself is not identified with another point. Then, for the intervals remaining in $\mathcal{I}$, identify the first of these intervals with the second, the third of these intervals with the fourth, etc. We have now constructed f.

Note that although the original Kuperberg example does not have compact image, the image, I, can be made compact in the following way. First embed I in its one point compactification and then identify the new point added with any point of I. The composition of these two maps is a one-to-one continuous function from I onto a (compact) continuum and the composition of this composite function with Kuperberg's 2-to-1 function is again a 2-to-1 function with one discontinuity, but this time the image is a continuum.

4. COVERING MAPS.

Covering maps defined on compact spaces are the tamest of all k-to- 1 maps. Two-fold covering maps are related to *crisp* maps, i.e. maps that are not just point-wise 2-to-1 but are continuum-wise 2- to-1 in that, if C is a proper subcontinuum of the image, then the inverse of C consists of two disjoint continua each of which is mapped homeomorphically onto C. Every crisp map is a two-fold covering map and every two-fold covering map has a crisp restriction to a subcontinuum, [25]. So far as I know, this relationship has not been studied for integers greater than two, so a natural question is:

Question 12. Define a map to be *k-crisp* if for each proper subcontinuum C of the image, the inverse of C consists of k disjoint continua, each of which is mapped homeomorphically onto C. What is the relationship, if any, between k-crisp maps and k-fold covering maps?

5. SIMPLE MAPS.

In [3] K. Borsuk and R. Molski define *simple* maps to be continuous functions whose point inverses all have exactly one or two points. Simple maps share some of the strength of exactly 2-to-1 maps and are much easier to construct. In fact, one instantly sees that the only space that does not support a simple map, that is not a homeomorphism, is the one point space. In case the simple map f, defined on a compactum, is open, J. W. Jaworowski [37] has shown that f is equivalent to a homeomorphism on the domain of period two. That is, the natural involution i on the compact domain defined by $i(x) = x$ if $f^{-1}f(x)$ has only one point and $i(x)$ is the other point of $f^{-1}f(x)$ otherwise, is a homeomorphism iff the simple map f is open. In contrast, if f is an open (exactly) 2-to-1 map, then f itself is locally one-to-one and is a local homeomorphism [25]; but this is not true of simple maps (a simple example of this is folding an arc in half).

Exactly k-to-1 maps never change the dimension [17], and Jaworowski [37] showed that open simple maps do not alter dimension; but in [3] Borsuk and Molski note that there is a simple map from the Cantor discontinuum onto an interval, so simple maps can raise dimension by one (but they point out that simple maps never change dimension other than to raise it by one). In a similar way, there is an natural simple map from the the Sierpiński universal curve into the plane that raises its dimension by one; however W. Dębski and J. Mioduszewski have proved [11] the surprising result that every simple map from the Sierpiński triangle into the plane has an image with empty interior (and so the image has dimension one). See [12] and [13] for other related results.

Borsuk and Molski [3] proved that every locally one-to-one map defined on a compactum is a finite composition of simple maps. So in this sense, simple maps are building blocks for locally one-to-one maps on compacta. In [46] Sieklucki showed even more: Every map of finite order defined on a finite dimensional compactum is a finite composition of simple maps. He also constructs an infinite dimensional counterexample. Since any finite composition of simple maps is necessarily of finite order, his theorem is the best possible for compacta. In response to the natural question of whether (or not) *open* maps of finite order (defined on a finite dimensional compacta) are finite compositions of *open* simple maps, John Baildon [2] proved that if f is an open simple map between 2-manifolds without boundaries, and if f is the composition of n open simple maps, then f has order 2^n. Hence, no such finite composition is possible for $w = z^3$, defined on the unit sphere, for instance. Note that Baildon adds to the definition of a simple map that it not be one-to-one.

These results do not extend, as is, to the exactly k-to-1 case. For instance, there is a 3-to-1 map defined the unit interval onto a simple closed curve, but it cannot be written as a composition of 2-to-1 maps and 1-to-1 maps because there is no 2-to-1 map defined on the unit interval at all, and finite compositions of one-to-one maps are homeomorphisms. But is there any kind of building block theory here?

Question 13. Under what circumstances are k-to-1 maps finite compositions of maps of lesser order?

6. DEFINITIONS.

Adjacency matrix. If V is a vertex set for a graph H, the *adjacency matrix* is a matrix indexed by $V \times V$ whose (v_1, v_2) entry is defined to be the number of edges in H between v_1 and v_2.

Arc-continuum. A continuum is an *arc-continuum* if each subcontinuum is either the whole continuum, a point or an arc.

Arc-like. A continuum is *arc-like* if for each positive number ϵ there is an ϵ-map from the continuum onto an arc, i.e. a continuous function from the continuum onto an arc such each point inverse has diameter less than ϵ.

Confluent. A function is *confluent* if for each continuum C in the image, each component of the preimage of C maps onto C.

Continuum. A topological space is a *continuum* if it is connected, compact, and metric.

Covering Map. A continuous function f from a space X onto a space Y is a *covering map* if for each point y in Y there is an open set U containing y such that

$f^{-1}(U)$ is the union of finitely many disjoint open sets, each of which is mapped homeomorphically by f onto U.

Crisp. A map f is *crisp* if, for any proper subcontinuum C of the image, the inverse of C is the union of two disjoint continua, each of which is mapped homeomorphically by f onto C.

Cut point. A point x in a continuum X is a *cut point* if $X \setminus \{x\}$ is not connected.

Decomposable. A continuum is *decomposable* if it is the union of two proper subcontinua.

Dendrite. A continuum is a *dendrite* if it is locally connected and contains no simple closed curve.

Euler number. The *Euler number* of a graph is the number of edges minus the number of vertices.

Finitely discontinuous. A function is *finitely discontinuous* if it has at most a finite number of discontinuities.

Finite order. A function has *finite order* if there is an integer k such that each point in the image has a preimage with no more than k points.

Graph. A continuum is a *graph* if it is homeomorphic to the finite union of straight arcs and points.

Hereditarily indecomposable. A continuum is *hereditarily indecomposable* if each subcontinuum is indecomposable.

Indecomposable. A continuum is *indecomposable* if it is not the union of two proper (unequal to the continuum) subcontinua.

Inverse Adjacency Matrix If f is a function from the vertex set of a graph G onto the vertex set V of a graph H, then the *inverse adjacency matrix* is indexed by $V \times V$ and its (v_1, v_2) entry is the number of edges in G that go from any point of $f^{-1}(v_1)$ to any point of $f^{-1}(v_2)$.

Involution. An *involution* is a fixed-point-free function f from a space into itself such that $f(f(x)) = x$ for every x. This condition implies that f is one-to-one and onto, but an involution need not be continuous.

k-to-1 A function is *k-to-1* if the preimage of each point in the image has exactly k points.

k-crisp. A map is *k-crisp* if for each continuum C in the image, the inverse of C consists of k disjoint continua each of which is mapped homeomorphically onto C.

Local Cantor Bundle A continuum is a *local Cantor bundle* if each point has a neighborhood homeomorphic to $C \times (0, 1)$, where C denotes the Cantor discontinuum.

Local Homeomorphism. A function f is a *local homeomorphism* if for each point p in the domain, there is an open set U containing p such that f is a homeomorphism on U and $f(U)$ is open.

Map. A function is a *map* if it is continuous.

Non-orientable Arc-continuum. See "Orientable Arc- continuum".

Non-unicoherent. A continuum is *non- unicoherent* if it is the union of two subcontinua whose intersection fails to be connected.

Orientable Arc-continuum. A general definition can be found in [1], but for arc-continua that are local Cantor bundles, the definition is equivalent to the following natural one. The arc-continuum is *orientable* if each separate arc component can be parameterized (given a direction) so that no sequence of arcs going one direction converges to an arc going the other direction.

Proper subcontinuum. A subcontinuum of a continuum C is *proper* if it is not equal to C.

Simple Map. A continuous function is *simple* if each of its point inverses has cardinality 1 or 2.

Tree. A graph is a *tree* if it is connected and contains no simple closed curves.

Tree-like. A continuum is *tree-like* if for each positive number ϵ there is an ϵ-map from the continuum onto a tree. (See "arc-like" for the definition of an ϵ-map.)

2-to-1 A function is *2-to-1* if the preimage of each point in the image has exactly two points.

Unicoherent. See "Non-unicoherent".

Weak Confluence. A function is *weakly confluent* if for each continuum C in the image, at least one component of the preimage of C maps onto C.

REFERENCES

1. J.M. Aarts and L.G. Oversteegen, *Flowbox Manifolds* Transactions AMS, **327** (1991) 449-463.
2. J. D. Baildon *Open simple maps and periodic homeomorphisms*, Proceedings AMS **39** (1973) 433-436.
3. K. Borsuk and R. Molski, *On a class of continuous maps*, Fund. Math. **45** (1957) 84-98.

4. A.V. Cernavskii, *Twofold Continuous Partitions of a Ball*, Sov. Math. Doklady **1**, 1960, p. 436.

5. A. V. Černavskii, *The impossibility of a strictly continuous 2-to-1 partition of a homology cube*, Doklady AN SSSR **144** (1962) 286-289 = Soviet Math. Dokl. **3** (1962) 726-729.

6. P. Civin, *Two-to-one mappings of manifolds*, Duke Math. J. **10** (1943), 49-57.

7. W. Dębski, *Two-to-one maps on solenoids and knaster continua*, Fund. Math. **141** (1992) 277-285.

8. W. Dębski, *A note on continuous k-to-1 maps for k even*, Preprint.

9. W. Dębski, J. Heath, J. Mioduszewski, *Exactly 2-to- 1 maps from continua onto some tree-like continua*, Fundamenta Mathematica **141** (1992) 269-276.

10. W. Dębski, J. Heath, J. Mioduszewski, *Exactly 2-to- 1 maps onto arc continua*, Preprint.

11. W. Dębski and J. Mioduszewski, *Simple plane images of the Sierpiński curve are nowhere dense*, Coll. Math. **59** (1990) 125-140.

12. W. Dębski and J. Mioduszewski, *Splitting property of dimension raising simple maps*, Coll. Math. to appear.

13. W. Dębski and J. Mioduszewski, *Conditions which assure that a simple map does not raise the dimension*, preprint.

14. P. Gilbert, *n-to-one mappings of linear graphs*, Duke Math J. **9** (1942), 475-486.

15. W.H. Gottschalk, *On k-to-1 transformations*, Bulletin AMS **53** (1947), 168-169.

16. O.G. Harrold, *The non-existence of a certain type of continuous transformation*, Duke Math J. **5** (1939), 789-793.

17. O.G. Harrold, *Exactly (k,1) transformations on connected linear graphs*, Amer. J. Math. **62** (1940), 823-834.

18. J. Heath, *Every exactly 2-to-1 function on the reals has an infinite number of discontinuities*, Proceedings AMS **98** (1986), 369- 373.

19. J. Heath, *K-to-1 functions on arcs for K even*, Proceedings AMS **101** (1987) 387-391

20. J. Heath, *There is no k-to-1 function from any continuum onto [0,1], or any dendrite, with only finitely many discontinuities*, Transactions AMS **306** (1988), 293-305.

21. J. Heath, *K-to-1 functions between graphs with finitely many discontinuities*, Proceedings AMS **103** (1988), 661-666.

22. J. Heath, *K-to-1 images of dendrites*, Topology and Its Applications **31** (1989) 73-82.

23. J. Heath, *Tree-like continua and exactly k-to-1 functions*, Proceedings AMS **105** (1989), 765-772.

24. J. Heath, *The structure of 2-to-1 maps on metric compacta*, Proceedings AMS **110** (1990) 549.

25. J. Heath, *2-to-1 maps with hereditarily indecomposable images*, Proceedings AMS **113** (1991) 839-846.

26. J. Heath, *2-to-1 Maps on Hereditarily Indecomposable Continua*, Transactions AMS **328** (1991) 433-443.

27. J. Heath, *Weakly confluent, 2-to-1 maps on hereditarily indecomposable continua*, Proceedings AMS **117** (1993) 569- 573.

28. J. Heath and A.J.W. Hilton, *Exactly k-to-1 maps between graphs*, Transactions AMS **331** (1992) 771-785.

29. J. Heath and A.J.W. Hilton, *Trees that admit 3-to-1 maps onto the circle*, Journal of Graph Theory **14** (1990) 311-320.

30. J. Heath and A.J.W. Hilton, *At most k-to-1 continuous mappings between graphs*, Contemporary methods in Graph Theory, BI Wessenschatsverlag, (1990) 383-398.

31. J. Heath and A.J.W. Hilton, *At most k-to-1 continuous mappings between graphs, II.*, To appear in Discrete Math.

32. J. Heath and A.J.W. Hilton, *Extensions of k- to-1 maps between graphs*, Houston J. Math. **20** (1994) 129-143.

33. G. W. Henderson *The mod 2 homology of the image of an exactly 2-to-1 map from a sphere*, Proceedings AMS **18** (1967) 723-726.

34. A.J.W. Hilton *The existence of k-to-1 continuous maps between graphs when k is sufficiently large*, Graph Theory **17** (1993) 443-461.

35. A.J.W. Hilton *The initial and threshold values for exactly k-to-1 continuous maps between graphs*, Congressus Numerantium **91** (1992), 254-270.

36. A.J.W. Hilton, J.P. Liu, and C. Zhao *Graphs that admit 3-to-1 or 2-to-1 maps onto the circle*, Discrete Applied Math., to appear.

37. J. W. Jaworowski *On simple regular mappings*, Fund. Math. **45** (1958) 119-129.

38. H. Katsuura and K. Kellum, *K-to-1 functions of an arc*, Proceedings AMS **101** (1987) 629-633.

39. H. Katsuura, *K-to-1 functions on (0,1)*, Real Analysis Exchange, bf 12 (1987), 516-527.

40. S. Miklos *Exactly (n,1) mappings on generalized local dendrites*, Topology Appl. **31** (1989) 47-53.

41. S. Miklos *Exactly (n,1) mappings on graphs* Period. Math. Hungar. **20** (1989) 35-39.

42. J. Mioduszewski, *On two-to-one continuous functions*, Dissertationes Math. (Rozprawy Mat.) **24** (1961), 42.

43. R. Myers *Contractible open 3-manifolds which are not covering spaces*, Topology **27** (1988) 27-35.

44. S.B. Nadler, Jr. and L.E. Ward, Jr., *Concerning exactly (n,1) images of continua*, Proceedings AMS **87** (1983), 351-354.

45. J.H. Roberts, *Two-to-one transformations*, Duke Math. J. **6** (1940), 256-262.

46. K. Sieklucki *On superpositions of simple mappings,* Fund. Math. **48** (1960) 217-228.

47. D. G. Wright *Contractible open manifolds which are not covering spaces.* Preprint. Mathematics Department; Brigham Young University; Provo, Utah 84602.

E-mail address: heathjw@mail.auburn.edu "Jo Heath"

6
A BRIEF HISTORY OF INDECOMPOSABLE CONTINUA

JUDY A. KENNEDY Department of Mathematical Sciences, University of
Delaware, Newark, Delaware 19716, USA

ABSTRACT. The history of indecomposable continua is surveyed from their dis-
covery in 1910 by L. E. J. Brouwer to modern times and their increasing role in
dynamical systems.

A *continuum* is a compact, connected, metric space. A *subcontinuum* X' of the
continuum X is a closed, connected subset of X. If there exist two nonempty
subcontinua H and K of the continuum X such that $H \neq X$ and $K \neq X$, but
$H \cup K = X$, then X is a *decomposable* continuum. Any continuum that is not
decomposable is *indecomposable*. The first question that might arise, if one is not
familiar with many continua, is "Do indecomposable continua really exist?" After
all, the continua that automatically jump to mind are probably manifolds of some
description. It follows from the definition that an indecomposable continuum would
have to be one that is unbreakable in some sense of the word: such a continuum
would have to have the property that if it were broken, it would shatter into an
uncountable number of "pieces" each nowhere dense in the original continuum. It
would be a little like spilling a box of toothpicks.

Obviously, if the title of this paper means anything, indecomposable continua do
exist. But how were they discovered, and why should mathematicians care? The
story of indecomposable continua has a theme that has occurred in mathematics
many times. These objects first arose as examples of extreme pathology, and, each

1991 *Mathematics Subject Classification*. primary 54F20, 54F50, 58F12.
Key words and phrases. indecomposable continua, hereditarily indecomposable continua, irreducible cotinua, composant, subcontinua, complicated invariant sets, chaotic behavior, inverse
limits, pseudoarc, pseudocircle, Lake of Wada continua, Knaster continua.
This research was supported by NSF grant DMS-9208201

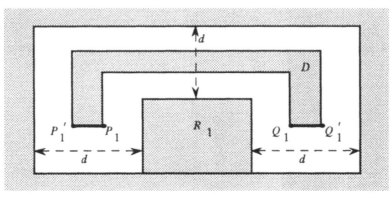

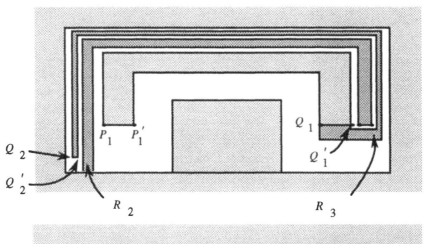

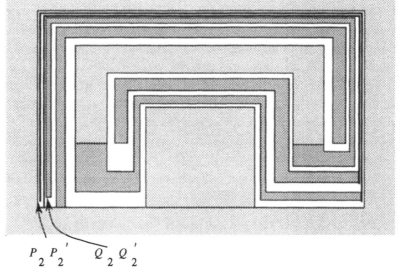

FIGURE 1. **Brouwer's original construction:** Form a rectangle R_0 in the plane of length a and height b (Brouwer's principle rectan-

gle). Now remove an open rectangle R_1 intersecting the boundary of R_0 from R_1, and remove another connected open set D (whose closure doesn't intersect the boundary of R_0 nor the boundary of R_1, and which is composed of the union of three rectangles, each with sides that are either horizontal or vertical) from the interior of $R_0 \backslash R_1$. Choose R_1 so that $R_0 \backslash R_1$ is the union of three rectangles each with the same width d, and so that the ratio of the length of its baseline to that of R_1 is $1/(2c + 1)$ for some $c > 0$. Then $d = [c/(2c + 1)]a$. Draw the open set D so that each of the rectangles composing D has width $[c/(c+1)^2]a$, and the boundary of D is composed of 8 line segments each parallel to one of the 8 line segments composing the boundary of $R_0 \backslash R_1$, and at a distance of $[c/(2c + 1)^2]a$ from the boundary of $R_0 \backslash R_1$. Now construct R_2 and remove it from $R_0 \backslash (R_1 \cup D)$: R_2 is a union of three rectangles and intersects the boundary of R_0, but has closure disjoint from D and R_1. The boundary of R_2 is composed of 8 line segments (all horizontal or vertical) and one horizontal line segment lies on the boundary of R_0 on the bottom side and to the left of R_1. Then R_2 goes around and above D and ends to the right of R_1 at the height of the horizontal segment $Q_1 Q_1'$ in the boundary of D. The width of each of these rectangles composing R_2 is $1/(2c + 1)$ of the width of the portion of $R_0 \backslash (R_1 \cup D)$ in which it lies. Then an open connected set R_3, consisting of the union of 5 rectangles, all with sides that are either horizontal or vertical, is removed from $R_0 \backslash (R_1 \cup D \cup R_2)$. The closure of R_3 does not intersect the closures of R_1 or R_2 or the boundary of R_0, but one side of R_3 coincides with one side of D. Starting at the lower right side of D, R_3 then wraps around R_2, going up, over, and then down again, before ending in the interior of R_0 to the left of that side of R_2 that is on the boundary of R_0, but ending at a height above that of the bottom side of R_0. It, too, removes rectangles of width $1/(2c + 1)$ of the width of that portion of $R_0 \backslash (R_1 \cup D \cup R_2)$ containing it (except for the rectangle adjoining D). Let $R_3' = R_3 \cup D$. In general, R_n', for n odd, $n > 1$ is extended from both ends: a continuation from $Q_r Q_r'$ to $Q_{r+1} Q_{r+1}'$ is followed in the next step by a continuation from $P_r P_r'$ to $P_{r+1} P_{r+1}'$ in R_0. Each R_n has width $1/(2c+1)$ of that portion of $R_0 \backslash (R_1 \cup D \cup R_2 \cup ... R_{n-1})$. For even n, R_n has one side lying on the bottom side of R_0. For each n, R_n is composed of rectangles, each with each side either horizontal or vertical. The boundary of each R_n' (for n odd and greater than 1) does not intersect the boundary of any R_m for m even, and does not intersect the boundary of any R_m for m even, and does not intersect the boundary of D or R_m, for m odd, unless $m = n-2$ or $n-4$ (and then it only intersects one of those). The horizontal segments $P_r P_r'$ have lengths approaching 0, and approach the lower left point of R_0, as do the horizontal segments $Q_r Q_r'$. The set $A = R_0 \backslash (D \cup (\cup_{n \geq 1} R_n))$ is a non-empty continuum, each point of which is arbitrarily close to some point of the connected open set $G = (R^2 \backslash R_0) \cup (\cup_{n \geq 1} R_{2n}) \cup R_1$, and is arbitrarily close to the connected open set $H = D \cup (\bigcup_{n \geq 1} R_{2n+1})$.

Further, G does not intersect H.

time they arise in a different context, it seems that, once again, they represent pathology. A. D. Wallace [Wa] summed up the feeling of many mathematicians about these objects when he said, "We commonly think of indecomposable spaces as being monstrous things created by set theoretic topologists for some evil (but purely mathematical) purpose." But then the unexpected, the bizarre, turns out to be common, and what happens "most" of the time. To understand indecomposable continua, a mathematician must be willing to develop an intuition that is quite non-Euclidean in nature. However, as these objects become more familiar, it becomes apparent that rather than totally wild, random, unfathomable objects, they have a rich, interesting structure in common, and appear in many different forms.

The reader should keep in mind that what follows is not an exhaustive history of indecomposable continua, but a brief one that represents this author's perspective. An attempt is made here to introduce non-expert readers to these objects, and for that reason the discussion is kept on an elementary level and many definitions are given. (After all, for those who are already experts, there is no need for an article such as this one.) For readers interested in learning more about indecomposable continua and their applications to dynamics, a fairly extensive bibliography follows. An important resource for this paper has been F. L. Jones's 1971 dissertation [J], *A History and Development of Indecomposable Continua*.

The first indecomposable continuum was discovered in 1910 by L.E.J. Brouwer [Br] as a counterexample to a conjecture of Schoenflies that the common boundary between two open, connected, disjoint sets in the plane had to be "decomposable", i.e., the union of two proper, closed, connected sets. The first steps in Brouwer's construction are indicated in Figure 1. Between 1912 and 1920, the Polish mathematician Janiszewksi [Ja1,Ja2] produced more examples of such continua, and was the first to produce an example in the plane that did not separate it. This example of Janiszewski is topologically equivalent to the continuum now widely known as the Knaster bucket handle. (Knaster later gave a simpler description of this example using semicircles. This description appeared in a paper written by Kuratowski [Ku1]. See Figure 2.) Janiszewski makes no mention of the indecomposability of his examples. He seems rather to be interested in their irreducibility, which is not surprising since his dissertation was on this subject. A continuum is *irreducible* between two of its points if no proper subcontinuum contains both points. The concepts of irreducibility and indecomposability are related both mathematically and historically. An indecomposable continuum is always irreducible between two of its points.

Another early example of an indecomposable continuum came from the other side of the world. In 1917, K. Yoneyama [Y] wrote an extensive English paper, "Theory of continuous sets of points". That paper is most known today for its beautiful description of an example known as the Lakes of Wada. (See Figure 3 for illustrations and description.) To mathematicians contemporary to Yoneyama

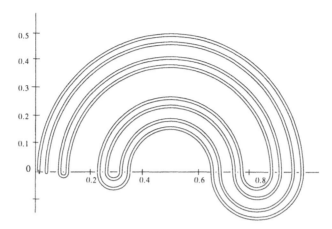

FIGURE 2. **Kuratowski's description of the Knaster bucket handle:** Suppose that C denotes the Cantor middle-thirds set sitting on the unit interval $[0, 1] \times \{0\}$ in the plane. Connect the points with semicircles as follows: (1) For each pair p, q of points of C that are equidistant from $(1/2, 0)$, connect p and q with a semicircle sitting above the x-axis. (2) For each pair p, q of points of points of C equidistant from $(5/6, 0)$, connect p and q with a semicircle extending below the x-axis. (3) For each pair p, q of points equidistant from the midpoint $(5/18, 0)$ of $(2/9, 0)$ and $(1/3, 0)$, connect p and q with a semicircle that extends below the x-axis. Continue this process. (The nth step consists of connecting each pair p, q of points equidistant from the midpoint $(5/3^n, 0)$ of the points $(2/3^n, 0)$ and $(1/3^n, 0)$ with a semicircle that extends below the x-axis.)

the example was nothing more than a novelty, and its indecomposability is not only not mentioned, it probably was not a concept known to the Japanese at that time. Again the purpose was to demonstrate irreducibility, and the original example formed the common boundary between only two plane regions. P. Urysohn [U] was actually the first to prove that the example given by Wada is indecomposable "for a convenient distribution of canals", and to generalize it to allow any countable number of lakes. Most mathematicians today know the example from the exposition given in the text, **Topology**, by Hocking and Young [HY]. However, it should be noted that Brouwer, in that very first paper in 1910, remarked that his construction could be modified so that the resulting continuum forms the common boundary for any finite $(n > 2)$ or countable number of disjoint, connected, simply connected sets in the plane.

By 1920, the Polish mathematicians began to study these objects as interesting entities in themselves, and not just as examples of pathology. The volumes of **Fundamenta Mathematicae** that appeared between 1920 and 1940 are full of papers laying the groundwork of knowledge about these continua. The journal was founded in 1920 by Z. Janiszewski, S. Mazurkiewicz, and W. Sierpiński. Its purpose was to present set-theoretic problems and results in papers written in English,

French, German, or Italian. Some mathematicians of the time feared that this restriction of topic would cause the journal to fail. It not only survived even World War II, but continues to prosper and to be a highly respected journal for this area of mathematics, although many of its main contributors did not survive the war and the Nazi occupation of Poland. Its existence has much to do with the development of the theory of indecomposable continua: it gave researchers in continuum theory a wonderful forum for the presentation of their results and problems. Sadly, the first volume also contained the obituary of Janiszewski. He died at the age of 32 as a result of a long illness, but had already been instrumental in starting the journal and the fine Polish school of mathematics, as well as contributing much to the early development of continuum theory.

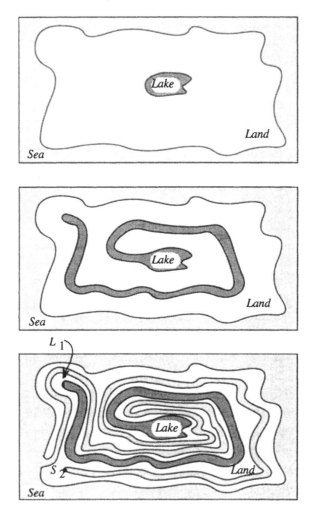

FIGURE 3. **The original Lakes of Wada construction (quoted)** [Y]: "Suppose that there is a land surrounded by sea and that in this land there is a fresh lake. Also suppose that from the lake and

sea canals are built to introduce the waters of them into the land according to the following scheme. Let $E_1, E_2, \ldots, E_n, E_{n+1}, \ldots$ be a sequence of positive numbers monotonously [sic] converging to zero; namely let $E_1 > E_2 > \ldots > E_n > E_{n+1} > \ldots$ and $\lim_{n \to \infty} E_n = 0$."

"On the first day a canal is built from the lake such that it does not meet the sea water and such that the shortest distance from any point on the shore of the sea to that of the lake and canal does not exceed E_1. The end point of this canal is denoted by L_1."

"On the second day a canal is built from the sea, never meeting the fresh water of the lake and canal constructed the day before, and the work is continued until the shortest distance from any point on the shore of the lake and canal filled with fresh water to that of the sea and canal filled with salt water does not exceed E_2. The end point of this canal is denoted by S_2."

"On the third day the work is begun from L_1 never cutting the canals already built, and the work is continued until the shortest distance from any point on the shore of the sea and canal filled with salt water to that of the lake and canal filled with freshwater does not exceed E_3. The endpoint of this canal is denoted by L_3."

"Now it is clear that we can continue the work day by day in the above way, by adequately narrowing the breadth of the canals, since the land is always semi–continuous [i.e, connected] at the end of the work every day. If we proceed in this way indefinitely, we get at last an everywhere dense set of waters fresh and salt, which never mingle together at any place."

"Now denote by M_L the shore of the lake and canal filled with fresh water, and by M_s that of the sea and canal filled with salt water, and by M_p the limiting points of M_L and M_s not contained in them. Then the sum of M_L, M_s, M_P forms a continuous set [continuum], and any three points, each taken from the above different sets form a system of three points, every two of which form a pair of principal points of the set [i.e.., the continuum is irreducible between any two of those three points]."

"...If we suppose that there are many such lakes in the land, we may obtain by the similar method a continuous set having the property" [Y, p. 60–62].

The first paper devoted to indecomposability was due to Mazurkiewicz [M1]. He was the first to use the term "indecomposable" and to restrict his study to closed and bounded subsets of $\mathbb{R}^n$, and therefore, compact sets. (Previous authors had been more interested in closed, connected sets in $\mathbb{R}^n$, although their examples had been bounded.) Mazurkiewicz proved in his paper that an indecomposable continuum in $\mathbb{R}^2$ has three points such that the continuum is irreducible between any two of them.

The paper, "Sur les continus indécomposables" [JK], by Janiszewski and Kuratowski, was the most influential paper on indecomposable continua to appear in the early 20's. In that paper the terms "composant", "boundary set", and "continuum of condensation" first appeared as well as several fundamental structure theorems and familiar characterizations of indecomposability.

Suppose X is a space. A subset A of the continuum X is a *boundary set* in X if $A \subseteq \overline{X \setminus A}$. A subcontinuum K of X is a *continuum of condensation* if $K \subseteq \overline{X \setminus K}$. Thus, a continuum of condensation K is nowhere dense in a continuum X. If x is a point in X, let $\mathrm{Com}(x) = \{y \text{ in } X \mid \text{ there is a proper subcontinuum in } X \text{ that contains both } x \text{ and } y\}$. The set $\mathrm{Com}(x)$ is called the *composant of x* in X. If X is a continuum and x is a point in X, then $\mathrm{Com}(x)$ is dense in X. It follows from the Janiszewski-Kuratowski paper that the continuum X is indecomposable if and only if $\{\mathrm{Com}(x) \mid x \text{ is in } X\}$ forms a partition of X into an uncountable collection of first category, connected sets each of which is dense in X. In fact, they proved that the continuum X is indecomposable and nondegenerate if and only if it possesses two disjoint composants. (A set is *first category* in X if it can be written as the union of a countable number of nowhere dense subsets of X. If a set is not first category, then it is *second category*. It follows from the Baire Category Theorem that no complete, separable metric space is first category in itself, and that no complete, separable metric space can be written as the union of a countable number of first category sets.) It then follows that if x is a point in an indecomposable continuum X, then the composant $\mathrm{Com}(x)$ is a boundary set in X, and if K is a proper subcontinuum of the indecomposable continuum X, then K is a continuum of condensation in X. They also proved that if X is an indecomposable continuum, then there exist three points such that X is irreducible between any two of them. It then follows as a corollary to all this that the continuum X is *indecomposable* if and only if for each point x in X there is a dense, second category subset $\mathrm{D}(x)$ in X such that X is irreducible between x and each point d in the set $D(x) = X \setminus Com(x)$.

A continuum X is *hereditarily indecomposable* if every subcontinuum of X is indecomposable. While the existence of indecomposable continua is amazing enough, that hereditarily indecomposable exist at all is rather remarkable. By 1922, Knaster [Kn1] described in his thesis a hereditarily indecomposable continuum that would later be called the pseudoarc. (Knaster did not use either the term "hereditarily indecomposable" or "pseudoarc." Most research on these objects was not done until after World War II.) Suppose we define a *Lakes of Wada continuum* to be a continuum that forms the common boundary between three or more open, disjoint, connected, simply connected sets in a 2-manifold. (Thus, the original example given by Wada is not a Lakes of Wada continuum, at least technically.) In 1924, Kuratowski [Ku2] proved that a Lakes of Wada continuum is either indecomposable or the union of two indecomposable continua, and Knaster [Kn2] gave examples to show that the second possibility can occur. R. L. Wilder [Wi] gave an example of a locally connected continuum in $\mathbb{R}^3$ to show that the common boundary between three open sets homeomorphic to the interior of a sphere need not be indecomposable or anything close to it. Thus, Lakes of Wada continua are only guaranteed for this situation in the 2-dimensional case.

Suppose X is a space, and K is a subset of X. If x is a point in K, then x is *accessible* from $X \backslash K$ if there is an arc P (i.e., a continuum homeomorphic to [0,1]) such that $P \cap K = \{x\}$, and x is an endpoint of P. If a point is not accessible from $X \backslash K$, then it is *inaccessible*. Mazurkiewicz [M2,M3] proved that if K is an indecomposable continuum in the plane, then (1) the union of all composants in K that contain more than a point accessible from $\mathbb{R}^2 \backslash K$ is a first category subset of K (with respect to the space K), and (2) the union of composants of K that contain more than one point accessible from $\mathbb{R}^2 \backslash K$ is at most countable. Thus, in some sense of the word, indecomposable continua in the plane behave as if they have "interior": they have a sort of pseudo-interior, namely all their inaccessible points. Again, this is a highly plane specific result: the situation is completely different for an indecomposable continuum sitting in a higher dimensional space. In that case every composant may be accessible from outside the continuum.

In 1932, G. D. Birkhoff [Bi] discovered his "remarkable curve". He constructed on the annulus A a homeomorphism with the property that there is an invariant continuum K contained in the interior of A that separates A into two connected open sets, one containing the inside boundary circle and the other containing the outside boundary circle. Further, the rotation number of points accessible from the component of $A \backslash K$ that contained one boundary component was different from the rotation number of points accessible from the other boundary component of $A \backslash K$. (This could not happen if the continuum K were anything close to a simple closed curve. We do not define the term "rotation number" here precisely. Loosely, what it measures is the average rotation of accessible points on K.) Now Birkhoff was a researcher in dynamics and not a topologist, but he apparently found the topology of his invariant continuum K interesting, because two years later, Marie Charpentier [C], a post-doctoral student working with Birkhoff at Harvard, proved that his continuum was indecomposable. Thus, we have (probably) the first example of an indecomposable continuum appearing in dynamical systems. However, once again, Birkhoff's example was one of pathology.

After World War II continuum theory underwent a change in direction and a change in personnel. Before the war most of the researchers working on indecomposable continua and related topics had been Europeans, primarily the Poles. After the war much of the research has been done by Americans, with the great majority of those being students of R. L. Moore or descendants of that tradition. (As mentioned before, at least one major factor here was the fact that a number of the Polish researchers died during the war.) Before the war research had been directed mainly towards developing a general theory of indecomposable continua. The only results on hereditarily indecomposable continua had been in Knaster's thesis [Kn1], with Knaster's goal being to show the existence of these continua. After the war much of the research was on hereditarily indecomposable continua, and, even more specifically, on certain examples of these continua, the pseudoarc and the pseudocircle.

The motivation for this change was some long outstanding problems in continuum theory, and the realization that their solutions involved hereditarily indecomposable continua. Many of these problems are still not completely solved. A topological space X is *homogeneous* if for each x and y in X, there is a homeomorphism f from X onto itself such that $f(x) = y$. A continuum X is *hereditarily equivalent* if each of its nondegenerate subcontinua is homeomorphic to the entire space X. A continuum X has the *fixed point property* if for each continuous map F from X to itself, there is a point x in X such that $F(x) = x$. A set X contained in a connected space Y is *nonseparating* if $Y \backslash X$ is connected. Three old, intriguing problems stated in the early volumes of **Fundamenta** were the following: (1) Does there exist another homogeneous plane continuum (besides the simple closed curve and the point)? (2) Is the arc the only hereditarily equivalent continuum? (3) Does every nonseparating plane continuum have the fixed point property?

In 1948, E. E. Moise [Mo1] constructed the pseudoarc in the plane and proved that it was hereditarily equivalent. (It followed from later results that Moise's continuum was homeomorphic to the one described in Knaster's thesis. See Figure 4 for a description of Moise's construction. Also see Figure 5 for a picture of how minimally complicated this construction must necessarily be, even at low levels of the construction.) At this time, Moise and R. H. Bing were at the University of Texas. Bing remarked to Moise then that he believed that this continuum had the additional property that it was homogeneous! Moise didn't believe it for a while: after all it looks from the construction as if this object has two endpoints, just as an arc does. Shortly thereafter, both Bing [B2] and Moise [Mo2], in independent papers, proved that the pseudoarc is indeed homogeneous: the pseudoarc is a chainable continuum and every point of it is an endpoint. (The finite cover $\{o_1, o_2, ... o_n\}$ of the continuum X is a *chain cover* if $o_i \cap o_j \neq \emptyset$ if and only if $|i - j| \leq 1$. A continuum is *chainable* or *arclike* or *snakelike* if for each positive number ϵ, it possesses an open chain cover of mesh less than ϵ. The point x of a chainable continuum X is an *endpoint* of X if for each $\epsilon > 0$, there is a chain cover $O = \{o_1, o_2, ..., o_n\}$ such that x is in o_1.) Thus, this one bizarre little continuum had shown that the answer to the first two of the old questions was no. It does not answer the third question: although the pseudoarc is a nonseparating plane continuum, it follows from results of O. H. Hamilton [Ha] that the pseudoarc has the fixed point property.

The race was on, so to speak. What other amazing properties did this and other hereditarily indecomposable continua have? Plenty, it turned out. R. H. Bing [B3] proved that all chainable, hereditarily indecomposable continua are homeomorphic to Moise's pseudoarc (from which it followed that Knaster's 1922 continuum is also homeomorphic to a pseudoarc, and today topologists use the term "pseudoarc" to mean any chainable, hereditarily indecomposable continuum). However, not all hereditarily indecomposable continua are pseudoarcs: Bing [B1, B5] proved that there are as many topologically distinct hereditarily indecomposable continua as there are real numbers, and for each positive integer $n > 1$, he constructed an n-dimensional hereditarily indecomposable continuum in $\mathbb{R}^n$ as well as an infinite-dimensional hereditarily indecomposable continuum in the Hilbert cube. However, Bing [B1] also proved that if X is either Euclidean n-space or the Hilbert cube, then most continua in X are pseudoarcs, in the sense that if the collection $F(X)$ of all continua in X is given the Hausdorff metric, then the subcollection of $F(X)$ that consists of the pseudoarcs is a dense, second category subset of $F(X)$ (and, is, in fact, a dense G_δ-subset of this complete, separable metric space). Thus, pseudoarcs aren't just most hereditarily indecomposable continua, or most indecomposable continua, they are most continua! Not arcs, or circles, or disks, or tori, but pseudoarcs! Does it really make sense to think of these objects as "bizarre", or "pathological", or "monstrous and evil"?

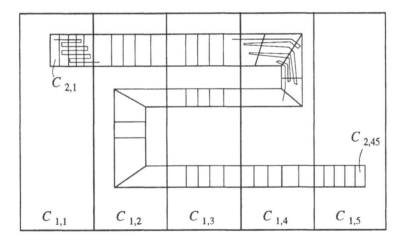

FIGURE 4. The first steps in Moise's construction. Portions of C_3 are indicated in C_2. Note that adding **one** link to a chain **exponentially increases** the number of links that a very crooked (refining) chain necessarily must have. **Moise's pseudoarc construction:** Suppose X is a metric space. A **chain** is a finite collection $C = \{c_1, c_2, \ldots, c_n\}$ of sets which have nonempty interior dense in the sets (i.e., $\overline{c_i^o} = \overline{c_i}$) such that $c_i \cap c_j \neq \emptyset$ if and only if $|i - j| < 2$. The members c of the chain C are called **links** of C. If p is a point in c_1 and q is a point in c_n, then c is a **chain from** p **to** q. The links c_1 and c_n are the **endlinks** of C. A **subchain** C' of the chain C is a subcollection of C which is itself a chain. Thus, C' can be written as $C' = \{c_k, c_{k+1}, \ldots, c_l\}$ for some $0 < k \leq l \leq n$. The chain D **refines** the chain C if each link of D is contained in some link of C. If the chain D refines the chain C in such a way that for each link c of C, the set of all links of D that lie in c is

a subchain of D, then D is **straight** with respect to C (or just straight in C). Suppose that $C = \{c_1, c_2, \ldots, c_n\}$ is a chain from p to q, and $D = \{d_1, d_2, \ldots d_m\}$ is a chain from p to q that refines C. Then D is **very crooked** in C if (1) $n < 5$ and D is straight in C, or (2) $n > 4$ and D is the union of (a) a chain D_1 from p to some x in c_{n-1}; (b) a chain D_2 from x to some y in c_2; (c) a chain D_3 from y to q; (d) each of $D_1, D_2,$ and D_3 is very crooked in $C \setminus \{c_{n-1}\}, C \setminus \{c_1, c_{n-1}\}$, and $C \setminus \{c_1\}$, respectively; and (e) no two of D_1, D_2 and D_3 have in common any link not an end link of both of them. Define the pseudoarc as follows: let $C_2, C_2, \ldots,$ be a sequence of chains in the plane from the point p to the point q such that

(1) $C_i = \{c_{i,1}, c_{i,2}, \ldots, c_{i,n_i}\}$ and $\overline{\cup C_i} = \overline{\cup_j c_{i,j}}$ is a compact set;

(2) for each i, C_{i+1} is very crooked in C_i and $\overline{\cup C_{i+1}}$ is a subset of the interior of $\cup C_i$;

(3) C_1 has five links;

(4) if c is a link of C_i and C' is a subchain of C_{i+1} whose union is contained in c and which is not contained properly in any other subchain of C_i whose union is contained in c, then C' consists of five links; and

(5) for each i, the diameter of each link in C_i is less than $1/i$ (i.e., the mesh of C_i is less than $1/i$).

Then $M = \bigcap_i (\cup C_i)$ is a **pseudoarc**. (Note that M is a nonempty continuum.)

Bing proved in [B3] that the pseudoarc can also be characterized as a homogeneous, chainable continuum. The pseudoarc has so many incredible properties tht it is really not possible to list them all here. The interested reader can consult [B2], [B4], [Hen], [K], [Le], [L1], [L2], [MT], and [OT] for more information on this continuum; and W. Lewis [L3] has written a survey listing the known properties of this continuum with an extensive bibliography.

The pseudoarc is the only chainable, hereditarily indecomposable, homogeneous continuum, but are there other homogeneous, hereditarily indecomposable continua? Bing [B1] had a candidate in 1951. Define a *circular chain cover* to be a finite open cover $C = \{c_0, c_1, \ldots, c_n\}$ of the continuum X such that $c_i \cap c_j \neq \emptyset$ if an only if $|i - j| < 2$ or $i = 0$ and $j = n$ (or vice versa). The continuum X is *circlelike* or *circularly chainable* if, for each $\epsilon > 0$, X has a circular chain cover of mesh less than ϵ. In a manner quite similar to Moise's pseudoarc construction, but using circular chains rather than (rectangular) chains, Bing constructed his pseudocircle in the plane. This continuum is circlelike, and hereditarily indecomposable. Bing asked whether or not it is homogeneous. His question wasn't answered until 1969, and then it was answered independently by two young researchers, J. T. Rogers, Jr. [R1] and L. Fearnley [F]. The answer was no, it is not homogeneous. They also proved that each circlelike, hereditarily indecomposable continuum in the plane is a

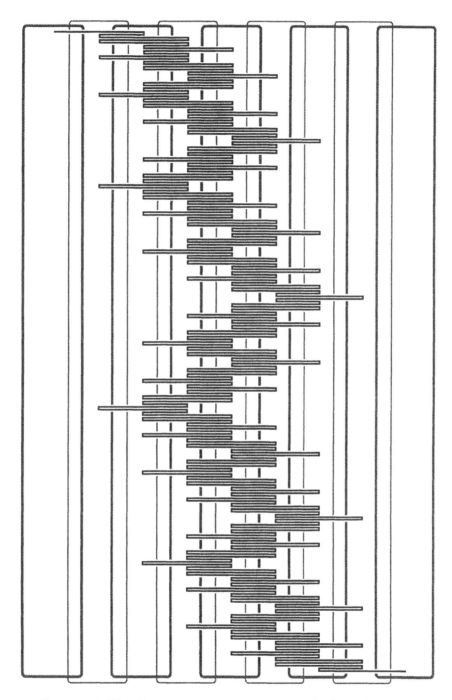

FIGURE 5. This figure was computer generated by Piotr Minc and
shows a very crooked chain (see definition in Figure 4, with the
5-link crookedness requirement changed to a 4-link requirement)
in a nine-link chain. Note that the figure illustrates the minimal
crookedness possible, i.e., no chain with fewer turns can be very
crooked (with respect to the 4-link definition) in a nine-link chain.

pseudocircle. However, circlelike, hereditarily indecomposable continua that cannot be realized as subsets of the plane exist in three space, and there are infinitely many topologically distinct ones of these [Mi]. If X is a continuum and x is a point in X, define the *orbit* $G(x)$ of x (under the action of the homeomorphism group) to be

$$G(x) = \{y \text{ in } X \mid \text{ for some homeomorphism } h \text{ from}$$

$$X \text{ onto itself, } h(x) = y\}.$$

For X a pseudocircle, it was proved in [KR] that not only is $G(x) \neq X$ for any x, but in fact each $G(x)$ is dense, but not second category in X. Thus, the pseudocircle is not even close to being homogeneous, in spite of the fact that it is locally exactly like the pseudoarc [KR]. However, the pseudocircle does behave somewhat like an abelian topological group! It has a nice group of "rotations" associated with it. (Note: Since the pseudocircle is not homogeneous, it cannot possibly be a topological group. The only abelian topological groups that are indecomposable are the solenoids. These were studied very early. See [VD], [VH], and [V].)

Sometime around 1980, another major direction appeared in the study of continua, and particularly in the study of indecomposable continua, but this time the new direction came about because of developments in another area of mathematics. In the 60's Stephen Smale [S1, S2] introduced one of the prime motivating examples of modern dynamics: the horseshoe map. This diffeomorphism takes a stadium shaped region of the plane into itself (see Figure 6), by stretching the region lengthwise, contracting it vertically, folding it once, and then placing it back into itself. The map has an invariant unstable set, its global attractor, and also an invariant Cantor set, which lies in the attractor. Smale was attempting, with this example, to construct a simplified model of the sort of behavior that occurs with the Poincaré return map for the forced van der Pol equations (for some parameter values). M. L. Cartwright and J. E. Littlewood [CL1, CL2] had earlier investigated these differential equations, and had even conjectured that an invariant continuum admitted by such a return map contained an indecomposable continuum. (It was not until 1989 that it was proved that, in fact, this continuum is indecomposable. See [BG].)

The dynamicists were mostly interested in the invariant Cantor set associated with the horseshoe map, for on that set the map is "chaotic". (That is, it admits a dense set of periodic points, the orbit of some point under the map is dense in the set, and the map is sensitive to initial conditions. This is Devaney's definition of "chaotic" (see [D1]). This word has a number of meanings used currently. The horseshoe map (on the Cantor set) satisfies all of them.) However, the global attractor for the map is just the Knaster bucket handle (on its side) that had been around since before 1920. (See Figures 2 and 6.) The points of the attractor that are not in the Cantor set are all wandering points, with the exception of one attracting fixed point. (A fixed point x is *attracting* for the map F if there is some open set u that contains x such that $\overline{F(u)} \subseteq u$ and $\bigcap_{n \geq 0} F^n(u) = \{x\}$. A point x is *wandering* under F if there is some open set u containing x such that the collection $\{F^n(u) | n \geq 0\}$ is mutually disjoint.) The type of stretch-contract-fold behavior demonstrated in the horseshoe map is typical of nonlinear dynamical systems.

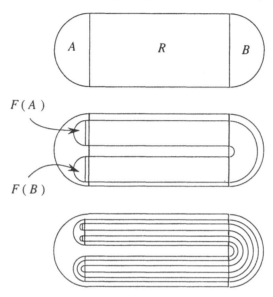

FIGURE 6. **The Smale horseshoe map.**

FIGURE 6. **The Smale horseshoe map:** Consider the stadium shaped region D, which consists of a rectangle R with interior and two semicircles A and B (interiors included), that are sewn onto the shorter sides of the rectangle. Now D is a subset of the plane and the horseshoe map F maps D into itself, as pictured. Think of F as having the following effect on D: the map F shrinks R vertically, stretches R horizontally, contracts the semicircles, and then folds D once and places the acted-upon D back into itself so that $F(A)$ and $F(B)$ are in the interior of A and $F(R)$ is in the interior of D.

One of the main tools used in understanding dynamical systems is that of symbolic dynamics. In its traditional form this has meant attempting to understand the dynamics on the "interesting" set by associating those dynamics with those on a map of the Cantor set to itself, with this map usually being a shift map or some generalization of a shift map. This is how one comes to understand the dynamics of the horseshoe map, for example. But suppose that the interesting dynamics happens on a set that is not even close to being a Cantor set? How does one reduce the behavior in this situation? Another dynamicist, Robert Williams [W1,W2,W3], developed a tool that allowed for a more general version

of symbolic dynamics. Instead of using a Cantor set as his underlying set for the
support of the interesting dynamics, he used what he calls "branched manifolds",
and maps of these branched manifolds to themselves, and then took inverse limits with these manifolds and maps. (Recall that if X is a compact, metric space
and $f : X \to X$ is continuous, then the *inverse limit*, denoted $\lim(X, f)$, is defined to

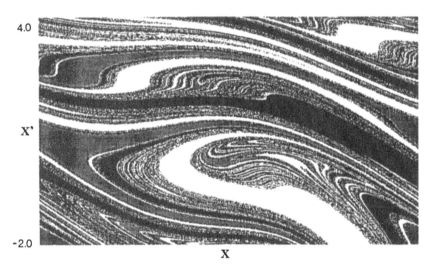

FIGURE 7. If p is an attracting point for a homeomorphism f,
then the **basin of attraction** for p is the set of all points that
are attracted to p under iteration by f. The figure shows three
basins of attraction for the Poincare return map of the differential
equation

$$x'' + 0.2x' + \sin(x) = 1.66\cos(t),$$

which models a forced, damped pendulum. This map is a diffeomorphism from the cylinder $S \times R$ onto itself. The points colored
black represent the basin of attraction for one attracting point,
while the points colored gray represent the basin of attraction for
another attracting point, and the points colored white represent
the basin of attraction for a third attracting point. Note that the
left boundary is identified with the right boundary, since the x-axis
is really a circle. There is strong evidence (mostly numerical) that
the boundary of the basins is common, and that when the cylinder
is compactified in the natural way, the common basin boundary
is a Lakes of Wada continuum. It is also probable that the basin
boundary is the closure of the stable manifold for any saddle point
in the basin boundary, and that the closure of the unstable manifold for such a saddle point is also an indecomposable continuum.
(The unstable manifold is not shown in the figure. The picture was
made with the software *Dynamics* [NY] developed by J. Yorke.)

be $\lim(X, f) = \{(x_1, x_2, ...) \text{ in } X^\infty | \text{ for each } i > 0, f(x_{i+1}) = x_i\}$. Thus, $\lim(X, f)$ is a closed subset of X^∞ with *bonding map* f. Further, the map f induces a homeomorphism $\widetilde{f}$, which is known as the *shift map* on $\lim(X, f)$. (We won't define the term "branched manifold" precisely here. However, the circle and the arc are important examples of branched manifolds.) Well known examples of inverse limits on branched manifolds used to model dynamical systems include the solenoids (which can be obtained both as a result of an embedding of solid tori into themselves, or as inverse limits over circles with bonding maps $f : S \to S$ defined by $f(x) = ax$ mod 1, where a is a positive integer greater than one, and $S = \mathbb{R} \bmod 1$), the Plykin attractor, and the geometric Lorenz attractor. Often these inverse limits turn out to be indecomposable continua. In the examples mentioned above the solenoids and the Plykin attractor are indecomposable, while the Lorenz attractor is not, although it fails to be locally connected at any point.

FIGURE 8. The Ikeda map F is a dissipative diffeomorphism on the plane that models laser optics. It admits a global attractor (which is pictured in the figure for parameter values stated) and at least two saddle points. For parameters r, a, b, and c, define for a point (x, y) in the plane

$$T(x, y) = a - b/(1 + x^2 + y^2), \text{ and}$$

$$F(x, y) = (r + b(x \cos(T) - y \sin(T)), b(x \sin(T) + y \cos(T))).$$

The figure pictures the global attractor for the parameter values $r = 1$, $a = 0.4$, $b = 0.9$, $c = 6$. (The picture was made with the software *Dynamics* [NY] developed by J. Yorke.) It follows from Barge's theorem [B] that this global attractor is an indecomposable continuum.

The Lakes of Wada continua described previously, though beautiful seem quite artificial, and perhaps it is hard to believe that such an object could occur in nature. However, there is strong numerical evidence that for many parameter values, the Poincaré map associated with the differential equation

$$x'' + a\,x' + b\sin(x) = c\,\cos(t)$$

that models the forced, damped pendulum admits attracting fixed points, and the basin boundaries for these fixed points is actually one common boundary, which with an appropriate compactification of the infinite cylinder, is a Lakes of Wada continuum [KY1]. (See Figure 7.) It is very difficult to prove directly that these Poincaré maps admit basin boundaries that are Lakes of Wada continua, so as is

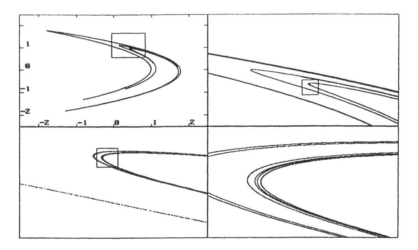

FIGURE 9. The Hénon map H is a plane diffeomorphism that exhibits many different kinds of dynamical behavior. It is not yet fully understood for many parameter values, but it does follow from Barge's theorem [B] that for many parameter values, the Hénon attractor is an indecomposable continuum. (This attractor is not a global attractor – an unbounded open set of points diverges to infinity.) Define for parameters r and c, the Hénon map H as follows:

$$H(x, y) = (r - x^2 + cy, x).$$

The figure shows the attractor, and several blowups for the map when $r = 1.4$ and $c = .3$. The figure in the upper right is a blowup of the portion of the attractor contained in the box drawn in the picture at the upper left, while the bottom picture is a blowup of the portion of the attractor contained in the box drawn in the picture at the upper right. (The picture was made with the software *Dynamics* [NY] developed by J. Yorke.)

commonly done in dynamics, in [KY2] an idealized model of such a map was constructed. For this model of the Poincaré map it was proved there that, with the appropriate compactification, there is a common basin boundary for three attracting fixed points, and that the common boundary is a Lakes of Wada continuum. Recently, J. Hubbard and R. Oberste-Vorth [HO] have proved that for certain parameter values the basin boundary associated with the Hénon map is a Lakes of Wada continuum. It is probably the case that whenever three or more disjoint basins of attraction occur in dynamical systems on 2-manifolds, the likely result is a Lakes of Wada continuum, or so says J. Yorke. Also, I. Kan [Ka] has obtained them in connection with flows on a 2-manifold.

Indecomposable continua have occurred in so many nonlinear dynamical systems, natural or otherwise, that it would seem probable that they are to be "expected" (although that has not been proven or made precise). Perhaps that is not too surprising when one thinks about the way fractal structures and chaotic dynamics go hand in hand. Indecomposable continua are really just a type of fractal structure. What follows is a list of some of the contexts in which these continua have arisen so far. The list grows longer every day:

(1) R. Devaney [D2] has shown that, with an appropriate compactification, an indecomposable continuum forms the Julia set in the complex plane for values of λ larger than e in the family of exponential maps $E_\lambda(x) = \lambda \exp(x)$. It is still not known whether they can occur in connection with Julia sets of polynomial maps in the complex plane, although there are many partial results here, and a number of continuum theorists have been working on this problem. See, for example, [MO1], [MO2], [MR], [R2], [R3], [R4], and [R5].

(2) M. Barge and J. Martin [BM1], [BM2], [BM3], [BM4] have shown that they must occur in inverse limits on intervals in a large number of cases. S. Shumann [Sc] has given necessary and sufficient conditions for them to occur in inverse limits on intervals, and Barge and B. Diamond [BD] have further explored this, (Inverse limits on intervals are often used in symbolic dynamics á la Williams.)

(3) M. Barge [B] has shown, that under mild conditions, when the stable and unstable manifolds of a fixed point for a diffeomorphism on the plane intersect (at a point other than the fixed point), then the closure of one branch of the unstable manifold is an indecomposable continuum. It follows that, for certain paramenter values, the Ikeda and Hénon attractors are indecomposable. (See Figures 8 and 9.) In fact, Barge's theorem leads immediately to a large number of examples, including the global attractor for the Smale horseshoe map. (A fixed point p for a diffeomorphism F on the plane is a *saddle point* if the Jacobian matrix at p has one eigenvalue with norm greater than one and one eigenvalue with norm less than one. The set of points that are attracted to p when F is iterated forms the *stable manifold* of p. The set of points that are attracted to p when F^{-1} is iterated forms the *unstable manifold* of p.)

(4) M. Handel [H] and M. Herman [He] have produced C^∞ diffeomorphisms on the plane with pseudocircles as invariant sets. For some of the maps the pseudocircles are attractors, and for others the maps are area-preserving (and therefore cannot

have attractors). The diffeomorphisms are minimal on the pseudocircles (i.e., the orbit of each point in the pseudocircle is dense in the pseudocircle).

(5) J. Yorke and the author [KY3,KY4,KY5] have produced examples of C^∞ maps on manifolds with invariant sets that are continuous Cantor sets of continua, most (a dense, second category subset) of which are pseudocircles. None of the pseudocircles is invariant, but both maps have the property that for an open set of differentiable maps (C^1 -metric), all maps in the open set possess a similar invariant set. The first example is a C^∞ map on a 3-manifold, and it is not a diffeomorphism, but the invariant set is a repellor for the map. The second example is a C^∞ diffeomorphism on a 7- manifold, and the invariant set is "somewhat" hyperbolic (in the sense that for an open set containing the set all points not in the invariant set are either attracted to it, repelled away from it, or come into the open set and then leave it forever.) However, the diffeomorphism is not hyperbolic. In both examples the maps are chaotic on the invariant Cantor set of continua (by nearly any definition of this term). Thus, these maps are perturbable in the sense that nearby maps possess similar invariant sets, and it is not possible to make the "pathology" go away by moving the map a little.

(6) While the pseudoarc has not appeared at this time as an invariant set for a differentiable dynamical system, it can be realized as a global attractor for a homeomorphism on the plane. This follows from results of Barge and Martin [BM4] and P. Minc and W. Transue [MT]. Barge and Martin showed how to realize any inverse limit on the interval with one bonding map as the global attractor in the plane for a homeomorphism, with the homeomorphism essentially being an extension of the homeomorphism induced by the bonding map on the interval used to make the inverse limit. Minc and Transue proved that there is a map from the interval to itself such that the resulting inverse limit is a pseudoarc and the induced homeomorphism is chaotic on the pseudoarc. (Henderson [Hen] had earlier shown that the pseudoarc can be realized as an inverse limit on an interval with one bonding map, but his induced homeomorphism was not chaotic.)

(7) In 1993, Krystyna Kuperberg [Ky] gave an example of a C^∞ flow on a 3-manifold that has an invariant indecomposable continuum (which is not a solenoid, but is somewhat like a solenoid) and which gives a counterexample to the C^∞ Seifert conjecture. Krystyna and her son, Greg [KK], have since given a real analytic counterexample to the C^∞ Seifert conjecture.

(8) A *cofrontier* is a continuum Λ that irreducibly separates the plane into exactly two complementary domains D_i and D_e. Suppose that F is an orientation preserving homeomorphism of the plane and Λ is a cofrontier with complementary domains D_i and D_e such that $F(\Lambda) = \Lambda$. Then it is possible (using prime end theory) to talk about the rotation numbers of points in Λ accessible from D_i and D_e, respectively. When these rotation numbers are different, M. Barge and R. Gillette [BG] have proved that it is often the case that Λ is indecomposable. Recall that this situation also arose in the first example of an indecomposable continuum in a dynamical system, with Birkhoff's annulus homeomorphism [Bi], which M. Charpentier [C] showed admitted an invariant indecomposable continuum (which was also a cofrontier). See also [BGM], [R5] and [Ru] for related results.

The introduction of dynamical systems into continuum theory has been somewhat controversial. Not all continuum theorists have welcomed it, even some of
those who have already made contributions to this interaction. This new and important direction in continuum theory has definitely changed the area, probably
irrevocably. However, on the plus side, it has meant that a whole new and un-
explored frontier has opened up to continuum theorists and it has revitalized the
subject. And there do still remain three very old, venerable, and important problems which have yet to be completely solved: (1) Does every map of a nonseparating
plane continuum to itself admit a fixed point? (2) Is there yet another homogeneous
plane continuum? (The point, the simple closed curve, the pseudoarc, and the circle of pseudoarcs are all homogeneous plane continua, but is this a complete list?)
(3) Is there still another hereditarily equivalent continuum (besides the point, arc
and pseudoarc?) Many partial answers have been obtained, and it is known that
if there is a nonseparating plane continuum without interior which does not admit
a fixed point, then that continuum has some really strange properties and it must
be a treelike, but not chainable, hereditarily indecomposable continuum; and that
if there is another homogeneous plane continuum, then it must be a treelike,but
not chainable, hereditarily indecomposable continuum, or a circle of such continua.
Similar results hold for the third question. These questions have been around since
the 20's and they have been worked on by nearly every continuum theorist.

REFERENCES

[B]   M. Barge, Homoclinic intersections and indecomposability, *Proc.AMS* (3)
       1010(1987) 541.
[BD]   M. Barge and B. Diamond, The dynamics of continuous maps of finite
       graphs through inverse limits, *Trans AMS* 344 (1994) 773.
[BG]   M. Barge and R. Gillette, Indecomposability and dynamics of invariant
       plane separating continua, *Contemp. Math.* 117 (1991) 13.
[BM1] M. Barge and J. Martin, Chaos, periodicity and snakelike continua, *Trans
       AMS* 289 (1985) 355.
[BM2] M. Barge and J. Martin, Dense orbits on the interval, *Michigan Math. J.*
       34 (1987) 3.
[BM3] M. Barge and J. Martin, Dense periodicity on the interval, *Proc.AMS* (4)
       94 (1985) 731.
[BM4] M. Barge and J. Martin, The construction of global attractors, *Proc.AMS*
       110 (1990) 523.
[B1]   R. H. Bing, Concerning hereditarily indecomposable continua, *Pacific J.
       Math.* 1 (1951) 43.
[B2]   R. H. Bing, A homogeneous indecomposable plane continuum, *Duke Math.
       J.* 15 (1948) 729.
[B3]   R. H. Bing, Each homogeous nondegenerate chainable continuum is a pseu-
       doarc, *Proc.AMS* 10 (1959) 345.
[B4]   R. H. Bing, Snakelike continua, *Duke Math. J.* 18 (1951) 653.
[B5]   R. H. Bing, Higher-dimensional hereditarily indecomposable continua,
       Trans AMS 71 (1951) 267.

[Bi]   G. D. Birkhoff, Sur quelques courbes fermées remarquable, *Bull. Soc. Math. France* 60 (1932) 1.

[BGM]   B. L. Brechner, M. D. Guay, and J. C. Mayer, Rotational dynamics on confrontiers, *Contemp. Math.* 117 (1991) 39.

[Br]   L. E. J. Brouwer, Zur analysis situs, *Math. Ann.* 68 (1910) 422.

[CL1]   M.L. Cartwright and J.E. Littlewood, On non-linear differential equations of the second order: 1. The equation $y'' - k(1-y^2)y' + y = b\lambda k\cos(t+\alpha), k$ large, *J. London Math. Soc.* 20 (1945) 180.

[CL2]   M.L. Cartwright and J.E. Littlewood, Some fixed point theorems, *Ann. Math.* 54 (1951) 1.

[C]   M. Charpentier, Sur quelques propriétés des courbes de M. Birkhoff, *Bull. Soc. Math. France* 62 (1934) 193.

[D1]   R. L. Devaney, **An Introduction to Chaotic Dynamical Systems**, 2nd ed. Addison-Wesley (1989) Redwood City, CA.

[D2]   R. L. Devaney, Knaster-like continua and complex dynamics, Ergodic Theory and Dynamical Systems 13 (1993) 627.

[F]   L. Fearnley, The pseudocircle is not homogeneous, *Bull. Amer. Math. Soc.* 75 (1969) 554.

[Ha]   O. H. Hamilton, A fixed point theorem for the pseudoarc and certain other plane continua, *Proc. AMS* 2 (1951) 173.

[H]   M. Handel, A pathological area preserving C^∞ diffeomorphism of the plane, *Proc.AMS* (1) 86 (1982) 163.

[Hen]   G. W. Henderson, The pseudo-arc as an inverse limit with one binding map, *Duke Math. J.* 31 (1964) 421.

[He]   M. Herman, Construction of some curious diffeomorphisms of the Riemann sphere, *J. London Math. Soc.* 34 (1986) 375.

[HY]   J. G. Hocking and G. S. Young, **Topology** ,Addison-Wesley,1961 (now available from Dover Publications, New York).

[HO]   J. H. Hubbard and R. Oberste-Vorth, Hénon mappings in the complex domain, preprint.

[Ja1]   Z. Janiszewski, Sur les continus irréductibles entre deux points, *J. de L'Ecole Polytechnique* 16 (1912) 79.

[Ja2]   Z. Janiszewski, Oeuvres choises, *Państwowe Wydawnictwo Naukowe* (1962).

[JK]   Z. Janiszewski and C. Kuratowski, Sur les continus indécomposables, *Fund. Math.* 1 (1920) 210.

[J]   F. L. Jones, **A History and Development of Indecomposable Continua,** Dissertation (1971), Michigan State University, East Lansing, Michigan.

[Ka]   I. Kan, Strange attractors of uniform flows, *Trans AMS* 293 (1986) 135.

[K]   J. Kennedy, Stable extensions of homeomorphisms on the pseudoarc, *Trans. AMS* 310 (1988) 167.

[KR]   J. Kennedy and J. T. Rogers, Jr., Orbits of the pseudocircle, *Trans. AMS* 296 (1986) 327.

[KY1]   J. Kennedy and J. A. Yorke, Basins of Wada, *Physica D* 51 (1991) 213.

[KY2]   J. Kennedy and J. A. Yorke, The forced damped pendulum and the Wada property, **Continuum Theory and Dynamical Systems, Lecture Notes in Pure and Applied Mathematics**, ed. by Thelma West, 149 (1993) 157.

[KY3]   J. Kennedy and J. A. Yorke, Pseudocircles in dynamical systems, *Trans. AMS* 343 (1994) 349.

[KY4]   J. Kennedy and J. A. Yorke, Pseudocircles, diffeomorphisms, and perturbable dynamical systems, to appear, *Ergodic Theory and Dynamical Systems.*

[KY5]   Kennedy and J. A. Yorke, Bizarre topology is natural in dynamical systems, to appear, *Bull. AMS.*

[Kn1]   B. Knaster, Un continu dont tout sous-continu est indécomposable, *Fund. Math.* 3 (1922) 247.

[Kn2]   B. Knaster, Quelques coupres singulières du plan, *Fund. Math.* 7 (1925) 264.

[KnKu] B. Knaster and C. Kuratowski, Sur les continus non-bornées, *Fund. Math* 5 (1924) 23.

[Kr]   J. Krasinkiewicz, Mapping properties of hereditarily indecomposable continua, *Houston J. Math.* 8 (1982) 507.

[KM]   J. Krasinkiewicz and P. Minc, Mappings onto indecomposable continua, *Bull. Acad. Pol. Sci.* 25 (1977) 675.

[Ky]   K. Kuperberg, *A* smooth counterexample to the Seifert conjecture, to appear, *Ann. Math.*

[KK]   G. Kuperberg and K. Kuperberg, Generalized counterexamples to the Seifert conjecture, preprint.

[Ku1]   C. Kuratowski, Theorie des continus irreductibles entre deux points I, *Fund. Math.* 3 (1922) 200.

[Ku2]   C. Kuratowski, Sur les coupures irreductibles du plan, *Fund. Math.* 6 (1924) 130.

[Le]   G. R. Lehner, Extending homeomorphisms on the pseudoarc, *Trans AMS* 98 (1961) 369.

[L1]   W. Lewis, Most maps of the pseudoarc are homeomorphisms, *Proc. AMS* 91 (1984) 147.

[L2]   W. Lewis, Stable homeomorphisms of the pseudo-arc, *Canadian J. Math.* 31 (1979) 363.

[L3]   W. Lewis, The pseudo-arc, *Contemp. Math.* 117 (1991) 103.

[MO1]   J. Mayer and L. Oversteegen, Denjoy meets rotation on an indecomposable frontier, **Continuum Theory and Dynamical Systems, Lecture Notes in Pure and Applied Mathematics**, ed. by Thelma West, 149 (1993) 183.

[MO2]   J. Mayer and L. Oversteegen, *Continuum Theory*, Technical report 1 (1992).

[MR]   J. Mayer and J. T. Rogers, Jr., Indecomposable continua and Julia sets of polynomials, *Proc. AMS* 117 (1993) 795.

[M1]   S. Mazurkiewicz, Un théorème sure les continus indécomposables, *Fund. Math.* 1 (1920) 35.

[M2]   S. Mazurkiewicz, Sur les points accessible des continus indécomposables, *Fund. Math.* 14 (1929) 107.

[M3]   S. Mazurkiewicz, Un théorème sur l'accessibilité des continus indécomposables, *Fund. Math.* 14 (1929) 271.

[MT]   P. Minc, and W.R.R. Transue, A transitive map on [0,1] whose inverse limit is the pseudoarc, *Proc. AMS* 111 (1991) 165.

[Mi]   J. Mioduszewski, On pseudocircles, **Theory of Sets and Topology**, Deutsch. Verlagwissensch. Berlin (1972) 363.

[Mo1]   E. E. Moise, An indecomposable plane continuum which is homeomorphic to each of its nondegenerate subcontinua, *Trans. AMS* 63 (1948) 581.

[Mo2]   E. E. Moise, A note on the pseudoarc, *Trans. AMS* 67 (1949) 57.

[NY]   H. Nusse and J. A. Yorke, **Dynamics: Numerical Explorations**, Spring-Verlag, New York (1994).

[OT]   L. G. Oversteegen and E. D. Tymchatyn, On hereditarily indecomposable continua, *Geometric and Algebraic Topology,* Banach Centre Publications, PWN, Warsaw,18 (1986) 403.

[R1]   J. T. Rogers, Jr., The pseudocircle is not homogeneous, *Trans. AMS* 148 (1970) 417.

[R2]   J. T. Rogers, Jr., Intrinsic rotations of simply connected regions and their boundaries, *Complex Variable Theory Appl.*

[R3]   J. T. Rogers, Jr., Is the boundary of a Siegel disk a Jordan curve?, *Bull. AMS* 27 (1992) 284.

[R4]   J. T. Rogers, Jr., Singularities in the boundary of local Siegel disks, *Ergodic Theory and Dynamical Systems*, 12 (1992) 803.

[R5]   J. T. Rogers, Jr., Rotations of simply connected regions and circle-like continua, *Contemp. Math.* 117 (1991) 139.

[Ru]   N. E. Rutt, Prime ends and indecomposability, *Bull. AMS* 41 (1935) 265.

[Sc]   S. Schumann, Dissertation, Unviversity of Wyoming,

[S1]   S. Smale, Diffeomorphisms with many periodic points, in: **Differential and Combinatorial Topology - A Symposium in Honor of Marsden Morse,** ed. S. Cairns, Princeton Univ. Press, Princeton, (1965) 63.

[S2]   S. Smale, Differentiable dynamical systems, *Bull. Amer. Math. Soc.* 73 (1967) 747.

[U]   P. Urysohn, Memoire sur les multiplicities Cantoriennes, *Fund. Math.* 8 (1926) 225.

[VD]   D. Van Dantzig, Homogene Kontinua, *Fund. Math.* 15 (1930) 102.

[VH]   A. Van Heemert, Topologische Gruppen und Unzerlegbare Kontinua, *Comp. Math.* 5 (1938) 319.

[V]   L. Vietoris, Uber den Hoheren Zusammenhang Kompakter Raume und eine Klasse von Zusammenhangstreuen Abbildungen, *Math. Ann.* 97 (1927) 454.

[Wa]   A. D. Wallace, The structure of topological semi-groups, *Bull. AMS* 61 (1955) 95.

[Wi]   R. L. Wilder, Domains and their boundaries, *Math. Ann.* 109 (1933) 273.

[W1]   R.F. Williams, Expanding attractors, *Publ. Math. IHES* 43 (1974) 169.

[W2]   R.F. Williams, One-dimensional wandering sets, Topology 6 (1967) 473.

[W3]   R.F. Williams, The structure of Lorenz attractors, *Publ. Math. IHES* 50 (1979) 101.

[Y]   K. Yoneyama, Theory of continuous sets of points, *Tohoku Math. J.* 11-12 (1917)43.

E-mail address: jkennedy@brahms.udel.edu "Judy Kennedy"

7
SPANS OF CONTINUA AND THEIR APPLICATIONS

ANDREW LELEK   Department of Mathematics, University of Houston, Houston, Texas 77204, USA

THELMA WEST   Department of Mathematics, University of Southwestern Louisiana, Lafayette, Louisiana 70504, USA

ABSTRACT. The concept of the span of a continuum was introduced in 1964. Various versions of span have been defined since then. The relationships between these spans and other concepts in topology have been extensively investigated. In this paper, we give an overview of the work done on these topics.

1. INTRODUCTION

In 1964, the concept of the span of a metric space was introduced by the first named author ([L1], p.209). Since that time, some modified versions of the span have been introduced (see [L5]). Later, the concept of the span of a map was introduced by Ingram in [I1]. The concept of symmetric span was defined by Davis in [Da2]. Recently, other less closely related variations on span have been defined ([BS] and [W7]).

Since the original concept of span was introduced, the relationships between these various versions of span have been explored. Very basic geometric conditions have been found to be related to the spans. The interactions between the spans and other concepts in metric topology, particularly in the theory of continua, have been investigated. The purpose of this paper is to review the literature on these topics. For other reviews see [L8] and [OF].

1991 *Mathematics Subject Classification.* primary 54F15, secondary 54F20.
Key words and phrases. span.

2. DEFINITIONS

If X is a non-empty metric space, we define the *span* $\sigma(X)$ of X to be the least upper bound of the set of real numbers α which satisfy the following condition: there exists a connected space C and continuous mappings f_1, $f_2 : C \to X$ such that

$$(\sigma) \qquad\qquad\qquad f_1(C) = f_2(C)$$

and $\alpha \leq \text{dist}\,[f_1(c), f_2(c)]$ for $c \in C$. Equivalently (see[L1], p.209), the span $\sigma(X)$ is the least upper bound of numbers α for which there exist connected subsets C_α of the product $X \times X$ such that

$$(\sigma)' \qquad\qquad\qquad p_2(C_\alpha) = p_2(C_\alpha)$$

and $\alpha \leq \text{dist}\,(x, y)$ for $(x, y) \in C_\alpha$, where p_1 and p_2 denote the projections of $X \times X$ onto X, i.e., $p_1(x, y) = x$ and $p_2(x, y) = y$ for $x, y \in X$. We note that, for compact spaces X, the sets C_α in the latter definition of $\sigma(X)$ can be assumed to be closed in $X \times X$.

The definition does not require X to be connected, but to simplify our discussion we will now consider X to be connected. The surjective span $\sigma^*(X)$, the semispan $\sigma_0(X)$, and the surjective semispan $\sigma_0^*(X)$ are defined as above, except we change conditions (σ) and $(\sigma)'$ to the following (see [L5]):

$$
\begin{array}{llll}
(\sigma^*) & f_1(C) = f_2(C) = X, & (\sigma^*)' & p_1(C_\alpha) = p_2(C_\alpha) = X, \\
(\sigma_0) & f_1(C) \subseteq f_2(C), & (\sigma_0)' & p_1(C_\alpha) \subseteq p_2(C_\alpha), \\
(\sigma_0^*) & f_1(C) \subseteq f_2(C) = X, & (\sigma_0^*)' & p_1(C_\alpha) \subseteq p_2(C_\alpha) = X.
\end{array}
$$

The following inequalities follow immediately from the definitions.

$$
\begin{array}{ccccccccc}
0 & \leq & \sigma^*(X) & \leq & \sigma(X) & \leq & \sigma_0(X) & \leq & \text{diam}\, X, \\
0 & \leq & \sigma^*(X) & \leq & \sigma_0^*(X) & \leq & \sigma_0(X) & \leq & \text{diam}\, X.
\end{array}
$$

The span can be thought of as a connectedness type analogue of the diameter of a space.

Suppose X and Y are spaces, Y is a metric space, and f is a mapping of X into Y. The *span of* f, denoted by σf, is the least upper bound of the set of numbers α for which there is a connected subset Z of $X \times X$ such that

$$(\sigma)_f \qquad\qquad\qquad p_1(Z) = p_2(Z),$$

and $\text{dist}\,(f(x), f(y)) \geq \alpha$ for each (x, y) in Z.

The surjective span $\sigma^*(f)$, the semispan $\sigma_0(f)$, and the surjective semispan $\sigma_0^*(f)$ are defined by changing the condition (σ_f) to the following:

$$(\sigma^*)_f \; p_1(Z) = p_2(Z) = X, \quad (\sigma_0)_f \; p_1(Z) \subseteq p_2(Z), \quad (\sigma_0^*)_f \; p_1(Z) \subseteq p_2(Z) = X.$$

For $\tau = \sigma, \sigma^*, \sigma_0, \sigma_0^*$, the corresponding spans $\tau(X)$ of a space X are the spans $\tau(id_X)$ of the identity map on X.

It has been shown by Lelek (see [L9] and [L10]) that in the above definitions, the inequalities

$$\alpha \leq \text{dist}\,(x, y) \quad \text{and} \quad \alpha \leq \text{dist}\,(f(x), f(y))$$

can actually be replaced by

$$\alpha = \text{dist}\,(x, y) \quad \text{and} \quad \alpha = \text{dist}\,(f(x), f(y))$$

respectively, in the case of compact spaces.

3.   REVIEW

Many of the questions in span theory deal with the relationships between the various versions of span (see [L5]), the spans of related spaces, and spans and topological properties. Two particularly interesting span problems are:

For a continuum X are $\sigma(X) = 0$ and chainability of X equivalent? ([L2])

If S_1 and S_2 are two simple closed curves in $\mathbb{R}^2$, where S_1 is contained in the bounded component of $\mathbb{R}^2 - S_2$, then is $\sigma(S_1) \leq \sigma(S_2)$? ([CIL2], Problem 173)

See [CIL1] and [CIL2] for other problems.

Calculating the spans of a space and establishing the relationships between the various spans, in general, are difficult. Surprisingly, a concept as simple as convexity has been used in some of these calculations. Convexity has been used to obtain a partial answer to whether or not $\sigma(X) = \sigma_0(X)$, when X is a simple closed curve ([T1]). Another paper on this topic is [W3]. Convexity has also been used in obtaining a partial solution to the problem on simple closed curves, which is mentioned above ([T2]). Span has been used in estimating the area of a convex disc in the plane ([Kw5]). It has been used to establish lower bounds for the ratios $\sigma^*(X)/\sigma_0^*(X)$ and $\sigma^*(X)/\sigma_0(X)$, where X is homeomorphic to S^n or R^n ([W5] and [W6]).

Another, much more complicated property, that of width of a continuum, has been found to be related to span. It has been shown that $\sigma(X) = w(X)$, where $w(X)$ is the width of X and X is a convex set in $\mathbb{R}^2$ ([Kw5]). Various relationships between $\sigma(X)$ and $w(X)$, where X is a simple triod or a tree are derived in [W2], [W3], and [L6].

It is a trivial result that $\sigma^*(X) = \sigma(X)$ when X is a simple closed curve. Also, all the spans are equal when X is chainable. However, none of these equalities necessarily hold when X is a simple triod. It has been shown in [W1], that for $t \in [\frac{1}{2}, 1]$, there is a simple triod such that $\sigma^*(X)/\sigma(X) = \sigma_0^*(X)/\sigma_0(X) = t$. Also, see [L3].

Using inverse limits, Ingram has constructed examples of continua which are atriodic, tree-like, and with chainable proper subcontinua, but which have positive span (see[I1-5], and [I7]). In [I6], there is an example of such a continuum, except that the proper subcontinua have positive span. Another example, similar to these, was constructed in [DI]. In [RSS], it is shown that, if there exists in $\mathbb{R}^2$ an uncountable collection of pairwise disjoint copies of a tree-like continuum X, then the symmetric span of X is zero. It is not known whether or not $\sigma(X)$ must be zero. An unpublished result of H. Cook says that the dyadic solenoid has symetric span equal zero, even though the span is positive (see [Da2]). The relationships between span and various types of maps have also been studied (see [Kw4], [L1], [L4], [LM], [CD2], [O1], and [Kw2]).

Much work has been done on spans equal to zero. Davis ([Da2]) has shown that for a continuum X, $\sigma(X) = 0$ is equivalent to $\sigma_0(X) = 0$. In [Ky2] it is shown that zero span is a sequential strong Whitney-reversible property. Later in [Ka], it was shown that zero semispan also has this property. In [KKT] it is shown that if $\sigma^*(X) = 0$, where X is a continuum, then X is tree-like and atriodic. It is also known, that an almost chainable homogeneous continuum has span zero ([Da1]). Zero surjective span and zero surjective semispan have been shown to be preserved by refinable maps (see [Ky1]). Cuervo and Duda have characterized span zero for a certain class of continua ([CD1]). Also, this was done for semispan ϵ, $\epsilon \geq 0$ ([Du2]). Kawamura has shown that open maps preserve span zero ([Kw2]).

Relationships between span and dimension of the space, also between span and essential maps, have been investigated ([L7] and [L1]). Limits have been found for spans of X, where X can be written as $A \cup B$ and certain conditions are met by the subcontinua A and B ([DK], [Du1], and [W3]).

The question, given above, about the equivalence of chainability and span zero, has prompted much work in span theory. Ovesteegen has found conditions on a continua, such that these are equivalent ([O2]). Oversteegen and Tymchatyn have extensively studied how weak-chainability is related to span ([OT2] and [OT4]). For related results see ([OT1], [OT3], and [OT5]).

REFERENCES

[BS]   M. Barge and R. Swanson, The essential span of a simple closed curve, Marcel Dekker, this volume.

[CIL1] H. Cook, W.T. Ingram and A. Lelek, Eleven annotated problems about continua, in: Open problems in Topology, J. van Mill and G.M. Reed, eds., North-Holland, Amsterdam 1990, 295-302.

[CIL2] H. Cook, W.T. Ingram and A. Lelek, A list of problems known as Houston Problem Book, Marcel Dekker, this volume.

[CD1] Maria Cuervo and Edwin Duda, A characterization of span zero, Houston J. Math., Vol. 12, No. 2 (1986), 177-182.

[CD2] Maria Cuervo and Edwin Duda, A finite to one open mapping preserves span zero, Topology Proc., Vol. 13 (1988), 181-187.

[Da1] James F. Davis, The span of almost chainable homogeneous continua, Ill. J. Math., Vol. 25, No. 4, Winter 1981, 622-625.

[Da2] James F. Davis, The equivalence of zero span and zero semispan, Proc. Amer. Math. Soc., Vol. 90, No. 1, January 1984, 133-138.

[DI]   James F. Davis and W.T. Ingram, An atriodic tree-like continuum with positive span which admits a monotone mapping to a chainable continuum, Fund. Math., 131 (1988), 13-24.

[Du1] Edwin Duda, A sum theorem for semispan of continua, General Topology and its Relations to Modern Analysis and Algebra V Proc. Fifth Prague Topol. Symp., 1981, J. Novak (ed.) 1982, 162-165.

[Du2] Edwin Duda, Acharacterization of semispan of continua, Proc. Amer. Math. Soc., Vol. 96, No. 1, January 1986, 171-174.

[DK] Edwin Duda and James Kell, III, Two sum theorems for semispan Houston
      J. Math., Vol. 8, No. 3 (1982), 317-321.

[I1] W.T. Ingram, An atriodic tree-like continuum with positive span, Fund.
      Math., 77 (1972), 99-107.

[I2] W.T. Ingram, An uncountable collection of mutually exclusive planar atri-
      odic tree-like continua with positive span, Fund. Math., 85 (1974), 73-78.

[I3] W.T. Ingram, Atriodic tree-like continua and the span of mappings, Topology
      Proc., Vol. 1 (1976), 329-332.

[I4] W.T. Ingram, Tree-like continua and span, Proc. Eighth Annual Univ. South-
      western Louisiana Conference (1977), 62-66.

[I5] W.T. Ingram, Atriodic tree-like continua: problems and recent results, Proc.
      Eleventh Annual Univ. Southwestern Louisiana Conference (1980), 15-19.

[I6] W.T. Ingram, Hereditarily indecomposable tree-like continua, II, Fund.
      Math., 111 (1981), 95-106.

[I7] W.T. Ingram, Positive span in inverse limits, Topology Proc., Vol. 9 (1984),
      313-317.

[Ka] Hisao Kato, On some sequential strong Whitney-reversible properties, Bul-
      letin of the Polish Academy of Sciences Mathematics, Vol. 37, No. 7-12
      (1989), 517-523.

[KKT] Hisao Kato, Akira Koyama, and E.D. Tymchatyn, Mappings with zero sur-
      jective span, Houston J. Math., Vol. 17, No. 3 (1991), 325-333.

[Kw1] Kazuhiro Kawamura, On span and inverse limits, Tsukuba J. Math., Vol.
      12, No. 2 (1988), 333-340.

[Kw2] Kazuhiro Kawamura, On some properties on span, J. Math. Soc. Japan, Vol.
      40, No. 4 (1988), 605-613.

[Kw3] Kazuhiro Kawamura, Span zero continua and the pseudo-arc, Tsukuba J.
      Math., Vol. 14, No. 2 (1990), 327-341.

[Kw4] Kazuhiro Kawamura, Some productive classes of maps which are related to
      confluent maps, Fund. Math., 138 (1991), 175-191.

[Kw5] Kazuhiro Kawamura, Spans, widths and areas of convex bodies in the plane,
      preprint.

[Ky1] Akira Koyama, A note on span under refinable maps, Tsukuba J. Math.,
      Vol. 9, No. 2 (1985), 237-240.

[Ky2] Akira Koyama, Zero span is a sequential strong Whitney-reversible property,
      Proc. Amer. Math. Soc., Vol. 101, No. 4, December 1987.

[L1] A. Lelek, Disjoint mappings and the span of spaces, Fund. Math., Vol. 55
      (1964), 199-214.

[L2] A. Lelek, Some problems concerning curves, Colloq. Math., Vol. 23 (1971),
      93-98.

[L3] A. Lelek, An example of a simple triod with surjective span smaller than
      span, Pacific J. Math., Vol. 64, No. 1 (1976), 207-215.

[L4] A. Lelek, Sets of distances and mappings of certain continua, Topology Proc.,
      Vol. 1 (1976), 325-328.

[L5] A. Lelek, On the surjective span and semispan of connected metric spaces,
      Colloq. Math., Vol. 37 (1977), 35-45.

[L6] A. Lelek, The span and the width of continua, Fund. Math., Vol. 98 (1978),
      181-199.

[L7]   A. Lelek, The span of mappings and spaces, Topology Proc., Vol. 4 (1979), 631-633.

[L8]   A. Lelek, Report on spans, Topology Proc., Vol. 6 (1981), 195-200.

[L9]   A. Lelek, Continua of constant distances related to the spans, Topology Proc., Vol. 9 (1984), 193-196.

[L10]   A. Lelek, Continua of constant distances in span theory, Pacific J. Math., Vol. 123, No. 1 (1986), 161-171.

[LM]   A. Lelek and L. Mohler, Real valued continuous functions and the span of continua, Colloq. Math., Vol. 32 (1975), 207-209.

[OF]   A.A. Odincov and V.V. Fedorčuk, Theory of continua I, Itogi Nauki i Tehniki, Sez. Algebra, Topology, Geometry, Vol. 29 (1991), 63-119.

[O1]   Lex G. Oversteegen, On products of confluent and weakly confluent mappings, Houston J. Math., Vol. 12, No. 1 (1986), 109-116.

[O2]   Lex G. Oversteegen, On span and chainability of continua, Houston J. Math., Vol. 15, No. 4 (1989), 573-593.

[OT1]   Lex G. Oversteegen and E.D. Tymchatyn, Plane strips and the span of continua, (I), Houston J. Math., Vol. 8, No. 1 (1982), 129-142.

[OT2]   Lex G. Oversteegen and E.D. Tymchatyn, On the span of weakly chainable continua, Fund. Math., Vol. 119, No. 2 (1983), 151-156.

[OT3]   Lex G. Oversteegen and E.D. Tymchatyn, Plane strips and the span of continua, (II), Houston J. Math., Vol. 10, No. 2 (1984), 255-266.

[OT4]   Lex G. Oversteegen and E.D. Tymchatyn, On span and weakly chainable continua, Fund. Math., Vol. 122, No. 2 (1984), 159-174.

[OT5]   Lex G. Oversteegen and E.D. Tymchatyn, On span and chainable continua, Fund. Math., Vol. 123, No. 2 (1984), 137-148.

[RS]   Dusan Repovś and E.V. Sčepin, On the symetric span of continua, Abstr. Amer. Math. Soc. 14 (1993) 319 No. 93T -54-42.

[RSS]   Dusan Repovś, Arkadij B. Skopenkov, and Evgenij V. Sčepin, On uncountable collections of continua and their span, preprint.

[T1]   Katarzyna Tkaczyńska, The span and semispan of some simple closed curves, Proc. Amer. Math. Soc., Vol. 111, No. 1, January 1991.

[T2]   Katarzyna Tkaczyńska, On the span of simple closed curves, Houston J. Math., Vol. 20, No. 3 (1994), 507-528.

[W1]   Thelma West, Spans of an odd triod, Topology Proc., Vol. 8 (1983), 347-353.

[W2]   Thelma West, Spans of simple triods, Proc. Amer. Math Soc., Vol. 102, No.2, February 1988, 407-415.

[W3]   Thelma West, On the spans and width of simple triods, Proc. Amer. Math. Soc., Vol. 105, No. 3, March 1989, 776-786.

[W4]   Thelma West, Spans of simple closed curves, Glasnik Mathematički, Vol. 24 (44) (1989), 405-415.

[W5]   Thelma West, Relating spans of some continua homeomorphic to S^n, Proc. Amer. Math. Soc., Vol. 112, No. 4, August 1991, 1185-1191.

[W6]   Thelma West, The relationships of spans of convex continua in R^n, Proc. Amer. Math. Soc., Vol. 111, No.1, January 1991, 261-265.

[W7]   Thelma West, On the surjective semispan of abstract graphs, to appear in the Journal of Combinatorial Mathematics and Combinatorial Computing.

E-mail address: asl@ math.uh.edu "Andrew Lelek"

8

COMPLEX DYNAMICS AND CONTINUUM THEORY

JOHN C. MAYER Department of Mathematics, University of Alabama at Birmingham, Birmingham, Alabama 35294, USA

ABSTRACT. Our point of departure in this survey is the question "What is there in complex analytic dynamics that would be of interest to continuum theorists?" We will present a collection of results and questions arising in the discrete dynamics of complex analytic maps of the complex plane $\mathbf{C}$ and the Riemann sphere $\mathbf{C}_\infty$. Our focus will be on the topological structure of the Julia sets of entire functions on the complex plane, and the tools used to study this structure. As it happens, there are a number of interesting open topological problems concerning Julia sets. The difficulty appears to be identifying those problems whose solutions are likely to be obtainable by topological methods.

A LITTLE HISTORY

During the decades of the 30's to the 50's there was considerable interaction between dynamics and topology. In the 60's and 70's and into the early 80's the interaction seemed to be at a minimum. In complex dynamics, a fruitful period of research by analysts and geometers began around 1981 with the advent of improved computer technology (for generating dramatic and complicated pictures) and the seminal articles by Douady and Hubbard [DH1-3] and Mañé, Sad, and Sullivan [MSS]. In the past few years indications are that another period of exciting interchange between dynamics and topology is again underway. A number of professional meetings, particularly the annual Spring Topology Conferences, have brought together dynamicists and topologists. Recent Midwest Dynamical Systems

1991 *Mathematics Subject Classification*. Primary: 30C35, 54F20.

Key words and phrases. Julia set, Fatou set, prime end, complex analytic dynamics, continuum theory, conformal dynamics, Mandelbrot set, quadratic map, indecomposable continuum, Siegel disk, Cantor bouquet, exponential map, topological equivalence, locally connected.

133

Conferences at Emory and Montana State had significant numbers of topologists in attendance. One of the Joint Summer Research Conferences for 1989 was devoted to continuum theory and dynamical systems. A Southeast Dynamical Systems Conference had its first meeting at Auburn in 1989, and has continued to have annual meetings well-attended by topologists. In the period 1990–93 there were three Special Sessions on Continuum Theory and Dynamical Systems scheduled at AMS meetings, and in 1994, the Special Session on Modern Methods in Continuum Theory had dynamics as one important focus. Most recently, the 1994 Spring Topology Conference at Auburn University was a joint meeting with the Southeast Dynamical Systems Conference.

During the period 1989–94, significant, though initial, progress has been made on the topological classification of non-locally-connected Julia sets: work by Aarts, Bula, Fokkink, Oversteegen, and Mayer [AO,BO,FO,M] have contributed to understanding the Cantor bouquet Julia sets of the exponential and related families; work by Rogers, Brechner, Guay, Oversteegen, and Mayer [Rg1-5,MR,BGM1,BGM2,-MO2] have contributed to understanding under what conditions polynomial Julia sets, and the boundaries of Siegel disks, might be indecomposable continua. Moreover, in [Rg5] Rogers has recently used topological methods to provide the most complete answer to date of what had generally been considered an analytical question: When does the boundary of a Siegel disk for a polynomial map contain a critical point?

An early, but still useful, introduction to complex dynamics on the Riemann sphere is the Bulletin article by Paul Blanchard [Bl]. Another excellent introduction to complex analytic dynamics may be found in John Milnor's lecture notes [Mi1]. Recently, two good graduate textbooks, one by A. F. Beardon and the other by L. Carleson and T. W. Gamelin, on the iteration of rational functions have appeared [Br,CG]. Also useful is the master's thesis of D. E. K. Sørenson [Sn].

JULIA SETS OF ENTIRE FUNCTIONS

In order to state theorems and questions in the simplest possible setting, we will confine our attention in this survey mainly to questions about the Julia sets of two one-parameter families of complex analytic functions: the quadratic family $f_c : z \to z^2 + c$ and the exponential family $f_\lambda : z \to \lambda e^z$. Our focus will be the topological structure of certain non-locally-connected Julia sets in these families: accessibility of (periodic) points, indecomposability, and ultimately, classification up to topological equivalence. Many of the results and questions can be stated in more general settings: e.g., rational functions on the Riemann sphere $\mathbf{C}_\infty$, and transcendental entire functions and meromorphic functions on $\mathbf{C}$.

The Julia set $J(h)$ of a complex analytic map $h : \mathbf{C} \to \mathbf{C}$ is the set of "unstable" points under iteration of h. The complementary stable set is now often called the Fatou set. The Julia set of h is defined as the set of points of $\mathbf{C}$ at which the family of iterates $\{h^n | n \in \mathbf{Z}^+\}$ of h is not a normal family. A family $\mathcal{F}$ of functions is *normal* at a point z iff there is an open set U about z on which each sequence of functions selected from $\mathcal{F}$ has a subsequence which converges uniformly (including converging to ∞) on compact subsets.

The *orbit* of a point z under h is $\{h^n(z) | n \in \mathbf{N}\}$. A point is *pre-periodic* iff its orbit is a finite set. A point z is said to be a *periodic point* of h iff for some $n \in \mathbf{N}$, $h^n(z) = z$. A point is *fixed* iff it is periodic for $n = 1$. Let z be a periodic point with n least such that $h^n(z) = z$, and let $\lambda = (h^n)'(z)$. We call λ the *eigenvalue* of the (periodic) orbit of z. (By the chain rule, λ is independent of which point in the periodic orbit we take for z.) We say the periodic point z is *repelling* iff $|\lambda| > 1$, *attracting* iff $|\lambda| < 1$, and *indifferent* or *neutral* iff $|\lambda| = 1$.

The *basin of attraction* of an attractive periodic orbit is the set of points whose orbits accumulate on the periodic orbit. It follows that the basin of attraction of an attractive periodic orbit is an open set, and a subset of the Fatou set [Mi1]. Since normality at a point is an open condition, the Fatou set is open and its complement, the Julia set is closed. For a polynomial (which has an analytic extension to $\mathbf{C}_\infty$ fixing ∞), the Julia set is also the boundary of the basin of attraction of ∞, and so compact. For polynomials (indeed, rational functions) of degree ≥ 2, the Julia set is nonempty, perfect, and fully invariant. Moreover, the set of preimages of each point in the Julia set is dense in the Julia set. Proofs of these basic facts may be found in any of the introductions cited above.

By a theorem of Baker [B], generalizing theorems of Julia [J] and Fatou [F], the Julia set is the closure of the set of repelling periodic points of h. A proof, at least for rational maps, may be found in most of the surveys cited above (e.g., [Mi1, Theorem 11.1]).

A sampling of the questions and problems which we discuss more fully in the subsequent sections includes the following:

(1) There are non-locally-connected Julia sets in the quadratic family of complex analytic maps of the plane. Describe them topologically. Are any of these Julia sets indecomposable continua?

(2) Are all the periodic points in the Julia set of a quadratic map accessible from ∞; in particular, is an irrationally indifferent fixed point?

(3) Suppose a Julia set J (in the quadratic family) contains the boundary of a Siegel disk S. Must ∂S be homeomorphic to a circle? May ∂S and/or J be an indecomposable continuum?

(4) Describe topologically the Julia sets in the exponential and related families. In these families, do topologically equivalent Julia sets occur unexpectedly in non-topologically-conjugate settings?

(5) Are topologically equivalent Julia sets always equivalently embedded in the plane?

(6) Are the indecomposable connected invariant subsets, which are found when

the Julia set of the exponential function is the entire complex plane, chainable?

THE QUADRATIC FAMILY

Let us specialize now to the family of quadratic maps $f_c(z) = z^2 + c$, for a complex parameter c, denoting the Julia set by J_c. The "roadmap" to quadratic Julia sets is the Mandelbrot set. The Mandelbrot set M is the set of parameter values $c \in \mathbf{C}$ for which the critical point 0 of f_c has a bounded orbit.

An important analytic theorem is that each attractive periodic orbit of a polynomial h must attract a (finite) critical point of h [Mi1, Theorem 6.6]. This means that f_c can have at most one attractive orbit, and that if f_c has an attractive orbit, then the orbit of the critical point is bounded, so in $c \in M$. Having an attractive orbit of a given minimal period n is stable under small perturbations of the parameter c, so if f_c has an attractive orbit of minimal period n so do all the functions $f_{c'}$ for c' in a neighborhood of c. A *hyperbolic component* of order n of M is a maximal connected open subset U of M such that for all parameter values $c \in U$, f_c has an attractive orbit of minimal period n. The collection of all hyperbolic components is clearly a subset of $\text{Int}(M)$.

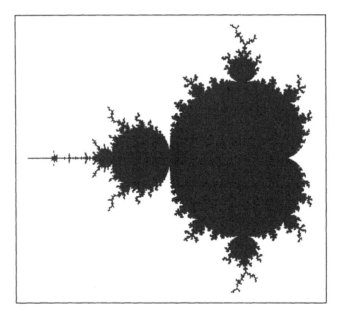

FIGURE 1. The Mandelbrot set M. The window on $\mathbf{C}$ is
$[-2.1, 0.6] \times [-1.2, 1.2]i$.

Except for $c = 1/4$, for which f_c has only one (indifferent) fixed point, f_c has two fixed points, at least one of which must be repelling. Let the fixed points be denoted z_0 and z_1, with z_0 denoting the one that is most repelling. For example, in the familiar picture of M (Figure 1), the interior of the large cardioid is the hyperbolic component M_1. More precisely,

$$M_1 = \{c \in M : |f_c'(z_1)| < 1\}$$
$$\partial M_1 = \{c \in M : |f_c'(z_1)| = 1\}$$

For $c \in \mathbf{C} - M$, the Julia set is homeomorphic to the Cantor set [Br, Section 9.8]; for $c \in M$, J_c is a continuum [Mi1]; for $c \in M_1$, J_c is a Jordan curve [Br, Section 9.9]. After that, it gets complicated.

We restrict our attention to $c \in M$, and pay particular attention to $c \in \partial M_1$. By a theorem of Douady and Hubbard [DH2], M is connected, in fact, cellular. It is unknown if M is locally connected. (This and other questions about M are most interesting, but we will confine ourselves to questions about Julia sets here.) A general question of long-standing interest is the following: For which $c \in M$ is J_c locally connected? In this connection, Douady and Hubbard prove the following structural theorem:

Theorem A. *For c in any hyperbolic component of $\mathrm{Int}(M)$, the Julia set J_c is locally connected, and is topologically a quotient space of S^1 with finite equivalence classes.*

The equivalence classes, and consequently, the topological structure of J_c, depend sharply upon the particular hyperbolic component of $\mathrm{Int}(M)$ in which c is found. Roughly speaking, the size of the equivalence classes, i.e. how many points are "pinched" together, determines the local cut point order of the "pinch" point in J_c. This is the single most important factor in classifying such Julia sets up to homeomorphism [BK]. See Figure 4, below.

For c in a non-hyperbolic component of $\mathrm{Int}(M)$ (if there are any, an open question), or for $c \in \partial M$, the situation is somewhat murkier. Yoccoz, Douady, and others have recently made additional progress on this problem [Mi2]. However, we will not discuss further the delicate analytic question of which Julia sets are locally connected, but rather concentrate on the more topological problem

Problem 1. *Find topological descriptions for the non-locally-connected Julia sets in the quadratic family.*

The Filled-in Julia Set

To discuss this, and related questions, we will first have to establish some terminology. A connected open set in $\mathbf{C}_\infty$ is a *domain*. For h a polynomial map of $\mathbf{C}$ of degree $d \geq 2$, let $K(h)$ denote the "filled-in" Julia set: $J(h)$ together with all its bounded complimentary domains, if any. Let $U_\infty = \mathbf{C}_\infty - K(h)$. This notation is appropriate since the domain $\mathbf{C}_\infty - K(h)$ is the basin of attraction of ∞. It can be proved that $\partial K(h) = \partial U_\infty = J(h)$ [Mi1, Problem 4.1 and Lemma 17.1].

Prime end theory is a way of studying the approaches to the boundary of a simply connected domain U with nongegenerate boundary. The existence of prime ends flows from the Riemann mapping theorem which guarantees the existence of a conformal isomorphism ϕ from the unit disk $\mathbf{D}$ onto U. This theory, part analysis and part topology, is one of the most important tools for studying the topological structure of Julia sets. In case the domain $U_\infty = \mathbf{C}_\infty - K(h)$ is simply connected, there is a particular choice of ϕ which is convenient.

It is a consequence of a classical theorem of Bötkher [Mi1, Theorem 6.7] that if h is a polynomial map of $\mathbf{C}$ of degree $d \geq 2$, then there is a neighborhood N_∞ of infinity on which h is analytically conjugate to the map $z \to z^d$ near ∞. To be more precise, let $D_\infty = \mathbf{C}_\infty - \mathrm{Cl}(\mathbf{D}_r)$, for some $r > 1$, be a disk about ∞. Then there exists a conformal isomorphism $\psi : N_\infty \to D_\infty$ taking ∞ to ∞ such that $\psi(h(z)) = \psi(z)^d$. In general, ψ cannot be extended as a conformal isomorphism from the complement U_∞ of the filled-in Julia set $K(h)$ (to some disk), because the critical points of h in U_∞ are an obstruction to the extension. However, if all the critical points are in $K(h)$, then there is no obstruction, and ψ can be extended to a conformal isomorphism $\psi : U_\infty \to \mathbf{C}_\infty - \mathrm{Cl}(\mathbf{D})$, unique up to multiplication by a $(d-1)$st root of unity [Mi1, Theorem 17.3].

Let $\mathbf{D}_\infty = \mathbf{C}_\infty - \mathrm{Cl}(\mathbf{D})$, the "unit disk" about ∞. It follows from the above remarks that there is a canonical conformal map $\phi : \mathbf{D}_\infty \to U_\infty$ (namely ψ^{-1} from above) taking ∞ to ∞. This map is sometimes called the Bötkher *uniformization* of $U_\infty = \mathbf{C}_\infty - K(h)$. The map induced on the circle $S^1 = \partial\mathbf{D} = \partial\mathbf{D}_\infty$ by $\phi^{-1}h\phi$ is $z \to z^d$. It is convenient to view S^1 as $\mathbf{R}/\mathbf{Z}$; then the induced map on $\partial\mathbf{D}$ is $t \to dt \pmod 1$.

The radial rays $R_t = \{re^{i\theta} | \theta = t, r > 1\}$ from $\infty \in \mathbf{D}_\infty$ are carried by ϕ to rays $\phi(R_t)$ from $\infty \in U_\infty$ which accummulate on $\partial U_\infty = J(h)$. Such rays are usually called *external rays* in complex dynamics literature. Without going into an involved discussion of prime ends, it suffices for our purposes to agree that each point $t \in \partial\mathbf{D}$, via the ray $\phi(R_t)$, defines a unique *prime end* of U_∞ [CoLo]. The limit set $\mathrm{Cl}(\phi(R_t)) - \phi(R_t) \subset \partial U_\infty$ of $\phi(R_t)$ for a radial ray R_t of $\mathbf{D}_\infty$ is the *principal continuum* of the prime end corresponding to $t \in \partial\mathbf{D}$. (Since it is the limit set of a ray, it is easy to see that the principal continuum corresponding to t is indeed a continuum.)

Via the underlying conjugacy, one can see that $h(\phi(R_t)) = \phi(R_{dt})$. Thus, we can speak of rays being "periodic" and "pre-periodic," depending upon whether t is periodic or pre-periodic under multiplication by $d \pmod 1$.

LANDING THEOREMS

Let U be a domain and x a point of ∂U. We say x is *accessible* from U iff there is an embedding e of $[0,1]$ in $U \cup \{x\}$ such that $e(0) \in U$ and $e(1) = x$. It can be shown that the accessible points of $K(h)$ (accessible from the complementary domain U_∞) are precisely the points which are singleton limit sets (degenerate principal continua) of external rays. Such an external ray is said to "land" at the corresponding accessible point. The following remarkable theorem is proved in [DH2] (see Theorem 18.1 of [Mi1]):

Theorem B. *If R_t is a radial ray with a rational argument $t \in [0,1)$, then the limit set of the external ray $\phi(R_t)$ is a point, which is either periodic, and repelling or indifferent with rational argument, or preperiodic to such a periodic point.*

Theorem B establishes a functional (not necessarily continuous) extension of ϕ to a dense subset of $\partial\mathbf{D}$. If ϕ can be continuously extended to all of $\partial\mathbf{D}$, then J_c, as the image of the circle, is consequently locally connected. It is exactly this which Douady and Hubbard accomplish to prove Theorem A.

Remarkably, it is possible for every ray defining a prime end to land at a point of the boundary of a simply connected domain U without ∂U being locally connected. It is not known whether this phenomenon occurs for a quadratic Julia set, but it does for Julia sets of the exponential family, as we discuss in a subsequent section.

Even more remarkably, the converse to Theorem B is a theorem of Douady and Yoccoz ([Mi1, Theorem 18.2 and Lemma 18.3]):

Theorem C. *If z is a repelling or rationally indifferent periodic point in $J(h) = \partial K(h)$, then at least one external ray lands on z, necessarily periodic, and at most finitely many rays do, all of the same period.*

Non-Locally-Connected Quadratic Julia Sets

We will first address the topological structure of J_c for c in the boundary of the hyperbolic component M_1 of Int(M) for which the corresponding quadratic maps $f_c(z) = z^2 + c$ have an attractive fixed point. Let K_c denote the filled-in Julia set and $U_\infty = \mathbf{C}_\infty - K_c$. For $c \in \partial M_1$, the indifferent fixed point z_1 has derivative $f_c'(z_1) = e^{2\pi i\alpha}$ for some $\alpha = \alpha(c) \in [0,1)$. If α is rational, then z_1 is in J_c, and J_c is locally connected [DH2]. This is the *parabolic* case: J_c is a "fatter" version of some corresponding hyperbolic case. The Julia set is locally connected, and, indeed, homeomorphic to a hyperbolic Julia set (one with an attractive orbit in the corresponding Fatou set).

For irrational α, there are two possibilities:

(1) *Cremer* case: $z_1 \in J_c$.
(2) *Siegel* case: $z_1 \notin J_c$

Cremer showed the first case could occur [Cr] and Siegel showed the second could occur [Si]. Roughly, the Siegel case occurs when α is badly approximated by rationals and the Cremer case occurs when α is well-approximated by rationals [Mi1, page 55]. The irrationals for which a Siegel disk occurs in the quadratic family have been shown by Yoccoz [Yo] to be precisely the Brjuno irrationals.

CREMER JULIA SETS

Call all non-Brjuno irrationals, Cremer irrationals. Douady (and independently Sullivan [Su3]) obtain the following result:

Theorem D. *For a Cremer irrational $\alpha = \alpha(c)$, J_c does not separate* **C** *(so $J_c = K_c$) and is not locally connected.*

The proof [Mi1, Corollary 18.6] of Theorem D reveals little else about the structure of J_c. It is not clear how badly non-locally-connected it might be. (Paul Blanchard originally brought this issue to my attention.) There are a number of speculations about the topology of J_c for Cremer irrationals. In this direction, the following questions are of interest (assume $c \in \partial M_1$ and $\alpha = \alpha(c)$ is a Cremer irrational):

Question 2. *Is the indifferent fixed point z_1 accessible? Is every periodic point in J_c accessible?*

Question 3. *Is the critical point 0 accessible.*

Question 4. *Does J_c have points of local connectivity as well as points where local connectivity fails, or is J_c nowhere locally connected?*

Question 5. *Is J_c arcwise connected? In particular, is there an arc from z_1 to z_1', the other preimage of z_1?*

Question 6. *Is J_c an indecomposable continuum?*

A continuum X is *indecomposable* iff X is not the union of two proper subcontinua iff no proper subcontinuum of X has interior; otherwise it is *decomposable*. If a point x in a continuum X is a separating point of X (meaning, $X - \{x\}$ is the union of two disjoint open subsets of X), then X is decomposable.

One thing that emerges from the proof of Theorem D is that, if z_1 is accessible, then it must be accessible with respect to infinitely many different prime ends, i.e. infinitely many external rays land on z_1. Indeed, z_1 will be a separating point of J_c. Thus, $J_c - \{z_1\}$ will have infinitely many components (trapped between the "wedges" defined by the infinitely many external rays landing on z_1).

Question 7. *Assuming z_1 is accessible, is the number of components of $J_c - \{z_1\}$ countably or uncountably infinite?*

Let us suppose for the sake of argument that for at least some Cremer Julia sets the indifferent fixed point z_1 and the critical point 0 are accessible and that the Julia set is arcwise connected. What kind of continua will such Julia sets be? Since the preimages of any point in J_c are dense in J_c, there will be a dense set of separating points, the pre-images of z_1 and 0. There are examples of non-locally-connected planar dendroids with a dense set of separating points. Some are nowhere locally connected, and others have points (even, a dense set of points) of local connectivity.

Oversteegen and Mayer take these assumptions as a starting point for a plausible topological model for such Julia sets, constructing a non-locally-connected planar dendroid J with a dense set of separating points, and a dense set of points of local connectivity, such that the dendroid is invariant under a two-to-one covering map f of the plane branched over a single "critical" point in the dendroid. A version of this model was presented in a talk by the author at the 1994 Spring Topology Conference, but has not yet been published. The building block is the Cantor bouquet described herein in the section on the exponential family.

The next step will be to show that f can be constructed so that there is a complex structure on the plane with respect to which f is analytic fixing ∞ [DH4]. A necessary condition is that f induce on the prime end compactification of the complement of J in the compactified plane a mapping which is at least topologically conjugate to z^2 on $\mathbf{D}_\infty$. We can construct f so that this is true. (The difficult step will be improving the conjugacy to an analytic one.)

Mayer's and Oversteegen's model, even if homeomorphic to some Cremer Julia set, cannot account for all of them. It is a theorem of Douady that, for some Cremer irrationals, there are prime ends whose principal continuum is nondegenerate and contains both z_1 and its other preimage z_1'. (A proof of Douady's "non-landing" theorem can be found in Sørenson's MSc thesis [Sn].) Since at least one external ray does not land, it follows immediately that J_c is not locally connected. The proof does not apply to all Cremer irrationals, but only a dense G_δ subset, unlike the more elementary proof of Theorem D. This raises the questions:

Question 8. *Is it the case for all Cremer irrationals that J_c has a prime end whose principal continuum is nondegenerate?*

Question 9. *Is there any Cremer irrational for which J_c has a prime end whose principal continuum is all of J_c?*

An affirmative answer to Question 9 implies that J_c is an indecomposable continuum, as we discuss below.

INDECOMPOSABLE JULIA SETS

In the quadratic family, for $c \in M - \mathrm{Cl}(M_1)$, there is always a repelling fixed point z_1, which, by the landing theorems and the fact that $t \to 2t$ has only 0 as a fixed point, has two or more periodic external rays landing on it. A point at which more than one external ray lands is clearly a separating point of K_c, and hence of $J_c = \partial K_c$. Hence, J_c cannot be indecomposable for $c \in M - \mathrm{Cl}(M_1)$. So, if there are indecomposable quadratic Julia sets hiding somewhere, it is for $c \in \partial M_1$.

There is a relationship between a plane continuum being indecomposable and the prime ends of its complement first elaborated by Rutt [Ru]. To state the relationship, we need some additional prime end terminology, which we have specialized to the case of the uniformization of the complement $U_\infty = \mathbf{C}_\infty - K(h)$ of the filled-in

Julia set $K(h)$ for a polynomial h of degree $d \geq 2$. Let $\phi : \mathbf{D}_\infty \to U_\infty$ be the uniformizing map. Let $P(t)$ denote the principal continuum corresponding to $t \in \partial \mathbf{D}$. A prime end such that $P(t) = \partial U_\infty = J(h)$ is said to be a *simple dense canal*. (The terminology is truly descriptive if you are familiar with a "single lake-of-Wada continuum.")

Suppose S_t is the image of a *ray*, an embedding e of [0,1) in $\mathbf{D}_\infty$ extendible to $\mathrm{Cl}(\mathbf{D}_\infty) = \mathbf{C}_\infty - \mathbf{D}$, such that $e(0) = \infty$ and $e(1) = t \in \partial \mathbf{D}_\infty = \partial \mathbf{D}$. Previously R_t has designated a radial ray, but S_t need not be radial; it just must go from $\infty \in \mathbf{D}_\infty$ to $t \in \partial \mathbf{D}$. It can be shown [CoLo] that the limit set of $\phi(S_t)$ in ∂U_∞ is a continuum which always contains the principal continuum $P(t)$ corresponding to t. The smallest continuum in ∂U_∞ containing each limit set of $\phi(S_t)$, taken over all such rays S_t, is called the *impression* of the prime end corresponding to t. Let $I(t)$ denote the impression of the prime end corresponding to $t \in \partial \mathbf{D}$.

Rutt's theorems may now be stated in our case as (1) if U_∞ has a simple dense canal, then ∂U_∞ is an indecomposable continuum, and (2) if U_∞ has a prime end whose impression is all of ∂U_∞, then ∂U_∞ is the union of at most two indecomposable continua. Toward the goal of identifying indecomposable Julia sets (if there are any), the author and Jim Rogers [MR] have proved the following theorems, partly based on Rutt's results:

Theorem E. *Suppose the continuum $J = J(h)$ is the Julia set of a polynomial h of degree $d \geq 2$. Let $U_\infty = \mathbf{C}_\infty - K(h)$. The following are equivalent:*

(1) *J is indecomposable,*

(2) *Some indecomposable subcontinuum of J has nonempty interior in J,*

(3) *The impression of some prime end of U_∞ has nonempty interior in J, and*

(4) *The impression of some prime end of U_∞ is equal to J.*

Theorem F. *Suppose $f_c(z) = z^2 + c$ with Julia set $J = J_c$ connected. Let U_∞ be the domain $\mathbf{C}_\infty - K_c$. Suppose that $\phi : \mathbf{D}_\infty = \mathbf{C}_\infty - \mathrm{Cl}(\mathbf{D}) \to U_\infty$ is a uniformization such that $\phi(\infty) = \infty$ and ϕ conjugates f_c on U to f_0 on $\mathbf{D}_\infty$. Then the following are equivalent:*

(1) *The Julia set J is indecomposable.*

(2) *For some $t \in \partial \mathbf{D}$, $I(t) = J$.*

(3) *For a dense subset $\mathcal{D}$ of $\partial \mathbf{D}$, for each $t \in \mathcal{D}$, $I(t) = J$.*

(4) *For every $t \in \partial \mathbf{D}$, $I(t) = J$.*

(5) *The set of simple dense canals of U_∞ is a residual set in $\partial \mathbf{D}$.*

(6) *For some $t \in \partial \mathbf{D}$, the prime end corresponding to t is a simple dense canal in U_∞.*

Theorem G. *Suppose $f_c(z) = z^2 + c$ with J_c an indecomposable continuum. Then no composant of J_c can contain the principal continuum of more than one prime end of the domain $U_\infty = \mathbf{C}_\infty - K_c$.*

The proofs of Theorems E–G follow from a close examination of what can happen to external rays and their limit sets under f_c and the branches of f_c^{-1}, making use of the Bötkher uniformization, Rutt's theorems, and some more subtle theorems regarding composants of indecomposable plane continua.

Theorem E, particularly condition 3, may give us a practical test for detecting indecomposable Julia sets using prime end techniques, while Theorems F and G afford a description of what an indecomposable Julia set would have to look like.

Most of Questions 2–9 above were raised and discussed by Douady and others at the 1990 Symposium on Conformal Dynamics at Stony Brook [Bd]. Several have been the subject of active research for some time. Questions 4 and 9 are due to the author.

SIEGEL DISKS

A *Siegel disk* is a component of the Fatou set on which f_c is conformally conjugate to a rigid rotation of the unit disk. Now suppose for $c \in \partial M_1$, the indifferent fixed point z_1 has derivative $f_c'(z_1) = e^{2\pi i\alpha}$ for some $\alpha = \alpha(c) \in [0,1)$ which is a Brjuno number, that is, is not too well approximated by rationals [Mi1, page 55; Si,Z,Bj,Yo]. In this case, the Fatou set of f_c contains a Siegel disk S_c about z_1, so $z_1 \notin J_c$ and $\partial S_c \subset J_c$. The long-standing Siegel disk problem (usually stated for rational functions in general) is the following:

Question 11. *Is the boundary ∂S_c of the Siegel disk S_c always homeomorphic to the circle S^1?*

If not, then ∂S_c might still be decomposable, but not locally connected, or it might even be indecomposable or hereditarily indecomposable. Other candidates for the boundary of a Siegel disk include the pseudocircle, the hairy circle [FO], and the members of the family of indecomposable, but not hereditarily indecomposable, cofrontiers described by the author and his collaborators in [BGM1,BGM2,MO2].

For Brjuno irrationals, the situation is complicated by the fact, demonstrated by Herman [Hr2], that there are quadratic Siegel disks whose boundary is homeomorphic to S^1, but which do not have the critical point in their boundary. In this case, as Douady has argued, the proof of Theorem D may be modified to show that J_c is again not locally connected. Thus, even if all Siegel disks have an S^1 boundary, the Julia set containing that boundary may still be quite complicated topologically.

Working now inside the Siegel disk, rather than in the basin of attraction of infinity, Rogers [Rg2] has proved theorems similar to E–G for the boundary of a local Siegel disk. Let R_α be a rigid rotation of $\mathbf{D}$ through angle $2\pi i\alpha$ for irrational argument $\alpha \in [0,1)$. Let $\phi : \mathbf{D} \to U$ be a conformal isomorphism onto a simply connected domain U. We say U is a *local Siegel disk* iff $h_\alpha = \phi R_\alpha \phi^{-1} : U \to U$ can be extended continuously to ∂U. Rogers' work [Rg1-5] on local Siegel disks has been inspired by examples and theorems of Moeckel [Me], Rodin [Ro,PoR], and

Herman [Hr1-3]. In addition to the parallel theorems to E–G, he has proved the following remarkable theorem greatly narrowing down the kinds of continua that could appear as boundaries of Siegel disks of polynomials [Rg3].

Theorem H. *The boundary ∂U of a bounded irreducible local Siegel disk U satisfies exactly one of the following properties:*

(1) *The inverse ϕ^{-1} of the Riemann map $\phi : \mathbf{D} \to U$ extends continuously to a map (also called) $\phi^{-1} : \partial U \to \partial \mathbf{D} = S^1$, or*

(2) *∂U is an indecomposable continuum.*

Independently, inspired by Handel [Ha] and Barge [BG1-2], the author has worked with Beverly Brechner and Merle Guay [BGM1-2] on a family of even more "abstract" Siegel disks. If U is a simply connected domain with nondegenerate boundary, $h_\alpha : \partial U \to \partial U$ is a homeomorphism extendible to a homeomorphism (also called h_α) of $\mathrm{Cl}(U)$, and the homeomorphism $H|\partial \mathbf{D}$ on the circle of prime ends of U induced by extending $H = \phi^{-1} h_\alpha \phi$ to $\mathrm{Cl}(\mathbf{D})$ is conjugate to a rotation R_α of $S^1 = \partial \mathbf{D}$, then we say h_α is a *one-sided pseudorotation* of $\mathrm{Cl}(U)$. (By comparison, for a local Siegel disk, one requires that the induced homeomorphism on the disk $\mathrm{Cl}(\mathbf{D})$ of prime ends be conformally conjugate to a rotation, but does not assume that the map h_α on ∂U is a homeomorphism.) If Λ is a continuum which irreducible separates the plane into two components, we call it a *cofrontier*; if the homeomorphism h_a is defined on the entire plane leaving Λ invariant, and the induced homeomorphism on both the inner and outer circles of prime ends is conjugate to the same rotation, then we say that h_a is a *pseudorotation* of Λ.

Though we obtain theorems similar to E–G in [BGM1-2] for cofrontiers under (one-sided) pseudorotations, our efforts have been directed more toward the construction of additional examples which might exemplify conditions (1) and (2) of Theorem H. In this connection, the author has an example of an indecomposable cofrontier, invariant under a homeomorphism of the plane, such that from one side the induced prime end map is conjugate to a rotation, while on the other it is conjugate to a Denjoy-type map of the circle [MO2].

Using Theorem H and topological techniques, in work to appear [Rg5], Rogers has recently proved the following theorem, answering a question first raised by Douady [Do1] and partially answered by Herman [Hr2]:

Theorem I. *If the polynomial f has a Siegel disk U whose rotation number satisfies a Diophantine condition, then the boundary ∂U of the Siegel disk contains a critical point.*

It is of interest that more than a few dynamicists would likely have regarded the question "When does the boundary of a Siegel disk contain a critical point?" as one not amenable to topological methods.

Evidently, one can work on the question of indecomposable Julia sets (or indecomposable subcontinua of Julia sets) from the "inside" as well as the "outside."

One would like to have further information on the topological properties of Siegel Julia sets, and to that end the following questions are of interest (assume $c \in \partial M_1$, $\alpha = \alpha(c)$ is a Brjuno irrational, and ∂S_c is the boundary in J_c of a Siegel disk S_c):

Question 12. *Does J_c (respectively, ∂S_c) have points of local connectivity as well as points where local connectivity fails, or is J_c (respectively, ∂S_c) nowhere locally connected?*

Question 13. *Is J_c (respectively, ∂S_c) arcwise connected?*

Question 14. *Is J_c (respectively, ∂S_c) an indecomposable continuum?*

Question 15. *For which Brjuno irrationals does J_c (respectively, ∂S_c) have a prime end whose principal continuum is nondegenerate?*

Question 16. *Is there any Brjuno irrational for which J_c (respectively, ∂S_c) has a prime end whose principal continuum is all of J_c (respectively, ∂S_c)?*

Question 17. *Is the critical point 0 of f_c ever accessible from U_∞ in a Siegel disk case?*

Rogers [Rg3] has also proved the very interesting theorem

Theorem J. *If ∂S_c contains a periodic point, then ∂S_c is an indecomposable continuum.*

which makes the following question (pointed out by John Milnor) much more intriguing:

Question 18. *Does ∂S_c ever contain a periodic point?*

Questions 12–18 are inspired by Rogers' work; he asks related, and additional, questions in [Rg3].

Assuming as a starting point that the boundary of some Siegel disk is a circle, that the critical point is in its boundary and accessible from U_∞, and that the Julia set is locally connected, Mayer and Oversteegen have presented a plausible model of what such a Siegel Julia set would look like topologically, in an as-yet-unpublished talk given at the 1994 Spring Topology Conference.

A Possible Model for a Siegel Julia Set

Fix an irrational number $\alpha \in [0,1)$, and regard S^1 as $\mathbf{R}/\mathbf{Z}$. Let

$$X_0 = S^1 \times \{0\}$$
$$X_1 = (S^1 \times \{0,1\})/\sim$$
$$\vdots$$
$$X_n = (S^1 \times B_n)/\sim$$

where

(1) B_n is the set of binary sequences of length n ($B_0 = \{0\}$), and

(2) $\sim$ is an equivalence relation that attaches circles, one to another, at certain points (determined by α).

The desired model is a *direct limit* (increasing union plus completion points) X_∞ in $\mathbf{C}$ of the X_n's:

$$X_0 \hookrightarrow X_1 \hookrightarrow X_2 \hookrightarrow \cdots \hookrightarrow X_\infty$$

See Figure 2. The copies of S^1 in the X_n's are identified in the figure by their binary coordinates, beginning with the finite strings B_n and ending with a string of 0's. To see X_n in the figure, truncate binary sequences at the nth coordinate.

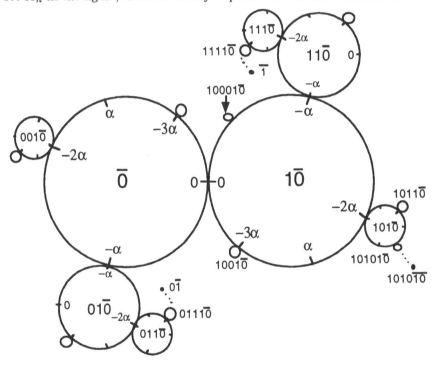

FIGURE 2. A possible model for a Siegel Julia set.
The Siegel disk is the inside of the circle labelled $\bar{0}$.

Let (t, b_n) denote a point in $S^1 \times B_n$. Let $\sigma_n : B_n \to B_{n-1}$ be the shift defined by

$$\sigma_n(i_1, i_2, \ldots, i_n) = (i_2, \ldots, i_n)$$

Consider the map

$$\sigma_{\alpha,n} : S^1 \times B_n \to S^1 \times B_{n-1}$$

given by

$$(t, b_n) \longrightarrow (t + \alpha, \sigma_n(b_n))$$

which is clearly a two-to-one covering map. The above map induces a map $f_{n-1}^n :$ $X_n \to X_{n-1}$, provided $\sigma_{\alpha,n}$ is equivariant with respect to $\sim$.

$$
\begin{array}{ccc}
S^1 \times B_n & \xrightarrow{\ \sigma_{\alpha,n}\ } & S^1 \times B_{n-1} \\
\sim\Big\downarrow & & \Big\downarrow\sim \\
X_n & \xrightarrow[\ f_{n-1}^n\ ]{} & X_{n-1}
\end{array}
$$

This can be accomplished in the manner suggested in the Figure 2, where $\sim$ "attaches" circles at certain multiples (mod 1) of α. The map $f_{n-1}^n : X_n \to X_{n-1}$ is a two-to-one branched covering, ramified over a single point $\{(0,\overline{0}) \sim (0,1\overline{0})\} \to$ $(\alpha,\overline{0})$.

Let B_∞ denote the set of all infinite binary sequences $(i_1, i_2, \dots)$. Let $(t, b_\infty) \in$ $S^1 \times B_\infty$ (but with the topology induced by X_∞). The model X_∞ consists of

(1) A countable collection of topological circles in $\bigcup_{n=0}^\infty X_n$, where (t, b_∞) is in such a circle iff $b_\infty = b_n\overline{0}$ for some $n = 0, 1, \dots$ and $b_n \in B_n$.
(2) A set of "completion points" (t, b_∞), where b_∞ does not end in $\overline{0}$. Each such point has many names: $(r, b_\infty) \equiv (s, b'_\infty)$ iff $b_\infty = b'_\infty$.
(3) X_∞ is locally connected.
(4) The set of "completion points" is a dense G_δ of endpoints in X_∞.

Let $\sigma : B_\infty \to B_\infty$ be the shift that "forgets" the first coordinate. The maps $f_{n-1}^n : X_n \to X_{n-1}$ induce a map $f_\infty : X_\infty \to X_\infty$ given by

$$
f_\infty(t, b_\infty) = (t + \alpha, \sigma(b_\infty))
$$

The part of the map given by $t \to t + \alpha$ "matters" just in case b_∞ ends in $\overline{0}$. It can be shown that f_∞ is a two-to-one branched covering map ramified over the point $(\alpha, \overline{0})$. The set of attachment points consists of all the preimages under f_∞ of $(\alpha, \overline{0})$. All periodic and preperiodic points of f_∞ are contained in the set of "completion points" of X_∞, and are dense in X_∞. The set of preimages of any point in X_∞ is dense in X_∞. The map f_∞ can be extended to a branched covering map of the plane (though this is not obvious).

THE EXPONENTIAL FAMILY

Consider the complex exponential function f_λ which maps $z \to \lambda e^z$, for a positive real parameter λ. The Julia set for $\lambda > 1/e$ is all of $\mathbf{C}$ [DK,D2]. For $\lambda < 1/e$, it is known [DG] that the Julia set $J(f_\lambda)$ is the boundary of the basin of attraction U_λ (which is an unbounded simply connected domain in this case) of the unique attracting fixed point q_λ of f_λ. There is, consequently, a prime end uniformization $\phi : \mathbf{D} \to U_\lambda$.

Cantor Bouquet Julia Sets

For $\lambda < 1/e$, the fine structure of the Julia set of the exponential function f_λ delineated by Devaney [DK,D1-2,DG] is quite interesting: the set A_λ of points of $J(f_\lambda)$ accessible from U_λ is a set of endpoints; $J(f_\lambda)$ itself is what he calls a *Cantor bouquet*, a set of rays extending from A_λ to ∞. The set A_λ contains all the repelling periodic points of f_λ, so it is dense in $J(f_\lambda)$. Though $J(f_\lambda)$ can be viewed as the closure of an increasing union of sets of the form (Cantor set) $\times [0, \infty)$, it is not itself homeomorphic to (Cantor set) $\times [0, \infty)$, nor is $J(f_\lambda) \cup \{\infty\}$ homeomorphic to the cone over the Cantor set. Precisely what it is, was not made clear until recently.

Devaney and Goldberg [DG] show that all radial limit points exist for the prime end uniformization ϕ; i.e. every image ray $\phi(R)$ of a radial ray R in **D** lands at a point of A_λ or at ∞. The author [M] has shown that this implies that the set of accessible points A_λ of $J(f_\lambda)$ is totally disconnected, but upon the addition of the point at infinity, becomes connected. Such a space has long been known topologically as an *explosion* (or *dispersion*) space, and the added point the explosion (or dispersion) point. It is surprising that such spaces occur in this "natural" setting.

It appeared to the author and others that this Cantor bouquet Julia set was homeomorphic to a certain topological model: a fan with a dense set of endpoints constructed by Lelek [L], minus its vertex. In this direction, Bula and Oversteegen [BO] (and independently, Charatonik) showed that all smooth fans having a dense set of endpoints are topologically equivalent. (Here "smooth" is meant in a point set topological sense.) The Cantor bouquet Julia set $J(f_\lambda) \cup \{\infty\}$ is a fan with vertex ∞ and a dense set of endpoints. It remained to be shown that it was smooth.

Apparently similar Cantor bouquet Julia sets occur for other entire functions as well [D1,DT,GK], such as in the family $s_\mu : z \to \mu \sin(z)$, $(0 < \mu < 1)$, or the family $f_\lambda : z \to \lambda e^z$ with complex parameter. The evident question here is the following:

Question 19. *Are the Cantor bouquets occurring in the Julia sets of various entire functions homeomorphic? Specifically, are the Julia sets of f_λ for fixed $0 < \lambda < 1/e$ and s_μ for fixed $(0 < \mu < 1)$ homeomorphic?*

Aarts and Oversteegen [AO] have shown that not only the Cantor bouquet Julia sets in the real-parameter exponential family, but also in certain sine, and hyperbolic sine families, are all homeomorphic to the conjectured model: Lelek's fan with a dense set of endpoints, minus its vertex. This answers the specific part of Question 19, and tends to confirm the, otherwise unsubstantiated, intuition that the Cantor bouquets known to occur in various non-conjugate settings are indeed representatives of one and the same topological space. This is somewhat surprising, particularly since some Cantor bouquet Julia sets have different measure-theoretic properties [Mc,EL]. It appears likely that Aarts' and Oversteegen's techniques can be applied to all the Cantor bouquet Julia sets that occur for the class (discussed in [DT,GK]) of entire functions of finite type satisfying an appropriate growth condition, and for the class of meromorphic functions which have an asymptotic value which lands on a pole after finitely many iterations.

Bergweiler's excellent survey article [Bg] of the dynamics of meromorphic functions (with a very complete bibliography) devotes only a short section to Cantor bouquets, but it does suggest to the author the possibility of their ubiquity in the Julia sets of transcendental meromorphic functions.

COMPLETE ERDÖS SPACE

The set A_λ of endpoints of the Cantor bouquet Julia set of the exponential function is an interesting set in its own right: it is totally disconnected, but 1-dimensional, and an example of an *almost zero-dimensional space* [OT]. Kawamura, Oversteegen and Tymchatyn [KOT] show that A_λ, the set of endpoints of Lelek's fan, and the set of endpoints of the universal separable **R**-tree [MNO], are all homeomorphic to *complete Erdös space*, the set of points in Hilbert space all of whose coordinates are irrational. Oversteegen and Tymchatyn have conjectured [MO3] that complete Erdös space is homeomorphic to the group of homeomorphisms of the Menger universal curve. Complete Erdös space, like the Cantor set, is probably ubiquitous.

COMPLEX-PARAMETER EXPONENTIAL FAMILY

Now consider the family $f_\lambda : z \to \lambda e^z$ with complex parameter λ. Ultimately, one would like to describe topologically the Julia sets of this family for all those cases for which it is not all of **C**, or at least obtain a partial classification of them like the Douady-Hubbard classification for the hyperbolic components of the Mandelbrot set M. The bifurcation set B for the exponential family, the analog of the Mandelbrot set, is described in [DGH] and [D1]: it is the set of parameter values for which the orbit of the singular point 0 is bounded. (See Figure 3.)

The topological conjugacy of f_{λ_1} and f_{λ_2} implies the topological equivalence of the corresponding Julia sets. Goldberg, for example, has shown in unpublished work that the maps corresponding to λ-values in the cardioid-like hyperbolic component B_1 of B, the component for which the dynamics has an attractive fixed point, are all quasiconformally conjugate. It follows that the corresponding Julia sets are homeomorphic. Nevertheless, one might still lack a clear idea of their topological structure, as well as be unsure of whether or not the very same topological object appears in non-conjugate settings. Therefore, one would like to accomplish two further objectives: "pin down" these Julia sets topologically, and produce a topological model for various Julia sets that "cuts across" conjugacy, much like the Cantor set appears ubiquitously in non-conjugate dynamical settings. It is in this way that the Cantor bouquet, which by the new results of Aarts and Oversteegen is a topologically unique object, not only forms the model for all the Julia sets corresponding to $\lambda \in B_1$, but also cuts across conjugacy, being the topological model for Julia sets in other (non-conjugate) families.

Computer pictures, and theoretical considerations [D1,DGH], of the bifurcation set B, indicate that there are topological models for the Julia sets of f_λ in those cases where f_λ has an attractive orbit of period greater than 1. Roughly, the Julia sets appear to be Cantor bouquets with certain endpoints identified, thus separating $\mathbf{C} - J(f_\lambda)$ into countably infinitely many domains. (See Figures 5 and 6.) This structural notion, well-known to researchers in the field, has yet to be made precise in a topological sense.

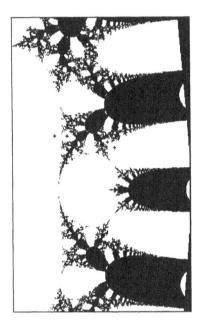

FIGURE 3. The bifurcation set B (in white) for the exponential family. The window on $\mathbf{C}$ is $[-5.7, 5.6] \times [-7.2, 10.0]i$.

The following problems require further work:

Problem 20. *Describe a topological model for the Julia set of $f_\lambda : z \to \lambda e^z$ with complex parameter λ for each hyperbolic component of the bifurcation set B of f_λ.*

Problem 21. *To what extent do topologically equivalent Julia sets appear* **unexpectedly** *in non-conjugate settings, in the exponential (and related) families?*

Example: Is $2 \times 3 \neq 3 \times 2$? An example will make clear why we believe expectations regarding the topological equivalence of Julia sets are sometimes subject to false intuitions. Consider two of the period 6 hyperbolic components of the Mandelbrot set marked with white crosses (+) in Figure 1. Typical Julia sets corresponding to c-parameter values (in $f_c(z) = z^2 + c$) for c in the left (respectively, right) marked "bulb" are shown in Figure 4 at the bottom (respectively, top). Given that the pictures accurately represent the (differing) patterns of local cut point order, the two

Julia sets in Figure 4 are not homeomorphic; the reason is tied to the fact that the bifurcation "routes" (doubling, then tripling, versus tripling, then doubling) differ.

The author has conjectured that the situation is otherwise with the Julia sets of the exponential and related functions. Consider the four black crosses marking points in the bifurcation set B for the exponential family in Figure 3. The leftmost cross lies in the 2-bulb, the next right in a 6=2x3-bulb; the rightmost cross in a 3-bulb, and the next left in a 6=3x2-bulb; the first pair of Julia sets are pictured in Figure 5, the second pair in Figure 6. We conjecture that the two "different" period 6 Julia sets are topologically equivalent. In addition to the dynamics, what distinguishes them, we believe, is how they are embedded in the plane. That is, we conjecture, that while there is a homeomorphism that takes one Julia set to the other, no homeomorphism of the plane can do so.

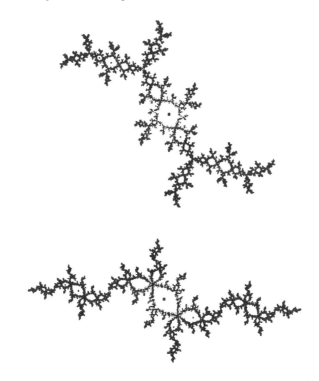

FIGURE 4. Nonhomeomorphic quadratic Julia sets whose corresponding Fatou sets each have a period 6 attractive orbit.

A UAB graduate student, Mark Widener, is presently making progress on the following topological problem, motivated by Problem 21 and the above example, with the idea of ultimately addressing Problem 20:

Problem 22. *Classify the spaces obtained by identifying subcollections of endpoints of Lelek's fan.*

In particular, one may specify any two disjoint, countably infinite, dense, subsets of endpoints $A = \{a_1, a_2, \dots\}$ and $B = \{b_1, b_2, \dots\}$ of the fan and identify a_i to b_i for each i. We conjecture that if the decomposition is continuous, then the same quotient space is obtained whatever the choices made for A and B.

EMBEDDINGS OF JULIA SETS

Recall that all Cantor sets in the plane are equivalently embedded. It can be shown that there are inequivalent embeddings of Lelek's fan in the plane. However, Aarts and Oversteegen [AO] show that the Cantor bouquet Julia sets they consider (of the exponential, sine, and hyperbolic sine) are all equivalently embedded in $\mathbf{C}_\infty$. In view of these considerations, the following question, to which we conjecture the answer is yes, is of interest:

Question 23. *Are there homeomorphic Julia sets in the exponential family which are not equivalently embedded in the plane?*

INDECOMPOSABLE INVARIANT SUBSETS OF JULIA SETS

Recall that for the complex exponential function f_λ mapping $z \to \lambda e^z$, for a positive real parameter $\lambda > 1/e$, the Julia set is all of $\mathbf{C}$ [D2]. Recently, Devaney [D3] has shown that for such parameter values there is a connected subset Λ_λ of $\mathbf{C}$, invariant under f_λ, which, upon compactification to a continuum Λ_λ^* by the addition of a countable set, becomes an indecomposable tree-like planar continuum. Let us agree to call a connected set, which **countably** compactifies into an indecomposable continuum, *indecomposable* itself. This is the first instance in which a topologically indecomposable set has been found as an invariant subset of a Julia set.

In addition to the obvious questions about how widespread the phenomenon of indecomposable continua appearing as invariant subsets of Julia sets is, there are a number of interesting topological questions about the particular subsets identified by Devaney.

Question 24. *For different parameter values $\lambda_1, \lambda_2 > 1/e$, are the continua $\Lambda_{\lambda_1}^*$ and $\Lambda_{\lambda_2}^*$ homeomorphic?*

Question 25. *Is Λ_λ^* chainable?*

Question 26. *Is every proper subcontinuum of Λ_λ^* chainable? Is every proper subcontinuum of Λ_λ^* an arc?*

Devaney asked Question 24 and has conjectured that the answer is no. An affirmative answer to either part of Question 26 would imply that Λ_λ^* is atriodic. We conjecture that every proper subcontinuum of Λ_λ is an arc, so that Λ_λ is an atriodic continuum, but that Λ_λ is not chainable.

FIGURE 5. Exponential Julia sets from 2-bulb (left)
and from 6=2x3-bulb.

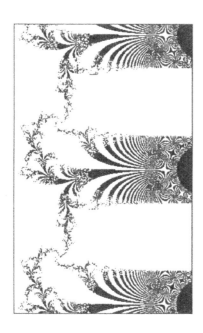

FIGURE 6. Exponential Julia sets from 3-bulb (left)
and from 6=3x2-bulb.

FINAL REMARKS

It is impossible to be complete in a survey of this sort. The topics selected, and the list of references, reflect my own interests rather than an attempt to be exhaustive. Several of the general references on complex dynamics have excellent bibliographies. Other topics that would be of interest to continuum theorists include how J_c and K_c change with the continuous change in the parameter c and expansive and continuum-wise expansive maps on Julia sets.

REFERENCES

[AO]   J. Aarts and L. G. Oversteegen, *The geometry of Julia sets*, Trans. AMS **338** (1993).

[B]   I. N. Baker, *Wandering domains in the iteration of entire functions*, Proc. London Math. Soc. **49** (1984), 563–576.

[BK]   C. Bandt and K. Keller, *Symbolic dynamics for angle-doubling on the circle: I. The topology of locally connected Julia sets*, Proc. 3rd Conference on Ergodic Theory and Related Topics, Gütrow (1990), 1–23.

[BG1]   M. Barge and R. M. Gillette, *Indecomposability and dynamics of invariant separating continua* (preprint).

[BG2]   _____, *Zero prime end rotation number implies a fixed point* Preliminary report, Abstracts AMS **10** (1989), 339.

[Br]   A. F. Beardon, *Iteration of Rational Functions*, Springer-Verlag, New York, NY, 1991.

[Bg]   W. Bergweiler, *Iteration of meromorphic functions*, Bulletin AMS **29** (1993), 151–188.

[Bd]   B. Bielefeld, *Conformal Dynamics Problem List, Preprint #1*, Institute for Mathematical Sciences, SUNY-Stony brook, 1990.

[Bi1]   G. D. Birkhoff, *Sur quelques courbes fermees remarquables*, Bull. Soc. Math. France **60** (1932), 1–26.

[Bn]   P. Blanchard, *Complex analytic dynamics on the Riemann sphere*, Bull. Amer. Math. Soc. **11** (1984), 85–141.

[BGM1] B. L. Brechner, J. C. Mayer, and M. D. Guay, *Rotational dynamics on cofrontiers*, Contemporary Mathematics **117** (1991), 39–48.

[BGM2] _____, *The rotational dynamics of cofrontiers*, in Continuum Theory and Dynamical Systems, Lecture Notes in Pure and Applied Mathematics **149** (Thelma West, ed.), Marcel Dekker, New York, 1993, pp. 59–82.

[Bj]   A. D. Brjuno, *Analytical form of differential equations*, Trans. Moscow Math. Soc. **25** (1971), 131–288; **26** (1972), 199–239.

[BO]   W. Bula and L. G. Oversteegen, *A characterization of smooth Cantor bouquets*, Proc. AMS **108** (1990), 529–534.

[Ca]   C. Caratheodory, *Über de Begrenzung einfach zusammenhängender Gebiete*, Math. Ann. **73** (1913), 323–370.

[CG]   L. Carleson and T. W. Gamelin, *Complex Dynamics*, Springer-Verlag, New York, NY, 1993.

[CoLo] E. F. Collingwood and A. J. Lohwater, *Theory of Cluster sets*, Cambridge Tracts in Math. and Math. Physics **56**, Cambridge University Press, Cambridge, 1966.

[Cr]   H. Cremer, *Zum Zentrumproblem*, Math. Ann. **98** (1928), 151–163.

[D1]   R. L. Devaney, e^z: *Dynamics and bifurcations*, International J. of Bifurcation and Chaos **1** (1991), 287–308.

[D2]   _____, *An Introduction to Chaotic Dynamical Systems*, Benjamin/-Cummings, Menlo Park, CA, 1986.

[D3]   _____, *Knaster-like continua and complex dynamics*, Ergodic Theory and Dynamical Systems **13** (1993), 627–634.

[DG]   R. L. Devaney and L. R. Goldberg, *Uniformization of attracting basins*, Duke Math. J. **55** (1987), 253–266.

[DGH]   R. L. Devaney, L. R. Goldberg, and J. H. Hubbard, *A dynamical approximation of the exponential by polynomials* (preprint).

[DK]   R. L. Devaney and M. Krych, *Dynamics of* $\exp(z)$, Ergodic Theory and Dynamical Systems **4** (1984), 35–52.

[DT]   R. L. Devaney and F. Tangerman, *Dynamics of entire functions near the essential singularity*, Ergodic Theory and Dynamical Systems **6** (1986), 498–503.

[Do1]   A. Douady, *Systèmes dynamiques holomorphes*, Séminar Bourbaki, exposé 599, Asterique **105–106** (1983–84), 39–63.

[Do2]   _____, *Disques de Siegel et anneaux de Herman*, Séminar Bourbaki, exposé 677, Asterique **108–109** (1986–87).

[DH1]   A. Douady and J. H. Hubbard, *Analysye complexe - Iteration des polynômes quadratiques complexes*, C. R. Acad. Sc. Paris **294** (1982), 123–126.

[DH2]   _____, *Étude dynamique des polynômes complexes (première partie)*, Publications Mathematiques D'Orsay **2** (1984), 1–75.

[DH3]   _____, *Étude dynamique des polynômes complexes (deuxième partie)*, Publications Mathematiques D'Orsay **4** (1985), 1–154.

[DH4]   _____, *A proof of Thurston's topological characterization of rational functions* (preprint).

[EL]   A. È. Eremenko and M. Yu. Ljubich, *Iterates of entire functions*, Soviet Math. Dokl. **30** (1984), 592–594.

[F]   P. Fatou, *Sur l'itération des fonctions trancendantes entieres*, Acta Math. **47** (1926), 337-370.

[FO]   R. J. Fokkink and L. G. Oversteegen, *A recurrent non-rotational homeomorphism on the annulus*, Trans. AMS **333** (1992), 865–875.

[GK]   L. M. Goldberg and L. Keen, *A finiteness theorem for a dynamical class of entire functions*, Ergodic Theory and Dynamical Systems **6** (1986), 183–192.

[GH]   W. H. Gottschalk and G. A. Hedlund, *Topological Dynamics*, vol. 36, AMS, Providence, RI, 1955.

[Ha]   M. Handel, *A pathological area-preserving diffeomorphism of the plane*, Proc. Amer. Math. Soc. **86** (1982), 163–168.

[Hr1]   M. R. Herman, *Construction of some curious diffeomorphisms of the Riemann sphere*, J. London Math. Soc. (2) **34** (1986), 375–384.

[Hr2]   _____, *Are there critical points in the boundary of singular domains*, Comm. Math. Phys. **99** (1985), 593–612.

[Hr3]   _____, *Recent results and some open questions on Siegel's linearization theorem of germs of complex analytic diffeomorphisms of* $\mathbf{C}^n$ *over a fixed point*, Proc. of the Eighth Int. Cong. Math. Phys., World Scientific, 1986, pp. 138–198.

[J]   G. Julia, *Memoire sur l'itération des fonctions rationelles*, J. Math. Pures Appl. **8** (1918), 47–225.

[KOT]   K. Kawamura, L. G. Oversteegen, and E. D. Tymchatyn, *On homogeneous totally disconnected 1-dimensional spaces* (preprint).

[MSS]   R. Mañé, P. Sad, and D. Sullivan, *On the dynamics of rational maps*, Ann. scient. Éc. Norm. Sup., 4e serie **16** (1983), 193–217.

[M]   J. C. Mayer, *An explosion point for the set of endpoints of the Julia set of* $\lambda \exp(z)$, Ergodic Theory and Dynamical Systems **10** (1990), 177–183.

[MMOT] J. C. Mayer, L. K. Mohler, L. G. Oversteegen, and E. D. Tymchatyn, *Characterization of separable metric R-trees*, Proc. AMS **115** (1992), 257–264.

[MNO]   J. C. Mayer, J. Nikiel, and L. G. Oversteegen, *On universal* **R**-*trees*, Trans. AMS **334**, 411–432.

[MO1]   J. C. Mayer and L. G. Oversteegen, *A topological characterization of R-trees*, Trans. AMS **320** (1990), 395–415.

[MO2]   _____, *Denjoy meets rotation on an indecomposable cofrontier*, in Continuum Theory and Dynamical Systems, Lecture Notes in Pure and Applied Mathematics **149** (Thelma West, ed.), Marcel Dekker, New York, 1993, pp. 183–200.

[MO3]   J. C. Mayer and L. G. Oversteegen, *Continuum Theory*, Recent Progress in General Topology (M. Hušek and J. van Mill, ed.), Elsevier (North-Holland), Amsterdam, 1992, pp. 454–492.

[MR]   J. C. Mayer and J. T. Rogers, Jr., *Indecomposable continua and the Julia sets of polynomials*, Proceeding AMS **117** (1993), 795–802.

[Mc]   C. McMullen, *Area and Hausdorff dimension of Julia sets of entire functions*, Trans. AMS **300** (1987), 329–342.

[Mi1]   J. Milnor, *Dynamics in one complex variable: introductory lectures*, Preprint #1990/5, IMS, SUNY-Stony Brook.

[Mi2]   _____, *Local connectivity of Julia Sets: Expository Lectures*, Preprint #1992/11, IMS, SUNY-Stony Brook.

[Me]   R. Moeckel, *Rotations of the closures of some simply connected domains*, Complex Variables **4** (1985), 285–294.

[OT]   L. G. Oversteegen and E. D. Tymchatyn, *On the dimension of some totally disconnected sets* (preprint) (1992).

[P]   G. Piranian, *The boundary of a simply connected domain*, Bulletin AMS **64** (1958), 45–55.

[PoR]   Ch. Pommerenke and B. Rodin, *Intrinsic rotations of simply connected regions, II*, Complex Variables **4** (1985), 223–232.

[Rg1]   J. T. Rogers, Jr., *Rotations of simply connected regions and circle-like continua*, Contemporary Mathematics **117** (1991), 139–148.

[Rg2]   _____, *Intrinsic rotations of simply connected regions and their boundaries*, Complex Variables Theory and Appl. **23** (1993), 17–23.

[Rg3]   _____, *Singularities in the boundaries of local Siegel disks*, Ergodic Theory and Dynamical Systems **12** (1992), 803–821.

[Rg4]   _____, *Is the boundary of a Siegel disk a Jordan curve?*, Bulletin AMS **27** (1992), 284–287.

[Rg5]   _____, *Diophantine conditions imply critical points on the boundaries of Siegel disks of polynomials I* (preprint).

[R]   B. Rodin, *Intrinsic rotations of simply connected regions*, Complex Variables **2** (1984), 319–326.

[Ru]   N.E. Rutt, *Prime ends and indecomposability*, Bull. Amer. Math. Soc. **41** (1935), 265–273.

[Si]   C. L. C. Siegel, *Iteration of analytic functions*, Math. Ann. **43** (1942), 607–612.

[So]   D. E. K. Sørenson, *Local conectivity of Julia sets*, M.Sc. Disseration, Mathematical Institute, Technical University of Denmark (1992).

[Su1]   D. Sullivan, *Seminar on conformal and hyperbolic geometry, Notes by M. Baker and J. Seade*, IHES, 1982.

[Su2]   _____, *Quasiconformal homeomorphisms and dynamics I*, Annals of Math. **122** (1985), 401–418.

[Su3]   _____, *Conformal dynamical systems*, Springer Lecture Notes Vol. 1007, Springer Verlag, New York, 1983, pp. 725–752.

[Wa]   R. Walker, *Basin boundaries with irrational rotations on prime ends*, Trans. AMS **324** (1991), 303–317.

[Yo]   J.-C. Yoccoz, *Theoreme de Siegel, polynomes quadratiques et ombres de Brjuno* (preprint).

[Z]   E. Zehnder, *A simple proof of a generalization of a theorem by Siegel*, Lecture Notes in Mathematics, Springer-Verlag **597** (1977), 855–866.

E-mail address: mayer@vorteb.math.uab.edu "John C. Mayer"

9

CONTINUA ON WHICH 2-TO-1 MAPS INDUCE CONTINUOUS INVOLUTIONS

JERZY MIODUSZEWSKI Instytut Matematyki, Uniwersytet Śląski, ul. Bankowa 14, 40-007 Katowice, Poland

ABSTRACT. A class of metric continua, described in terms of continuous involutions defined on neighbourhoods of points, is distinguished in order to clarify problems concerning 2-to-1 maps. This class is, even in dimension 1, somewhat wider than the class of connected compact manifolds. A list of examples is given in an attempt to estimate how wide this class is.

If $f : X \to Y$ is an (exactly) 2–to–1 map then it induces a fixed point free involution on X assigning to each x from X the point in $f^{-1}(f(x))$ distinct from x. Assuming that f is continuous, that X is metric and compact and that Y is Hausdorff, the involution corresponding to f is continuous on a dense and open subset of X. In the case when X is a manifold (author's paper, 1961) or even a homological manifold (Černavskii, 1962), this involution can be extended continuously from the set of continuity points by redefining it to be the identity outside of this set. Thus, as we shall say, 2–to–1 maps from homological manifolds *induce continuous involutions*.

However, the class of compacta on which 2–to–1 maps induce continuous involutions is somewhat wider, although only sporadic 1-dimensional non-manifolds are known as examples. The problem arises to describe this class of compacta.

1991 *Mathematics Subject Classification.* 54F15, 55M99.
Key words and phrases. 2-to-1 maps, continuous involutions, solenoids, Knaster continua, pseudoarc.
Supported by the Polish Scientific Grant

The interest in 2–to–1 maps goes back to papers by Harrold (1939) and Civin (1943), and then to papers by the author and Černavskii (loco cit.); for recent developments see papers by J. Heath, e.g. a 1994 survey.

Recently, W. Wójcik (doctoral dissertation) showed that the result of Černavskii remains valid if the compact metric space X has only a particular property of homological manifolds, assuming namely that its points cannot lie on boundaries of interiors of the fixed point sets of continuous involutions defined on open subsets of X. This property of spaces will be called here *the Newman property*. Even without assuming compactness, manifolds (Newman, 1930) and homological manifolds (Smith, 1941) have this property, because the set of fixed points of non–identical continuous involutions defined on them are nowhere dense.

Obviously, if a continuum has the Newman property then the Newman–Smith theorem, quoted above, is valid for it. The converse is not true.

Example: solenoids and Knaster continua. To explain this, observe that solenoids and Knaster continua are nowhere Newman. Points of solenoids (Knaster continua) have open Cantor bundles (open bent Cantor bundles) as neighbourhoods. There are continuous involutions moving "a half" of the bundle on the boundary of which (with respect to the bundle) the point lies, leaving "the remaining half" fixed. None the less (Dębski, 1992) these continua have the Newman–Smith property since the non–identical continuous involutions on solenoids and Knaster continua have at most two fixed points.

However, as was shown by Dębski (1992, loco cit.), 2–to–1 maps from solenoids and Knaster continua induce continuous involutions. Thus, the converse to the Wójcik result is not valid.

The Sierpiński carpet serves as an example of a non–manifold having the Newman property that is not far from being homogeneous (two kinds of non–equivalent points; Krasinkiewicz, 1969).

No homogeneous non–manifold serving as such example is known.

1. The Menger universal curve does not have the Newman property.
2. The Hilbert cube does not have the Newman property.

In both cases this is a consequence of failure of the Newman–Smith theorem for these spaces.

1. From the Anderson (1958) characterization of the Menger universal curve it follows that it is homeomorphic to the union of two of its copies which are sewn along an edge segment. If the Menger curve is embedded in the unit cube $[0,1]^3$ in the usual way then a suitable "edge segment" is all of the points in $[0,1]^3$ with first two coordinates equal to zero. The identity on one "half" and symmetry on the "remaining half", leaving the common segment fixed, is a non–Newman involution.

For these purposes, see also the thesis of N. Benakli (1992), where the Menger universal curve is regarded as a growth in a compactification of an appropriate hyperbolic space which allows (T. Januszkiewicz, oral communication) the production of "non–Newman" involutions on the growth.

2. Having the Hilbert cube in the form $Y \times I \times I \times \dots$, where the factor Y is a triod (Anderson, 1964), we get easily a non–Newman involution on it.

   In view of homogeneity, the Menger universal curve and the Hilbert cube are nowhere Newman. In contrast to solenoids and Knaster continua (which are also nowhere Newman), there are 2–to–1 maps from the Hilbert cube which do not induce continuous involutions (write the Hilbert cube in the form "an appropriate infinite dendrite" $\times I \times I \times \dots$ and consider a 2–to–1 map on the first factor).

The Newman property is closely related to the *Brouwer property* of the invariance of domain (homeomorphic subsets are both open if one of them is open).

The Hilbert cube and the Menger universal curve do not have these properties, because they can be embedded into themselves as nowhere dense subspaces.

However, these properties overlap.

The Sierpiński carpet has – as was mentioned – the Newman property, but does not have, for the same reasons as the Menger curve, the Brouwer property.

The Sierpiński triangular curve does have the Newman property (at the local separating points), but does not have the Brouwer property after a non–essential modification which removes a singular situation at three angles; see a paper by W. Dębski and the author, 1990, and a paper by W. Bandt and T. Retta, 1992.

For the pseudo–arc, the Newman–Smith theorem (as stated) holds; Fugate and McLean, 1981. However, it is not known if the pseudo–arc has the Newman property.

REFERENCES

1. J. Mioduszewski, *On two–to–one continuous functions, Dissertationes Mathematicae* **24** (1961).
2. A. V. Černavskii, *Nevozmoshnost' strogo dvukratnogo nepreryvnogo razbijenija gomologičeskogo shara*, Doklady AN SSSR **144** (1962), 286–289.
3. O. G. Harrold, *The non–existence of certain type of continuous transformations*, Duke Math. J. **5** (1939), 789–793,
4. P. Civin, *Two–to–one mappings on manifolds*, Duke Math. J. **10** (1943), 49–57.
5. J. Heath, *Exactly k–to–1 maps : from pathological functions with finitely many discontinuities to well–behaved covering maps*, this volume.
6. W. Wójcik, Dissertation.
7. W. Dębski, *Two–to–one maps on solenoids and Knaster continua*, Fund. Math. **141** (1992), 277–285.
8. M. H. A. Newman, *A theorem on periodic transformations of spaces*, Journal of Math., Oxford Series, **2** (1931), 1–8.

9. J. Krasinkiewicz, *On homeomorphisms of the Sierpiński curve*, Prace Matematyczne (Commentationes Mathematicae) **13** (1969), 255–257 (in Polish).

10. R. D. Anderson, *A characterization of the universal curve and a proof of its homogeneity*, Ann. of Math. **67** (1958), 313–324.

11. R. D. Anderson, Notices AMS, 1964.

12. V. B. Fugate, T. B. McLean, *Compact groups of homeomorphisms on tree–like continua*, Trans. AMS **267** (1981), 609–620.

13. N. Benakli, *Polyedres hyperboliques. Passage du local au global*, Thèse, 1992, Paris Sud.

14. W. Dębski, J. Mioduszewski, *Simple plane images of the Sierpiński triangular curve are nowhere dense*, Coll. Math. **49** (1990), 125–140.

15. C. Bandt, T. Retta, *Topological spaces admitting a unique fractal structure*, Fund. Math. **141** (1992), 257–268.

E-mail address: mioduszewski@gate.math.us.edu.pl "Jerzy Mioduszewski"

Part II

Research Papers

10
ENDPOINTS OF INVERSE LIMIT SPACES
AND DYNAMICS

MARCY BARGE Department of Mathematics, Montana State University, Bozeman, Montana 59717, USA

JOE MARTIN Department of Mathematics (Emeritus), University of Wyoming, Laramie, Wyoming 82071 , USA

ABSTRACT. In this paper, endpoints of inverse limits of interval maps are charcterized and this characterization is used to relate dynamical properties of such maps with the existence of endpoints in their inverse limit spaces. In particular, it is shown that the inverse limit of a core map in the tent family has endpoints if and only if the critical point is recurrent.

INTRODUCTION

Besides being of intrinsic interest, inverse limits of one-dimensional maps have been studied for their relevance to dynamical systems. The topology of an inverse limit space carries a considerable amount of information about the dynamical properties of the bonding map used to create the space. These spaces, and the natural shift homeomorphisms on them, are also used to model attractors of dynamical systems in higher dimensions. The situation that seems most relevant to dynamics occurs when the bonding map is piecewise monotone and the inverse limit space is indecomposable. In this case, the inverse limit space is locally, at most points, the product of a Cantor set and an arc and it can be quite difficult to decide whether, for two different bonding maps, the corresponding spaces are, or are not, homeomorphic. For example, it is unknown whether there are two distinct parameter

1991 *Mathematics Subject Classification.* 58F13, 58F22, 55F20, 54H20.
Key words and phrases. Endpoints, inverse limits.
Research of the first author was supported in part by NSF-OSR grant #9350546.

values in the tent family (see section 3) for which the corresponding inverse limit spaces are homeomorphic.

The cardinality of the set of endpoints is a topological invariant of inverse limit spaces that proves to be, at least in the case of piecewise monotone bonding maps, relatively accessible. We find in this paper that the existence of endpoints is related to the recurrence of a turning point. This relationship is developed in section 2 and applied to the tent family in section 3. As a consequence of the results of section 3, there are, for any nondegenerate interval of parameters in the tent family, infinitely many corresponding inverse limit spaces with no endpoints, infinitely many with a finite and nonzero number of endpoints, and infinitely many with infinitely many endpoints. (We do not know if there are uncountably many topologically distinct inverse limits in the tent family.)

For techniques that can sometimes be used to distinguish between inverse limit spaces with the same number of endpoints, see [W], [D], and [B-D].

1. ENDPOINTS

In this article a continuum is a compact connected metrizable space. If X is a continuum and f is a continuous map from X into itself, then the inverse limit space with bonding map f,

$$\varprojlim f = \{(x_0, x_1, \cdots) : x_n \epsilon X, \ f(x_{n+1}) = x_n, n = 0, 1, 2, \cdots\},$$

with the product topology, is also a continuum. In this situation there is an induced shift homeomorphism $\hat{f} : \varprojlim f \hookleftarrow$ given by $\hat{f}((x_0, x_1, \cdots)) = (f(x_0), x_0, x_1, \cdots)$. We will let $\pi_n : \varprojlim f \to X$ denote projection onto the nth coordinate and when X is a real interval we will use the metric

$$d((x_0, x_1, \cdots), (y_0, y_1, \cdots)) = \sum_{n=0}^{\infty} \frac{|x_n - y_n|}{2^n}.$$

A point x in the continuum X is an *endpoint* of X provided for every pair A and B of subcontinua of X with $x \in A \cap B$, either $A \subset B$ or $B \subset A$. Clearly any homeomorphism from one continuum onto another must carry endpoints to endpoints. Thus the cardinality of the set of endpoints of a continuum is a topological invariant, one that can sometimes be easily employed to distinguish between continua.

For example, let f and g be the maps of the interval $[0, 1]$ whose graphs are pictured in $Fig.$ 1. Then $\varprojlim f$ and $\varprojlim g$ are not homeomorphic since $(0, 0, 0, \cdots)$ is the only endpoint of $\varprojlim f$, while $\varprojlim g$ has two endpoints: $(0, 0, 0, \cdots)$ and $(1, 1, 1, \cdots)$.

The reader can easily check that the points claimed to be endpoints in the previous example are indeed endpoints. But for other maps of the interval it can be fairly difficult to find endpoints of the associated inverse limit space. To begin our investigation of endpoints in inverse limit spaces we make the following definition:

If $f : [c, d] \to [a, b]$ is a continuous surjection of the interval $[c, d]$ onto the interval $[a, b]$, $p \in [c, d]$, and $\epsilon > 0$, then f is ϵ-*crooked with respect to* p provided p does not separate $f^{-1}([a, a + \epsilon])$ from $f^{-1}([b - \epsilon, b])$ in $[c, d]$.

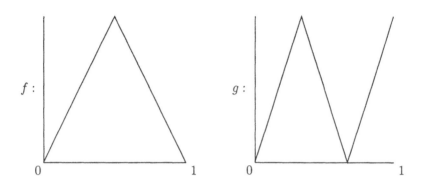

FIGURE 1.

Lemma 1.1: *Suppose that J and L are closed intervals, that $\epsilon > 0$, that $f : J \to L$ is continuous, that $p \in J$, and that f fails to be ϵ-crooked with respect to p. Then there is a unique subinterval K of J such that*

    (1) $p \in K$
    (2) $f(K) = L$
    (3) $f : K \to L$ *is not ϵ-crooked with respect to p, and*
    (4) K *is irreducible with respect to 1, 2, and 3.*

Proof: Assume the notation of the hypothesis. It is easy to check that if each of $R_1, R_2, \cdots$ is a closed subinterval of J satisfying 1, 2, and 3, and $R_{i+1} \subset R_i$, then $X = \cap_{i=1}^{\infty} R_i$ satisfies 1, 2, and 3. It then follows from the Brouwer Reduction Theorem, that there is a subinterval K of J which is irreducible with respect to 1, 2, and 3. It is clear that K is unique. ∎

Notice that if the endpoints of L are a and b, then $f^{-1}(a) \cap K$ is one endpoint of K and $f^{-1}(b) \cap K$ is the other. This is because of the irreducibility of K.

Lemma 1.2: *Suppose that I is a closed interval, $f : I \to I$ is continuous, $R_1, R_2,$ and R_3 are subintervals of I, $f(R_3) = R_2$, $f(R_2) = R_1$, $p_i \in R_i$, and $f(p_{i+1}) = p_i$. Suppose further that $f : R_2 \to R_1$ and $f^2 : R_3 \to R_1$ both fail to be ϵ-crooked with respect to p_2 and p_3 respectively. Let $K_2 \subset R_2$ and $K_3 \subset R_3$ be the irreducible intervals for f and f^2 promised by Lemma 1.1. Then $f(K_3) = K_2$.*

Proof: Let $R_1 = [a, b]$. Then it follows from the remark following Lemma 1.1 that $f^{-1}(a) \cap K_2$ and $f^{-1}(b) \cap K_2$ are the endpoints of K_2 and $f^{-2}(a) \cap K_3$ and $f^{-2}(b) \cap K_3$ are the endpoints of K_3. Then it is clear that $f(K_3) = K_2$. ∎

Lemma 1.3: *Suppose that I is a closed interval, $f : I \to I$ is continuous, $\epsilon > 0$, $J_0, J_1, J_2, \cdots$ is a sequence of subintervals of I, $p_0, p_1, p_2, \cdots$ is a sequence of points of I, $p_i \in J_i$, $f(p_{i+1}) = p_i$, and $f(J_{i+1}) = J_i$. Further, suppose that for each n, $f^n : J_n \to J_0$ is not ϵ-crooked with respect to p_n. Let $K_1, K_2, K_3, \cdots$ be the irreducible intervals promised by Lemma 1. Then $f(K_{i+1}) = K_i$. Furthermore, for each i, there are proper subintervals A_i and B_i of K_i such that*

    (1) $p_i \in A_i \cap B_i$

    (2) $A_i \cup B_i = K_i$

    (3) $f(A_{i+1}) = A_i$, and

    (4) $f(B_{i+1}) = B_i$.

Proof: It follows from Lemma 1.2 that $f(K_{i+1}) = K_i$.

We define A_i and B_i as follows: let $J_0 = [a, b]$ and consider the sets $f^{-i}([a, a+\epsilon])$ and $f^{-i}([b-\epsilon, b])$. Since f^i is not ϵ-crooked with respect to p_i, p_i separates these two sets. Let L_i be the lower of the two sets, and U_i the upper of the two sets. Let $l_i = \sup(L_i)$ and $u_i = \inf(U_i)$. Then $l_i < p_i < u_i$ and either $f(l_{i+1}) = l_i$ and $f(u_{i+1}) = u_i$, or $f(l_{i+1}) = u_i$ and $f(u_{i+1}) = l_i$.

If $K_i = [a_i, b_i]$, then the two sets we want are $C_i = [a_i, u_i]$ and $D_i = [l_i, b_i]$. Then either $f(C_{i+1}) = C_i$ and $f(D_{i+1}) = D_i$, or $f(C_{i+1}) = D_i$ and $f(D_{i+1}) = C_i$.

So, let $A_1 = C_1$, $B_1 = D_1$, and inductively define $A_{i+1} \in \{C_{i+1}, D_{i+1}\}$ and $B_{i+1} \in \{C_{i+1}, D_{i+1}\}$ so that $f(A_{i+1}) = A_i$ and $f(B_{i+1}) = B_i$. ∎

Theorem 1.4: *Let I be a closed interval, and let $f : I \to I$ be continuous. Then $(p_0, p_1, \cdots)$ is an endpoint of $\varprojlim f$ if and only if for each integer i, each interval J_i with $p_i \in \operatorname{int} J_i$, and each $\epsilon > 0$, there is a positive integer N such that if $p_{N+i} \in J_{N+i}$ and $f^N(J_{N+i}) = J_i$, then f^N is ϵ-crooked with respect to p_{N+i}.*

Proof: First, suppose that $\underline{p} = (p_0, p_1, p_2, \cdots)$ fails to be an endpoint of $\varprojlim f$. Then there are proper subcontinua A and B of $\varprojlim f$ such that $\underline{p} \in A \cap B$, $A - B \neq \emptyset$, and $B - A \neq \emptyset$. For each i, let $A_i = \pi_i(A)$ and $B_i = \pi_i(B)$. Let $J_i = A_i \cup B_i$. Since $p_i \in A_i \cap B_i$ we see that J_i is an interval. Let $J_i = [x_i, y_i]$.

Now, since $A - B \neq \emptyset$ and $B - A \neq \emptyset$, there is an integer i such that if $j \geq i$, then $A_j - B_j \neq \emptyset$, and $B_j - A_j \neq \emptyset$. Choose ϵ such that $[x_i, x_i + \epsilon]$ lies in one of $A_i - B_i$, $B_i - A_i$ and $[y_i - \epsilon, y_i]$ in the other. Since, for each N, $f^N(A_{i+N}) = A_i$, $f^N(B_{i+N}) = B_i$, and $p_{i+N} \in A_{i+N} \cap B_{i+N}$, we see that p_{i+N} separates $f^{-N}([x_i, x_i + \epsilon])$ and $f^{-N}([y_i - \epsilon, y_i])$. Thus for each N, $f^N : J_{N+i} \to J_i$ fails to be ϵ-crooked with respect to p_{i+N}. This establishes the first half of the theorem.

Next, suppose that the condition in the hypothesis fails. Then there is an integer i, an interval J_i with $p_i \in \operatorname{int} J_i$ and an $\epsilon > 0$ such that for each integer N, there is an interval J_{N+i} containing p_{N+i} such that $f^N(J_{i+N}) = J_i$ and $f^N : J_{i+N} \to J_i$ fails to be ϵ-crooked with respect to p_{i+N}.

It now follows from Lemma 1.3, that for each $K > i$, there are intervals A_K and B_K such that $p_K \in A_K \cap B_K$, $f(A_{K+1}) = A_K$, $f(B_{K+1}) = B_K$, $A_K - B_K \neq \emptyset$, and $B_K - A_K \neq \emptyset$. Now, for $0 \leq K \leq i$, define $A_K = f(A_{K+1})$ and $B_K = f(B_{K+1})$. Now, let $A = \{(x_0, x_1, \cdots) \in \varprojlim f \mid x_K \in A_K\}$ and $B = \{(x_0, x_1, \cdots) \in \varprojlim f \mid x_K \in B_K\}$. Then A and B are subcontinua of $\varprojlim f$, $(p_0, p_1, p_2, \cdots) \in A \cap B$, $A - B \neq \emptyset$ and $B - A \neq \emptyset$. Then $(p_0, p_1, p_2, \cdots)$ is not an endpoint of $\varprojlim f$. ∎

While the statement of this theorem seems somewhat unwieldy, it can be used with some utility after a bit of practice. We proceed by presenting some examples.

Example 1: If $f : [0, 1] \to [0, 1]$ and $f(0) = 0$, let $\underline{p} = (0, 0, 0, \cdots)$. Then since there is no interval J with $0 \in$ int J we may choose $N = 1$ for any ϵ and so $\underline{p}$ is an endpoint.

Example 2: See Figure 2. Let $\underline{p} = (\frac{1}{2}, \frac{1}{2}, \cdots)$. Suppose that $\frac{1}{2} \in$ int $[a, b]$. If $b < 1$ then there is no interval J with $\frac{1}{2} \in J$, and $f(J) = [a, b]$. If $b = 1$, $\frac{1}{2} \in J$, and $f(J) = [a, 1]$ then $\frac{1}{2}$ does not separate $f^{-1}(a)$ from $f^{-1}(1)$. So for any $\epsilon > 0$ we may choose $N = 1$ and we see that $\underline{p}$ is an endpoint.

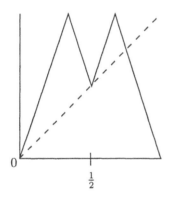
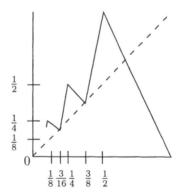

FIGURE 2.   FIGURE 3.

Example 3: See Figure 3. Let $\underline{p} = (\frac{1}{2}, \frac{1}{4}, \frac{1}{8}, \cdots)$. We will argue that $\underline{p}$ is an endpoint. We will check the condition for $\frac{1}{2}$. Checking on the other coordinates is analogous.

Suppose that $\frac{1}{2} \in$ int $[a, b]$ and that $a > 0$. If $a > \frac{3}{8}$, there is no interval J containing $\frac{1}{4}$ with $f(J) = [a, b]$. We may choose $N = 1$ for any ϵ. If $a > \frac{3}{16}$, there is no interval J containing $\frac{1}{8}$ with $f^2(J) = [a, b]$. We may choose $N = 2$ for any ϵ. Continue this argument using $N = j$ if $a > \frac{3}{2^{j+2}}$.

So we may assume $a = 0$. Now choose $\epsilon > 0$. If $\epsilon > \frac{3}{8}$ choose $N = 1$ and f is ϵ-crooked with respect to $\frac{1}{4}$. If $\epsilon > \frac{3}{16}$ choose $N = 2$ and f^2 is ϵ-crooked with respect to $\frac{1}{8}$. Continue this argument using $N = j$ if $\epsilon > \frac{3}{2^{j+2}}$.

As an exercise the reader can use similar arguments to show that $\underline{q} = (\frac{3}{8}, \frac{3}{8}, ...)$ is an endpoint.

Example 4: See Figure 4. Let $\underline{p} = (\frac{1}{2}, \frac{1}{2}, \cdots)$. We will argue that $\underline{p}$ is not an endpoint. Consider $\frac{1}{2} \in [\frac{1}{4}, \frac{3}{4}]$ and let $\epsilon = \frac{1}{12}$. Then if $\frac{1}{2} \in J$ and $f^N(J) = [\frac{1}{4}, \frac{3}{4}]$, $\frac{1}{2}$ separates $f^{-N}([\frac{1}{4}, \frac{1}{3}])$ from $f^{-N}([\frac{2}{3}, \frac{3}{4}])$. Thus f^N is not ϵ-crooked with respect to $\frac{1}{2}$.

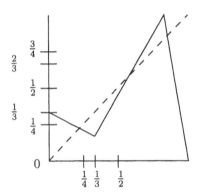

FIGURE 4.

Example 5: See Figure 4. Let $\underline{p} = (\frac{1}{4}, \frac{1}{4}, \cdots)$. Arguments similar to the ones previously given can be used to show that $\underline{p}$ is an endpoint.

2. ENDPOINTS AND DYNAMICS

Let $f : [0, 1] \to [0, 1]$ be continuous. The statement that f has *finitely many turning points* means that there is a partition $0 = t_0 < t_1 < \cdots < t_n = 1$ such that f is strictly monotone on each of $[t_i, t_{i+1}]$ and is increasing on $[t_i, t_{i+1}]$ if and only if it is decreasing on $[t_{i-1}, t_i]$. The set $T = \{t_0, t_1, \cdots t_n\}$ is called the set of turning points for f. Notice that if t is a turning point, $t \notin \{0, 1\}$, then there is an open interval U containing t such that $f(U)$ is a half open interval with endpoint $f(t)$.

In general if $f : [0, 1] \to [0, 1]$ is continuous, the statement that t is a *turning point* for f means that $t = 0$, $t = 1$, or there is an open interval U containing t such that $f(U)$ is a half open interval with endpoint $f(t)$.

Lemma 2.1: *Suppose that t is a turning point for f and $t \notin \{0, 1\}$. Then there is a positive δ such that if J and L are intervals, $f(t) \in \text{int } L$, $t \in J$, and $f(J) = L$, then diam $J > \delta$.*

Proof: This follows directly from the definition of turning point. ∎

Lemma 2.2: *Suppose that $f : [0,1] \to [0,1]$ is continuous and that f^2 has a dense orbit. Then given $\epsilon > 0$ and $\delta > 0$, there is an N such that if $K > N$, J is a subinterval and diam $J > \delta$, then $[\epsilon, 1 - \epsilon] \subset f^K(J)$.*

Proof: Suppose to the contrary that there is an $\epsilon > 0$, a $\delta > 0$, a sequence $J_1, J_2, \cdots$ of subintervals of $[0,1]$, and a sequence $K_1, K_2, \cdots$ of integers such that

(1) diam $(J_i) > \delta$
(2) $K_i > i$, and
(3) $f^{K_i}(J_i)$ does not contain $[\epsilon, 1 - \epsilon]$.

Then if $J_i = [a_i, b_i]$, there is a subsequence $i_1, i_2, \cdots$ and points a and b such that $a_{i_j} \to a$ and $b_{i_j} \to b$. Since diam $J_i > \delta$ it follows that $a < b$. Let $J = [a, b]$.

Now, since f^2 has a dense orbit, there is an integer N such that if $K > N$, then $[\frac{\epsilon}{2}, 1 - \frac{\epsilon}{2}] \subset f^K(J)$. (Theorem 6, [B-M2]). Thus, there is an M such that if $l > M$ and $K > N$, then $[\frac{\epsilon}{2}, 1 - \frac{\epsilon}{2}] \subset f^K[a_{i_l}, b_{i_l}]$. Now, choose $i_l > M$ and $i_l > N$. Then $f^{i_l}[a_{i_l}, b_{i_l}] \supset [\epsilon, 1 - \epsilon]$. But since $K_{i_l} > i_l$, we have $f^{K_{i_l}}(J_{i_l}) \supset [\epsilon, 1 - \epsilon]$. This is a contradiction and the result follows. ∎

Theorem 2.3: *Suppose that $f : [0,1] \to [0,1]$ is continuous, and that f has a dense orbit. Suppose that t is a turning point for f and that $f(t) = t$. Then $\underline{t} = (t, t, t, \cdots)$ is an endpoint of $\varprojlim f$.*

Proof: If $t = 0$ or $t = 1$ it is clear that $\underline{t}$ is an endpoint. We assume that $0 < t < 1$. In addition, we will first make the argument assuming that f^2 has a dense orbit.

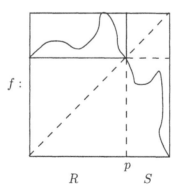

FIGURE 5.

Suppose then that H is a proper subcontinuum of $\varprojlim f$ containing $\underline{t}$. Let $H_n = \prod_n(H)$. Assume that $H_0 \neq [0,1]$ and further assume that for infinitely many values of K, $t \in$ int H_K. It follows from Lemma 2.1 that there is a $\delta > 0$ such that if $t \in$ int H_K, then diam $H_{K+1} > \delta$. Since H_0 is proper, there is an $\epsilon > 0$ such that either $[0, \epsilon] \cap H_0 = \emptyset$, or $[1 - \epsilon, 1] \cap H_0 = \emptyset$. We assume the former.

It follows from Lemma 2 of [B-M1] that there is an N such that if $K > N$ and diam $J > \delta$ then $[\epsilon, 1 - \epsilon] \subset f^K(J)$. Now there is an l such that $l > N$ and diam $H_l > \delta$. Thus $[\epsilon, 1 - \epsilon] \subset f^l(H_l) = H_0$. This contradicts the fact that $H_0 \cap [0, \epsilon] = \emptyset$.

Hence there is a j such that if $K > j$, then t is an endpoint of H_K. This implies that if A and B are proper subcontinua of $\varprojlim f$, both containing $\underline{t}$, then either $A \subset B$ or $B \subset A$. Hence $\underline{t}$ is an endpoint of $\varprojlim f$.

We now deal with the case in which f has a dense orbit, but f^2 does not. It follows from Lemma 2.2 that there are subintervals R and S of $[0, 1]$ such that

(1) $R \cup S = [0, 1]$,
(2) $R \cap S = \{p\}$,
(3) $f(R) = S$, and
(4) $f(S) = R$.

In this case p is the only fixed point for f, but $t \neq p$ since p is not a turning point for f. It follows that this case cannot occur. This establishes Theorem 2.3. ∎

Notice that Theorem 2.3 does not require that f have finitely many turning points.

Lemma 2.4: *Suppose f has finitely many turning points $T = \{t_0, t_1, \cdots, t_n\}$. Let T_n be the collection of turning points for f^n. Then T_n is finite, and $T_n = T \cup f^{-1}(T) \cup \cdots \cup f^{-(n-1)}(T)$. Consequently, $f^n(T_n) = f^n(T) \cup f^{n-1}(T) \cdots \cup f(T)$ and so $\bigcup_{n=1}^{\infty} f^n(T_n) = \bigcup_{K=1}^{\infty} f^K(T)$.*

Proof: Left to the reader. ∎

If $f : [0, 1] \to [0, 1]$ is continuous and $x \in [0, 1]$, then the statement that x is *eventually periodic* means that there is an $n > 0$ such that $f^n(x)$ is a periodic point of f.

Lemma 2.5: *Suppose that f has finitely many turning points and that each turning point is eventually periodic. Let T_n be the set of turning points for f^n. Then $\bigcup_{n=1}^{\infty} f^n(T_n)$ is finite.*

Proof: By Lemma 2.4 we have that $\bigcup_{n=1}^{\infty} f^n(T_n) = \bigcup_{K=1}^{\infty} f^K(T_1)$. Now T_1 is finite and since each turning point is eventually periodic it follows that if $t \in T_1$, then $\bigcup_{K=1}^{\infty} \{f^K(t)\}$ is finite. Hence $\bigcup_{K=1}^{\infty} f^K(T_1)$ is finite, and the conclusion follows. ∎

Theorem 2.6: *Suppose that f has finitely many turning points, and that each turning point is eventually periodic. Suppose that $(p_0, p_1, \cdots) \in \varprojlim f$ and there is* an i such that if $K > i$, p_K is not a turning point of f. Then $(p_0, p_1, \cdots)$ is not an endpoint of $\varprojlim f$.

Proof: Since $(p_0, p_1, \cdots)$ is an endpoint if and only if $(p_{i+1}, p_{i+2}, \cdots)$ is an endpoint, we may assume that no element of $\{p_0, p_1, \cdots\}$ is a turning point.

Using Lemma 2.5, there is an interval $J = [a, b]$ such that

(1) $p_0 \in$ int J and
(2) $J \cap [\cup_{k=1}^\infty f^K(T) - \{p_0\}] = \emptyset$.

This last condition means that no point of J different from p_0 is the image of a turning point of f^n.

Let $\epsilon < \min\{|a - p_0|, |b - p_0|\}$ and let $n > 0$ and consider f^n. p_n is not a turning point for f^n, for otherwise $f^n(p_n) = p_0$ would be a turning point for f.

Since $J \cap [f^n(T_n) - \{p_0\}] = \emptyset$ it follows that there is an interval L_n such that $p_n \in$ int L_n, $f^n(L_n) = J$, and f^n is monotone on L_n. that if Then p_n separates $f^{-n}([a, a + \epsilon])$ from $f^{-n}([b - \epsilon, b])$ in L_n and hence $f^n : L_n \to J$ is not ϵ-crooked with respect to p_n.

Thus, it follows from Theorem 1.4 that $(p_0, p_1, \cdots)$ is not an endpoint of $\varprojlim f$. ∎

Lemma 2.7: *Suppose that $g : [0, 1] \to [0, 1]$ is continuous. Suppose that R and S are subintervals of $[0, 1]$ such that*

(1) $R \cup S = [0, 1]$
(2) $R \cap S = \{p\}$
(3) $g(R) = R$, $g(S) = S$ *and*
(4) $g^2 : R \to R$ *has a dense orbit.*

Let t be a point of R such that $g(t) = t$ and t is a turning point for g. Then $\underline{t} = (t, t, t, \cdots)$ is an endpoint of $\varprojlim g$.

Proof: We will use Theorem 1.4. We must argue that if J is a subinterval of I, $t \in$ int J and $\epsilon > 0$, then there is an n such that if $t \in L$ and $g^n(L) = J$ then $g^n : L \to J$ is ϵ-crooked with respect to t.

Notice that $t \neq p$ since p is not a turning point for g. Assume the notation is chosen so that $0 \in R$.

Let J be a subinterval of $[0, 1]$ with $t \in$ int J and let $\epsilon > 0$. Let $J = [a, b]$. If $J \subset R$ we use the fact that $(t, t, \cdots)$ is an endpoint of $\varprojlim g|_R$ to conclude that there is an n such that if $\underline{g}^n(L) = J$ and $t \in J$ then $\underline{g}^n$ is ϵ-crooked with respect to t.

So we may assume that $J \cap$ int $S \neq \emptyset$. Now it follows that if $t \in L$ and $g^n(L) = J$ then $[t, p] \subset L$.

Suppose $a > 0$. Since g^2 has a dense orbit on R, there is an integer j and a point x, $t < x < p$ such that $g^j(x) < a$. This means that there is no interval L such that $t \in L$ and $g^n(L) = J$. This takes care of the case when $a > 0$.

So, we may assume that $J = [0, b]$ where $p < b$.

Now, as before, there is an integer K and a point x, $t < x < p$, such that $g^K(x) < \epsilon$. Also, there is a point y, $p < y$ such that $g^K(y) = b$. Then if $g^K(L) = J$ we have that $g^K : L \to J$ is ϵ-crooked with respect to t. Thus, $(t, t, \cdots)$. is an endpoint of $\varprojlim g$. ∎

Theorem 2.8: *Suppose that $f : [0, 1] \to [0, 1]$ is continuous. Suppose that f has a dense orbit, f has finitely many turning points and every turning point is eventually periodic.*

Then a point $(p_0, p_1, \cdots)$ is an endpoint of $\varprojlim f$ if and only if

(1) *there is an n such that, for each i, $p_i = p_{i+n}$, and*
(2) *at least one of p_0, $p_1, \cdots$ is a turning point.*

Proof: We remark that this theorem would be a fairly straightforward consequence of Theorems 2.3 and 2.6 were it not for the fact that f^2 might not have a dense orbit. This is where Lemma 2.7 is needed.

First, suppose that 1 and 2 hold. Let $g = f^n$. Then $(p_0, p_0, \cdots) \in \varprojlim g$ and p_0 is a turning point for g. Now if f^2 has a dense orbit then it follows from [B-M1], Lemma 2, that $g = f^n$ has a dense orbit and hence, by Theorem 2.3, $(p_0, p_0, \cdots)$ is an endpoint for $\varprojlim g$. Since $\varprojlim g$ and $\varprojlim f$ are homeomorphic with homeomorphism $(a_0, a_1, a_2, \cdots) \leftrightarrow (a_0, a_n, a_{2n}, \cdots)$ it follows that $(p_0, p_1, \cdots)$ is an endpoint of $\varprojlim f$.

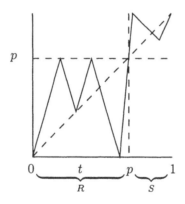

FIGURE 6.

Suppose that f^2 does not have a dense orbit. Then it follows from [B-M1], Lemma 2, that there are subintervals R and S of $[0, 1]$ such that

(1) $R \cup S = [0, 1]$
(2) $R \cap S = \{p\}$
(3) $f(R) = S$

(4) $f(S) = R$

and if $h = f^2$, then

(5) $h^2 : R \to R$ has a dense orbit and

(6) $h^2 : S \to S$ has a dense orbit.

Now, since $g(p_0) = p_0$, $p_0 \neq p$ it follows that $g = h^j$. Since h^2 has a dense orbit on each of R and S it follows from [B-M1], Lemma 2, that g^2 has a dense orbit on each of R and S. Hence Lemma 2.7 applies and we conclude that $(p_0, p_0, \cdots)$ is an endpoint for $\underleftarrow{\lim} g$. As before $(p_0, p_1, \cdots p_n, \cdots)$ is an endpoint for $\underleftarrow{\lim} f$.

Next, suppose that $(p_0, p_1, \cdots)$ is an endpoint of $\underleftarrow{\lim} f$. It follows from Theorem 2.6 that there is an increasing subsequence $K_1, K_2, \cdots$ such that for each i, p_{K_i} is a turning point for f. Since there are only finitely many turning points, it follows that there is a turning point t and an increasing subsequence $m_1, m_2, \cdots$ such that for each i, $p_{m_i} = t$. Hence t is periodic under f. Let n be the period of t.

Fix m_i. Then for each $j < m_i$, $f^n(p_j) = p_j$ or $p_{j-n} = p_j$. So for each $K \geq m_i - n$, $p_K = p_{K+n}$. Since this is true for each m_i it follows that for each j, $p_j = p_{j+n}$. Furthermore, one of $p_0, p_1, \cdots p_n$ is t and hence 2 holds. ∎

Let I be the unit interval $[0, 1]$. If $f : I \to I$ and $x \in I$, x is *recurrent* under f provided there is a sequence of positive integers $n_i \to \infty$ such that $f^{n_i}(x) \to x$.

Theorem 2.9: *Suppose that $f : I \to I$ is continuous and that either 0 or 1 is recurrent under f. Then $\underleftarrow{\lim} f$ has at least one endpoint.*

Proof: Suppose that 0 is recurrent under f (the argument for 1 is the same). We will construct an endpoint in $\underleftarrow{\lim} f$. For $t \in I$ and $\delta > 0$, we use the notation $B_\delta(t) = \{s \in I : |s - t| < \delta\}$. Let $n_1 = 1$ and $\epsilon_1 = \frac{1}{2}$. Let n_2 be such that $f^{n_2}(0) \in [0, \epsilon_1)$ and $f^{n_1 + n_2}(0) \in B_{\frac{1}{4}}(f^{n_1}(0))$. Let $\epsilon_2 > 0$ be small enough so that $f^{n_2}([0, \epsilon_2]) \subset [0, \epsilon_1)$ and $f^{n_1 + n_2}([0, \epsilon_2]) \subset B_{\frac{1}{4}}(f^{n_1}(0))$.

Proceeding inductively, assume that positive integers $n_1, n_2, \cdots, n_k$ and positive numbers $\epsilon_1, \epsilon_2, \cdots, \epsilon_k$ have been defined. Let n_{k+1} be such that:

$$f^{n_{k+1}}(0) \in [0, \epsilon_k),$$

$$f^{n_k + n_{k+1}}(0) \in [0, \epsilon_{k-1}) \cap B_{\frac{1}{2^{k+1}}}(f^{n_k}(0)),$$

$$f^{n_{k-1} + n_k + n_{k+1}}(0) \in [0, \epsilon_{k-2}) \cap B_{\frac{1}{2^{k+1}}}(f^{n_{k-1} + n_k}(0)),$$

$$\vdots$$

$$f^{n_1 + \cdots + n_{k+1}}(0) \in B_{\frac{1}{2^{k+1}}}(f^{n_1 + \cdots + n_k}(0)).$$

Let $\epsilon_{k+1} > 0$ be small enough so that

$$f^{n_{k+1}}([0, \epsilon_{k+1}]) \subset [0, \epsilon_k),$$

$$f^{n_k + n_{k+1}}([0, \epsilon_{k+1}]) \subset [0, \epsilon_{k-1}) \cap B_{\frac{1}{2^{k+1}}}(f^{n_k}(0)),$$

$$\vdots$$

$$f^{n_1 + n_2 + \cdots + n_{k+1}}([0, \epsilon_{k+1}]) \subset B_{\frac{1}{2^{k+1}}}(f^{n_1 + \cdots + n_k}(0)).$$

Let $N_k = n_1 + n_2 + \cdots + n_k$ and, for each $k \geq 1$, let x_{N_k} be the limit of the Cauchy sequence $\{f^{N_i - N_k}(0) | i = k, k+1, \cdots\}$. It is clear from the construction that $x_{N_k} \in [0, \epsilon_k]$, $f^{N_i - N_k}(x_{N_i}) = x_{N_k}$, and $f^{N_i - N_k}([0, \epsilon_i]) \subset B_{\frac{1}{2^{i-1}}}(x_{N_k})$ for $k = 1, 2, \cdots$ and $i \geq k$. Let $\underline{x} \in \varprojlim f$ be the point defined by $\pi_{N_k}(\underline{x}) = x_{N_k}$ for $k = 1, 2, \cdots$. We claim that $\underline{x}$ is an endpoint of $\varprojlim f$. Indeed, suppose that $J_{N_k} = [a, b]$ is a subinterval of I with $x_{N_k} \in \overset{o}{J}_{N_k}$. Let $\epsilon > 0$ be given; we may as well suppose that ϵ is small enough so that $x_{N_k} \in (a + \epsilon, b - \epsilon)$. There is then an $i > k$ large enough so that $B_{\frac{1}{2^{i-1}}}(x_{N_k}) \subset (a + \epsilon, b - \epsilon)$. Suppose that J_{N_i} is an interval with $f^{N_i - N_k}(J_{N_i}) = J_{N_k}$ and $x_{N_i} \in J_{N_i}$. Now $f^{N_i - N_k}([0, \epsilon_i]) \subset B_{\frac{1}{2^{i-1}}}(x_{N_k}) \subset (a - \epsilon, b + \epsilon)$. Since $x_{N_i} \in [0, \epsilon_i]$, we see that x_{N_i} does not separate $f^{N_k - N_i}([a, a + \epsilon))$ from $f^{N_k - N_i}((b - \epsilon, b])$ in J_{N_i}. That is, $f^{N_i - N_k}|_{J_{N_i}}$ is ϵ-crooked with respect to x_{N_i}. By Theorem 1.4, $\underline{x}$ is an endpoint of $\varprojlim f$.   ∎

Lemma 2.10: *Suppose that the continuous map $f : I \to I$ has a dense orbit. Then*

(1) *there is an $\eta > 0$ so that for every nontrivial subinterval J of I there is an $N = N(J)$ with diam $(f^n(J)) \geq \eta$ for all $n \geq N$, and*

(2) *for each point $\underline{x}$ of the inverse limit space $\varprojlim f$ there is a nondegenerate subcontinuum A of $\varprojlim f$ such that $\underline{x} \in A$ and diam $(\pi_n(A)) \to 0$ as $n \to \infty$.*

Proof: If f^2 has a dense orbit this follows easily from Lemma 2.2. If f has a dense orbit but f^2 does not, then there are subintervals I_1, I_2 of I with $I_1 \cap I_2 = \{p\}$, p fixed, $I_1 \cup I_2 = I$, such that $f^2|_{I_i}$ and $f^4|_{I_i}$ have dense orbits for $i = 1, 2$ (see [B-M1], Lemma 2). The lemma at hand then follows by applying Lemma 2.2 to $f^2|_{I_i}$, $i = 1, 2$.   ∎

Given a continuous map $f : I \to I$ and a point $x \in I$, the ω-*limit set* of x is $\omega(x) = \bigcap_{N \geq 0} \mathrm{cl}\{f^n(x) | n \geq N\}$. If S is a subset of I, let $\omega(S) = \bigcup_{x \in S} \omega(x)$. Note that $\omega(S)$ is invariant $(f(\omega(S)) = \omega(S))$ and closed if S is finite.

Theorem 2.11: *Suppose the continuous map $f : I \to I$ has finitely many turning points and a dense orbit. In addition, suppose that the set T of turning points of f*

(0 and 1 are in T) satisfies $\omega(T) \cap T = \emptyset$. Then the inverse limit space $\varprojlim f$ does not have an endpoint.

Proof: Let $\underline{x} = (x_0, x_1, \cdots)$ be a point of $\varprojlim f$. We will show that $\underline{x}$ is contained in an arc in $\varprojlim f$ of which it is not an endpoint. We consider two cases.

Case 1: There is a $\delta > 0$ and an N such that $d(x_n, T) > \delta$ for all $n \geq N$.

In this case, let A be a nondegenerate subcontinuum of $\varprojlim f$ with $\underline{x} \in A$ and diam $(\pi_n(A)) \to 0$ as $n \to \infty$. Let η be as in the preceding Lemma 2.10 and let M be such that diam $(\pi_n(A)) < \delta' = \frac{1}{2}\min(\delta, \eta)$ for all $n \geq M$. Then $\pi_n(A) \cap T = \emptyset$ for all $n \geq K = \max(M, N)$ so that $f : \pi_{n+1}(A) \to \pi_n(A)$ is monotone for all $n \geq K$ and A is an arc. If x_n is not an endpoint of $\pi_n(A)$ for some $n \geq K$ then $\underline{x}$ is not an endpoint of A and hence not an endpoint of $\varprojlim f$.

Suppose then that $\pi_n(A) = \langle x_n, a_n \rangle$, the convex hull of $\{x_n, a_n\}$, for all $n \geq K$. We will fatten A to a larger arc of which $\underline{x}$ is not an endpoint. For each $n \geq K$, let L_n be the closed interval of length δ' such that $L_n \cap \pi_n(A) = \{x_n\}$ (note that $L_n \subset I$ since $d(x_n, T) > \delta'$ and $\{0, 1\} \subset T$). Since $T \cap L_n = \emptyset$ we see that either $f(L_{n+1}) \subset L_n$ or $f(L_{n+1}) \supset L_n$ and that $f(L_{n+1}) \cap \pi_n(A) = \{x_n\}$ for each $n \geq K$.

Choose a monotone sequence $\{s_i\} \subset L_K \backslash \{x_K\}$ with $s_i \to x_K$ as $i \to \infty$. For each i let $k(i) = \sup\{k|$ there is an $s_i^k \in L_{K+k}$ with $f^k(s_i^k) = s_i$ and $f^l(\langle s_i^k, x_{K+k} \rangle) \subset L_{K+k-l}$ for $l = \{0, 1, \cdots, k\}\}$. If, for some i, $k(i) = \infty$, let $B_{K+k} = \langle s_i^k, a_{K+k} \rangle$. Then $B = \{(y_0, y_1, \cdots)|y_{K+k} \in B_{K+k}$ for all $k \geq 0\}$ is an arc containing $\underline{x}$, of which $\underline{x}$ is not an endpoint. Suppose then that $k(i) < \infty$ for all i (the proof in case 1 will be complete when we show that this is not possible). From the continuity of f, $k(i) \to \infty$ as $i \to \infty$. Let $\alpha > 0$ be such that diam $(f(L_n)) \geq \alpha$ for all $n \geq K$ and let $s_i^{k(i)} \in L_{K+k(i)}$ be such that $f^{k(i)}(s_i^{k(i)}) = s_i$ and $f^l(\langle s_i^{k(i)}, x_{K+k(i)} \rangle) \subset L_{K+k(i)-l}$ for $l = 0, 1, \cdots, k(i)$ and each i.

From the definition of $k(i)$, $s_i^{k(i)} \notin f(L_{K+k(i)+1})$ so that diam $(\langle s_i^{k(i)}, x_{K+k(i)} \rangle) \geq \alpha$. Now choose a subsequence $\{i_j\}$ such that $s_{i_j}^{k(i_j)} \to s$ and $x_{K+k(i_j)} \to x$, some s and x, as $j \to \infty$. Then diam $(\langle s, x \rangle) \geq \alpha$ so that for a sufficiently large n, diam $(f^n(\langle s, x \rangle)) \geq \eta > \delta'$ (as in the preceding Lemma 2.10). But then for all sufficiently large j, diam $(f^n(\langle s_{i_j}^{k(i_j)}, x_{K+k(i_j)} \rangle)) > \delta'$, which is not possible since $f^n(\langle s_{i_j}^{k(i_j)}, x_{K+k(i_j)} \rangle) \subset L_{K+k(i_j)-n}$ (for j such that $k(i_j) > n$) and diam $(L_{K+k(i_j)-n}) \leq \delta'$.

Case 2: For each $\epsilon > 0$ there is an n such that $d(x_n, T) < \epsilon$.

Since T is finite and $\omega(T) \cap T = \emptyset$, there is an $\epsilon > 0$ and an N such that $|f^n(s) - t| > \epsilon$ for all $s, t \in T$. Let M be such that $d(x_M, T) < \epsilon/2$. Then $x_{M+N} \notin T$. Let $\delta > 0$ be small enough so that, letting $J_{M+N} = [x_{M+N} - \delta, x_{M+N} + \delta]$,

$f^N(J_{M+N}) \subset [x_M - \frac{\epsilon}{2}, x_M + \frac{\epsilon}{2}]$. Then $f^m(T) \cap J_{M+N} = \emptyset$ for all $m \geq 0$ so that each component of $f^{-m}(J_{M+N})$ is mapped monotonically onto J_{M+N} by f^m for all $m \geq 0$. For each $m \geq 0$ let J_{M+N+m} be the component of $f^{-m}(J_{M+N})$ that contains x_{M+N+m}. Then $f|_{J_{n+1}} : J_{n+1} \to J_n$ is a monotone surjection for each $n \geq M + N$ so that $J = \{(y_0, y_1, \cdots) \in \varprojlim f : y_n \in J_n \text{ for all } n \geq M + N\}$ is an arc. Clearly $\underline{x}$ is not an endpoint of J so $\underline{x}$ is not an endpoint of $\varprojlim f$. ∎

3. ENDPOINTS IN THE TENT FAMILY

As an application of Theorems 2.9 and 2.11 we consider the family of maps $g_s : I \to I$, $1 \leq s \leq 2$, given by

$$g_s(x) = \begin{cases} sx + 2 - s, & 0 \leq x \leq \frac{s-1}{s} \\ -sx + s, & \frac{s-1}{s} \leq x \leq 1 \end{cases}.$$

Let $c_s = \frac{s-1}{s}$ be the critical point of g_s.

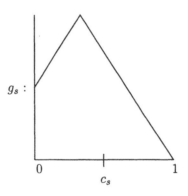

FIGURE 7.

Theorem 3.1: *For $1 \leq s \leq 2$, $\varprojlim g_s$ has an endpoint if and only if 0 is recurrent under g_s.*

Towards the proof of Theorem 3.1 we first establish the following lemma.

Lemma 3.2: *For $1 < s < \sqrt{2}$*

    (1) *$\varprojlim g_s$ has an endpoint if and only if $\varprojlim g_{s^2}$ has an endpoint and*
    (2) *0 is recurrent under g_s if and only if 0 is recurrent under g_{s^2}.*

Proof: (Part 1.) Let $p_s = \frac{s}{1+s}$ denote the unique fixed point of g_s ($1 < s < \sqrt{2}$) and note that $g_s(0) = 2 - s > p_s$. Thus, for s in this range, $g_s^2([0, p_s]) = [0, p_s]$ and

$g_s^2([p_s, 1]) = [p_s, 1]$. Moreover, $0 < c_s < g_s^2(0) < p_s < g_s^2(1) < 1$, and the intervals $[0, g_s^2(0)]$ and $[g_s^2(1), 1]$ are invariant under g_s^2.

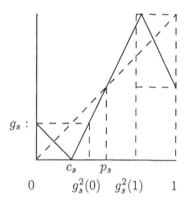

$$g_s :$$

$$0 \qquad c_s \qquad p_s \qquad 1$$
$$g_s^2(0) \quad g_s^2(1)$$

FIGURE 8.

It is easily checked that $g_s^2|_{[0, g_s^2(0)]}$ and $g_s^2|_{[g_s^2(1), 1]}$ are topologically conjugate (in fact, by affine homeomorphisms) with g_{s^2} on $[0, 1] = I$. It follows that the continua $H = \varprojlim g_s^2|_{[0, g_s^2(0)]}$, $K = \varprojlim g_s^2|_{[g_s^2(1), 1]}$, and $\varprojlim g_{s^2}$ are all homeomorphic. Suppose now that A is a subcontinuum of $\varprojlim g_s^2$ that contains a point $\underline{x} \in K$ and a point $\underline{y} \in \varprojlim g_s^2 \backslash K$. There is then an integer N such that $y_n = \pi_n(\underline{y}) \notin [g_s^2(1), 1]$ for all $n \geq N$. Let q_s denote the fixed point of $g_s^2|_{[g_s^2(1), 1]}$. Since $s^2 < 2$, there is a $k > 0$ such that $g_s^{2k}(g_s^2(1)) \geq q_s$.

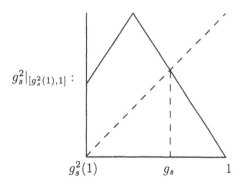

$$g_s^2|_{[g_s^2(1), 1]} :$$

$$g_s^2(1) \qquad\qquad g_s \qquad\qquad 1$$

FIGURE 9.

Now $y_n < g_s^2(1) \leq x_n$ for all $n \geq N$ so that $g_s^2(1) \in \pi_n(A)$ for all $n \geq N$ and thus $g_s^{2k}(1) \in \pi_n(A)$ for all $n \geq N$. From this we see that $[g_s^2(1), q_s] \subset \pi_n(A)$ for all $n \geq N$. Since $g_s^2([g_s^2(1), q_s]) \supset [q_s, 1]$, $[g_s^2(1), 1] \subset \pi_n(A)$ for all $n \geq N$ so that

$[g_s^2(1),1] \subset \pi_n(A)$ for all n and thus $K \subset A$. That is, any subcontinuum of $\varprojlim g_s^2$ that contains a point of K and a point not in K must contain K. In the same way, any subcontinuum of $\varprojlim g_s^2$ that meets both H and $\varprojlim g_s^2 \backslash H$ must contain H.

Now suppose that $\varprojlim g_s^2$ has an endpoint. Then K also contains an endpoint $(K \simeq \varprojlim g_{s^2})$. Let $\underline{x}$ be an endpoint of K and suppose that A and B are subcontinua of $\varprojlim g_s^2$ with $\underline{x} \in A \cap B$.

If A and B are both contained in K then $A \subset B$ or $B \subset A$ since $\underline{x}$ is an endpoint of K. If one of A or B, say A, is not contained in K, and B is, then $A \supset K \supset B$. If one of A or B, say A, contains a point of H then $A = \varprojlim g_s^2 \supset B$.

Finally, if $A \cap B \supset K$, $(A \cup B) \cap H = \emptyset$, and each of A and B contains a point of $\varprojlim g_s^2 \backslash (H \cup K)$, then there is an N such that $\pi_n(A) = [a_n,1]$, $\pi_n(B) = [b_n,1]$, and $\{a_n,b_n\} \subset [g_s^2(0),g_s^2(1)]$ for all $n \geq N$. Since $g_s^2|_{[g_s^2(0),g_s^2(1)]}$ is monotone increasing we see that $g_s^2(a_{n+1}) = a_n$ and $g_s^2(b_{n+1}) = b_n$ for all $n \geq N$. Now, if $a_N \leq b_N$ then $a_n \leq b_n$ for all $n \geq N$ so that $A \supset B$ and if $b_N \leq a_N$ then $B \supset A$. Thus $\underline{x}$ is an endpoint of $\varprojlim g_s^2$ which is homeomorphic with $\varprojlim g_s$ and one direction of part 1 has been proved.

For the other direction, suppose that $\varprojlim g_s$ has an endpoint. Then $\varprojlim g_s^2$, being homeomorphic with $\varprojlim g_s$, also has an endpoint, say $\underline{x}$. If $\underline{x} \in H$ or $\underline{x} \in K$ then $\varprojlim g_{s^2}$ has an endpoint as H, K, and $\varprojlim g_{s^2}$ are all homeomorphic. We will show that for $\underline{x}$ to be an endpoint of $\varprojlim g_s^2$, $\underline{x}$ must lie in $H \cup K$ and the proof of part 1 will be complete.

If $\underline{x} \notin H \cup K$ then there is an N so that $g_s^2(0) < \pi_n(\underline{x}) = x_n < g_s^2(1)$ for all $n \geq N$. Let a_N and b_N be such that $g_s^2(0) < a_N < x_N < b_N < g_s^2(1)$ and let $a_{N+i} = g_s^{-2i}(a_N)$, $b_{N+i} = g_s^{-2i}(b_N)$ for $i = 1,2,\cdots$ (the choices of inverse image are unique). Let $A_n = [a_n,x_n]$ and $B_n = [x_n,b_n]$ for $n \geq N$. Then $g_s^2(A_{n+1}) = A_n$ and $g_s^2(B_{n+1}) = B_n$ for all $n \geq N$ so that the subcontinua $A = \{(y_0,y_1,\cdots)|y_n \in A_n$ for $n \geq N\}$ and $B = \{(y_0,y_1,\cdots)|y_n \in B_n$ for $n \geq N\}$ of $\varprojlim g_s^2$ satisfy $\underline{x} \in A \cap B$, $A \not\subset B$, and $B \not\subset A$. Thus $\underline{x}$ is not an endpoint of $\varprojlim g_s^2$.

For the proof of *part 2*, suppose first that 0 is recurrent under g_s. Then 0 is recurrent under g_{s^2} (since $g_s^{2n+1}(0) \in [g_s^2(1),1]$ for $1 < s < \sqrt{2}$ and all $n \geq 0$). This implies that 1 is recurrent under g_s^2 since $g_s^2|_{[0,g_s^2(0)]}$ is topologically conjugate with g_s^2 on $[0,1]$ by a conjugacy that takes 0 to 1 (if $h : [0,g_s^2(0)] \rightarrow [0,1]$ is given by $h(x) = 1 - \frac{x}{g_s^2(0)}$ then $h \circ g_s^2|_{[0,g_s^2(0)]} = g_{s^2} \circ h$). Let $n_i \rightarrow \infty$ be a sequence such that $g_{s^2}^{n_i}(1) \rightarrow 1$. Then $g_{s^2}^{n_i+1}(1) \rightarrow g_{s^2}(1) = 0$, that is, $g_{s^2}^{n_i}(0) \rightarrow 0$ as $i \rightarrow \infty$ so that 0 is recurrent under g_{s^2}.

Conversely, suppose that 0 is recurrent under g_{s^2}. Then, as above, $g_s^2(0)$ is recurrent under $g_s^2|_{[0,g_s^2(0)]}$. Let $n_i \rightarrow \infty$ be such that $g_s^{2n_i}(g_s^2(0)) \rightarrow g_s^2(0)$. Then $g_s^2(g_s^{2n_i}(0)) \rightarrow g_s^2(0)$. Now for $1 < s < \sqrt{2}$ there is an $\epsilon > 0$ (namely, $\epsilon = g_s^2(0) -$

$g_s^4(0))$ such that if $|g_s^2(x) - g_s^2(0)| < \epsilon$ and $x \in [0, g_s^2(0)]$ then $x < \frac{\epsilon}{s^2}$. Thus $g_s^{2n_i}(0) \to 0$ as $i \to \infty$ and 0 is recurrent under g_s. ∎

The following is a fairly well-known fact.

Lemma 3.3: *For $\sqrt{2} \le s \le 2$, g_s has a dense orbit.*

Proof: Suppose first that $\sqrt{2} < s \le 2$ so that $s^2 > 2$. In this range, g_s^2 has either two or three interior critical points and in either case, if J is an interval containing two distinct interior critical points of g_s^2 then $g_s^2(J) = [0,1]$. Let U be a nonempty open interval in $[0,1]$. If U contains no more than one interior critical point of g_s^2 then $\text{length}(g_s^2(U)) \ge \frac{s^2}{2}\text{length}(U)$. Since $\text{length}(g_s^{2n}(U))$ is bounded, it must be that, for some $n \ge 0$, $g_s^{2n}(U)$ contains two interior critical points of g_s^2 so that $g_s^{2n+2}(U) = [0,1]$. Thus, for each nonempty open set U in $[0,1]$, there is an $n = n(U)$ such that $g_s^n(U) = [0,1]$. It follows that $\bigcup_{n\ge 0} g_s^{-n}(U)$ is dense in $[0,1]$ for each nonempty open U in $[0,1]$ so that (by [A-Y], Lemma 3, for example) g_s has a dense orbit.

For $s = \sqrt{2}$, let $p_{\sqrt{2}}(0) = \frac{\sqrt{2}}{1+\sqrt{2}}$ be the fixed point of $g_{\sqrt{2}}$. Then $g_{\sqrt{2}}^2|_{[0,p_{\sqrt{2}}]}$ is topologically conjugate with g_2 on $[0,1]$ (here $p_{\sqrt{2}} = g_{\sqrt{2}}^2(0)$ — see the proof of the preceeding lemma) so that $g_{\sqrt{2}}^2|_{[0,p_{\sqrt{2}}]}$ has a dense orbit, say $\{g_{\sqrt{2}}^{2n}(x)|n = 0,1,2\cdots\}$. But then $g_{\sqrt{2}}\{g_{\sqrt{2}}^{2n}(x)|n = 0,1,2\cdots\} = \{g_{\sqrt{2}}^{2n+1}(x)|n = 0,1,2\cdots\}$ is dense in $g_{\sqrt{2}}([0,p_{\sqrt{2}}]) = [p_{\sqrt{2}},1]$. That is, $\{g_{\sqrt{2}}^n(x)|n = 0,1,2\cdots\}$ is dense in $[0,1]$. ∎

Proof of Theorem 3.1: For $s = 1$, 0 is recurrent under g_1 and $\varprojlim g_1$ is an arc with endpoints $(0,1,0,1,\cdots)$ and $(1,0,1,0,\cdots)$. For $1 < s < \sqrt{2}$, $s^n \in [\sqrt{2},2]$ for some $n \ge 2$. Thus, by lemma 3.2, the problem is reduced to considering $\sqrt{2} \le s \le 2$. Now for any s, $1 < s \le 2$, the set of turning points of g_s is $\{0, c_s, 1\} = T_s$ and $g_s^2(c_s) = g_s(1) = 0$. Since $\omega(T_s)$ is invariant under g_s it follows that 0 is recurrent under g_s if and only if $\omega(T_s) \cap T_s \ne \emptyset$. For $\sqrt{2} \le s \le 2$, g_s has a dense orbit and T_s is finite so that, using Theorem 2.11, if 0 is not recurrent then $\omega(T_s) \cup T_s = \emptyset$ and $\varprojlim g_s$ has no endpoints. On the other hand, if 0 is recurrent then, by Theorem 2.10, $\varprojlim g_s$ has an endpoint. ∎

Note: According to [B-D-O-T], each of the following sets of parameter values is dense in $[1,2]$:
$$P = \{s : 0 \text{ is periodic under } g_s\};$$
$$R = \{s : 0 \text{ is recurrent but not periodic under } g_s\};$$
$$N = \{s : 0 \text{ is not recurrent under } g_s\}.$$

In case $s \in P$, $\varprojlim g_s$ has a nonempty and finite set of endpoints (equal in number to the period of 0). For $s \in R$, $\varprojlim g_s$ has infinitely many endpoints and for $s \in N$, $\varprojlim g_s$ has none. We see that the topology of $\varprojlim g_s$ is quite sensitive to s.

REFERENCES

[A-Y]   J. Auslander and J. Yorke. *Interval maps, factors of maps, and chaos*, Tohoku J. (2) 32 (1980) 177-188.

[B-D]   M. Barge and B. Diamond, *Homeomorphisms of inverse limit spaces of one-dimensional maps*, preprint.

[B-D-O-T] K.M. Brucks, B. Diamond, M.V. Otero-Espinar, and C. Tresser, *Dense orbits of critical points for the tent map*, Contemporary Math. 117 (1991) 57-61.

[B-M1]   M. Barge and J. Martin, *Chaos, periodicity, and snakelike continua*, Trans. AMS. (1) 289 (1985) 355-365.

[B-M2]   M. Barge and J. Martin, *Dense orbit on the interval*, Michigan Math.J. 34(1987) 3-11.

[D]   W. Dębski, *On topological types of the simplest indecomposable continua*, Colloq. Math. XLIX (1985), 203-211.

[W]   W.T. Watkins, *Homeomorphic classification of certain inverse limit spaces with open bonding maps*, Pacific J. Math. 103 (1982), 589-601.

E-mail address: barge@math.montana.edu "Marcy Barge"

11

THE ESSENTIAL SPAN OF A CLOSED CURVE

MARCY BARGE Department of Mathematics, Montana State University, Bozeman, Montana 59717, USA

RICHARD SWANSON Department of Mathematics, Montana State University, Bozeman, Montana 59717, USA

ABSTRACT: We prove in this article that for a plane topological annulus, the essential span of the inner boundary is less than the essential span of the outer boundary. In addition we provide various geometrical and dual characterizations of the essential span of a closed planar curve.

1. INTRODUCTION

Two iceskaters glide around the outside of an irregular frozen pond in the same direction. How *large* can their distance of closest approach be, if each skater makes a complete circuit of the pond in one minute? This metric invariant is called the "essential span" of the closed curve forming the perimeter. Suppose now that in the middle of the pond is an island, and that two other skaters perform the same experiment passing around the inner boundary of ice next to the shoreline of the island. They might be faster or slower than the first pair of skaters, but they still circle the island in one minute. Common sense might suggest that the essential span measured by the second (inner) pair of skaters is smaller than that recorded by the initial (outer) pair.

The original notion of span is due to Lelek [L1]. In that formulation (restricted to simple closed curves as above), the two skaters are required to make *at least* one full circuit (rather than *exactly* a full circuit) of the pond. For this formulation, Cook [C-I-L] (Problem 173 in the Houston Problem Book, cf. the present volume),

1991 *Mathematics Subject Classification.* primary 54F15.

Key words and phrases. span of a closed curve.

Research of both authors was supported in part by NSF-OSR grant #9350546 has asked whether the span of the perimeter of the island is less than or equal to the span of the outer shoreline of the pond. A host of other notions exist, including the surjective span, the semispan (Lelek [L2]), and the symmetric span, which we will compare in Section 4.

In the next section we will introduce and investigate another metric invariant called the "dual essential span". Returning for a moment to the iceskater illustration, we now require the two skaters to move in opposing directions about the pond, and we ask how *small* their maximal separation can be. Readers can verify for themselves that in simple examples (such as an ellipse) the essential span and dual essential span are equal (to the length of the minor axis).

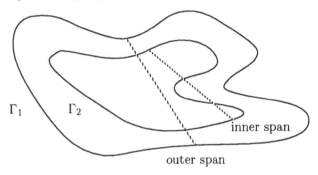

FIGURE 1. The Essential Span of Annulus Boundaries.

In the next section, we shall establish the equivalence of essential span and dual essential span, generally. This equality will be crucial for the proof of the main result (Theorem 3.2): that the inner essential span is less than the outer essential span.

Our argument will also depend on the celebrated "Mountain Climbers Theorem" . The latter asserts that two climbers can ascend opposite sides of a mountain, having no flat sections, in such a way that they maintain the same height all the way to the summit. Numerous backtrackings may be required. The version we need (Lemma 3.1) was proved by Mioduszewski [Mi] (and see also Keleti ([K])).

2. THE ESSENTIAL SPAN EQUALS THE DUAL ESSENTIAL SPAN

2.1. The Definition of the Essential Span.

Suppose that $f, g : \mathbb{S}^1 \to \Gamma$ are degree one maps, where Γ is an embedded circle in the plane and $\mathbb{S}^1$ is the unit circle. Then the *essential span* of Γ is the nonnegative number

$$\sigma_e(\Gamma) = \sup_{f,g} \inf_{\theta \in \mathbb{S}^1} \|f(\theta) - g(\theta)\|.$$

Observe that the essential span is never larger than the diameter of Γ.

Essential span has a useful continuity property. Suppose that $\mathcal{E}_1$ denotes the collection of all degree one continuous embeddings of $\mathbb{S}^1$ into $\mathbb{R}^2$ endowed with the uniform metric, $\sup \{\|f(\theta) - g(\theta)\| : \theta \in \mathbb{S}^1\}$

Proposition 2.1. *Let* $\sigma_e : \mathcal{E}_1 \to \mathbb{R}$ *where* $\sigma_e(\gamma) = \sigma_e(\gamma(\mathbb{S}^1))$. *Then* σ_e *is continuous. In fact,* σ_e *is Lipschitz with constant 2.*

Proof. Let $\gamma_1, \gamma_2 \in \mathcal{E}_1$, let $\epsilon > 0$ be given, and suppose that $f_1, g_1 : [0,1] \to \gamma_1(\mathbb{S}^1) = \Gamma_1$ are maps such that $\|f_1(t) - g_1(t)\| \geq \sigma_e(\Gamma_1) - \epsilon$, for all $t \in [0,1]$. Define $f_2 = \gamma_2 \circ \gamma_1^{-1} \circ f_1$ and $g_2 = \gamma_2 \circ \gamma_1^{-1} \circ g_1$.

Then

$$\|f_2(t) - g_2(t)\| = \|\gamma_2 \circ \gamma_1^{-1} \circ f_1(t) - \gamma_2 \circ \gamma_1^{-1} \circ g_1(t)\|$$
$$= \|\gamma_2 \circ \gamma_1^{-1} \circ f_1(t) + f_1(t) - \gamma_1 \circ \gamma_1^{-1} \circ f_1(t) - g_1(t) + \gamma_2 \circ \gamma_2^{-1} \circ g_1(t) - \gamma_2 \circ \gamma_1^{-1} \circ g_1(t)\|$$
$$\geq \|f_1(t) - g_1(t)\| - 2\,\mathrm{d}(\gamma_1, \gamma_2) \geq \sigma_e(\gamma_1) - \epsilon - 2\,\mathrm{d}(\gamma_1, \gamma_2).$$

Thus, $\sigma_e(\gamma_2) \geq \sigma_e(\gamma_1) - 2\,\mathrm{d}(\gamma_1, \gamma_2)$, and similarly $\sigma_e(\gamma_1) \geq \sigma_e(\gamma_2) - 2\,\mathrm{d}(\gamma_1, \gamma_2)$.

So $|\sigma_e(\gamma_1) - \sigma_e(\gamma_2)| \leq 2\,\mathrm{d}(\gamma_1, \gamma_2)$.

∎

2.2. Definition of the Dual Essential Span. Suppose that $f : \mathbb{S}^1 \to \Gamma$ and $g : \mathbb{S}^1 \to \Gamma$ are parameterizations of the closed curve Γ of degrees 1 and -1, respectively. Now define the *dual essential span* to be the nonnegative real number given by

$$\sigma_e^*(\Gamma) = \inf_{f,g} \sup_{\theta \in \mathbb{S}^1} \|f(\theta) - g(\theta)\|.$$

For $\sigma_e^*(\Gamma)$, it will be convenient to have, in addition, a lifted version of the definition on hand. Suppose $h : [0,1] \to \Gamma$ and $\pi : \mathbb{R} \to \Gamma$ is a covering projection (of period one). Then define

$$\deg(h) = \tilde{h}(1) - \tilde{h}(0),$$

where $\tilde{h} : [0,1] \to \mathbb{R}$ is a lift of h. If $f, g : [0,1] \to \Gamma$, with $f(1) = g(1)$, define the composite map (the homotopy product fg^{-1})

$$f * \overline{g}(t) = \begin{cases} f(2t), & 0 \leq t \leq 1/2 \\ g(2 - 2t), & 1/2 \leq t \leq 1 \end{cases}$$

Then we have the following

Proposition 2.2. *1) Suppose that $h, k : [0,1] \to \Gamma$ denote arbitrary maps having degrees 1 and -1, respectively, such that $h(0) = k(0)$. Then the dual essential span is given by*

$$\sigma_e^*(\Gamma) = \inf_{h,k} \sup_{0 \leq t \leq 1} \|h(t) - k(t)\|.$$

2) Suppose that $f, g : [0,1] \to \Gamma$ denote arbitrary maps such that $f * \overline{g}$ has degree $+1$. Then

$$\sigma_e^*(\Gamma) = \inf_{f,g} \sup_{0 \leq t \leq 1} \|f(t) - g(t)\|.$$

Proof. Assertion 1) is clear. For 2), notice that if $f * \overline{g}$ is defined and has degree 1, then $f(0) = g(0)$ and $f(1) = g(1)$. Suppose $h = f * \overline{g}$ and $k = g * \overline{f}$ so that h and k have degrees 1 and -1, respectively. Then by 1), $\sigma_e^*(\Gamma) \leq \inf_{f,g} \sup_{0 \leq t \leq 1} \|f(t) - g(t)\|$, with the infimum being over all such f, g. The reverse inequality is similar. ∎

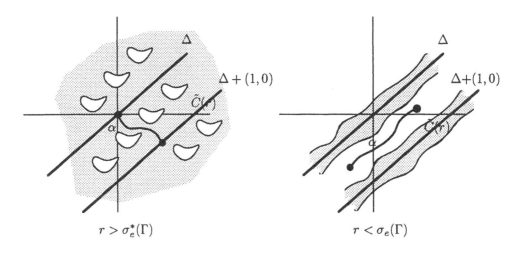

$$r > \sigma_e^*(\Gamma) \qquad\qquad\qquad r < \sigma_e(\Gamma)$$

FIGURE 2. The Proof that Essential Span equals Dual Span.

2.3. Dual Essential Span Equals Essential Span.
In this section, we will prove the following

Theorem 2.3. *If* Γ *denotes a simple closed curve in the plane, then*

$$\sigma_e(\Gamma) = \sigma_e^*(\Gamma).$$

Proof. Given $r > 0$, let

$$C(r) = \{(x, y) \in \Gamma \times \Gamma : \ d(x, y) < r\}.$$

Then $C(r)$ is open in $\Gamma \times \Gamma$, and if $r_1 < r_2$, we obtain $\overline{C}(r_1) \subset C(r_2)$.

Let $\Pi : \mathbb{R}^2 \to \Gamma \times \Gamma$ be the universal cover with fundamental domain $[0, 1] \times [0, 1]$ and coordinate projections $\pi_1, \pi_2 : \Gamma \times \Gamma \to \Gamma$ onto, respectively, the first and second factors. Finally, suppose that Δ denotes the diagonal subspace of $\mathbb{R}^2$.

Suppose that $f : [0, 1] \to \Gamma$ and $g : [0, 1] \to \Gamma$ are such that the composite $f * \overline{g}$ has degree $+1$, $f(0) = g(0)$, $f(1) = g(1)$, and $d(f(t), g(t)) < r$, for all t. Then there is a lift $\alpha : [0, 1] \to \mathbb{R}^2$ of the map $t \mapsto (f(t), g(t))$ which is a path from Δ to $\Delta + (1, 0)$, lying wholly in $\tilde{C}(r) = \Pi^{-1}(C(r))$. See the left side of Figure 2 for an illustration. From this, if we define

$$\rho \equiv \inf\{r > 0 : \ \Delta \text{ and } \Delta + (1, 0) \text{ are in the same path component of } \tilde{C}(r)\},$$

then $\rho \leq \sigma_e^*(\Gamma)$.

Conversely, if α is a path in $\tilde{C}(r)$ from Δ to $\Delta + (1,0)$, put $f = \pi_1 \circ \Pi \circ \alpha$ and $g = \pi_2 \circ \Pi \circ \alpha$. Then, $f(0) = g(0)$, $f(1) = g(1)$, $f * \overline{g}$ has degree 1, and $d(f(t), g(t)) < r$ for all $t \in [0,1]$. Thus, $\sigma_e^*(\Gamma) = \rho$.

If $f, g : [0,1] \to \Gamma$ are degree one maps with $d(f(t), g(t)) \geq r$ for all $t \in [0,1]$, there is a lift $\alpha : [0,1] \to \mathbb{R}^2$ of $t \mapsto (f(t), g(t))$ with $\alpha(0)$ lying in the strip between Δ and $\Delta + (1,0)$. Then $\bigcup_{n \in \mathbb{Z}} (\alpha([0,1]) + (n,n))$ lies in the complement of $\tilde{C}(r)$ and separates Δ from $\Delta + (1,0)$. It follows that $\sigma_e(\Gamma) \leq \rho \leq \sigma_e^*(\Gamma)$. See the right side of Figure 2.

Now we prove the reverse inequality. Suppose that $0 < r_1 < \sigma_e^*(\Gamma)$, and let r_2 be such that $0 < r_2 < r_1$. Then, since $\overline{C}(r_2) \subset C(r_1)$, there exists a finite collection of rectangles $R_i = I_i \times J_i$, for $i = 1, 2, \ldots n$, with I_i and J_i open arcs in Γ such that $\overline{C}(r_2) \subset U = \bigcup_i^n R_i \subset C(r_1)$. The boundary B of U is then made up of finitely many polygonal arc-components, and since Δ and $\Delta + (1,0)$ are in different path components of $\tilde{C}(r_1)$, every path from Δ to $\Delta + (1,0)$ must meet $\tilde{B} = \Pi^{-1}(B)$. Thus, some component B_0 of $\tilde{B}$ separates Δ from $\Delta + (1,0)$ and $B_0 \cap (B_0 + (1,1)) \neq \emptyset$.

So, there is a path $\alpha : [0,1] \to B_0$ with $\alpha(1) = \alpha(0) + (1,1)$. The maps $f = \pi_1 \circ \Pi \circ \alpha$ and $g = \pi_2 \circ \Pi \circ \alpha$ then have the properties: $f, g : [0,1] \to \Gamma$ are of degree one and $d(f(t), g(t)) \geq r_2$ for all t. In particular, $\sigma_e(\Gamma) \geq r_2$, and since $r_2 < r_1 < \sigma_e^*(\Gamma)$, we obtain $\sigma_e(\Gamma) \geq \sigma_e^*(\Gamma)$. $\blacksquare$

3. COOK'S INEQUALITY FOR ESSENTIAL SPAN

We will write $\Gamma_1 \prec \Gamma_2$ if Γ_1 lies in the bounded complementary domain to Γ_2.

Before proving our main theorem we will have need of the following result proved by Mioduszewski [Mi] (and see also Keleti [K].)

Lemma 3.1 (Mountain Climber's Theorem). *Suppose $f, g : [0,1] \to [0,1]$ are continuous maps with the properties that $f(0) = g(0) = 0$ and $f(1) = g(1) = 1$ and there are no intervals on which f or g is a constant. Then there are continuous maps $f', g' : [0,1] \to [0,1]$ such that $f'(0) = g'(0) = 0$, $f'(1) = g'(1) = 1$ and $f(f'(t)) = g(g'(t))$ for all $t \in [0,1]$.*

In this section, we shall prove

Theorem 3.2. *If $\Gamma_1 \prec \Gamma_2$, then $\sigma_e(\Gamma_1) < \sigma_e(\Gamma_2)$.*

Proof. Begin by setting $\delta = \inf\{|x - y| : x \in \Gamma_1, y \in \Gamma_2\}$. We will prove that $\sigma_e(\Gamma_1) \leq \sigma_e(\Gamma_2) - 2\delta$.

Since $\sigma_e(\gamma(S^1))$ varies continuously with the embedding γ, we may assume that Γ_1 and Γ_2 are C^∞ smooth. Suppose $\epsilon > 0$. Let $f_2, g_2 : [0,1] \to \Gamma_2$ be such that $f_2(0) = g_2(0)$, $f_2(1) = g_2(1)$, $f_2 * \overline{g}_2$ is degree one, and $\|f_2(x) - g_2(x)\| < \sigma_e^*(\Gamma_2) + \epsilon/3$

for all $x \in [0,1]$. By slightly perturbing f_2 and g_2, we may assume they are, in addition, smooth and satisfy $||f_2(x) - g_2(x)|| \leq \sigma_e^*(\Gamma_2) + \epsilon/3$ for all x.

Define $F : [0,1] \times [0,1] \to \mathbb{R}^2$ by $F(x,y) = (1-y)f_2(x) + yg_2(x)$. There is then a smooth $H : [0,1] \times [0,1] \to \mathbb{R}^2$ such that H is transverse to Γ_1 and $||H(x,y) - F(x,y)|| < \epsilon/6$ for all $(x,y) \in [0,1] \times [0,1]$. See Figure 3 for an illustration of this situation.

Now the set of smooth maps $G : [0,1] \times [0,1] \to \mathbb{R}^2$ with the property that any tangencies of arcs of the form $G(\{x\} \times [0,1])$ with Γ_1 are quadratic is dense (in the C^0 maps) by the jet transversality theorem (e.g. [G-G]). By the openness of transversality (in the C^1 maps), we may, by slightly perturbing H, assume that: (i) $||H(x,y) - F(x,y)|| < \epsilon/3$ for all (x,y); (ii) H is transverse to Γ_1; and (iii) all tangencies of $H(\{x\} \times [0,1])$ with Γ_1 are quadratic.

From (ii) and the implicit function theorem, it follows that $H^{-1}(\Gamma_1) = \bigcup_{i=1}^m \Lambda_i$ is the disjoint union of simple closed curves Λ_i. Also by (ii), it follows that $(\{x\} \times [0,1]) \cap \Lambda_i$ is finite for each $x \in [0,1]$ and $i = 1, 2 \ldots m$.

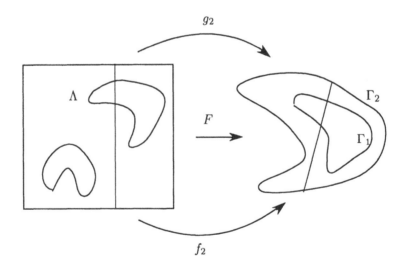

FIGURE 3. The Proof of Cook's Conjecture for Essential Span.

Pick a point z in the bounded domain complementary to Γ_1. Then $H^{-1}(\{z\}) \subset \bigcup_i U_i$, where each U_i denotes the bounded domain complementary to Λ_i. Since $\deg(f_2 * \overline{g}_2) = 1$, the index of F restricted to the boundary $\beta = \partial([0,1] \times [0,1])$ and, hence, of $H|_\beta$, about z, is $+1$. Thus, $\sum_i \mathrm{ind} H|_{\Lambda_i} = 1$, and for some i, the index of $H|_{\Lambda_i}$ is positive. In other words, the degree of $H|_{\Lambda_i} : \Lambda_i \to \Gamma_1$ is positive. Set $\Lambda = \Lambda_i$ for this particular i.

Let $P_1, P_2 : [0,1] \times [0,1] \to [0,1]$, denote projections onto the first or second coordinate, respectively. Let $P_1(\Lambda) = [a,b]$ and choose $p \in P_1^{-1}(a) \cap \Lambda$ and $q \in$

$P_1^{-1}(b) \cap \Lambda$. There are then simple arcs Λ^- and Λ^+ with $\Lambda^- \cup \Lambda^+ = \Lambda$ and $\Lambda^- \cap \Lambda^+ = \{p, q\}$. Since, from the quadratic tangency, $P_1|_\Lambda$ is a finite-to-one map, there are, by the Mountain Climber's Theorem (Lemma 3.1), surjective maps $\alpha^\pm : [0,1] \to \Lambda^\pm$ such that

$$
\begin{aligned}
\alpha^+(0) &= \alpha^-(0) = p; \\
\alpha^+(1) &= \alpha^-(1) = q; \text{ and} \\
P_1 \circ \alpha^+ &= P_1 \circ \alpha^-.
\end{aligned}
$$

Note that $\alpha^- * \overline{\alpha}^+$ is a degree one map.

Now let $f_1 = H \circ \alpha^-$ and $g_1 = H \circ \alpha^+$. Since $H|_\Lambda$ has positive degree, $f_1 * \overline{g}_1$ has positive degree. Thus, $\sigma_e^*(\Gamma_1) \le \sup_t \|f_1(t) - g_1(t)\|$ and for each $t \in [0,1]$,

$$
\begin{aligned}
|f_1(t) - g_1(t)| &= \|H \circ \alpha^-(t) - H \circ \alpha^+(t)\| \\
&\le \|F \circ \alpha^-(t) - F \circ \alpha^+(t)\| + 2\epsilon/3.
\end{aligned}
$$

The points $F \circ \alpha^-(t)$ and $F \circ \alpha^+(t)$ both lie on the intersection of Γ_1 with the straight line segment between $f_2(x) \in \Gamma_2$ and $g_2(x) \in \Gamma_2$, where $x = P_1(\alpha^-(t)) = P_1(\alpha^+(t))$. Since the minimum distance between points of Γ_1 and Γ_2 is δ, we see that

$$
\begin{aligned}
\|F \circ \alpha^-(t) - F \circ \alpha^+(t)\| &\le \|f_2(x) - g_2(x)\| - 2\delta \\
&\le \sigma_e^*(\Gamma_2) + \epsilon/3 - 2\delta.
\end{aligned}
$$

Combining this with the estimate above, we obtain $\|f_1(t) - g_1(t)\| \le \sigma_e^*(\Gamma_2) - 2\delta + \epsilon$. Thus, $\sigma_e^*(\Gamma_1) \le \sigma_e^*(\Gamma_2) - 2\delta + \epsilon$ with ϵ arbitrary, and, hence, $\sigma_e^*(\Gamma_1) \le \sigma_e^*(\Gamma_2) - 2\delta$. Finally, by Theorem 2.3, we get $\sigma_e(\Gamma_1) \le \sigma_e(\Gamma_2) - 2\delta$, as desired. ∎

4. Comparison with Other Notions of Span

4.1. Definitions. Suppose K denotes a compact connected metric space. Let $\pi_i : K \times K \to K$ denote the ith coordinate projection. As in Lelek [L2], define the *surjective span* $\sigma_{surj}(K)$ (respectively, the *surjective semi-span* $\sigma_{ss}(K)$), of K to be the least upper bound of the set of real numbers $r \ge 0$ for which there exists a connected set $C_r \subset K \times K$ such that $d(x,y) \ge r$ for $(x,y) \in C_r$ and $\pi_1(C_r) = K = \pi_2(C_r)$ (respectively, $\pi_1(C_r) = K$).

The *span* of K, written $\sigma(K)$, [respectively, *semispan* $\sigma_{semi}(K)$] is given by

$$
\sigma(K) = \sup\{\sigma_{surj}(A) \mid A \subset K, \ A \ne \emptyset, \ A \text{ connected}\}
$$

[resp., $\sigma_{semi}(K) = \sup\{\sigma_{ss}(A) \mid A \subset K, \ A \ne \emptyset, \ A \text{ connected}\}$].

When $K = \Gamma$, a simple closed curve, these definitions have a simpler form:

$$
sigma(\Gamma) = \max_{f,g:[0,1]\overset{onto}{\to}\Gamma} \{ \min_{t\in[0,1]} \|f(t) - g(t)\| \}.
$$

$$sigma_{semi}(\Gamma) = \max_{\substack{f:[0,1]\overset{onto}{\to}\Gamma \\ g:[0,1]\to\Gamma}} \{\min_{t\in[0,1]} \|f(t) - g(t)\|\}.$$

Another concept of span is the *symmetric span*, which, in the case of a simple closed curve Γ is given by

$$\sup_{f,g} \inf_{0\le t\le 1} \|f(t) - g(t)\|,$$

where $f, g : [0, 1] \to \Gamma$ are continuous maps such that

$$\{(f(t), g(t)) \mid 0 \le t \le 1\} = \{(g(s), f(s)) \mid 0 \le s \le 1\}.$$

We begin with a lemma.

Lemma 4.1. *Suppose that $f, g : [0, 1] \to \Gamma$ are one-to-one, with $f(0) = g(0)$, $f(1) = g(1)$, and $deg(f * \overline{g}) = \pm 1$. Suppose also that $\|f(t) - g(t)\| \le r$ for all $t \in [0, 1]$. Then $\sigma(\Gamma) \le \sigma_{semi}(\Gamma) \le r$.*

Proof. For each $t \in [0, 1]$, define $L(t) = f[0, t] \cup g[0, t]$ and $R(t) = f[t, 1] \cup g[t, 1]$. Given $h, k : [0, 1] \to \Gamma$ with h onto, let

$$H(t) = \inf\{t' \mid h(t) \in L(t')\} = \sup\{t' \mid h(t) \in R(t')\}, \text{ and}$$

$$K(t) = \inf\{t' \mid k(t) \in L(t')\} = \sup\{t' \mid k(t) \in R(t')\}.$$

Then, $H, K : [0, 1] \to [0, 1]$ are continuous maps, and H is onto. Thus, there exists $t \in [0, 1]$ such that $H(t) = K(t)$. So, for some t', $\{h(t), k(t)\} \subset \{f(t'), g(t')\}$ and $|h(t) - k(t)| \le r$. This yields the desired result, since h, k are arbitrary, and it is evident that $\sigma(\Gamma) \le \sigma_{semi}(\Gamma)$. ∎

4.2. Comparisons. $\mathbb{A}$ is an *ϵ-annular neighborhood* of Γ if there exists an embedding $H : \mathbb{S}^1 \times (-1, 1) \to \mathbb{R}^2$ with $H(\mathbb{S}^1 \times \{0\}) = \Gamma$ and $\text{diam}\,(H(\{t\} \times (-1, 1))) < \epsilon$, for all $t \in \mathbb{S}^1$.

Let $\overline{\sigma}$ denote one of σ, σ_e or σ_{semi}. Then $\overline{\sigma}$ will be termed *essentially continuous* at Γ if for every $\epsilon > 0$, there exists a $\delta > 0$ such that if $\mathbb{A}$ is a δ-annular neighborhood of Γ containing the essential simple closed curve Γ', then $|\overline{\sigma}(\Gamma) - \overline{\sigma}(\Gamma')| < \epsilon$.

Let us say that Cook's Inequality holds for a given version $\overline{\sigma}$ of the span if the span of a simple closed curve Γ_1 is strictly smaller than the span of a simple closed curve Γ_2, whenever $\Gamma_1 \prec \Gamma_2$. We know that Cook's Inequality holds for essential span (Theorem 3.2). Our aim in this section is to prove some results relating Cook's Inequality to the equivalence of various competing notions of the span. Toward this end, we prove

Lemma 4.2. *Suppose Cook's Inequality holds for $\overline{\sigma} = \sigma$, σ_e or σ_{semi}, then $\overline{\sigma}$ is essentially continuous.*

Proof. We will prove the contrapositive. Suppose that $\overline{\sigma}$ is *not* essentially continuous at Γ. Then for $n = 1, 2, \ldots$, there exist closed curves $\Gamma_n \subset \mathbb{A}_n$, with each $\mathbb{A}_n$ a $(1/n)$-annular neighborhood of Γ, such that $|\overline{\sigma}(\Gamma_n) - \overline{\sigma}(\Gamma)| > \delta > 0$, Γ_n essential in $\mathbb{A}_n$. Suppose that $\overline{\sigma}(\Gamma_n) > \overline{\sigma}(\Gamma) + \delta$ for all n. For each n there is a Γ_n' in $\mathbb{A}_n$ obtained by pushing Γ_n along radial arcs until $\Gamma_n' \prec \Gamma$ and $\Gamma_n' = \psi(\Gamma_n)$ for an embedding ψ that moves each point of Γ_n by an amount less than $2/n$. But then $|\overline{\sigma}(\Gamma_n') - \overline{\sigma}(\Gamma_n)| < 4/n$.

Taking n large enough that $4/n < \delta$ implies $\overline{\sigma}(\Gamma_n') > \overline{\sigma}(\Gamma)$. So Cook's Inequality is false. ∎

The next result reveals the importance of Cook's Problem in sorting out different notions of span.

Theorem 4.3. *The following hold:*

  a) *If Cook's Inequality holds for* <u>semi-span</u> $\sigma_{semi}(\Gamma)$, *then* $\sigma = \sigma_{semi} = \sigma_e$.
  b) *If Cook's Inequality holds for* <u>span</u> $\sigma(\Gamma)$, *then* $\sigma(\Gamma) = \sigma_e(\Gamma)$.

Proof. (a) From the definitions, $\sigma_{semi} \geq \sigma \geq \sigma_e$, and, hence, it suffices to show that $\sigma_e \geq \sigma_{semi} - \epsilon$ for all $\epsilon > 0$.

Fix a simple closed curve Γ. By Lemma 4.2, σ_{semi} is essentially continuous at Γ. Let $f, g : [0, 1] \to \Gamma$ be such that $f(0) = g(0)$; $f(1) = g(1)$; $f * \overline{g}$ has degree ± 1; and $\sup_{t \in [0,1]} \|f(t) - g(t)\| \leq \sigma_e(\Gamma) + \epsilon/3$.

Now let $f', g' : [0, 1] \to \mathbb{R}^2$ be such that $f'(0) = g'(0)$; $f'(1) = g'(1)$; f', g' are 1-1; $f' * \overline{g}'$ has degree ± 1; $\|f' - f\| < \epsilon/6$, $\|g' - g\| < \epsilon/6$; and $\Gamma' = f'([0, 1]) \bigcup g'([0, 1])$ is a simple closed curve that is essential in the δ-annular neighborhood of Γ.

By Lemma 4.2, choose δ small enough that $|\sigma_{semi}(\Gamma') - \sigma_{semi}(\Gamma)| < \epsilon/3$.

Now $\forall t$, $\|f'(t) - g'(t)\| \leq \|f(t) - g(t)\| + \epsilon/3 \leq \sigma_e(\Gamma) + 2\epsilon/3$. Then by Lemma 4.1, $\sigma_{semi}(\Gamma') \leq \sigma_e(\Gamma) + 2\epsilon/3$. Thus, $\sigma_{semi}(\Gamma) \leq \sigma_e(\Gamma) + \epsilon$, as desired.

(b) The proof is very similar to that of (a): Since Cook's Inequality is assumed for σ, Lemma 4.2 implies $|\sigma(\Gamma') - \sigma(\Gamma)| \leq \epsilon/3$. By Lemma 4.1, $\sigma(\Gamma') \leq \sigma_e(\Gamma) + 2\epsilon/3$. This time, we conclude that $\sigma(\Gamma) \leq \sigma_e(\Gamma) + \epsilon$, completing the proof. ∎

We conclude with a brief discussion of symmetric span $\sigma_{sym}(\Gamma)$ for a simple closed curve Γ, defined in Section 4.1.

Proposition 4.4. *If Γ is a simple closed curve in the plane, then*

$$\sigma_{sym}(\Gamma) \leq \sigma_e(\Gamma).$$

Proof. Suppose that $0 < r < \sigma_{sym}(\Gamma)$. Let f, g be as in the definition of symmetric span, such that $\inf_{0 \leq t \leq 1} \|f(t) - g(t)\| \geq r$. We will argue that $r \leq \sigma_e(\Gamma)$. To this end, let X be the continuum $X = \{(f(t), g(t)) : 0 \leq t \leq 1\}$. Then X is a symmetric set about $x = y$ which does not meet the diagonal in $\Gamma \times \Gamma$. We claim that some component of the lift $\tilde{X}$ of X to the universal cover $\mathbb{R} \times \mathbb{R}$ of $\Gamma \times \Gamma$ separates the diagonal Δ in $\mathbb{R} \times \mathbb{R}$ from $\Delta + (1, 0)$.

To verify this, suppose K is a component of $\tilde{X}$ lying in the strip $\{(x, y) : y < x < y + 1\}$ and suppose that K is bounded. Let K^s denote the reflection of K about $x = y$. Then K^s must also be a component of $\tilde{X}$, as X is symmetric. Thus, $K^s = K + (m, n)$ for some integers m and n. We have $K = (K^s)^s = (K + (m, n))^s = (K + (m, n)) + (n, m) = K + (m + n, m + n)$ so that $n = -m$, since K is bounded. But if $(x, y) \in K$, $y < x < y + 1$, then the translated point $(x + m, y + n) = (y', x')$ satisfies $x + m < y + n < x + m + 1$, since it is a reflection. It follows that $2 > n - m > 0$. But for integers, this inequality contradicts $m = -n$. So, K must be unbounded.

Thus, we may conclude that K does separate Δ from $\Delta + (1, 0)$. Now with $C(r)$ as in the proof of Theorem 2.3, we see that Δ and $\Delta + (1, 0)$ must sit in different path components in $\tilde{C}(r)$. It follows that $r \leq \sigma_e^*(\Gamma) = \sigma_e(\Gamma)$ and, hence, $\sigma_{sym}(\Gamma) \leq \sigma_e(\Gamma)$. ∎

One question raised in [C-I-L] is whether span and symmetric span agree on all plane continua. If so, then essential span and span must coincide for simple closed curves and, by Theorem 3.2, Cook's inequality would hold for span.

REFERENCES

[C-I-L] H. Cook, T. Ingram, and A. Lelek, Eleven Annotated Problems about Continua. In: *Open Problems in Topology*, North-Holland, Amsterdam, 1990.

[G-G] M. Golubitsky and V. Guillemin, *Stable mappings and their singularities*, Springer-Verlag, New York, 1974.

[K] T. Keleti, The Mountain Climber's Problem, *Proc. of the Amer. Math. Soc.* **117**, no. 1, (1993), 89-97.

[L1] A. Lelek, Disjoint mappings and the span of spaces, *Fund. Math.* **LV** (1964), 199-214.

[L2] _____ , On the surjective span and semi-span of connected metric spaces, *Colloq. Math.* **37** (1977), 35-45.

[Mi] J. Mioduszewski, On a quasi-ordering in the class of continuous mappings of a closed interval into itself, *Colloq. Math.* 9, (1962), 233-240.

[M-O] J. Mayer and L. Oversteegen, Continuum Theory, *UAB: Technical Report*, 1992.

[O-T] L. Oversteegen and E.D. Tymchatyn, Plane strips and the span of continua (I), *Houston Jour. of Math.*, Vol. **8**, No. 1,(1982) 129-142.

E-mail address: barge@math.montana.edu "Marcy Barge"
E-mail address: dswanson@math.montana.edu "Richard Swanson"

12

INVERTIBILITY OF THE PSEUDO–ARC

DAVID P. BELLAMY Department of Mathematical Sciences, University of Delaware, Newark, Delaware 19716, USA

ABSTRACT. A topological space X is called invertible if and only if, given any open $O \subseteq X$, if $O \neq \phi$, there is a homeomorphism $h : X \to X$ such that $h(X - O) \subseteq O$. A simple proof is given that the pseudo–arc has this property.

Invertibility has been investigated in [3], [4], [5], [6], [7] and [8]. The only invertible finite dimensional manifolds are spheres. Some other examples of invertible spaces are the Hilbert cube and the solenoid groups.

A *continuum* is a compact connected metric space. A continuum is *indecomposable* if and only if it is not the union of two of its proper subcontinua, and is *hereditarily indecomposable* provided every one of its subcontinua is indecomposable. A *pseudo–arc* is a continuum which is hereditarily indecomposable and can be expressed as an inverse limit of closed intervals and continuous maps.

The literature on pseudo–arcs is extensive and will not be summarized here. The facts which we need are stated in the following three theorems.

Theorem 1. *(R. H. Bing [1]) A pseudo arc P is homogeneous; that is, given any $x, y \in P$, there is a homeomorphism $h : P \to P$ such that $h(x) = y$.*

A shorter proof than Bing's can be found in [11].

Theorem 2. *(R. H. Bing [2]) All pseudo–arcs are homeomorphic.*

Hence we sometimes say "the" pseudo–arc.

1991 *Mathematics Subject Classification.* primary 54F15, secondary 54C10.
Key words and phrases. pseudo-arc, invertibility.

Theorem 3. *(G. W. Henderson [9]) There is a continuous function $f : [0,1] \to [0,1]$ with $f(x) < x$ for all x except 0 and 1; $f(0) = 0$; $f(1) = 1$; and such that the inverse limit of intervals with each one-step bonding map equal to f is a pseudo–arc.*

In what follows, a pseudo–arc of this sort will be called a Henderson pseudo–arc. The points $< 1 >$, with all coordinates, equal to one, and $< 0 >$, with all coordinates equal to zero, exist in any Henderson pseudo–arc. The other points are of the form $< t_n >_{n=1}^{\infty}$, where $f(t_n) = t_{n-1}$ for each $n > 1$; in particular $\lim_{n \to \infty} t_n = 1$.

The question of whether the pseudo–arc is invertible was posed by P. H. Doyle and H. S. Davis in the 1960's. It has apparently never appeared in print, but has been a folklore question for some time. It could no doubt be settled by the traditional chaining arguments, but the following proof will probably be accessible to a broader audience.

Lemma A. *Let $f : [0,1] \to [0,1]$ be any continuous function such that $f(0) = 0$; $f(1) = 1$; and $f(x) < x$ otherwise. Then for all $p, q \in (0,1)$ there exists a positive integer m such that $f^m([0,p]) \subseteq [0,q]$.*

Proof. Let f be such a function, and let $p, q \in (0,1)$. The collection $(f^n)^{-1}[0,q), n$ a positive integer, is an open cover of $[0,1)$. Hence some finite number of these sets covers $[0,p]$, as $p < 1$. Since they are nested, a single one does.

Theorem B. *The pseudo–arc is invertible.*

Proof. Let P be a pseudo–arc and let U be any nonvoid open subset of P. Let $x \in U$, and choose homeomorphisms g and k from P onto a Henderson pseudo arc $H = \lim\{[0,1]; f\}$ such that $g(x) =< 1 >$ and $k(x) =< 0 >$. These homeomorphisms exist by Theorems 1 and 2. Define $F : H \to H$ to be the shift map $F(\langle t_i \rangle_{i=1}^{\infty}) = \langle f(t_i) \rangle_{i=1}^{\infty}$.

Now, since $g(U)$ and $f(U)$ are open in H, using straightforward arguments it is possible to find a positive integer n and an $\epsilon > 0$ such that $\{\langle t_i \rangle_{i=1}^{\infty} \in H | t_n < \epsilon\} \subseteq k(U)$ and $\{\langle t_i \rangle_{i=1}^{\infty} \in H | t_n > 1 - \epsilon\} \subseteq g(U)$. By the Lemma, there is a positive integer m such that $f^m(t) < \epsilon$ for all $t \leq 1 - \epsilon$. Define h be the composition $k^{-1} \circ F^m \circ g : P \to P$. Now, if $y \in P - U$, $g(y) = \langle x_i \rangle_{i=1}^{\infty}$ for some $\langle x_i \rangle_{i=1}^{\infty} \in H$ with $x_n \leq 1 - \epsilon$. Thus $F^m(g(y)) = \langle f^m(x_i) \rangle_{i=1}^{\infty}$ and $f^m(x_n) < \epsilon$. Thus, $F^m(g(y)) \in \{\langle t_i \rangle \in H | t_n < \epsilon\} \subseteq k(U)$, and so $k^{-1}(F^m(g(y))) \in U$. Hence $h(P - U) \subseteq U$ and the proof is done.

The natural questions which follow here are whether a pseudo–circle is invertible; or more generally whether there exist any other invertible, hereditarily indecomposable continua. Is the pseudo–arc the only invertible tree–like indecomposable continuum?

REFERENCES

1. R. H. Bing, *A homogeneous indecomposable plane continuum*, Duke Math J. **15** (1948), 729–742.

2. R. H. Bing, *Snakelike continua*, Duke Math J. **18** (1951), 653–663.

3. H. S. Davis and P. H. Doyle, *Invertible continua*, Portugal Math. **26** (1967), 487–491.

4. P. H. Doyle and J. G. Hocking, *A characterization of Euclidean n-space*, Michigan Math J. **7** (1960), 199–200.

5. P. H. Doyle and J. G. Hocking, *Continuously invertible spaces*, Pacific J. Math. **12** (1962), 499–503.

6. P. H. Doyle and J. G. Hocking, *Dimensional Invertibility*, Pacific J. Math. **12** (1962), 1235–1240.

7. P. H. Doyle and J. G. Hocking, *Invertible spaces*, Amer. Math. Monthly **68** (1961), 959–965.

8. P. H. Doyle, J. G. Hocking and R. P. Osborne, *Local invertibility*, Fund. Math. **54** (1964), 15–25.

9. G. W. Henderson, *The pseudo-arc as an inverse limit with one binding map*, Duke Math. J. **31** (1964), 421–425.

10. E. E. Moise, *An indecomposable plane continuum which is homeomorphic to each of its nondegenerate subcontinua*, Trans. Amer. Math. Soc. **63** (1948), 581–594.

11. E. E. Moise, *A note on the pseudo-arc*, Trans. Amer. Math. Soc. **67** (1949), 57–58.

E-mail address: bellamy@math.udel.edu "David P. Bellamy"

13
INVERSE LIMIT SPACES, PERIODIC POINTS, AND ARCS

LOUIS BLOCK Department of Mathematics, University of Florida, Gainesville, Florida 32611, USA
SHANNON SCHUMANN   Department of Mathematics, University of Kentucky, Lexington, Kentucky 40506, USA

ABSTRACT. Much has been studied by way of the relationship between a function's dynamics and the inverse limit space obtained using the function as the sole bonding map. In 1985, Barge and Martin [BM] discovered relationships between the periods a function of the interval may have and the decomposability or indecomposabilty of its inverse limit space. These results have been generalized to maps of finite graphs by Barge and Diamond in [BD].

More recently, Holte [H] has found conditions for certain piecewise monotone functions to have homeomorphic inverse limit spaces, and questions are arising between the relationship between the kneading sequence of unimodal maps and the inverse limit space. This paper addresses a related question about unimodal maps – that of the possible periodic points when the inverse limit space is an arc.

1. INTRODUCTION

Results from [BM] indicate that functions with simple dynamics have uncomplicated inverse limit spaces. This paper deals with a specific case of this phenomenon – when the inverse limit is an arc. In particular, we give a complete characterization for the unimodal maps whose inverse limits are arcs with respect to periodic points. In section 3, we prove some specific results about periodic points and inverse limits. We prove that for a continuous map of the interval to itself, if the inverse limit space is an arc, then all periodic points have period one or two. Section 4 contains

1991 *Mathematics Subject Classification.* 58F03, 54B99, 54H20.
Key words and phrases. period of a periodic point, unimodal map, and inverse limit space.

the main result, which summarizes the work in section 3, as well as an example showing that this result cannot be generalized to piecewise monotone maps.

2. PRELIMINARIES

For what follows, let I denote the unit interval, and let $f : I \to I$ be continuous. Now define the inverse limit space of f by the following:

$$(I, f) = \{\underline{x} = < x_0, x_1, \cdots >| \ x_i \in I \text{ and } f(x_{i+1}) = x_i \text{ for } i = 0, 1, \ldots\},$$

with the metric $d(\underline{x}, \underline{y}) = \sum_{i=0}^{\infty} \frac{|x_i - y_i|}{2^i}$.

We say a topological space is an arc if it is homeomorphic to the interval $[0, 1]$. We say a continuous map $f : [0, 1] \to [0, 1]$ is unimodal if there is a point $c \in (0, 1)$ such that either f is strictly increasing on $[0, c]$ and strictly decreasing on $[c, 1]$ or f is strictly decreasing on $[0, c]$ and strictly increasing on $[c, 1]$.

The following theorems are well known or easy to prove and are given without proof

THEOREM 2.1. (I, f) *is a chainable continuum. Furthermore, if f is an onto homeomorphism, then (I, f) is an arc.*

THEOREM 2.2. *For $n \in \mathrm{N}$, (I, f) is homeomorphic to (I, f^n).*

THEOREM 2.3. *If h is a homeomorphism, and $f = h^{-1} \circ g \circ h$, then (I, f) is homeomorphic to (I, g).*

THEOREM 2.4. *If $< x_1, x_2, x_3, x_4, \cdots > \in (I, f)$, then for each natural number n, $x_n \in \bigcap_{i=0}^{\infty} f^i(I)$.*

The function f induces a canonical homeomorphism, $\hat{f} : (I, f) \to (I, f)$ defined by the following:

$$\hat{f}(< x_0, x_1, x_2, \cdots >) = < f(x_0), x_0, x_1, \cdots >.$$

It is easy to see that the dynamics of $\hat{f}$ mimic the dynamics of f.

For $i = 0, 1, 2, \ldots$, let π_i be projection of the i-th coordinate into I. Notice that π_i is a continuous function for each i.

3. INVERSE LIMITS, ARCS, AND PERIODIC POINTS

In this section we prove preliminary propositions which lead to our main result: When $f : I \to I$ is a unimodal map, then (I, f) is an arc if and only if one of the following two conditions is met: 1) f has more than one fixed point but no points of other periods; or 2) f has a single fixed point and a period two orbit, but no point of any other period.

PROPOSITION 3.1. *Let $f : I \to I$ be continuous. Then (I, f) is a point if and only if f has exactly one fixed point and no other periodic points.*

Proof. Since each periodic point of f yields a point $\underline{x} = < x_0, x_1, x_2, \cdots > \in (I, f)$, it is obvious that if f has more than one periodic point then (I, f) is not a point. Suppose that (I, f) has exactly one periodic point, p. Then $f(p) = p$, since $f(p)$ must be periodic.

We claim that f does not map I onto I. To prove this claim, suppose that f does map I onto I. First suppose that $p = 0$. Then $f(1) < 1$ and $f(x) < x$ for all $x \in (0, 1]$. Since f is onto, there is a point $< x_0, x_1, x_2, \cdots > \in (I, f)$ with $x_0 \neq 0$. Since $f(x_i) = x_{i-1}$, the sequence (x_n) is strictly increasing, and has a limit x with $f(x) = x$. This contradicts the fact that 0 is the unique fixed point of f. Thus, we may assume that $p \neq 0$. By a similar argument, we may also assume that $p \neq 1$. Then $f(0) > 0$ and $f(1) < 1$, and it follows that $f(x) > x$ for $x \in [0, p)$ and $f(x) < x$ for $x \in (p, 1]$.

Now, since f is onto, there are points $x \in [0, p)$ and $y \in (p, 1]$ such that $f(x) = 1$ and $f(y) = 0$. Also, there is a point $w \in (x, p)$ with $f(w) = y$. Then $0 < w < p$, and $f^2(w) = 0$. Since $f^2(0) > 0$, this contradicts the fact that p is the unique fixed point of f^2. This establishes our claim that f is not onto.

Let K denote the (nested) intersection of the sets $f^k(I)$ where $k = 1, 2, \ldots$. Then K is a compact, connected set containing the point p, and f maps K onto K. Since the restriction of f to K has exactly one periodic point, it follows from the claim above that K cannot be a nondegenerate interval. Thus, $K = \{p\}$.

Finally, note that if $\underline{x} = < x_0, x_1, x_2, \cdots > \in (I, f)$, then $x_j \in K$ for each $j = 0, 1, 2, \ldots$. Thus the only element of (I, f) is the point $< p, p, p, \cdots >$.

PROPOSITION 3.2. *Let $f : I \to I$ be continuous, and suppose that there are distinct points $w, y, z \in I$ such that w is not between y and z, $f(w) = w$, $f(y) = z$, and $f(z) = y$. Then (I, f) is not an arc.*

Proof. Without loss of generality, we can assume that $y < z$. First, assume that $w < y$. To show that (I, f) is not an arc, we find a connected subset, $\mathcal{A}$, of (I, f), such that $\overline{\mathcal{A}} \setminus \mathcal{A}$ contains more than two points. Let $S = \{x \in [w, y] \mid f(x) = x\}$, and let $p = \sup(S)$. Then $f(p) = p$, and $w \leq p < y$.

Since $f(y) = z > y$ and f has no fixed points in the interval $(p, y]$, it follows that $f(x) > x$ for $x \in (p, y]$. Also, since $f(y) > y$ and $f(z) < z$, there is a point $q \in (y, z)$ with $f(q) = q$.

Let $\mathcal{A} = \{\underline{x} = < x_0, x_1, x_2, \cdots > \in (I, f) \mid \lim_{i \to \infty} |x_i - p| = 0\}$. Then $\mathcal{A}$ is a connected subset of (I, f) by Lemma 2.1 and Theorem 2.2 of [M]. Let $\underline{q}$, $\underline{y}$, and $\underline{z}$ be the points in (I, f) given by $\underline{q} = < q, q, q, \cdots >$, $\underline{y} = < y, z, y, z, \cdots >$, and $\underline{z} = < z, y, z, y, \cdots >$. We will show that each of these three distinct points are in $\overline{\mathcal{A}} \setminus \mathcal{A}$. Obviously none of the three points are in $\mathcal{A}$.

We will show that $\underline{y} \in \overline{\mathcal{A}}$. Let $\epsilon > 0$ with $\epsilon < y - p$. We claim that there is a natural number k such that $f^k\left([p,p+\epsilon]\right)$ contains $[p,z]$. To see this, recall that $f(x) > x$ for each $x \in (p,y]$. Let $t \in (p,p+\epsilon)$. If $f^k(t) < y$, for every natural number k, then $\lim_{k\to\infty} f^k(t) = T \leq y$. Then $f(T) = f\left(\lim_{k\to\infty} f^k(t)\right) = \lim_{k\to\infty} f^{k+1}(T) = T$. Since $T \in S$ and $T > p$, this contradicts that $p = \sup S$. Thus $f^{k-1}(t) \geq y$ for some natural number k, and $f^k\left([p,p+\epsilon]\right)$ is an interval containing $[p, f(y)] = [p,z]$. This establishes the claim.

Next, choose an even natural number n with $\sum_{i=n+1}^{\infty} 2^{-i} < \epsilon$. By the claim, there is a point $r \in (p,p+\epsilon)$ such that $f^k(r) = y$. Since $f(t) > t$ for $t \in (p,p+\epsilon)$, there is a point $v \in (p,r)$ with $f(v) = r$. By applying this argument inductively, we obtain a point $\underline{r} \in \mathcal{A}$ with $\pi_{n+k}\left(\underline{r}\right) = r$. Notice that for $i \leq n$, if i is even, then $\pi_i\left(\underline{r}\right) = y = \pi_i\left(\underline{y}\right)$, and if i is odd, then $\pi_i\left(\underline{r}\right) = z = \pi_i\left(\underline{y}\right)$. Therefore,

$$d\left(\underline{r},\underline{y}\right) = \sum_{i=0}^{\infty} 2^{-i}\left|\pi_i\left(\underline{r}\right) - \pi_i\left(\underline{y}\right)\right| \leq \sum_{i=n+1}^{\infty} 2^{-i}\left|\pi_i\left(\underline{r}\right) - \pi_i\left(\underline{y}\right)\right| < \epsilon.$$

Thus, $\underline{y} \in \overline{\mathcal{A}}$. A similar argument will show that $\underline{q}$ and $\underline{z}$ are in $\overline{\mathcal{A}}$, which completes the proof in the case $w < y$.

Finally, if $w > z$, then conjugating f by the homeomorphism $h(x) = 1 - x$ yields a function which satisfies the requirements of the previous case. Since (I,f) is homeomorphic to $\left(I, h^{-1} \circ f \circ h\right)$, the proposition is proved.

COROLLARY 3.3. *If $f : I \to I$ is continuous and (I,f) is an arc, then all periodic points of f are either fixed or period two.*

Proof. Suppose that f has a point with period larger than two. By the theorem of Sharkovsky [S], f has a periodic point of period four. By theorem A of [BH], f has a periodic orbit of period 4, $\{y_1, y_2, y_3, y_4\}$, with $y_1 < y_2 < y_3 < y_4$, satisfying $f\left(\{y_1, y_2\}\right) = \{y_3, y_4\}$ and $f\left(\{y_3, y_4\}\right) = \{y_1, y_2\}$. This implies that $f^2(y_3) = y_4$, $f^2(y_4) = y_3$, and there is a point q with $y_2 < q < y_3$ and $f(q) = q$.

To summarize, we have $q < y_3 < y_4$ with $f^2(q) = f(q) = q$, $f^2(y_3) = y_4$, and $f^2(y_4) = y_3$. Thus, f^2 satisfies the hypotheses of Proposition 3.1, and hence $\left(I, f^2\right)$ is not an arc. But (I,f) is homeomorphic to $\left(I, f^2\right)$, and this completes the proof.

PROPOSITION 3.4. *Let $f : I \to I$ be continuous. Suppose there is a point $c \in (0,1)$ such that $f(c) > c$, $f|_{[0,c]}$ is strictly increasing, and $f|_{[c,1]}$ is strictly decreasing. Suppose further that there is a point $\gamma < c$ for which $f(\gamma) = \gamma$, and that f has no points of period two. Then (I,f) is an arc.*

Proof. Let $S = \{x \in I \mid x < c \text{ and } f(x) = x\}$. S is not empty and is bounded, so let $\alpha = \sup(S)$ and $\beta = \inf(S)$. Let $g = f|_{[0,\alpha]}$. If $\alpha = \beta$, then (I,g) is the single point $< \alpha, \alpha, \alpha, \cdots >$. If $\alpha > \beta$, then since f is strictly increasing on $[0,c]$, we know that g is a homeomorphism, and by theorem 2.1, (I,g) is an arc with endpoints $\hat{\beta} = < \beta, \beta, \beta, \cdots >$ and $\hat{\alpha} = < \alpha, \alpha, \alpha, \cdots >$.

We now restrict our attention to $f|_{[\alpha,1]}$.

CLAIM 1: We first claim that $f^2(c) > c$. To see this, suppose that $f^2(c) \leq c$. Then since $f(c) > c$, and $f(f(c)) = f^2(c) < c$, there exists $y \in (c, f(c))$ such that $f(y) = c$. Since $\sup(f(I)) = f(c)$, we have $f^2(y) = f(f(y)) = f(c) > y$, and $f^2(f(c)) \leq f(c)$. Thus there is a point $q \in (y, f(c)]$ such that $f^2(q) = q$. Since $f|_{[c,1]}$ is one to one, f has a unique fixed point, p, in $[c, 1]$. Also, $p < y < q \leq f(c)$, so $q \neq y$. Therefore, q has period two, which contradicts the hypotheses of the proposition.

Next, we show that if $x \in (p, f(c)]$, then $f^2(x) \in (p, f(c)]$. This is true because $f|_{[c,1]}$ is a homeomorphism, so f maps $[c, p]$ onto $[p, f(c)]$. Then, by the above claim, f maps $[p, f(c)]$ into $[c, p]$.

Notice that this remark implies the following:

(1) If $x \in (p, f(c)]$, then $f^2(x) < x$.

(2) If $x \in [c, p)$, then $f^2(x) > x$.

(3) If $< x_0, x_1, x_2, \cdots > \in \left(I, f|_{[\alpha,1]}\right)$ and $x_k > c$, then $x_j > c$ for all $j < k$.

(4) If $< x_0, x_1, x_2, \cdots > \in \left(I, f|_{[\alpha,1]}\right)$ and $x_k < c$, then $x_{k+n} < c$ for each natural number n.

(5) There is a unique thread, $< x_0, x_1, x_2, x_3, \cdots > \in \left(I, f|_{[\alpha,1]}\right)$ with $x_0 = c$. Call this thread $\hat{c}$.

CLAIM 2: Now we show that if $< x_0, x_1, x_2, \cdots > \in \left(I, f|_{[\alpha,1]}\right)$ and $x_i > c$ for each $i \in \{0, 1, 2, \ldots\}$, that for each $i \in \{0, 1, 2, \ldots\}$, $x_i = p$. To see this, suppose the contrary. Then $x_0 \neq p$, because $f|_{[c,f(c)]}$ is one to one. Thus, $x_0 > p$ or $x_0 < p$. If $x_0 < p$, then we have $x_1 \in (p, f(c))$, and so $x_2 \in (c, p)$. Also, by remark (2) above, $x_0 > x_2$. Continuing in this fashion we obtain $x_0 > x_2 > x_4 > \ldots$, a decreasing sequence which is bounded below by c. Let $x = \lim_{n \to \infty} x_{2n}$. Then $c \leq x < p$; also, $f^2(x) = f^2(\lim_{n \to \infty} x_{2n}) = \lim_{n \to \infty} f^2(x_{2n}) = \lim_{n \to \infty} x_{2n-2} = x$. Since f has no point of period two, we know that x must be a fixed point. This is a contradiction since p is the unique fixed point of f in $[c, 1]$. Similarly, if $x_0 > p$, remark (1) above yields contradiction.

Now, let $A_0 = \{< x_0, x_1, x_2, \cdots > \in (I, f) \mid x_i \leq \alpha$ for $i = 0, 1, 2, \ldots\}$. As earlier noted, A_0 is either an arc or a point. Let $A_1 = \{< x_0, x_1, x_2, \cdots > \in (I, f) \mid x_1 \leq c$ and $x_i \geq \alpha$ for $i = 0, 1, 2, \ldots\}$. A_1 is an arc because $f|_{[\alpha,c]}$ is a homeomorphism, and the point $\hat{\alpha}$ is an endpoint of A_1. Furthermore, $A_0 \cap A_1 = \{\hat{\alpha}\}$. Next, let $B_1 = A_0 \cup A_1$. B_1 is an arc with endpoints $\hat{\beta}$ and $\hat{f}(\hat{c})$. Similarly, let $A_2 = \{< x_0, x_1, x_2, \cdots > \in (I, f) \mid \alpha \leq x_0, x_1 \geq c$, and $x_2 \leq c\}$. A_2 is an arc because $f|_{[0,c]}$ and $f|_{[c,f(c)]}$ are homeomorphisms. The endpoints of A_2 are $\hat{f}^2(\hat{c})$ and $\hat{f}(\hat{c})$, and $A_2 \cap B_1 = \{\hat{f}(\hat{c})\}$. Let $B_2 = B_1 \cup A_2$. B_2 is and arc with endpoints $\hat{\beta}$ and $\hat{f}^2(\hat{c})$.

Inductively, define

$$A_k = \{< x_0, x_1, x_2, \cdots > \in (I, f) \mid \alpha \leq x_0, x_{k-1} \geq c, \text{ and } x_k \leq c\},$$

and $B_k = A_k \cup B_{k-1}$. Note that A_k is an arc with endpoints $\hat{f}^k(\hat{c})$ and $\hat{f}^{k-1}(\hat{c})$; $A_k \cap B_{k-1} = \{\hat{f}^{k-1}(\hat{c})\}$; and therefore, B_k is an arc with endpoints $\hat{\beta}$ and $\hat{f}^k(\hat{c})$.

Let $B = \bigcup_{k=1}^{\infty} B_k$. By Claim 2, $B \cup \{\hat{p}\} = (I, f)$. Now we show that (I, f) is an arc.

CLAIM 3: If $\{\hat{z}(n)\}$ is a sequence in (I, f) with $\hat{z}(n) \in A_n$ for each n, then $\{\hat{z}(n)\}$ converges to $\hat{p}$. To see this, suppose that $\hat{z}(n)$ is described as above. Since (I, f) is a compact metric space, the set $\{\hat{z}(n) \mid n \in \mathbb{N}\}$ has a limit point. Also, if $\hat{z}(n) = <\hat{z}(n)_1, \hat{z}(n)_2, \hat{z}(n)_3, \cdots>$, we have that $\hat{z}(n)_k \geq c$ for $k \leq n-1$. Now let $\hat{x}$ be any limit point of $\{\hat{z}(n) \mid n \in \mathbb{N}\}$. By the above remarks, $x_k \geq c$ for each k, and by Claim 2, it follows that $\hat{x} = \hat{p}$. This means that $\{\hat{z}(n)\}$ converges to $\hat{p}$.

Finally, since for each n, B_n is an arc and $B_n \subseteq B_{n+1}$, there are homeomorphisms $f_n : [0, 1 - 1/n] \to B_n$ such that

$$f_{n+1}|_{[0,1-1/n]} = f_n.$$

The sequence $\{f_n\}$ induces a homeomorphism $h : [0, 1) \to B$, which can be extended to $f^* : [0, 1] \to \overline{B}$. The function f^* is clearly one to one and onto because h is a homeomorphism. By Claim 3, f^* is a homeomorphism from $[0, 1]$ onto (I, f). This completes the proof of Proposition 3.4.

PROPOSITION 3.5. *Let $f : I \to I$ be continuous. Suppose there is a point $c \in (0, 1)$ such that $f(c) = c$, $f|_{[0,c]}$ is strictly increasing, and $f|_{[c,1]}$ is strictly decreasing. Then if there is a point $x < c$ such that $f(x) = x$, (I, f) is an arc, and if there is no such point, (I, f) is a point.*

Proof. By Proposition 2.4, we know that $(I, f) = (K, f|_K)$, where $K = \bigcap_{i=0}^{\infty} f^i(I)$. Let $S = \{x < c \mid f(x) = x\}$. If S is empty, then for each point $x < c$, $c > f(x) > x$, and therefore the sequence $f^n(x)$ converges. Let $T = \lim_{x \to c} f^n(x)$. By continuity, $f(T) = f(\lim_{x \to c} f^n(x)) = \lim_{x \to c} f^{n+1}(x) = \lim_{x \to c} f^n(x) = T$, and so we have that $K = \{c\}$, and that (I, f) is a single point, $\{<c, c, \cdots>\}$. If S is not empty, then let $\beta = \inf(S)$. As before, if $x < \beta$, then $\lim_{x \to \infty} f^n(x) = \beta$, and so in this case, $K = [\beta, c]$. But then $f|_K$ is an onto homeomorphism, so by Theorem 2.1, (I, f) is an arc.

PROPOSITION 3.6. *Let $f : I \to I$ be continuous. Suppose there is a point $c \in (0, 1)$ such that $f(c) < c$, $f|_{[0,c]}$ is strictly increasing, and $f|_{[c,1]}$ is strictly decreasing. Then if f has a single fixed point, (I, f) is a point, and if f has more than one fixed point, (I, f) is an arc.*

Proof. Since $f(c) < c$ and f is decreasing on $[c, 1]$, f has no fixed points to the right of c. Furthermore, we have that f has no points of period 2. To see this, suppose that x_1 and x_2 form a period two orbit with $x_1 < x_2$. Then $f(x_1) = x_2 > x_1$, and $f(x_2) = x_1 < x_2$. Thus there is a fixed point p with $x_1 < p < x_2$. Since all fixed points of f are to the left of c, we know that $x_1 < c$. Now, suppose that $x_2 < c$. The function f is increasing on $[0, c]$, so then $x_2 = f(x_1) < f(x_2) = x_1$. This is a contradiction, and we know that $x_2 > c$. But since f is increasing on $[0, c]$, we have the following inequality: $x_2 = f(x_1) < f(c) < c$. This contradicts that $x_2 > c$, so f cannot have a period 2 orbit.

Next, let $S = \{x \mid f(x) = x\}$. If S is a single point, then the absence of period two orbits and Proposition 3.1 give that (I, f) is a point. Suppose that S contains two points. As in Proposition 3.5, we have (I, f) is homeomorphic to $(K, f|_K)$ where $K = \bigcap_{i=0}^{\infty} f^i(I)$. But $K \subseteq f(I) \subseteq [0, c]$, and $f|_{[0,c]}$ is one to one, so $f|_K$ is an onto homeomorphism, and by Theorem 2.1, (I, f) is an arc.

THEOREM 3.7. *If $f : I \to I$ is continuous and unimodal and f has no points of period two, then (I, f) is either an arc or a point.*

Proof. Let c be the point for which f is one to one on each of $[0, c]$ and $[c, 1]$. When f is increasing on $[0, c]$, there are two cases to consider. In the case that $f(c) > c$ and f has no fixed points to the left of c, we have that f has a unique fixed point, p. In addition, since f has no point of period 2, the theorem of Sharkovsky guarantees that f has exactly one fixed point and no other periodic points. Thus, by Proposition 3.1, (I, f) is a point. When $f(c) > c$ and f has a fixed point to the left of c or when $f(c) \leq c$, Propositions 3.4, 3.5, and 3.6 give that (I, f) is either an arc or a point. When f is decreasing on $[0, c]$, conjugation of f by the homeomorphism $h(x) = 1 - x$ produces a unimodal function which is increasing on $[0, c]$ and decreasing on $[c, 1]$. The conclusion then follows from Theorem 2.3.

PROPOSITION 3.8. *Let $f : I \to I$ be continuous. Suppose there is a point $c \in (0, 1)$ such that $f|_{[0,c]}$ is strictly increasing, and $f|_{[c,1]}$ is strictly decreasing. Suppose further that f has a point of period two but no point of period four, and that f has no fixed point less than c. Then (I, f) is an arc.*

Proof. Notice that f has exactly one fixed point, p, and $p > c$. Furthermore, by remarks made in the proof of Proposition 3.6, $f(c) > c$.

By Theorem 2.4, if $< x_0, x_1, x_2, \cdots > \in (I, f)$, then $x_i \in \bigcap_{i=0}^{\infty} f^i(I) = K$. In particular, $a = f^2(c) \leq x_i \leq f(c) = b$ for each $i = 0, 1, 2, \ldots$. We will consider two cases.

CASE 1: $f^2(c) \geq c$. In this case, $K \subseteq [c, 1]$, and $f|_K$ is one to one. Therefore, (I, f) is either an arc or a point. Since f has a period two orbit, we know that (I, f) contains more than one point, and so (I, f) is an arc.

CASE 2: $f^2(c) < c$. In this case we see by Theorem 2.4 that (I, f) is homeomorphic to $([a, b], f|_{[a,b]})$. We claim that $f(a) > c$. To see this, note that if $f(a) \leq c$, then since $f^3(c) \leq c$, $f^3(0) \geq 0$, and f has no fixed points less than c, f has a point of period 3. By the theorem of Sharkovsky [S], f has a point of period four, which contradicts the hypothesis of the theorem.

Now, since f is one to one on $[a, c]$ and $[c, b]$, and $f([a, c]) \subseteq [c, b]$, we know that f^2 is one to one on $[a, c]$. Furthermore, since $f^2(a) > a$ and $f^2(c) = a$, we know that f^2 is strictly decreasing on $[a, c]$. Also, there is a unique point $r \in (c, b)$ for which $f(r) = c$, and furthermore, $c < p < r$.

Similarly, since f is one to one on $[c, r]$ and on $[c, b]$ and $f([c, r]) = [c, b]$, we know that f^2 is one to one on $[c, r]$. Also, since $f^2(c) = a$ and $f^2(r) = b$, we know that f^2 is strictly increasing on $[c, r]$.

Next, notice that since f is one to one on $[r, b]$ and $[a, c]$ and $f([r, b]) = [a, c]$, we know that f^2 is one to one on $[r, b]$. Furthermore, since $f^2(r) = b$ and $f^2(b) < b$, we know that f^2 is strictly decreasing on $[r, b]$.

We next claim that $f^2(b) > p$. To see this suppose not. Then $[p, b]$ is contained in both $f^2([p, r])$ and $f^2([r, b])$. This means that f^2 is turbulent and hence has periodic points of all periods [BC, Lemma II.3, page 26]. But this contradicts the hypotheses of the theorem. Similarly, $f^2(a) < p$, for if $f^2(a) \geq p$, then $[a, p]$ is contained in both $f^2([a, c])$ and $f^2([c, p])$, which means that f^2 is turbulent.

By the above claims, we see that $f^2([a, p]) = [a, p]$ and $f^2([p, b]) = [p, b]$. Therefore, $([a, b], f^2|_{[a,b]})$ is the union of the two inverse limit spaces, $([a, p], f^2|_{[a,p]})$ and $([p, b], f^2|_{[p,b]})$. But we have shown that $f^2|_{[a,p]}$ and $f^2|_{[p,b]}$ satisfy the conditions of Theorem 3.7, and so these two inverse limit spaces are arcs or points.

Finally, (I, f^2) is the union of two arcs or points, but is not a point by Proposition 3.1. Since (I, f^2) is a chainable continuum, (I, f^2) is an arc, and (I, f^2) is homeomorphic to (I, f) by Theorem 2.2.

COROLLARY 3.9. *Let $f : I \to I$ be unimodal and have a single fixed point, a period two orbit, but no other periods. Then (I, f) is an arc.*

Proof. When f is increasing on $[0, c]$, then this is Proposition 3.8. When f is decreasing on $[0, c]$, then conjugation by the homeomorphism $h(x) = 1 - x$ produces a function which satisfies the conditions of Proposition 3.8. By Theorem 2.3, (I, f) is homeomorphic to $(I, h^{-1} \circ f \circ h)$.

4. A CHARACTERIZATION OF UNIMODAL MAPS WHOSE INVERSE LIMITS ARE ARCS

We now see our main result:

THEOREM 4.1. *If f is a unimodal map, then (I, f) is an arc if and only if one of the following two conditions are met:*

1) *f has more than one fixed point but no points of other periods.*
2) *f has a single fixed point, a period two orbit, but no points of period greater than two.*

Proof. By Proposition 3.1 and Theorem 3.7, condition 1) implies that (I, f) is an arc. By Corollary 3.9, condition 2) implies that (I, f) is an arc.

Suppose that (I, f) is an arc, but condition 1) is not met. Then either f has a single fixed point, or f has points of period greater than one. If f has a single fixed point but no other periodic points, then (I, f) is a point by Proposition 3.1. Therefore, f has a point of some other period. By the theorem of Sharkovsky, [S], f has a periodic orbit $\{y, z\}$ of period two. Furthermore, by Corollary 3.3, f has no points of period greater than two, because (I, f) is an arc. Since f is unimodal, it is easy to verify that f has a unique fixed point between y and z. Then by Proposition 3.2, (I, f) has a single fixed point, and condition 2) is met.

We end the paper with an example which shows that this result cannot be extended to piecewise monotone maps.

EXAMPLE 4.2. *There is a piecewise monotone function $f : I \to I$ such that f has no points of period two, but (I, f) is not an arc or a point.*

Let $g : I \to I$ be the function defined by $g(x) = 2x$ for $0 \leq x \leq 1/2$, and $g(x) = 3/2 - x$ for $1/2 \leq x \leq 1$. Notice that g satisfies the hypotheses of Proposition 3.2, and therefore, (I, g) is not an arc. (In fact, it is commonly known that (I, g) is a $\sin(1/x)$ continuum). Furthermore, g has only points with periods one and two. Consider the function $f = g^2$. This function is piecewise linear and has no point of period two, but (I, f) is homeomorphic to (I, g).

References

[BC] L.S. Block and W.A. Coppel, *Dynamics in One Dimension*, Lecture Notes in Mathematics, Volume 1513, Springer-Verlag, Berlin, 1992.

[BD] M. Barge and B. Diamond, *The Dynamics of Continuous Maps on Finite Graphs through Inverse Limits*, (preprint).

[BH] L. Block and D. Hart, *Stratification of the space of unimodal interval maps*, Ergod.Th. and Dynam. Sys. **3** (1983), 533–539.

[BM] M. Barge and J. Martin, *Chaos, Periodicity, and Snakelike Continua*, Trans. Amer. Math. Soc. **289** (1985), 355–365.

[H]   S. Holte, *Generalized Horseshoe Maps and Inverse Limits*, Pacific J. Math. **156** (1992), 297–305.

[M]   R. Munasinghe, *Composants, Unstable Sets, and Minimal Sets of Inverse Limit Spaces*, Doctoral dissertation, University of Wyoming, 1991.

[S]   A. Sharkovsky, *Coexistence of Cycles of a Continuous Map of the Line into itself*, Ukrain. Mat. Zh. **16** (1964), 61–71.

E-mail address: block@math.ufl.edu "Louis Block"

14

A SYMBOLIC REPRESENTATION OF INVERSE LIMIT SPACES FOR A CLASS OF UNIMODAL MAPS

KAREN M. BRUCKS Department of Mathematical Sciences, University of Wisconsin at Milwaukee, Milwaukee, Wisconsin 53201, USA
BEVERLY DIAMOND Department of Mathematics, University of Charleston, Charleston, South Carolina 29424, USA

ABSTRACT. An algorithm which provides a planar embedding of inverse limit spaces for a class of unimodal maps (including certain maps from the tent family) is presented. The algorithm is based primarily on kneading theory for unimodal maps.

1. INTRODUCTION

Inverse limit spaces tend to be difficult to represent geometrically. The intent of this paper is to make use of kneading theory for unimodal maps and coding for Markov partitions to provide a symbolic representation of inverse limit spaces formed by using certain unimodal maps as single bonding maps, and to provide an algorithm which draws in the plane segments of arbitrary length of the composants of these inverse limit spaces. For a unimodal map whose second iterate has a dense orbit, we prove that the shift map on the inverse limit space is conjugate to the shift map on a set of symbol sequences (where the topology on the set of symbol sequences is a modification of the usual determined by identifying certain sequences). In other cases, the set of symbol sequences is a homeomorphic representation of the inverse limit space. If the critical point of the map satisfies certain additional conditions, then composants of elements of the inverse limit space can be identified symbolically. A planar embedding of the inverse limit space motivated by work of Barge and Holte on embedding inverse limit spaces of unimodal maps with periodic critical point as

1991 *Mathematics Subject Classification.* primary 58F03, 54F13, 54H20, secondary 58F13.
Key words and phrases. inverse limit space, unimodal map, kneading theory.

attractors of planar homeomorphisms is presented. Since this type of coding is well-known to dynamicists, this paper is aimed primarily at continuum theorists who work with inverse limit spaces.

In the remainder of this section, we introduce terminology, notation and preliminary results. In §2, we present kneading theory and the symbolic coding of points in the inverse limit space and identify composants for certain bonding maps. In §3, we give an algorithm for representing the inverse limit space in the plane.

A *continuum* is a compact connected metric space. For any nondegenerate continuum X and any point $x \in X$, the *composant* of x in X is the union of all proper subcontinua of X containing x, and the *arc component* of x in X is the union of all arcs in X containing x. Given a continuous function $f : I \to I$ of the interval $I = [0, 1]$, the associated *inverse limit space* (I, f) with single bonding map f is defined by

$$(I, f) = \{\underline{x} = (\dots, x_1, x_0) : x_n \in I, f(x_{n+1}) = x_n, n = 0, 1, 2, \dots\}$$

and has metric

$$\underline{d}(\underline{x}, \underline{y}) = \sum_{i \geq 0} \frac{|x_i - y_i|}{2^i}.$$

The *shift homeomorphism* $\hat{f} : (I, f) \to (I, f)$ is defined by

$$\hat{f}((\dots, x_1, x_0)) = (\dots, x_1, x_0, f(x_0)).$$

For $n \in Z^+ \cup \{0\}$, $\Pi_n : (I, f) \to I$ is the nth projection map defined by $\Pi_n(\underline{x}) = x_n$. If a map $f : I \to I$ is not onto, then $(I, f) = (J, f|_J)$ where $J = \cap_{n=1}^{\infty} f^n(I)$; since $f|_J$ is a surjective map, it is enough to prove our results for surjective maps.

As usual, f^n will denote the nth iterate of f, i.e., $f^n = f \circ f \circ \cdots \circ f$ n times. A continuous map $f : I \to I$ is *Markov* if there is a finite collection of points $0 = c_0 < c_1 < \cdots < c_n = 1$ such that $f(\{c_0, c_1, \dots, c_n\}) \subseteq \{c_0, c_1, \dots, c_n\}$ and each interval $[c_{i-1}, c_i]$, $i = 1, 2, \dots, n$, can be written as a finite union of subintervals $[a, b]$ such that $f|_{[a,b]}$ is monotone and $f(\{a, b\}) \subseteq \{c_0, c_1, \dots, c_n\}$. Any such collection $\{c_0, c_1, \dots, c_n\}$ will be called a *Markov partition* for f. If f is Markov, there is a standard way to assign a code or itinerary to $x \in I$ representing the behavior of the forward iterates of x under f (which can be extended to $\underline{x} \in (I, f)$). The assigning of itineraries in kneading theory, which we present in the next section, is very similar. Although we modify the itineraries provided by kneading theory slightly in a manner consistent with the standard coding associated with a Markov partition in order to state the theorems more easily, we feel the modification is simple enough so that we do not present full details here of the coding for points under a Markov map.

A map $f : I \to I$ is *unimodal* if there is $c \in (0, 1)$ such that f is strictly increasing on $[0, c)$ and strictly decreasing on $(c, 1]$; the point c is the *critical point* of f. If f is a surjective unimodal map, then the interval $[f^2(c), f(c)]$ is invariant under f.

That is, $f([f^2(c), f(c)]) = [f^2(c), f(c)]$. Let f restricted to the interval $[f^2(c), f(c)]$ be called the *core map* of f. The inverse limit space for the original map is identical to that of the core map except in possibly having an additional arc (if $f(1) \neq 0$) which is either compact, if $f(1)$ is fixed, or an infinite ray entwined with the inverse limit space of the core map, if $f(1)$ is not fixed. If the critical point of the core map is periodic, then both the original map and the core map are Markov.

A continuous map $f : I \to I$ is *locally eventually onto* if for every $\varepsilon > 0$, there is a positive integer N such that if U is an open subinterval of I with $diam(U) \geq \varepsilon$, then for every $n \geq N$, $f^n(U) = I$. (Since I is compact, this is equivalent to simply requiring that for every open subinterval U of I, there is a positive integer N such that for $n \geq N$, $f^n(U) = I$.) If f is a unimodal map, then f is locally eventually onto if and only if f^2 has a dense orbit (1.2). If a unimodal map f is locally eventually onto, then f must be onto and f must be its own core map. That is, $f(1) = 0$.

Define the family of tent maps $f_\lambda : I \to I$, for $0 \leq \lambda \leq 2$, as

$$f_\lambda(x) = \begin{cases} \lambda x, & 0 \leq x \leq .5 \\ \lambda(1 - x), & .5 \leq x \leq 1. \end{cases}$$

It is easy to prove that if $\lambda \in [\sqrt{2}, 2]$, the core of f_λ is locally eventually onto. For instance, see the proof of [3, Lemma 2] or [8, Proposition 2.5.5]. Examples in §3 will be drawn from the tent family. Results of this paper can be applied to more general unimodal families of maps, for example the logistic family $f_\lambda(x) = 4\lambda x(1 - x)$ with $\lambda \in [0, 1]$, using the work of Holte [7] (see the remarks following 2.5).

The properties of locally eventually onto maps that are important for our results are contained in the next lemma. To anyone unfamiliar with kneading theory, the significance of the first property may not be clear until the next section where we symbolically code points in the inverse limit space.

Lemma 1.1. *Suppose that a unimodal map $f : I \to I$ is locally eventually onto. Then*

(1) if x, y are distinct points of I, then for some nonnegative integer n, c lies between $f^n(x)$ and $f^n(y)$; and

(2) if H is a proper connected closed subset of (I, f), then as $n \to \infty$,

$$diam(\Pi_n(H)) \to 0.$$

Proof: (1) Suppose that $x, y \in I$ with $x < y$, and let n be the least nonnegative integer for which $f^n([x, y])$ contains the critical point c.

(2) If H is connected, then $\Pi_n(H)$ is connected for each $n \in Z^+$. Suppose that there is $\varepsilon > 0$ such that for infinitely many n, $diam(\Pi_n(H)) > \varepsilon$. Since f is locally eventually onto, there is a positive integer m such that any subinterval of I of width ε maps over I under f^m. Then for infinitely many coordinates, $\Pi_k(H) = I$, a contradiction to the fact that H is a proper closed subset of (I, f). $\square$

Lemma 1.2. *Suppose that f is unimodal. Then f is locally eventually onto if and only if f^2 has a dense orbit.*

Proof: Since f^2 has a dense orbit if and only if given nonempty open sets U, V of I, there is an integer n such that $f^{2n}(U) \cap V \neq \emptyset$ (see, for example, [10, Theorem 5.9]), it follows easily from the definition that if f is locally eventually onto, then f^2 has a dense orbit.

According to [2, Lemma 2.2], if for a continuous map $g : I \to I$, g^2 has a dense orbit, then given $\varepsilon, \delta > 0$, there is an integer N such that if $n > N$, J is a subinterval of I, and $\operatorname{diam}(J) > \delta$, then $[\varepsilon, 1 - \varepsilon] \subseteq f^n(J)$. If ε is chosen small enough so that $c \in [\varepsilon, 1 - \varepsilon]$, then $c, 1 \in f^{n+1}(J)$ and $0, 1 \in f^{n+2}(J)$. $\square$

2. THE KNEADING THEORY

We recall some basic notions of kneading theory; for more details see [4]. Let f be a unimodal map with critical point c. For each $x \in [0, 1]$ the *itinerary* of x under the map f is given by $I(x) = b_0 b_1 b_2 \ldots$, where $b_i = 2$ if $f^i(x) > c$, $b_i = 1$ if $f^i(c) < c$, and $b_i = C$ if $f^i(x) = c$, with the usual convention that the itinerary stops after the first C. The *kneading sequence* of the map f, denoted $K(f)$, is defined to be the itinerary of $f(c)$. If f is surjective, then $K(f)$ begins $21 \ldots$. The *parity-lexicographical ordering* is put on the set of itineraries as follows. Set $1 < C < 2$. Let $w = w_0 w_1 \ldots$ and $v = v_0 v_1 \ldots$ be two distinct itineraries and let k be the first index where the itineraries differ. If $k = 0$, then $w < v$ iff $w_0 < v_0$. If $k > 0$ and $w_0 \ldots w_{k-1} = v_0 \ldots v_{k-1}$ has an even number of 1's, i.e., has *even parity*, then $w < v$ iff $w_k < v_k$; if $w_0 \ldots w_{k-1}$ has an odd number of 1's, then $w < v$ iff $v_k < w_k$. It is an elementary fact that the map $x \mapsto I(x)$ is monotone, i.e., $x < y$ implies that $I(x) \leq I(y)$ [4, Lemma II.1.3]. As $f(c)$ is the maximum value of the function, it follows that if $A = a_0 a_1 \ldots = K(f)$, then any shift of A ($a_j a_{j+1} \ldots$ for $j \geq 0$) is less than or equal to the kneading sequence itself in the parity-lexicographical order. If the kneading sequence of f is finite, say $K(f) = a_0 a_1 \ldots a_n C$, then one can show that exactly one of $P = (a_0 a_1 \ldots a_n 2)^\infty$, $Q = (a_0 a_1 \ldots a_n 1)^\infty$ is less than $K(f)$ and exactly one is greater than $K(f)$ in the parity-lexicographical order. Moreover, the only itinerary falling between P and Q in the parity-lexicographical order is $K(f)$. For convenience, when $K(f)$ is finite, let $I'(f(c))$ denote the smaller of P and Q defined above. The following lemma is a rephrasing of [4, Theorem II.3.8] and can be proven as in [4].

Lemma 2.1. *Let f be a unimodal map. Suppose that A is a sequence of 1's and 2's, either infinite or finite and ending in C, such that all shifts of A are less than $K(f)$, or $I'(f(c))$ if $K(f)$ is finite, and greater than or equal to $I(f^2(c))$. Then there exists an $x \in [0, 1]$ such that $I(x) = A$.*

It follows from 2.1 and the preceeding paragraph that if f is a unimodal map and if $w = w_1 w_2 \ldots w_n$, with $w_i \in \{1, 2\}$ for $1 \leq i \leq n$, is such that $I(f^2(c)) \leq$

$w_j \ldots w_n \le I(f(c))$ for $1 \le j \le n$, then for some $x \in [0,1]$, the initial segment of $I(x)$ of length n is precisely w.

We say that the unimodal map f has *uniqueness of itineraries* provided that $x \ne y$ implies that $I(x) \ne I(y)$. A unimodal map f which is locally eventually onto has uniqueness of itineraries (1.1, part (1)).

For the remainder of this paper, unless otherwise stated, all maps are unimodal and locally eventually onto.

We make use of the forward itinerary $I(x)$ defined for $x \in I$ under f to define a two-sided itinerary for $\underline{x} \in (I, f)$, but modify the itinerary of c and any point eventually mapping to c. We wish the nth coordinate of the forward itinerary of a point x to reflect in which of the closed intervals, $J_1 = [0, c]$ or $J_2 = [c, 1]$, the nth forward iterate of x falls. The point c belongs to both intervals, and points below c and above c have codes 1 and 2 respectively; the point c is assigned (temporarily) two codes having 1 and 2 as 0th coordinates.

Suppose that c is periodic, and let $K(f) = 21 \ldots C$. One can prove that if f has uniqueness of itineraries, neither of the sequences $(21 \ldots 1)^\infty$, $(21 \ldots 2)^\infty$ are realized for the map f. As was pointed out above, one is too large; which depends on the parity of the segment of code preceding the first C. Since no sequence lies between $(21 \ldots C)^\infty$ and the remaining sequence in the parity lexicographical ordering, if itineraries are unique for f, this second sequence can not be realized. On the other hand, the map f has a turning point at c, hence there is an open interval U containing c such that all x in $f(U) \setminus \{1\}$ have itinerary beginning with the same finite segment of length n, namely, whichever of $21 \ldots 1$ and $21 \ldots 2$ is less than $K(f)$. We define the *modified forward itinerary* of $f(c)$ to be $(21 \ldots *)^\infty = I'(f(c))$. (For those familiar with coding for Markov partitions, this is the code obtained by taking intersections of images and preimages of interiors of rectangles to define codes— the second code defines a single point rather than an interval of points having a given finite itinerary.) The forward itinerary of any $x \in I$ eventually mapping to c now becomes modified in the obvious way, and will be denoted by $I'(x)$. Note that itineraries are still unique for the map f. For ease in the following discussion, if $x \ne c$ and either c is not periodic or c is periodic but x does not map onto c, then $I'(x) = I(x)$.

In either the periodic or non-periodic case, for $\underline{x} \in (I, f)$, a *two-sided itinerary* for $\underline{x}$, $E(\underline{x})$, is defined to be a sequence $S = (\ldots s_{-n} s_{-n+1} \ldots s_{-1} s_0 . s_1 \ldots s_n \ldots)$ of 1's and 2's with the property that (a) $x_{|i|} \in J_{s_i}$ for $i \le 0$, (b) $f^i(x_0) \in J_{s_i}$ for $i \ge 0$, and (c) if x_j (or $f^j(x_0)$) $= f(c)$ for any j, then $s_{-j} s_{-j+1} s_{-j+2} \ldots$ ($s_j s_{j+1} s_{j+2} \ldots$ respectively) equals $(21 \ldots *)^\infty$. If c is periodic with period n_0, then the two-sided itinerary of the point $\ldots f^{n_0-1}(c)c . f(c)f^2(c) \ldots f^{n_0-1}(c)c \ldots \in (I, f)$ (or any shift of this element) is, by the above, $\ldots 21 \ldots * . 21 \ldots * \ldots$ (or the appropriate shift, respectively). Every $\underline{x}$ has at least one two-sided itinerary and no $\underline{x}$ has more than two two-sided itineraries (2.3).

The one-sided infinite sequence $\ldots s_{-n}s_{-n+1}\ldots s_{-1}s_0$. will be called the *backward itinerary* of $\underline{x} \in (I, f)$. We occasionally refer to a backward itinerary as an interval itinerary since one such itinerary usually represents an interval of points in (I, f) (see the remark following the proof of 2.2).

We put a *reverse order*, $<_r$, on the set of backward itineraries as follows. Let $A = (\ldots a_{-n}a_{-n+1}\ldots a_{-1}a_0)$, and $B = (\ldots b_{-n}b_{-n+1}\ldots b_{-1}b_0)$ be distinct backward itineraries. We say $A <_r B$ provided reading from right to left, $A < B$ in the parity-lexicographical ordering, similarly for $A >_r B$. Likewise we say $A = a_{-n}a_{-n+1}\ldots a_{-1}a_0 <_r B = b_{-n}b_{-n+1}\ldots b_{-1}b_0$ provided reading from right to left, $A < B$ in the parity-lexicographical ordering, same for $A >_r B$.

Let Σ denote the set of all two-sided sequences of 1's and 2's with metric given by

$$d((\ldots s_{-n}s_{-n+1}\ldots s_0 . s_1 \ldots s_n \ldots), (\ldots t_{-n}t_{-n+1}\ldots t_0 . t_1 \ldots t_n \ldots)) =$$

$$\sum_{i \in Z} \frac{|s_i - t_i|}{2^{|i|}}.$$

The space Σ is a compact T_2 space. The *shift map* $\sigma : \Sigma \to \Sigma$ is defined by

$$\sigma((\ldots s_{-n}s_{-n+1}\ldots s_0 . s_1 \ldots s_n \ldots)) = (\ldots s_{-n}s_{-n+1}\ldots s_0 s_1 .. s_2 \ldots s_n \ldots).$$

Define $\Sigma' \subseteq \Sigma$ as follows: $S \in \Sigma'$ if every one-sided infinite segment of S of the form $s_{-k}s_{-k+1}\ldots s_0 . s_1 \ldots s_n \ldots$, $k \in Z$, is realized for f as a (possibly modified) forward itinerary. The first easy result simply states that Σ' is precisely the set of two-sided itineraries for (I, f).

Lemma 2.2. *If $S \in \Sigma'$, there is exactly one $\underline{x} \in (I, f)$ witnessing S.*

Proof: Suppose that $S \in \Sigma'$, and for $n \in Z^+$, let S_n denote the subword of S of length $2n + 1$ covering coordinates $-n$ to n, that is, $s_{-n}s_{-n+1}\ldots s_0 . s_1 \ldots s_n$. For each n, $\{x \in I : I'(x)$ begins with $S_n\}$ is a nonempty compact subset of I. Since f is onto, it follows that $K_n = \{\underline{x} : I'(x_n)$ begins with $S_n\}$ is a nonempty compact subset of (I, f), hence $\{K_n : n \in Z^+\}$ is a decreasing collection of compact sets. Thus $\bigcap_{n \in Z^+} K_n \neq \emptyset$, so that for at least one $\underline{x} \in (I, f)$, $E(\underline{x}) = S$.

Suppose that $\underline{x} \neq \underline{y}$, so that $x_n \neq y_n$ for some coordinate n. Then $I'(x_n) \neq I'(y_n)$, so $E(\underline{x}) \neq E(\underline{y})$. $\square$

Note that the proof of the above lemma can be modified easily to show that Σ' is a closed subset of Σ, thus Σ' is compact. A different modification proves that for any backward itinerary $\ldots s_{-n}s_{-n+1}\ldots s_0 .$, $K = \{\underline{x} \in (I, f) \mid \underline{x}$ has backward itinerary $\ldots s_{-n}s_{-n+1}\ldots s_0 ..\}$ is a connected nonempty set, since by the definition of the parity-lexicographical ordering, $\{\{\underline{x} : I'(x_n)$ begins with $s_{-n}s_{-n+1}\ldots s_0\}\}_{n \in Z^+}$ is a decreasing collection of compact connected sets.

In the following, $orb(x) = \{f^n(x) : n \in Z^+ \cup \{0\}\}$.

Lemma 2.3. *If $\underline{x} \in (I, f)$, then $\underline{x}$ witnesses at most two two-sided codes; $\underline{x}$ witnesses exactly two two-sided codes if and only if there exist positive integers n, m such that $x_m \notin orb(c)$ while $f^n(x_m) = c$.*

Proof: Suppose that there exist n, m such that $x_m \notin orb(c)$ while $f^n(x_m) = c$. For $i \geq m$, which of J_1, J_2 contains x_i is well-defined. Choose j to be the smallest integer for which $f^j(x_m) = c$; for $i < j$, $f^i(x_m)$ is in a unique interval. There are exactly two choices for the interval containing $f^j(x_m)$. Since the forward itinerary of $f(c)$ is uniquely defined, there are exactly two choices for the two-sided itinerary for $\underline{x}$.

Suppose that there do not exist n, m such that $x_m \notin orb(c)$ while $f^n(x_m) = c$. If $x_n \in orb(c)$ for every n, then c must be periodic, and the code for $\underline{x}$ is some shift of the code $(21 \ldots *)^\infty$. If for some m, $x_m \notin orb(c)$ and for every $n \in Z^+ \cup \{0\}$, $f^n(x_m) \neq c$, then the interval containing x_m or $f^n(x_m)$ is well-defined and $\underline{x}$ has a single itinerary. $\square$

We identify any two sequences in Σ' having the property that they disagree in exactly one coordinate and the code following the coordinate on which they disagree (that is, representing a forward itinerary) is the modified code for $f(c)$. (Note that two sequences representing itineraries can disagree in precisely one place but not have the code following the point of disagreement be the modified code for $f(c)$.) The following lemma indicates that this identification is well defined and provides a decomposition $Q\Sigma$ of Σ'.

Lemma 2.4. *Each sequence in Σ' is identified with at most one other sequence.*

Proof: Let R, S be sequences identified according to the above rule and disagreeing in coordinate k. Suppose that S, T are also sequences identified and disagreeing in coordinate l. Without loss of generality, $l < k$. Then

$$s_{l+1} s_{l+2} \cdots s_k s_{k+1} \cdots$$

and

$$s_{k+1} s_{k+2} \cdots$$

equal $I'(f(c))$. Since f has uniqueness of itineraries, c is periodic and for some n, $k - l = n n_0$ where n_0 is the period of c. But then

$$s_{k-n_0+1} s_{k-n_0+2} \cdots s_k,$$

$$r_{k-n_0+1} r_{k-n_0+2} \cdots r_k$$

are of the form $w1$ and $w2$, where wC is the initial segment of $K(f)$. This contradicts the fact that only one of the two words $w1$ and $w2$ is allowed. $\square$

We can now define a $1 - 1$ map from this quotient space $Q\Sigma$ of Σ' onto the inverse limit space (I, f). Maps $f : X \to X$, $g : Y \to Y$ are *conjugate* if there is a homeomorphism $h : X \to Y$ such that $h(f(x)) = g(h(x))$ for each $x \in X$. The map h is a (topological) *conjugacy*.

Theorem 2.5. *Suppose that f is a unimodal map which is locally eventually onto. Let $Q\Sigma$ denote the collection of all allowed sequences for f as defined above with identifications as indicated. The map $h : Q\Sigma \to (I, f)$ which assigns to $S \in Q\Sigma$ the unique element of (I, f) having S as two-sided code is a conjugacy.*

Proof: We have already mentioned that h is $1 - 1$ and onto. Let $q : \Sigma' \to Q\Sigma$ denote the quotient map. According to [9, Theorem 9.4], to show that h is continuous, it suffices to show that the composition $h \circ q$ is continuous.

Suppose that $\varepsilon > 0$; we wish to show that for some $\delta > 0$, $d(E(\underline{x}), E(\underline{y})) < \delta$ implies that $\underline{d}(\underline{x}, \underline{y}) < \varepsilon$. Choose N so that $\sum_{i \geq N} 1/2^i < \varepsilon/2$. According to 1.1, there is $M \geq N$ such that for $a, b \in I$, if $|a - b| \geq \varepsilon/4$ then $I'(a)$ and $I'(b)$ disagree by the Mth place. Choose $\delta > 0$ so that for $S, T \in \Sigma'$, if $d(S, T) < \delta$ then S and T agree in the $2M + 1$ places from M to $-M$. Suppose that $E(\underline{x})$ and $E(\underline{y})$ agree in the $2M + 1$ places. Then $|x_i - y_i| < \varepsilon/4$ for $0 \leq i \leq M$ and so

$$\underline{d}(\underline{x}, \underline{y}) < \Sigma_{i=1}^{M}(\varepsilon/4)(1/2^i) + \varepsilon/2$$
$$< \varepsilon/2 + \varepsilon/2 < \varepsilon.$$

Since the composition $h \circ q : \Sigma' \to (I, f)$ is continuous, and $\Sigma, (I, f)$ are compact T_2 spaces, the map $h \circ q$ is closed, hence is a quotient map. Again by [9, Theorem 9.4], h^{-1} is continuous. The fact that h is a conjugacy follows from the fact that $E(\hat{f}(\underline{x})) = \sigma(E(\underline{x}))$. $\square$

Suppose that g is any unimodal map with periodic critical point and kneading sequence equal to that for some map f_λ in the tent family for $\lambda \in [\sqrt{2}, 2]$. By going to the core of each of these maps, we can assume that g, f_λ are surjective and f_λ is locally eventually onto. (Recall from §1 that we may lose an arc from the inverse limit space by going to the core.) According to [7, Corollary 1], (I, g) and (I, f) are homeomorphic, and so $Q\Sigma$ is a model for (I, g) even though g is not necessarily locally eventually onto. On the other hand, if the kneading sequence for g is not realized by a map in the tent family, as with the period doublings in the quadratic family, for example, then 2.5 does not provide a model for (I, g). Often there are subsets of (I, g) which can be collapsed to provide a conjugacy between the resulting quotient space and an appropriate set of symbol sequences, but we are not providing a general theory for such maps.

Suppose that $A = \ldots a_{-n}a_{-n+1}\ldots a_0.$ is an interval itinerary for f and that $w = w_{-n+1}\ldots w_0$ is a word of length n. Let A/w denote the infinite backward symbol code obtained by replacing the initial segment of A of length $|w|$ (i.e., $a_{-n+1}\ldots a_0$) by w. We say that w is *allowed for the tail of A* (under f) or, more simply, *allowed for A*, if every finite subword of A/w is realized as a finite forward orbit itinerary for f. The following states that in this case, A/w is actually an interval itinerary for (I, f); the proof is essentially identical to that of 2.2.

Lemma 2.6. *Suppose that the one-sided infinite code* $\ldots s_{-n}s_{-n+1}\ldots s_{-1}s_0$. *has the property that every finite subword is realized as a finite orbit itinerary for f. Then, for some $\underline{x} \in (I, f)$, $\ldots s_{-n}s_{-n+1}\ldots s_{-1}s_0$. is the backward itinerary of $\underline{x}$.*

We abuse terminology slightly and think of an interval itinerary A as the set of elements in the inverse limit space having A as backward itinerary. Two infinite interval codes A and B are *joined in* (I, f) if $A \cup B$ is connected in the inverse limit space, that is, if A and B contain identified points. The original identifications made on the set of symbol sequences tell us when two distinct interval itineraries contain identified points; their codes must agree in all but one place, and the code following the coordinate of disagreement equals the modified code for $f(c)$. The next result makes it easier to describe the algorithm generating joined interval itineraries.

Theorem 2.7. *Let $A = \ldots a_{-n}a_{-n+1}\ldots a_0$. and $B = \ldots b_{-n}b_{-n+1}\ldots b_0$. be two interval codes. Then A and B are joined in (I, f) if and only if for each $n \geq 0$, the segments $S_n = a_{-n}a_{-n+1}\ldots a_0$ and $S'_n = b_{-n}b_{-n+1}\ldots b_0$ either agree or are adjacent in the parity-lexicographic ordering on both all words of length n allowed for f and all words of length n allowed for the tail of A.*

Proof: Let W_n and $W_{n,A}$ denote the sets of all words of length n that are allowed for f and the tail of A, respectively. For each word $w = w_{-n+1}\ldots w_0 \in W_n$, define the subinterval $J_{w_{-n+1}\ldots w_0}$ of I to be the set of those $x \in I$ whose forward itinerary under f has initial segment of length n equal to $w_{-n+1}\ldots w_0$.

By the definition of the parity-lexicographical ordering, if $w, z \in W_n$, the two intervals J_w and J_z are adjacent in I if and only if w and z are adjacent in W_n. Suppose that there are finite words $w < u < z$ of length n allowed for f such that $w, z \in W_{n,A}$. Then every finite subword of A/w or A/z is realized as a finite forward orbit itinerary for f. Since a finite subword of A/u beginning with some a_j lies between the corresponding subwords of A/w and A/z in the parity-lexicographical ordering, and every subword of u is realized as a finite forward itinerary for f, it follows from 2.1 that each finite subword of A/u is realized as a finite forward orbit itinerary for f. Then by 2.6, $u \in W_{n,A}$, so $W_{n,A}$ is a consecutive set of words from W_n and $\bigcup_{w \in W_{n,A}} J_w$ is a connected subset of I. That is, words are adjacent in $W_{n,A}$ if and only if they are adjacent in W_n.

For $n \in N$, let S_n and S'_n be as defined in the statement of the lemma. Suppose that the interval codes A and B are joined, so that $A \cup B$ is a connected set. Then for each n, $\Pi_n(A \cup B)$ is connected and is contained in $J_{S_n} \cup J_{S'_n}$. If S_n and S'_n were not equal or adjacent, $\Pi_n(A \cup B)$ would contain J_w for at least one additional $w \in W$, a contradiction.

Conversely, suppose that for each n, S_n and S'_n either agree or are adjacent.

Claim: For each n, S_n and S'_n either agree entirely or disagree in exactly one place, and are of the form $w_1 1 w_2$ and $w_1 2 w_2$, where w_2 is the initial segment of $I'(f(c))$ of the length required to make $w_1 2 w_2$ of length n.

Proof of claim: Suppose that w and z are adjacent in the set of words of length n allowed for f, and that they disagree in at least two places. Let m and k be the first two coordinates on which w and z disagree, so that (without loss of generality) w, z are of the form

$$w = w_1 w_2 \ldots w_{m-1} 1 w_{m+1} \ldots w_{k-1} w_k w_{k+1} \ldots w_n$$

and

$$z = w_1 w_2 \ldots w_{m-1} 2 w_{m+1} \ldots w_{k-1} z_k z_{k+1} \ldots z_n$$

where $w_k \neq z_k$. Either $w_k = 2, z_k = 1$, or $w_k = 1, z_k = 2$. Consider the first case. Depending on the parity of $w_{m+1} \ldots w_{k-1}$, either

$$s = w_1 w_2 \ldots w_{m-1} 1 w_{m+1} \ldots w_{k-1} 1 w_{k+1} \ldots w_n$$

or

$$t = w_1 w_2 \ldots w_{m-1} 2 w_{m+1} \ldots w_{k-1} 2 w_{k+1} \ldots w_n$$

is between w and z and is allowed for f by 2.1. This contradicts the fact that w and z are adjacent. The other case is similar and the claim is proved.

Then A and B must also disagree in exactly one place and have the code following equal a finite segment of the modified code for $f(c)$. If each interval code is extended to a two-sided itinerary by continuing with the code of $f(c)$, then the resulting two two-sided itineraries are identified in the inverse limit space. It follows that $A \cup B$ is a connected subset of (I, f). $\square$

We now describe the composants of (I, f) symbolically for certain unimodal maps.

Lemma 2.8. *Suppose that $\underline{x}$ and $\underline{y}$ have backward itineraries differing in at most finitely many coordinates. Then $\underline{x}$ and $\underline{y}$ are in the same arc component of (I, f).*

Proof: Let $-n$ be the smallest index for which the backward itineraries for $\underline{x}$ and $\underline{y}$ differ (or 1 if they agree), and define H to be the set of all $\underline{z} \in (I, f)$ such that the backward itinerary of $\underline{z}$ agrees with that of $\underline{x}$ and $\underline{y}$ past the $-n^{th}$ index. Then H is an arc in (I, f) parametrized by the actual coordinate in I_{n+1}. (If two points have the same actual coordinate in I_{n+1}, then there is some $m > n + 1$ for which the two points have differing coordinates. These differing coordinates must be associated with different interval codes.) $\square$

Lemma 2.9. *Suppose that for a unimodal, locally eventually onto map f, the critical point is either periodic or not recurrent. If $\underline{x}$ and $\underline{y}$ have backward itineraries differing infinitely often, then $\underline{x}$ and $\underline{y}$ are not in the same composant of (I, f).*

Proof: Let $\underline{x}$ and $\underline{y}$ be as above. Suppose that H is a proper connected closed subset of (I, f) containing $\underline{x}$ and $\underline{y}$.

Case 1: f has periodic critical point.

Let n_0 be the period of the critical point c, and choose $\varepsilon > 0$ so that for $i \neq j$ and $0 \leq i, j < n_0$, $d(f^i(c), f^j(c)) \geq \varepsilon$. According to 1.1, there is N such that for $j \geq N$, $diam(\Pi_j(H)) < \varepsilon$, thus $\Pi_j(H)$ contains at most one element of the periodic orbit of c. Choose m, n such that $N \leq m < n$, the codes of $\underline{x}$ and $\underline{y}$ differ on both the $-m^{th}$ and $-n^{th}$ coordinates, and m, n are consecutive such integers. Then $c \in \Pi_n(H)$, $c \in \Pi_m(H)$, $c \in \Pi_{n-pn_0}(H)$ for all p, and $c \notin \Pi_j(H)$ for $N \leq j \leq n$ unless $j = n - pn_0$ for some p. It follows that $m = n - pn_0$ for some p, and that the codes of $\underline{x}$ and $\underline{y}$ are of the form:

$$\ldots (w*)^i w 2 \ldots$$

and

$$\ldots (w*)^i w 1 \ldots$$

where $wC \ldots$ is the original and unmodified kneading sequence of f. But of the two symbol sequences obtained by following w by either a 1 or 2, only one is allowed, a contradiction.

Case 2: the critical point of f is not recurrent.

Since c is not recurrent, there is $\epsilon > 0$ such that $d(c, f^n(c)) > \epsilon$ for $n \in Z^+$. According to 1.1, there is a positive integer N such that for $j \geq N$, $diam(\Pi_j(H)) < \varepsilon$. Choose $m > N+1$ such that the codes of $\underline{x}$ and $\underline{y}$ differ on the $-m^{th}$ coordinate. Let n be any integer with $N < n < m$. Since c lies between x_m and y_m, there is $\underline{z} \in H$ such that $z_m = c$. Now $d(z_n, c) > \epsilon$, the three points $x_n, y_n, z_n \in \Pi_n(H)$, and $diam(\Pi_n(H)) < \varepsilon$, thus x_n and y_n cannot lie on opposite sides of c. That is, the codes for $\underline{x}$ and $\underline{y}$ cannot differ on the nth coordinate. This contradicts the assumption that the codes for $\underline{x}$ and $\underline{y}$ differ infinitely often. $\square$

That is, the arc component of a point equals the composant of the point in this special case. The proof of 2.9 can be modified to show that if a unimodal, locally eventually onto map f does not have the "shadowing property" (see [5]), then the arc components of (I, f) are the composants of (I, f). It is unknown whether this holds for any unimodal, locally eventually onto map.

Corollary 2.10. *Suppose that for a unimodal, locally eventually onto map f, the critical point of f is either periodic or not recurrent. Then $\underline{x}$ and $\underline{y}$ are in the same composant of (I, f) if and only if their backward itineraries differ in at most finitely many coordinates.*

3. THE PLANAR REPRESENTATION

The symbolic representation of the inverse limit space of §2 is independent of any embedding in the plane. In this section, we present a particular embedding of (I, f) in the plane that makes a second use of kneading theory to place segments of composants.

Throughout this section assume that f is a locally eventually onto, unimodal map with critical point either periodic or nonrecurrent. According to the previous section, the composant of $\underline{x} \in (I, f)$ consists of all points having backward itinerary agreeing with that of $\underline{x}$ except in finitely many coordinates (2.10). Also, a segment of the composant of $\underline{x}$ can be obtained from taking all words of a given length n allowed for the tail or backward itinerary of $\underline{x}$ and attaching each to the tail to obtain an interval itinerary or set of points with given backward itinerary (2.7 - 2.10). We conclude this section with an embedding of the entire inverse limit space, but lead up to the embedding by building segments of single composants. This reflects what one would program a computer to do, and the proof that the final placement is in fact an embedding relies on this construction.

First, we build something we call the *folded nth iterate of f*. For $n \in N$, let W_n be the set of codes of length n allowed as forward itineraries for f with $|W_n| = m_n$. Recall that $J_w = \{x \in I \mid w$ is the initial segment of the forward itinerary of $x\}$; each J_w is a subinterval of monotonicity for f^n. If $w \in W_n$ is in the ith position when W_n is reverse ordered, then a vertical interval representing w is placed over the point $(\frac{i-1}{m_n-1}, 0)$ extending from $(\frac{i-1}{m_n-1}, w_l)$ to $(\frac{i-1}{m_n-1}, w_h)$ where w_h, w_l are the upper and lower endpoints of $f^n(J_w)$. (The horizontal coordinates of the intervals are unimportant– what is important is that the intervals are placed in accordance with the reverse ordering on the words.) The upper and lower endpoints, w_h and w_l, can be determined in the following way.

If $w, z \in W_n$ are adjacent in the usual ordering, we wish to identify an endpoint of the vertical interval representing each to indicate adjacency and joining of backward or interval codes in (I, f). Suppose that

$$w = w_1 w_2 \ldots w_{j-1} 1 w_{j+1} \ldots w_n,$$

$$z = w_1 w_2 \ldots w_{j-1} 2 w_{j+1} \ldots w_n.$$

The points of (I, f) we wish to identify have two-sided itineraries of the form $(\ldots s_{-n} s_{-n+1} \ldots s_{-1} s_0 . s_1 \ldots)$ where $s_{-n+1} \ldots s_{-1} s_0$ equals either w or z, and

$$s_{-n+j+1} \ldots s_{-1} s_0 s_1 \ldots = I'(f(c)).$$

That is, if $\underline{x}, \underline{y} \in (I, f)$ are to be identified as endpoints of interval itineraries, then $f(x_0) = f(y_0) = f^{n-j+1}(c)$. One of the endpoints of each of the vertical intervals representing w, z has second coordinate $f^{n-j+1}(c)$; we draw a semicircle joining the intervals at these endpoints and say we have joined the intervals at height $f^{n-j+1}(c)$. (An easier way of describing the height of the joining ignores the indexing of coordinates; if, looking backwards, w and z disagree in the kth position, then w and z are joined at height $f^k(c)$.) If w is either the largest or smallest word in W_n, then the second endpoint of the vertical interval representing w has second coordinate $f^n(1)$ or $f^n(0)$ respectively. We call this object the *folded nth iterate of f* for reasons made clear below.

Note that the above algorithm does not require that the critical point of f be periodic; if the critical point is periodic so that the map is Markov, then the set of endpoints for the intervals representing words will be finite, and the determination of images of intervals may be simpler. Folded iterates of the unimodal map with kneading sequence 21C and modified itinerary of $f(c)$ equal to $(212)^\infty$ are drawn following the proof of Theorem 3.1.

For $a, b \in I$, $< a, b >$ will denote the smallest subinterval of I containing both a and b.

Theorem 3.1. *The folded n^{th} iterate of f can be drawn with no crossings.*

Proof: We prove this by induction on n. The folded first iterate of f can be drawn with no crossings, since there are precisely two codes of length one, 1 and 2, joined at $f(c)$.

The adjacency of words of length $n + 1$ in the usual ordering can be determined from adjacency in the set W_n of words of length n as follows. If $w \in W_n$ and both $w1$ and $w2$ are allowed as forward itineraries, then $w1$ and $w2$ are adjacent in W_{n+1}. If distinct words w_1, w_2 are adjacent in W_n, then for precisely one assignment of $* = 1, 2$ (whichever continues the modified code of $f(c)$ past the point of disagreement), the words w_1*, w_2* are adjacent in W_{n+1}. No other words are adjacent in W_{n+1}.

Suppose that a crossing is forced with words of length $n + 1$. A crossing can be forced in one of two ways. First, there may be words $w_1 <_r w_2 <_r w_3$ such that w_1, w_3 are joined at a height of $f^j(c)$ where $f^j(c)$ is in the interior of the vertical interval representing w_2 (or more precisely, in the interior of the projection onto the y-axis of the interval representing w_2). Second, there may be words $w_1 <_r w_2 <_r w_3 <_r w_4$ such that w_1, w_3 and w_2, w_4 are joined at the same height.

Suppose that a crossing of the first type occurs. If w_1, w_3 are of the form $w1, w2$ for some $w \in W_n$, then w_1 and w_3 are joined at $f(c)$; since $f(c)$ is not in the interior of any interval representing a word, this is impossible. Then w_1, w_3 are of the form $w_1'*, w_3'*$ where $*$ is either 1 or 2 and w_1', w_3' are adjacent in W_n. Since $w_1 <_r w_2 <_r w_3$, $w_2 = w_2'*$. It follows that either $w_1' <_r w_2' <_r w_3'$ or $w_3' <_r w_2' <_r w_1'$. If $f^{j-1}(c)$ is in the interior of the interval representing w_2', then there is a crossing at level n. Let I_w denote the projection onto the y-axis of the interval representing the word w. Then $I_{w_i} = f(I_{w_i'} \cap J_*)$. Since f is monotone on each of $I_{w_i'} \cap J_*$, if $f^{j-1}(c)$ is not in the interior of $I_{w_2'}$, then $f^j(c)$ is not in the interior of I_{w_2}, a contradiction. That is, a crossing of the first type at the $n + 1^{th}$ stage forces a crossing at the nth stage.

Suppose then that there is a crossing of the second type. Since the words w_1, w_3 are joined at the same height as w_2, w_4, either both pairs are joined at $f(c)$ or neither pair is. In the first case, w_1, w_3 and w_2, w_4 are of the form $w1, w2$ and $t1, t2$ for words $w, t \in W_n$. But then either

$$w1 <_r t1 <_r t2 <_r w2$$

or

$$t1 <_r w1 <_r w2 <_r t2,$$

a contradiction. In the second case, w_1, w_3 and w_2, w_4 are of the form $w_1'*, w_3'*$ and $w_2'*', w_4'*'$ where $w_i' \in W_n$ for $i = 1$ to 4 with w_1', w_3' and w_2', w_4' adjacent in W_n. We have that $* = *'$, since $w_1'* <_r w_2'*' <_r w_3'* <_r w_4'*'$. That is either $w_1' <_r w_2' <_r w_3' <_r w_4'$ or $w_4' <_r w_3' <_r w_2' <_r w_1'$, and the pairs w_1', w_3' and w_2', w_4' are joined at the same heights, so again there is a crossing at the nth stage. This proves the theorem. $\square$

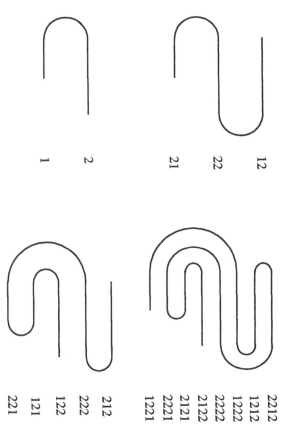

FIGURE 1

Example. Let f be the map from the tent family with critical point c periodic of period three, kneading sequence $21C$ and thus modified itinerary of $f(c)$ equal to $(212)^\infty$. Several folded iterates of f appear in Figure 1. The allowed words for f of any length are most easily described as those having no subword of the form 11. The allowed words of length 4, listed in terms of which have identified endpoints when representing intervals in the inverse limit space, that is, in the usual ordering, are: 1221, 1222, 1212, 2212, 2222, 2221, 2121, 2122. The set of words listed according to their placement in the plane from left to right, i.e., in the reverse ordering, are: 1221, 2221, 2121, 2122, 2222, 1222, 1212, 2212.

Work of [1] and [6] shows that for a surjective unimodal map f of the interval with periodic critical point, there is a disk homeomorphism F with global attractor homeomorphic to the inverse limit space (I, f). The action of the map F is indicated in Figure 2 for the period three map of the example, and can be described in terms similar to the usual horseshoe map. That is, one can think of contracting the disk in one direction, stretching it in the second, and placing the stretched version inside itself so that the 'pattern' of the original map is observed. From a heuristic point of view, one can think of generating the general shape of $F^n(D)$ by 'performing the action of f' on its own graph— by bending it around the critical point and stretching appropriately— $n - 1$ times (since we begin with the graph of f).

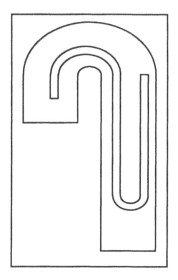

FIGURE 2

If the intervals of monotonicity for f^n are coded in the usual way, with words of length n indicating where forward iterates lie with respect to the critical point, then these words of length n are joined if and only if they are adjacent in W_n. One can show inductively that the placement of the intervals of monotonicity in the plane is that given by the drawing algorithm above. The bending around the critical point generates the codes at stage $n + 1$ from those at stage n by adding a 1 to those pieces of intervals below c and leaving them in the same relative position, adding a 2 to those pieces of intervals above c and reversing their relative positions, and leaving all codes ending with a 1 below all those ending with a 2. That is, if the codes are reverse ordered at stage n, they are reverse ordered at stage $n + 1$. In addition, joinings in this folded map are given by the determination of adjacency as described before Theorem 3.1. It is for this reason that the object constructed above is called the folded n^{th} iterate of f. According to [1] and [6], the shape of $F^n(D)$, which is the shape of the nth folded iterate of f, is an approximation of the shape of the inverse limit space in one planar embedding; according to the results of this paper (and as is made clear below), in the planar embedding being discussed, segments of composants have exactly the shape of pieces of the nth folded iterate.

We return to the question of drawing segments of composants with particular tails. Recall that A/w denotes the infinite backward symbol code obtained by replacing the initial segment of A of length $|w|$ by w. As indicated earlier, if A is a backward itinerary, then a segment of the composant associated with A can be obtained from taking all words of a given length n allowed for A, $W_{n,A}$, forming the set of corresponding interval itineraries $\{A/w \mid w \in W_{n,A}\}$, and considering A/w_1 and A/w_2 joined if w_1 and w_2 are adjacent in the usual ordering. For $w \in W_{n,A}$, we want to place a vertical interval representing A/w according to the reverse ordering on $\{A/w \mid w \in W_{n,A}\}$ (or $W_{n,A}$; they are identical). In particular, the projection onto the y-axis of a vertical interval representing A/w is to equal $f(\Pi_0(A/w))$. It may be the case that for certain composants, not all words of length n are allowed. As indicated in the proof of 2.7, the words of length n that are allowed for A form a consecutive set of words in the set of all words of length n allowed for f, that is, a set whose representation in the nth folded iterate of f is a connected set. As with the folded n^{th} iterate, then, the common second coordinate of one endpoint of two joined words A/w_1 and A/w_2 can be calculated as follows: if looking backwards, the codes A/w_1 and A/w_2 agree in precisely k places before the single point of disagreement, then the common second coordinate is $f^{k+1}(c)$. If the critical point of f is periodic of period n, then the codes $\hat{f}^k(\ldots cf(c)f^2(c)\ldots f^n(c)c\,.)$ for $0 \le k < n$ have as second coordinate of their other endpoint $f^{k+1}(c)$ for $0 \le k < n$, respectively. To use some algorithm involving adjacent words to determine the second coordinate of the other endpoint of a code A/w which is not of the above form but where w is either largest or smallest in the usual ordering on $W_{n,A}$, it is necessary to go to words of greater length to find the second interval itinerary joined in the composant since this second coordinate may differ from that for the word w in the folded n^{th} iterate. (For instance, in the folded iterates of the tent map with kneading sequence 2122C, the projection of the interval representing the word 22 is larger than that of the interval representing 122. However, if words of length k suffice to provide the second endpoint, the projection of the vertical interval representing the itinerary A/w is identical to that of the interval representing the initial segment of A/w of length k in the kth folded iterate of f.) To guarantee that for some k, words of length k suffice, we need the following, which is a special case of Theorems 2.3, 2.6 and 2.11 of [2].

Lemma 3.2. *If the critical point of f is not recurrent, then no point of (I, f) is an* *endpoint. If the critical point of f is periodic of period n, then (I, f) has precisely n endpoints: $\hat{f}^k(\ldots cf(c)f^2(c)\ldots f^n(c)c \cdot f(c)f^2(c)\ldots)$ for $0 \le k < n$.*

Each interval itinerary A is an arc in (I, f); that is, $f : \Pi_{n+1}(A) \to \Pi_n(A)$ is monotone for each n since $\Pi_n(A)$ does not contain c in its interior. The code A is either the largest or smallest in $\{A/w \mid w \in W_{n,A}\}$ in the usual ordering for every n if and only if an endpoint of A (in (I, f)) is actually an endpoint of (I, f). According to 3.2, this occurs if and only if A is of the periodic form described above, thus there is a k large enough so that both words joined to A are distinguished.

We have not specified horizontal coordinates in the above, since the placement according to the reverse ordering is all that is strictly necessary.

Segments of two composants associated with interval itineraries A and B can be drawn simultaneously by placing vertical intervals according to the reverse ordering on $W_{n,A} \cup W_{n,B}$. We indicate the drawing of segments of several composants following the next result.

Theorem 3.3. *For any two composants of (I, f), segments corresponding to the words of length n allowed for each can be drawn simultaneously with no crossings.*

Proof: Again we prove this by induction on n. Suppose that backward itineraries A, B are fixed. Segments of the composants associated with A and B corresponding to words of length 1 can be drawn with no crossings, since either

$$A/1 <_r B/1 <_r B/2 <_r A/2$$

or

$$B/1 <_r A/1 <_r A/2 <_r B/2,$$

regardless of whether 1 and 2 are allowed for each of A and B.

Suppose that a crossing is forced with $W_{n+1,A} \cup W_{n+1,B}$. A crossing can be forced in one of two ways. First, there may be words w_1, s, w_2 of length $n + 1$ such that s is allowed for, say, B, w_1, w_2 are allowed for A, $A/w_1 <_r B/s <_r A/w_2$, and $A/w_1, A/w_2$ are joined at a height of $f^j(c)$ where $f^j(c)$ is in the interior of the (projection of the) vertical interval representing B/s. Second, there may be words w_1, w_2, w_3, w_4 of length $n + 1$ such that w_1, w_3 are allowed for A, w_2, w_4 are allowed for B, $A/w_1 <_r B/w_2 <_r A/w_3 <_r B/w_4$, and $A/w_1, A/w_3$ and $B/w_2, B/w_4$ are joined at the same height.

Suppose that a crossing of the first type occurs. Without loss of generality, n is large enough so that the projections of the vertical intervals representing $A/w_1, B/s$ and A/w_2 agree with those of the words w_1, s and w_2, since the determination of the vertical interval representing a code A/w is independent of n, and any crossing appearing with the placing of $W_{n,A} \cup W_{n,B}$ will persist with the placing of $W_{n+k,A} \cup W_{n+k,B}$ for $k \in N$. It follows that $s \ne w_1$, $s \ne w_2$, for otherwise the endpoint $f^j(c)$ is not in the interior of the vertical interval representing s. The position of B/s relative to A/w_1 and A/w_2 is the same as that of s relative to w_1 and w_2 in the

$n + 1^{th}$ folded iterate of f, regardless of whether s is allowed for A, and thus can not lead to a crossing, a contradiction.

Suppose then that a crossing of the second type occurs. Since the codes A/w_1, A/w_3 are joined at the same height as B/w_2, B/w_4, either both pairs are joined at $f(c)$ or neither pair is. In the first case, A/w_1, A/w_3 and B/w_2, B/w_4 are of the form $A/w1$, $A/w2$ and $B/t1$, $B/t2$ for words $w, t \in W_n$. But then either

$$A/w1 <_r B/t1 <_r B/t2 <_r A/w2$$

or

$$B/t1 <_r A/w1 <_r A/w2 <_r B/t2,$$

a contradiction. In the second case, A/w_1, A/w_3 and B/w_2, B/w_4 are of the form A/w'_1*, A/w'_3* and B/w'_2*', B/w'_4*' where $w'_i \in W_n$ for $i = 1$ to 4 with A/w'_1, A/w'_3 and B/w'_2, B/w'_4 adjacent in W_n. We have that $* = *'$, since $w'_1* <_r w'_2*' <_r w'_3* <_r w'_4*'$. That is, either

$$A/w'_1 <_r B/w'_2 <_r A/w'_3 <_r B/w'_4$$

or

$$B/w'_4 <_r A/w'_3 <_r B/w'_2 <_r A/w'_1,$$

and the pairs A/w'_1, A/w'_3 and B/w'_2, B/w'_4 are joined at the same heights, so there is a crossing at the nth stage. This proves the theorem. $\square$

Returning to the example of the tent map with kneading sequence $21C$, let $A = \ldots 212212. = (212)^{\infty}.$, $B = (122)^{\infty}.$, and $C = (221)^{\infty}..$ The usual ordering on the allowed words of length 3 is

$$122 < 121 < 221 < 222 < 212.$$

The reverse ordering on $W_{3,A} \cup W_{3,B} \cup W_{3,C}$ is as indicated in Figure 3, where the segments of the three composants associated with A, B and C and corresponding to allowed words of length 3 appear.

Finally, we indicate how to place the entire inverse limit space in the plane. Let K be the standard middle thirds Cantor set. Assign codes to the 2^n intervals remaining at the nth stage of the usual construction of K of the 2^n words of $1's$ and $2's$ according to the reverse ordering of the latter. That is, the interval with code 22212 is to the left of that with code 22112 since $22212 <_r 22112$. (The reverse ordering is used because we wish to place backward itineraries.) Then any point in K has a unique code assigned in the obvious way. Given any backward itinerary A allowed for f, a vertical interval representing A (with second coordinates of endpoints defined as above) is placed over the point $(x, 0)$ where x is the element of K having code A. Note that the ordering of the placed words agrees with that defined earlier. According to 3.3, this placement is an embedding, since no two composants can intersect.

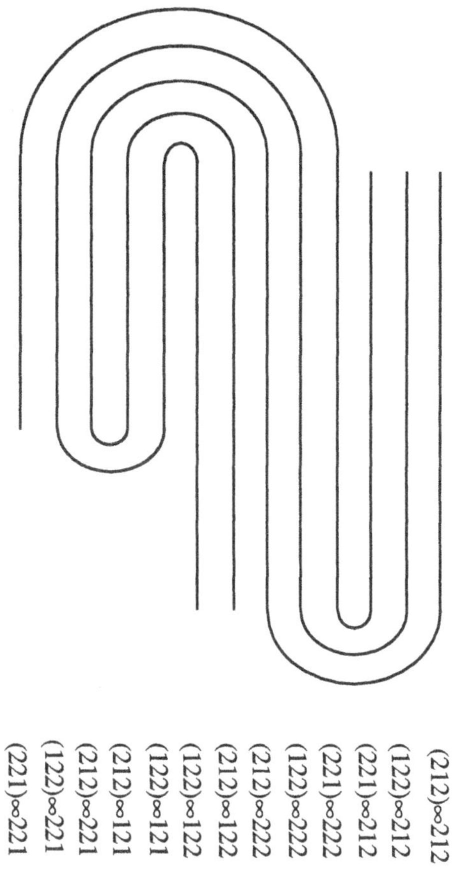
(221)∞221
(122)∞221
(212)∞221
(212)∞121
(122)∞121
(122)∞122
(212)∞122
(212)∞222
(122)∞222
(221)∞222
(221)∞212
(122)∞212
(212)∞212

FIGURE 3

REFERENCES

1. M. Barge, *Horseshoe maps and inverse limits*, Pacific Jour. Math. **121** (1986), 29-39.
2. M. Barge and J. Martin, *Endpoints of inverse limit spaces and dynamics*, preprint.
3. K.M. Brucks, B. Diamond, M.V. Otero-Espinar and C. Tresser, *Dense orbits of critical points for the tent map*, Contemporary Mathematics, **117** (1991), 57-61.
4. P. Collet and J.P. Eckmann, *Iterated maps on the interval as dynamical systems*, Birkhauser, Boston, 1980.
5. E.M. Coven, I. Kan and J.A. Yorke, *Pseudo-orbit shadowing in the family of tent maps*, Trans. Amer. Math. Soc. **308** (1988), 227-241.
6. S. Holte, *Generalized horseshoe maps and inverse limits*, Pacific Jour. Math., **156** (1992), 297-305.
7. S. Holte, *Inverse limits of Markov interval maps*, preprint.

8. S. van Strien, *Smooth dynamics on the interval*, New directions in dynamical systems, London Math. Soc. Lecture Note Ser. 127, Cambridge University Press, 1988.

9. S. Willard, General topology, Addison-Wesley, 1970.

10. P. Walters, *An introduction to ergodic theory*, Springer-Verlag, New York, 1985.

E-mail address: kmbrucks@csd4.csd.uwm.edu "Karen M. Brucks"
E-mail address: diamondb@cofc.edu "Beverly Diamond"

15

ON THE HOMOTOPIC STRUCTURE OF DYNAMICAL SYSTEMS CONTAINING GLOBAL ATTRACTORS

BERND GÜNTHER Fachbereich Mathematik, Johann Wolfgang Goethe-Universität, Robert-Mayer-Straße 6-10, 60054 Frankfurt, Germany

ABSTRACT. We dualize the standard approach to symbolic dynamics by replacing
inverse limits with direct limits and use this to analyze the relationship between
the shape of an attractor of a discrete dynamical system and the homotopy type
of its basin of attraction. When supplemented by the shape type of the motion on
the attractor and the homotopy type of the motion on the basin of attraction these
two perspectives are proved to contain exactly the same amount of information.
We show that an attractor has the shape of a (possibly infinite) polyhedron if and
only if its basin of attraction is homotopy dominated by a finite polyhedron if and
only if the inclusion map from the attractor into its basin of attraction is a shape
equivalence. The motion on the basin of attraction is of finite order up to homotopy
if and only if it is of finite order up to shape on the attractor. Furthermore a
classification scheme for attractors in the plane is given.

1. INTRODUCTION

By a (discrete) dynamical system we mean a pair (X, f) consisting of a manifold
X and a self-homeomorphism $f : X \approx X$, and we will assume that X contains a
global attractor A (definitions are given below). The purpose of this paper is to
analyze the relationship between X and A and to answer questions such as: Up to
what extent does X determine A, and which properties of X can be reconstructed
from A? Evidently, the homeomorphism type of X cannot be reconstructed: We
can replace X by $X' = X \times \mathbb{R}$ and f by $f'(x, t) = (f(x), t/2)$ without altering the

1991 *Mathematics Subject Classification.* 54C56, 54H20, 58F12.
Key words and phrases. homotopy, shape, attractor, dynamical system, symbolic dynamics..
The writing of this article was supported by a DFG grant.

attractor A. On the other hand X does not determine the homeomorphism type of A, because it can be altered by a small perturbation in f, but we will see that the shape of A remains unchanged. Therefore the properties under consideration will be the homotopy type of X and the shape of A, and this information will be supplemented by the homotopy type of the homeomorphism $f : X \approx X$ and the shape type of its restriction $f_{|A} : A \approx A$. Our main theorem, proved in section 4, states that properties of the manifold X and of the attractor A, which can be expressed in these terms, determine each other:

Theorem 1.1. *Let $f : X \approx X$ and $g : Y \approx Y$ be two homeomorphisms of manifolds. We assume that $A \subseteq X$ and $B \subseteq Y$ are global attractors of f, g respectively, and that $P \subseteq X$, $Q \subseteq Y$ are open domains of attraction with $f\left(\overline{P}\right) \subseteq P$, $g\left(\overline{Q}\right) \subseteq Q$. Then the following conditions are equivalent:*

(i) *There exists a homotopy equivalence $\Phi : X \simeq Y$ and an integer $n \geq 1$ with $g^n \Phi \simeq \Phi f^n$.*

(ii) *There exists a shape equivalence $\Phi \in \mathbf{sh}(A, B)$ and an integer $n \geq 1$ with $\eta\left(g_{|B}^n\right)\Phi = \Phi\eta\left(f_{|A}^n\right)$.*

$$
\begin{array}{ccc}
X \xrightarrow[\simeq]{\Phi} Y & \qquad & A \xrightarrow[\approx]{\Phi} B \\
f^n \Big\downarrow \approx \quad \approx \Big\downarrow g^n & \qquad & \eta(f_{|A}^n) \Big\downarrow \approx \quad \approx \Big\downarrow \eta(g_{|B}^n) \\
X \xrightarrow[\simeq]{\Phi} Y & \qquad & A \xrightarrow[\approx]{\Phi} B
\end{array}
$$

(iii) *There exist continuous maps $\varphi : P \to Q$ and $\psi : Q \to P$ and an integer $n \geq 1$ with $\varphi\psi \simeq g_{|Q}^n$ in Q and $\psi\varphi \simeq f_{|P}^n$ in P.*

If these conditions are satisfied, then the shape equivalence $\Phi_1 \in \mathbf{sh}(A, B)$, the map $\varphi : P \to Q$ and the homotopy equivalence $\Phi_2 : X \simeq Y$ can be chosen such that in addition the following diagram in $\mathbf{sh}$ is commutative, where the horizontal arrows are induced by inclusion maps:

$$
(1) \qquad
\begin{array}{ccccc}
A & \longrightarrow & P & \longrightarrow & X \\
\Phi_1 \Big\downarrow & & \eta(\varphi) \Big\downarrow & & \Big\downarrow \eta(\Phi_2) \\
B & \longrightarrow & Q & \longrightarrow & Y
\end{array}
$$

We observe that condition (iii) means that the maps $f_{|P}^n : P \to P$ and $g_{|Q}^n : Q \to Q$ are shift equivalent up to homotopy in the sense of R.F. Williams [10]. In [16] Tezer observed that topological conjugacy of the homeomorphisms $f_{|A} : A \approx A$ and $g_{|B} : B \approx B$ implies shift equivalence of the maps $f_{|P} : P \to P$ and $g_{|Q} : Q \to Q$; the implication (ii)⇒(iii) of our Theorem 1.1 strengthens this by replacing topological conjugacy with conjugacy up to shape. On the topological level the equivalence of conditions (ii) and (iii) for branched 1-manifolds has been proved in [10, p.342,Cor.]. Shape theory has been used by Robbin and Salamon in [12] to introduce an index for isolated invariant sets of dynamical systems; in case of an attractor the Robbin-Salamon index is the shape type of the attractor with an added discrete basepoint. Proposition 4.18 shows that even in simple cases our classification incorporating the shape type of the motion is strictly finer; we see no chance of establishing a 1-1-correspondence between attractor and basin of attraction as in our Theorem 1.1 without taking into account some information about the motion.

In our context a *manifold* means a topological manifold, i.e. a locally compact, separable Hausdorff space where each point has an open neighborhood homeomorphic to a Euclidean space. We observe that each manifold is metrizable. Sometimes we consider richer structures such as differentiable manifolds of class C^r, $1 \leq r \leq \infty$, or real analytic manifolds; they are defined as in standard text books.

An *ANR-space* is a metrizable space Y such that any map $f : A \to Y$ defined on a closed subspace A of a metrizable space X has an extension over a neighborhood of A in X. When it is always possible to extend over the entire space X, then Y is called an *AR-space*. An ANR-space is an AR-space if and only if it is contractible. Every compact polyhedron [6, Ch.III,Thm.7.8] and every manifold [6, Ch.V,Thm.7.1] is an ANR-space. Every closed inclusion map between metrizable spaces has the homotopy extension property with respect to ANR-spaces [6, Ch.IV,Thm.2.2].

Let's consider a dynamical system $f : X \approx X$. A compact subset $A \subseteq X$ with $f(A) = A$ is an *attractor* of f if there is an open neighborhood U of A in X with $f(\overline{U}) \subseteq U$ and $A = \bigcap_{n=1}^{\infty} f^n(U)$; U is called *domain of attraction*. Every domain of attraction is relatively compact. A is a *global attractor*, if at least for one (and hence for any) domain of attraction U we have $\bigcup_{n=1}^{\infty} f^{-n}(U) = X$. Every dynamical system can have at most one global attractor. The only global attractor on a *compact* manifold is the whole manifold itself. We observe that for any attractor the set $W := \bigcup_{n=1}^{\infty} f^{-n}(U)$ is an open subset of X invariant under f; it is called *basin of attraction* and consists exactly of those points of X converging to the attractor under forward iteration of f. Hence, if our main interest is the study of attractors, then we can always assume that our attractors are global by restricting the dynamical system to W.

2. Symbolic dynamics on inverse limits

Definition 2.1. A self-embedding $f : P \rightarrowtail P$ of a manifold P will be called *small*, if $f(P)$ is relatively compact in P.

A tower $\mathbf{P} = \left\{ P_1 \xleftarrow{f_1} P_2 \xleftarrow{f_2} P_3 \xleftarrow{f_3} \cdots \right\}$ of manifolds is called *homogeneous*, if all P_n are equal to the same manifold P and if all f_n are equal to the same map $f : P \rightarrowtail P$, which is supposed to be a small embedding. For any such homogeneous tower $\mathbf{P}$ we set $\overleftarrow{P} := \varprojlim \mathbf{P}$. The *shift map* $\overleftarrow{f} : \overleftarrow{P} \approx \overleftarrow{P}$ is defined by $\overleftarrow{f}(x_1, x_2, \dots) := (f(x_1), x_1, x_2, \dots)$. By $p : \overleftarrow{P} \to P$ we denote the projection map onto the first factor and observe $fp = p\overleftarrow{f}$; the projection map onto the n-th factor is $p\overleftarrow{f}^{-n+1}$. From [7, Ch.I,§6.1,Thm.2] we learn that the tower $\mathbf{P}$ together with the projection maps $p\overleftarrow{f}^{-n+1}$ is an ANR-expansion of $\overleftarrow{P}$. We say that the homogeneous tower $\mathbf{P}$, the compactum $\overleftarrow{P}$ and the shift map $\overleftarrow{f}$ are *associated* to the small self-embedding $f : P \rightarrowtail P$.

Occasionally we may encounter towers of the form $\mathbf{P} = \{ P \xleftarrow{f} P \xleftarrow{f} P \xleftarrow{f} \cdots \}$, where P has the homotopy type of a compact polyhedron[2] and $f : P \to P$ is an arbitrary self-map. Then the following lemma allows to replace $\mathbf{P}$ with a homogeneous tower without altering its shape properties:

Lemma 2.2. *For every self-map* $f : P \to P$, *where* P *is homotopy equivalent to a compact polyhedron, there exists a real analytic manifold* P', *an analytic small self-embedding* $f' : P' \rightarrowtail P'$ *and a homotopy equivalence* $\varphi : P \simeq P'$ *with* $f'\varphi \simeq \varphi f$.

Proof. We may suppose that P is a compact polyhedron contained in the open unit ball $B^n \subset \mathbb{R}^n$. We choose an open, regular neighborhood V of P whose closure is contained in B^n and which is small enough, that there exists a map $r : \overline{V} \to P$ with $r_{|P} = \mathrm{id}$; furthermore we choose $\varepsilon > 0$ such that the 2ε-neighborhood around P is contained in V. The inclusion map $i : P \hookrightarrow V$ is a homotopy equivalence with inverse $r_{|V} : V \to P$. The Stone-Weierstrass theorem provides us with a polynomial function $q : \overline{V} \to \mathbb{R}^n$ with $\|q(x) - fr(x)\| < \varepsilon$ for all $x \in \overline{V}$, and this means in particular $q(V) \subseteq V$ and $q_{|V} \simeq fr_{|V}$ in V. Now we set $P' := V \times B^n$ and define $f' : V \times B^n \rightarrowtail V \times B^n$ by

$$(2) \qquad\qquad f'(x,y) := (q(x) + \varepsilon y, x).$$

f' is a small analytic embedding because $f'(V \times B^n) \subseteq (P + 2\varepsilon\overline{B^n}) \times \overline{V} \subseteq V \times B^n$. The homotopy equivalence $\varphi : P \simeq V \times B^n$ is obtained as the inclusion map $P \hookrightarrow V \approx V \times \{0\} \hookrightarrow V \times B^n$. $\qquad\square$

Lemma 2.3. *We consider a small self-embedding* $f : P \rightarrowtail P$ *of a manifold and the associated shift map* $\overleftarrow{f} : \overleftarrow{P} \approx \overleftarrow{P}$, *and we suppose we are given a space* A, *a shape isomorphism* $\alpha \in \mathbf{sh}(A, A)$ *and a map* $\varphi : A \to P$ *with* $\eta(f\varphi) = \eta(\varphi)\alpha$. *Then there is a unique shape morphism* $\Phi \in \mathbf{sh}\left(A, \overleftarrow{P}\right)$ *with* $\eta\left(\overleftarrow{f}\right)\Phi = \Phi\alpha$ *and* $\eta(f)\Phi = \eta(\varphi)$.

Proof. By [7, Ch.1,§2.3,Thm.5] a shape morphism $\Phi \in \mathbf{sh}\left(A, \overleftarrow{P}\right)$ satisfying the

---

[2]We note that every compact CW-space has this property [7, App.1,§2.2,Thm.8].

conditions above corresponds bijectively to a sequence of maps $\Phi_n : A \to P$ with

$$
(3) \qquad\qquad\qquad f\Phi_{n+1} \quad \simeq \quad \Phi_n
$$

$$
(4) \qquad\qquad\qquad \eta\left(\Phi_n\right)\alpha \quad = \quad \eta\left(f\Phi_n\right)
$$

$$
(5) \qquad\qquad\qquad \Phi_1 \quad \simeq \quad \varphi
$$

This implies: $\eta\left(\Phi_{n+1}\right) = \eta\left(\Phi_{n+1}\right)\alpha\alpha^{-1} = \eta\left(f\Phi_{n+1}\right)\alpha^{-1} = \eta\left(\Phi_n\right)\alpha^{-1}$ and hence by induction $\eta\left(\Phi_n\right) = \eta(\varphi)\alpha^{-n}$. By [7, Ch.1,§2.3,Thm.4] this determines Φ_n up to homotopy, therefore Φ is unique. Defining Φ by $\eta\left(\Phi_n\right) := \eta(\varphi)\alpha^{-n}$ proves existence. $\qquad\qquad\qquad\qquad\qquad\qquad\qquad\qquad\qquad\qquad\qquad\qquad\square$

Lemma 2.3 characterizes the shape of $\overleftarrow{P}$ by a universal property. To take full advantage of this property we introduce the *category of attractors* **Atr**, whose objects are pairs (A,g) consisting of a compactum A and a homeomorphism $g : A \approx A$, such that there exists a manifold X and a homeomorphism $G : X \approx X$ with A as global attractor and $g = G_{|A}$. A morphism $\varphi : (A,g) \to (B,h)$ in **Atr** is a shape morphism $\varphi \in \mathbf{sh}(A,B)$ with $\eta(h)\varphi = \varphi\eta(g)$. Two morphisms $\varphi,\psi : (A,g) \to (B,h)$ in **Atr** will be identified if there is $n \in \mathbb{Z}$ with $\varphi\eta\left(g^n\right) = \psi$. Composition of morphisms in **Atr** is defined in an obvious way.

Two attractors are called *similar* if they are isomorphic as objects of **Atr**.

Lemma 2.4. *Two objects (A,g) and (B,h) of **Atr** are isomorphic in **Atr** if and only if there is a shape equivalence $\varphi \in \mathbf{sh}(A,B)$ with $\eta(h)\varphi = \varphi\eta(g)$.*

Proof. Only necessity must be proved, so we assume we are given shape *morphisms* $\varphi \in \mathbf{sh}(A,B)$ and $\psi \in \mathbf{sh}(B,A)$ with $\eta(h)\varphi = \varphi\eta(g)$ and $\eta(g)\psi = \psi\eta(h)$ and numbers $m,n \in \mathbb{Z}$ with $\varphi\psi = \eta\left(h^n\right)$ and $\psi\varphi = \eta\left(g^m\right)$. We have to show that φ is a shape *equivalence* and claim that $\psi' := \eta\left(g^{-m}\right)\psi$ is an inverse. The equation $\psi'\varphi = \mathrm{id}_A$ is evident. On the other hand the assumption implies: $\eta(h)^{2n} = \varphi\psi\varphi\psi = \varphi\eta(g)^m\psi = \eta(h)^m\varphi\psi = \eta(h)^{n+m} \;\Rightarrow\; \eta(h)^n = \eta(h)^m$ and therefore $\varphi\psi' = \varphi\psi\eta(h)^{-m} = \eta(h)^n\eta(h)^{-m} = \mathrm{id}_B$. $\qquad\qquad\qquad\square$

We also introduce the *category* **Sem** *of self-embeddings*: objects are pairs (P,f) consisting of a manifold P and a small embedding $f : P \rightarrowtail P$. A morphism $\varphi : (P,f) \to (Q,g)$ in **Sem** is a continuous map $\varphi : P \to Q$ with $g\varphi \simeq \varphi f$; two morphisms $\varphi,\psi : P \to Q$ are identified if there exist integers $m,n \geq 0$ with $\varphi f^m \simeq \psi f^n$.

Now we are able to extend our limit construction to a functor $\overleftarrow{F} : \mathbf{Sem} \to \mathbf{Atr}$ by $\overleftarrow{F}(P,f) := \left(\overleftarrow{P}, \overleftarrow{f}\right)$. The construction of a manifold containing $\overleftarrow{P}$ as global attractor will be given in Lemma 3.1. For $\varphi \in \mathbf{Sem}(P,f;Q,g)$ we apply Lemma 2.3 and define $\overleftarrow{F}(\varphi) \in \mathbf{Atr}\left(\overleftarrow{P}, \overleftarrow{f}; \overleftarrow{Q}, \overleftarrow{g}\right)$ to be the unique shape morphism fitting commutatively into the following diagram:

(6)
$$
\begin{array}{ccccc}
\overleftarrow{P} & \xrightarrow{\;\eta\left(\overleftarrow{f}\right)\;} & \overleftarrow{P} & \xrightarrow{\;\eta(p)\;} & P \\[4pt]
\overleftarrow{F}(\varphi)\Big\downarrow & & \Big\downarrow\overleftarrow{F}(\varphi) & & \Big\downarrow\eta(\varphi) \\[4pt]
\overleftarrow{Q} & \xrightarrow[\eta\left(\overleftarrow{g}\right)]{} & \overleftarrow{Q} & \xrightarrow[\eta(q)]{} & Q
\end{array}
$$

Proposition 2.5. *The functor* $\overleftarrow{F} : \mathbf{Sem} \to \mathbf{Atr}$ *is an equivalence of categories.*

Proof. By [9, Ch.IV,§4,Thm.1] we have to show:

a) Every attractor A of a homeomorphism $g : X \approx X$ on a manifold is similar to $\overleftarrow{F}(P, f)$ for a suitable small self-embedding $f : P \rightarrowtail P$: If P is an open domain of attraction for A with $g\left(\overline{P}\right) \subseteq P$, then $f := g_{|P} : P \rightarrowtail P$ has the necessary properties.

b) The functor $\overleftarrow{F}$ is faithful: Suppose we are given two morphisms $\varphi, \psi \in \mathbf{Sem}(P, f; Q, g)$ with $\overleftarrow{F}(\varphi) = \overleftarrow{F}(\psi)$. Then there exists an integer $n \geq 0$ such that the following diagram in $\mathbf{sh}$ is commutative:

(7)
$$
\begin{array}{ccccc}
P & \xleftarrow{\;\eta(p)\;} & \overleftarrow{P} & \xrightarrow{\;\eta(p)\;} & P \\[4pt]
\eta(\psi)\Big\downarrow & \overleftarrow{F}(\psi)=\Big\downarrow\overleftarrow{F}(\varphi)\overleftarrow{f}^{\,n} & & & \Big\downarrow\eta(\varphi f^{n}) \\[4pt]
Q & \xleftarrow[\eta(q)]{} & \overleftarrow{Q} & \xrightarrow[\eta(q)]{} & Q
\end{array}
$$

By [7, Ch.1,§2.3,Thm.4] this means $\psi p \simeq \varphi f^{n} p$, and since the homogeneous tower generated by $f : P \rightarrowtail P$ is an ANR-expansion of $\overleftarrow{P}$ there must be an integer $m \geq 0$ with $\psi f^{m} \simeq \varphi f^{m+n}$.

c) The functor $\overleftarrow{F}$ is full: Suppose we are given a shape morphism $\Phi \in \mathbf{Atr}\left(\overleftarrow{P}, \overleftarrow{f}; \overleftarrow{Q}, \overleftarrow{g}\right)$. We consider the function $\Phi : \mathbf{HTop}\left(\overleftarrow{Q}, Q\right) \to \mathbf{HTop}\left(\overleftarrow{P}, Q\right)$ and set $h := \Phi(q)$; the equation $\Phi\eta\left(\overleftarrow{f}\right) = \eta\left(\overleftarrow{g}\right)\Phi$ implies $gh \simeq hf$. Since the homogeneous tower generated by $f : P \rightarrowtail P$ is an ANR-expansion of $\overleftarrow{P}$ there is $n \geq 0$ and a map $\varphi : P \to Q$ with $h \simeq \varphi p\,\overleftarrow{f}^{\,-n}$. Then $g\varphi p\,\overleftarrow{f}^{\,-n} \simeq gh \simeq hf \simeq \varphi p\,\overleftarrow{f}^{\,-n+1} \simeq \varphi f p\,\overleftarrow{f}^{\,-n}$. Consequently there must be $m \geq 0$ with $g\varphi f^{m} \simeq \varphi f^{m+1}$, and this implies $\eta(q)\Phi = \eta\left(\varphi p\,\overleftarrow{f}^{\,-n}\right)$, hence $\eta(q)\Phi\eta\left(\overleftarrow{f}^{\,n+m}\right) = \eta(\varphi f^{m} p)$ and therefore $\overleftarrow{F}\left(\varphi f^{m}\right) = \Phi\eta\left(\overleftarrow{f}^{\,n+m}\right) \sim \Phi$. $\qquad\square$

At least up to shape we are able to determine the class of compacta that can be attractors of homeomorphisms:

Proposition 2.6. *We consider a homeomorphism* $g : A \approx A$ *of a compactum* A. *Then* (A, g) *is similar to an attractor in the sense of Lemma 2.4 if and only if there* *exists a map* $\varrho : A \to P$ *into a compact polyhedron* P *and a map* $f : P \to P$ *with* $f\varrho \simeq \varrho g$, *such that*

(i) *for every map* $\varphi : A \to Q$ *into an ANR-space* Q *there is* $\psi : P \to Q$ *and an integer* $r \geq 0$ *with* $\varphi g^r \simeq \psi \varrho$, *and*

(ii) *for any two maps* $\varphi, \psi : P \to Q$ *into an ANR-space* Q *with* $\varphi \varrho \simeq \psi \varrho$ *there is* $r \geq 0$ *with* $\varphi f^r \simeq \psi f^r$.

Proof. The conditions (i) and (ii) mean that the tower $\mathbf{P} = \{ P \xleftarrow{f} P \xleftarrow{f} P \xleftarrow{f} \cdots \}$ together with the maps $\varrho g^{-n} : A \to P$ forms an ANR-expansion of A. Therefore sufficiency follows from Lemma 2.2. To prove the converse we suppose A is an attractor of a homeomorphism $h : X \approx X$ on a manifold and choose an open domain of attraction $P' \subseteq X$ with $h\left(\overline{P'}\right) \subseteq P'$; by abuse of notation the restricted map $P' \to P'$ is again denoted by h. We embed P' in $\mathbb{R}^n$ as closed subspace and take an open neighborhood U of P' in $\mathbb{R}^n$ such that there exists a retraction map $\vartheta : U \to P'$. Let $P \subseteq \mathbb{R}^n$ be a compact polyhedron with $h\left(\overline{P'}\right) \subseteq P \subseteq U$ and set $f := h\vartheta$. By Morita's lemma [7, Ch.II,§2.2,Thm.5] the level morphism of towers equal to $r : P \to P'$ on each level is an equivalence $\mathbf{P} \simeq \mathbf{P}' = \left\{ P' \xleftarrow{h} P' \xleftarrow{h} \cdots \right\}$, and this finishes the proof. $\square$

Proposition 2.7. *We suppose that X is a manifold, that A is an attractor (not necessarily global) of $g : X \approx X$ and denote by $\varphi := \check{H}^*\left(g_{|A}\right) : \check{H}^*(A; R) \approx \check{H}^*(A; R)$ the induced automorphism of Čech cohomology groups with coefficients in a principal ideal domain R. Then $\check{H}^*(A; R)$ contains a finitely generated, φ-invariant submodule G with $\bigcup_{n=1}^{\infty} \varphi^{-n}(G) = \check{H}^*(A; R)$. If R is a field, then the vector space $\check{H}^*(A; R)$ is finite dimensional over R.*

Proof. We take a map $\varrho : A \to P$ into a compact polyhedron P with the properties of Proposition 2.6, consider the induced homomorphism $\varrho^* : \check{H}^*(P; R) \to \check{H}^*(A; R)$ of Čech cohomology groups and set $G := \operatorname{im} \varrho^*$. It is well known that $\check{H}^n(Z; R) = \mathbf{HTop}(Z; K(R, n))$ for every space Z (for CW-spaces this is [14, Ch.8,§1,Thm.10], for arbitrary spaces this carries over by continuity [7, Ch.II,§3.2,Thm.4]). Since the Eilenberg-MacLane space $K(R, n)$ can be assumed to be an ANR-space [7, App.1,§2.2,Thm.5], part (i) of Proposition 2.6 ensures that G has the desired properties. If R is a field, then for each $n \geq 0$ $\varphi^n : \varphi^{-n}(G) \rightarrowtail G$ is a vector space monomorphism and therefore $\dim \varphi^{-n}(G) = \dim G$. $\square$

Proposition 2.7 means that the submodule $G \subseteq \check{H}^*(A; R)$ is an "attractor" of the automorphism φ and should be viewed as a finiteness property of $\check{H}^*(A; R)$. It may also be expressed as $\check{H}^*(A; R) \approx G \otimes_{\mathbb{Z}[\varphi]} \mathbb{Z}\left[\varphi, \varphi^{-1}\right]$, where the $\mathbb{Z}[\varphi]$-module G is finitely generated as an abelian group. This structural property restricts the class of compacta that can be attractors [5, Lemma 1].

Corollary 2.8. *An attractor of a homeomorphism on a manifold can have only finitely many components.*

Proof. By Proposition 2.7 $\check{H}^0(A; \mathbb{Q})$ is finitely generated. □

If $A \subseteq X$ is an attractor of $f : X \approx X$, then $f_{|A} : A \approx A$ permutes the components of A. Taking n to be the order of this permutation f^n leaves each component fixed, hence each component is an attractor of f^n. Therefore there is no loss in generality if we restrict our attention to connected attractors.

The shape theoretic method, though powerful, also has its limitations. The characterization of attractors of continuous dynamical systems (i.e., manifolds equipped with a one-parameter group of homeomorphisms) in [4] raises the following question: Given two compacta A and B of identical shape such that A can be embedded in a manifold X and a homeomorphism $f : X \approx X$ can be constructed turning A into an attractor, is this also possible for B? In short: Is the property of being an attractor shape invariant? Example 2.9 shows that this conjecture is false; this property is not even invariant under homotopy.

Example 2.9. We denote by A the dyadic solenoid, i.e., the limit of an inverse sequence of circles with bonding maps wrapping the n+1-st circle twice around the n-th circle, and choose a dense sequence of points $a_1, a_2, a_3, \ldots \in A$. To the point a_n we attach n line segments of length $1/n$, the resulting compactum serves as B. In other words we set $C_n := \{ r \exp(2\pi i k/n) | 0 \le r \le 1/n, k = 0, \ldots n - 1 \} \subset D^2 = \{ z \in \mathbb{C} | |z| \le 1 \}$ and $B := A \times \{0\} \cup \bigcup_{n=1}^{\infty} \{a_n\} \times C_n \subset A \times D^2$. A is a strong deformation retract of B, and it is well known that A is an attractor of a homeomorphism on a 3-dimensional manifold.

Now we suppose the existence of a homeomorphism $g : B \approx B$ turning B into an attractor. Since A is not locally connected (every point has a fundamental system of neighborhoods homeomorphic to the Cartesian product of a line segment with Cantor's discontinuum) our homeomorphism g must leave A invariant and must permute the spaces C_n, and since no two of them are homeomorphic this means $g(A) = A$ and $g(\{a_n\} \times C_n) = \{a_n\} \times C_n$. We conclude $g(a_n) = a_n$ and therefore $g_{|A} = \mathrm{id}$. Hence Proposition 2.7 must be satisfied with $\varphi = \mathrm{id}$ and consequently $\check{H}^1(B) = \check{H}^1(A)$ should be finitely generated, which is not true.

3. Symbolic Dynamics on Direct Limits

We consider the *homogeneous direct sequence* $\mathbf{P} = \left\{ P \xrightarrow{f} P \xrightarrow{f} P \xrightarrow{f} \cdots \right\}$ determined by a small self-embedding $f : P \rightarrowtail P$ of a manifold and set $\overrightarrow{P} := \varinjlim \mathbf{P}$. This means that $\overrightarrow{P}$ is the quotient space obtained from $P \times \mathbb{N}$ by imposing the equivalence relation

$$(8) \qquad (x, m) \sim (y, n) :\Leftrightarrow y = f^{n-m}(x) \text{ if } n \ge m, \ x = f^{m-n}(y) \text{ if } n \le m.$$

By $\overrightarrow{f} : \overrightarrow{P} \approx \overrightarrow{P}$ we denote the map $\overrightarrow{f}(x,n) := (f(x),n)$; $\overrightarrow{f}$ is a homeomorphism with inverse $\overrightarrow{f}^{-1}(x,n) = (x,n+1)$ and is called the *shrinking map* determined by f. There is an embedding $i : P \rightarrowtail \overrightarrow{P}$ defined by $i(x) := (x,1)$, and we have $\overrightarrow{f}\, i = if$.

Lemma 3.1. $\overrightarrow{P}$ *is a manifold and* $\overrightarrow{f} : \overrightarrow{P} \approx \overrightarrow{P}$ *is a homeomorphism with a global attractor* $A \subseteq \overrightarrow{P}$, *such that there exists a homeomorphism* $A \approx \overleftarrow{P}$ *fitting commutatively into diagram* (9). *The subsets* $\overrightarrow{f}^{-n}i(P) \subseteq \overrightarrow{P}$ *are open and relatively compact and cover* $\overrightarrow{P}$.

(9)
$$
\begin{array}{ccc}
A & \xrightarrow{\ \approx\ } & \overleftarrow{P} \\
{\scriptstyle \overrightarrow{f}_{|A}}\downarrow & & \downarrow{\scriptstyle \overleftarrow{f}} \\
A & \xrightarrow{\ \approx\ } & \overleftarrow{P}
\end{array}
$$

Proof. Evidently the graph of the equivalence relation (8) is closed. The saturation of an open subset $U = \bigcup_{n=1}^{\infty} U_n \times \{n\}$ is given by $V = \bigcup_{n=1}^{\infty} V_n \times \{n\}$ with $V_n := \bigcup_{k=1}^{\infty} f^{n-k}(U_k)$, and since $f : P \rightarrowtail P$ is an open mapping by invariance of domain, V is open. Now [2, Ch.I,§8.3,Prop.8] and [2, Ch.I,§10.4,Prop.10] imply that $\overrightarrow{P}$ is a locally compact Hausdorff space. $\overrightarrow{f}^{-n}i(P)$ is the image of $P \times \{n+1\} \subseteq P \times \mathbb{N}$ and is therefore open, and these sets cover $\overrightarrow{P}$; on the other hand these sets are relatively compact because $\overrightarrow{f}^{-n}i(P) \subseteq \overrightarrow{f}^{-n-1}i\left(\overline{f(P)}\right)$. Every relatively compact, open subset of $\overrightarrow{P}$ is homeomorphic to an open subset of P, hence $\overrightarrow{P}$ is a manifold. $A := \bigcap_{n=1}^{\infty} \overrightarrow{f}^{\,n}i(P) = \bigcap_{n=1}^{\infty} \overrightarrow{f}^{\,n}\overline{i(P)}$ is a global attractor, and $h : \overleftarrow{P} \approx A$, $h(x_1, x_2, \ldots) := i(x_1)$ is a homeomorphism with $h\overleftarrow{f} = \overrightarrow{f}h$. $\square$

Remark 3.2. If the manifold P and the small embedding $f : P \rightarrowtail P$ are differentiable or real analytic[3], then the manifold $\overrightarrow{P}$ and the shrinking map $\overrightarrow{f}$ are of the same class.

We recall that two maps $\varphi, \psi : X \to Y$ are *pre-homotopic* (also called weakly homotopic [15, 9.22] and denoted $\varphi \simeq_w \psi$) if for any compact subset $K \subseteq X$ we have $\varphi_{|K} \simeq \psi_{|K}$. A map $\varphi : X \to Y$ is a *pre-homotopy equivalence*, if there is a map $\psi : Y \to X$ with $\psi\varphi \simeq_w \mathrm{id}_X$ and $\varphi\psi \simeq_w \mathrm{id}_Y$.

Lemma 3.3. *Let* $f : P \rightarrowtail P$ *be a small self-embedding,* $g : X \approx X$ *a self-homotopy equivalence on an arbitrary space and* $\varphi : P \to X$ *a map with* $g\varphi \simeq \varphi f$. *Then there exists a map* $\Phi : \overrightarrow{P} \to X$ *with* $g\Phi \simeq_w \Phi \overrightarrow{f}$ *and* $\Phi i \simeq \varphi$; Φ *is unique up to pre-homotopy.*

---

[3]Differentiable and real analytic embeddings are always assumed to be immersions.

Proof. For Φ as above we set $\varphi_n := \Phi \overrightarrow{f}^{-n} i : P \to X$ and get $\varphi_0 \simeq \varphi$, $g\varphi_{n+1} \simeq \varphi_n$, hence $\varphi_n \simeq g^{-n}\varphi$. Since the open sets $\overrightarrow{f}^{-n} i(P)$ cover $\overrightarrow{P}$ this determines the pre-homotopy class of Φ.

To prove existence we construct a sequence of maps $\varphi_n : P \to X$ with $\varphi_0 = \varphi$, $\varphi_n \simeq g^{-n}\varphi$, $\varphi_{n+1}f^2 = \varphi_n f$ by induction. If φ_n is already given we choose a homotopy $H : g^{-n-1}\varphi f \simeq \varphi_n$ and an Urysohn function $\lambda : P \to I$ with $\lambda \equiv 1$ on $f^2(P)$ and $\lambda \equiv 0$ on $P \setminus f(P)$ and set:

$$(10) \qquad \varphi_{n+1}(x) := \begin{cases} g^{-n-1}\varphi(x) & \text{for } \lambda(x) \leq 1/2 \\ H\left(f^{-1}(x), 2\lambda(x) - 1\right) & \text{for } \lambda(x) \geq 1/2. \end{cases}$$

Then we define $\Phi(x,n) := \varphi_n f(x)$; this means $\Phi \overrightarrow{f}^{-n} i = \varphi_{n+1} f$ and therefore all requirements are satisfied. $\qquad\square$

We introduce the *category* **Dyn** *of dynamical systems:* objects are pairs (X, g) formed of a manifold X and a homeomorphism $g : X \approx X$ containing a global attractor. A morphism $\varphi : (X, g) \to (Y, h)$ in **Dyn** is a map $\varphi : X \to Y$ with $\varphi g \simeq_w h\varphi$. Two morphisms $\varphi, \psi \in \textbf{Dyn}(X, g; Y, h)$ are considered equal if there exists $n \in \mathbb{Z}$ with $\varphi g^n \simeq_w \psi$. Two dynamical systems are *similar* if they are equivalent as objects of **Dyn**. Our limit construction yields a functor $\overrightarrow{F} : \textbf{Sem} \to \textbf{Dyn}$ defined by $\overrightarrow{F}(P, f) = \left(\overrightarrow{P}, \overrightarrow{f}\right)$, and for $\varphi \in \textbf{Sem}(P, f; Q, g)$ we define $\overrightarrow{F}(\varphi) \in \textbf{Dyn}\left(\overrightarrow{P}, \overrightarrow{f}; \overrightarrow{Q}, \overrightarrow{g}\right)$ to be the unique morphism making the following diagram commutative up to pre-homotopy:

$$(11) \qquad \begin{array}{ccccc} P & \xrightarrow{\ i\ } & \overrightarrow{P} & \xrightarrow{\ \overrightarrow{f}\ } & \overrightarrow{P} \\ {\scriptstyle \varphi}\downarrow & & \downarrow{\scriptstyle \overrightarrow{F}(\varphi)} & & \downarrow{\scriptstyle \overrightarrow{F}(\varphi)} \\ Q & \xrightarrow[\ j\ ]{} & \overrightarrow{Q} & \xrightarrow[\ \overrightarrow{g}\ ]{} & \overrightarrow{Q} \end{array}$$

Proposition 3.4. *The functor* $\overrightarrow{F} : \textbf{Sem} \to \textbf{Dyn}$ *is an equivalence of categories.*

Proof. a) Let $(X, g) \in \textbf{Dyn}$ be prescribed. Let $P \subseteq X$ be an open domain of attraction for A with $g\left(\overline{P}\right) \subseteq P$. Then $(X, g) = \overrightarrow{F}(P, f)$ with $f := g_{|P}$.

b) To show that $\overrightarrow{F}$ is faithful we consider two morphisms $\varphi, \psi \in \textbf{Sem}(P, f; Q, g)$ with $\overrightarrow{F}(\varphi) = \overrightarrow{F}(\psi)$. By definition this means $j\psi \simeq j\varphi f^n$ for suitable $n \geq 0$ (if n should turn up < 0 we interchange φ and ψ), and there exists $m \geq 0$ such that this homotopy runs in $\overrightarrow{g}^{-m} j(Q)$. Then $j\psi f^m \simeq \overrightarrow{g}^m j\psi \simeq \overrightarrow{g}^m j\varphi f^n \simeq j\varphi f^{m+n}$ in $j(Q)$ and that means $\psi f^m \simeq \varphi f^{m+n}$.

c) To show that the functor $\overrightarrow{F}$ is full we suppose we are given a morphism $\Phi \in \textbf{Dyn}\left(\overrightarrow{P}, \overrightarrow{f}; \overrightarrow{Q}, \overrightarrow{g}\right)$. There is $n \geq 0$ with $\Phi i(P) \subseteq \overrightarrow{g}^{-n} j(Q)$, and we take a map $\varphi : P \to Q$ with $j\varphi = \overrightarrow{g}^n \Phi i$. Let $m \geq n$ be such that $\overrightarrow{g}\Phi_{|P} \simeq \Phi \overrightarrow{f}_{|P}$

in $\overrightarrow{g}^{-m}j(Q)$. Then $g^{m-n+1}\varphi \simeq g^{m-n}\varphi f$, $jg^{m-n}\varphi \simeq \Phi \overrightarrow{f}^m i$ and consequently $\overrightarrow{F}(g^{m-n}\varphi) = \Phi \overrightarrow{f}^m \sim \Phi$. $\qquad\square$

Lemma 3.5. *If two dynamical systems (X,g), $(Y,h) \in \mathbf{Dyn}$ are similar, then there exists a homotopy equivalence $\Phi : X \simeq Y$ and an integer $n \geq 1$ with $\Phi g^n \simeq h^n \Phi$.*

Proof. The difficult part is to get rid of pre-homotopy, and for this purpose we need a few preparations from coherent homotopy theory: Two homotopies $G, H : a \simeq b$ connecting the same two maps $a, b : Z \to Z'$ are homotopic if they are related by a homotopy which is stationary on $Z \times \dot{I}$. A homotopy $H : a \simeq b$ can be composed with a map $c : Z' \to \cdot$ to yield a homotopy $cH : ca \simeq cb$ and with a map $d : \cdot \to Z$ to yield a homotopy $Hd := H\,(d \times \mathrm{id}_I) : ad \simeq bd$. For two homotopies $G : a \simeq b$, $H : b \simeq c$ we denote by $G \circ H : a \simeq c$ their composition defined in the obvious way. If maps $a, b : Z \to Z'$, $c, d : Z' \to Z''$ and homotopies $H : a \simeq b$, $G : c \simeq d$ are given, then the composed homotopies $cG \circ Hb : ca \simeq db$ and $Ha \circ dG : ca \simeq db$ are homotopic [17].

Now let's suppose we are given maps $\varphi' : X \to Y$ and $\psi' : Y \to X$ with $\varphi'g \simeq_w h\varphi'$, $\psi'h \simeq_w g\psi'$, $\varphi'\psi' \simeq_w h^k$ and $\psi'\varphi' \simeq_w g^m$ for suitable numbers $k, m \in \mathbb{Z}$. Then $g^{2m} \simeq_w \psi'\varphi'\psi'\varphi' \simeq_w \psi'h^k\varphi' \simeq_w g^{k+m} \Rightarrow \psi'\varphi' \simeq_w g^m \simeq_w g^k$. We choose *compact* domains of attraction $P \subseteq X$ and $Q \subseteq Y$ with $g(P) \subseteq P^\circ$ and $h(Q) \subseteq Q^\circ$. There exist numbers $a, b, c \geq 1$ with $c \geq a$, $b \geq -k$ such that

$$
(12) \qquad\qquad \varphi'(P) \quad \subseteq \quad h^{-a}(Q^\circ)
$$

$$
(13) \qquad\qquad h\varphi'_{|P} \quad \simeq \quad \varphi'g_{|P} \text{ in } h^{-a}(Q^\circ)
$$

$$
(14) \qquad\qquad \psi'h^{-a}(Q) \quad \subseteq \quad g^{-b}(P^\circ)
$$

$$
(15) \qquad\qquad \psi'\varphi'_{|P} \quad \simeq \quad g^k_{|P} \text{ in } g^{-b}(P^\circ)
$$

$$
(16) \qquad \varphi'g^b\psi'_{|h^{-a}(Q)} \quad \simeq \quad h^{b+k}_{|h^{-a}(Q)} \text{ in } h^{-c}(Q^\circ).
$$

We set $\varphi := h^c\varphi' : P \to Q$, $\psi := g^b\psi'h^{-a} : Q \to P$, $g_\diamond := g_{|P} : P \rightarrowtail P$, $h_\diamond := h_{|Q} : Q \rightarrowtail Q$, $r := k+b+c-a$ and conclude $\psi\varphi \simeq g^r_\diamond$, $\varphi\psi \simeq h^r_\diamond$. Here and in the remainder of the proof we assume that all maps and homotopies with target space P (resp. Q) take their values in P° (resp. Q°). This enables us to make use of the homotopy extension property of closed inclusion maps between metrizable spaces, because P° and Q° are ANR-spaces. We make a specific choice of homotopies $G : \varphi\psi \simeq h^r_\diamond$ and $F : \psi\varphi \simeq g^r_\diamond$ and define

$$
(17) \qquad\qquad \Sigma := G^{-1}\varphi \circ \varphi F : h^r_\diamond\varphi \simeq \varphi g^r_\diamond
$$

$$
(18) \qquad\qquad \Omega := F^{-1}\psi \circ \psi G : g^r_\diamond\psi \simeq \psi h^r_\diamond
$$

and observe

$$(19) \qquad\qquad \Omega\varphi \circ \psi\Sigma \circ Fg_\diamond^r \quad \simeq \quad g_\diamond^r F$$

$$(20) \qquad\qquad \Sigma\psi \circ \varphi\Omega \circ Gh_\diamond^r \quad \simeq \quad h_\diamond^r G.$$

Using the homotopy extension property we can define inductively sequences of maps $\varphi_n : P \to Q$, $\psi_n : Q \to P$, such that the following diagram is *strictly* commutative:

$$(21)$$

$$
\begin{array}{ccccc}
P & \xrightarrow{\ \varphi_n\ } & Q & \xrightarrow{\ \psi_n\ } & P \\
\downarrow{\scriptstyle g_\diamond^r} & & \downarrow{\scriptstyle h_\diamond^r} & & \downarrow{\scriptstyle g_\diamond^r} \\
P & \xrightarrow{\ \varphi_{n+1}\ } & Q & \xrightarrow{\ \psi_{n+1}\ } & P,
\end{array}
$$

and such that there exist homotopies $R_n : \varphi \simeq \varphi_n$ and $L_n : \psi \simeq \psi_n$ with

$$(22) \qquad\qquad R_{n+1}g_\diamond^r \quad = \quad \Sigma^{-1} \circ h_\diamond^r R_n$$

$$(23) \qquad\qquad L_{n+1}h_\diamond^r \quad = \quad \Omega^{-1} \circ g_\diamond^r L_n.$$

(21) allows us to define maps $\Phi : X \to Y$, $\Psi : Y \to X$ by setting

$$(24)$$
$$\Phi = h^{-nr}\varphi_n g^{nr} = h^{-(n+1)r}\varphi_{n+1}g^{(n+1)r} : g^{-nr}(P) \to h^{-nr}(Q) \quad \text{on } g^{-nr}(P)$$

$$(25)$$
$$\Psi = g^{-(n+1)r}\psi_n h^{nr} : h^{-nr}(Q) \to g^{-(n+1)r}(P) \quad \text{on } h^{-nr}(Q).$$

We define homotopies $\theta_n : \varphi_{n+1} \simeq \varphi_n$ by $\theta_n := R_{n+1}^{-1} \circ R_n$. (22) implies $\theta_{n+1}g_\diamond^r \simeq h_\diamond^r \theta_n$, and the homotopy extension property enables us to ensure $\theta_{n+1}g_\diamond^r = h_\diamond^r \theta_n$ by successive homotopic alterations in θ_{n+1}. This gives rise to a homotopy $h^r \Phi \simeq \Phi g^r$ equal to $h^{-nr}\theta_n g^{(n+1)r}$ on $g^{-nr}(P)$.

Furthermore there are homotopies

$$(26) \qquad \Gamma_n := L_n^{-1}\varphi_n \circ \psi R_n^{-1} \circ F : \quad \psi_n\varphi_n \simeq g_\diamond^r$$

$$(27) \qquad \Delta_n := R_{n+1}^{-1}\psi_n \circ \varphi L_n^{-1} \circ G : \varphi_{n+1}\psi_n \simeq h_\diamond^r;$$

they satisfy the relations

$$(28) \qquad\qquad \Gamma_{n+1}g_\diamond^r \quad \simeq \quad g_\diamond^r \Gamma_n$$

$$(29) \qquad\qquad \Delta_{n+1}h_\diamond^r \quad \simeq \quad h_\diamond^r \Delta_n.$$

Making use of the homotopy extension property of $g_\diamond^r$ and $h_\diamond^r$ we can alter the homotopies Γ_{n+1} and Δ_{n+1} by induction such that the relations (28) and (29) hold *strictly*. Then there are homotopies $\Psi\Phi \simeq \mathrm{id}_X$, $\Phi\Psi \simeq \mathrm{id}_Y$, which coincide with $g^{-(n+1)r}\Gamma_n g^{nr}$ on $g^{-nr}(P)$ and with $h^{-(n+1)r}\Delta_n h^{nr}$ on $h^{-nr}(Q)$. $\qquad\square$

Proposition 3.6. *We consider a homeomorphism $g : X \approx X$ on a manifold X containing a global attractor and the induced automorphism $\varphi := H_*(g) : H_*(X; R) \approx H_*(X; R)$ of singular homology groups with coefficients in a principal ideal domain*

R. Then $H_*(X; R)$ contains a finitely generated, φ-invariant submodule G with $\bigcup_{n=1}^{\infty} \varphi^{-n}(G) = H_*(X; R)$. If R is a field, then the vector space $H_*(X; R)$ is finite dimensional over R.

Proof. Let $P \subseteq X$ be a compact domain of attraction with $g(P) \subseteq P$. Since the inclusion map $P \hookrightarrow X$ factors over a compact polyhedron up to homotopy, the submodule $G := \text{im}\{H_*(P; R) \to H_*(X; R)\}$ is finitely generated. The property of compact supports [14, Ch.4,§8,Ax.11] implies $H_*(X; R) = H_*\left(\bigcup_{n=1}^{\infty} g^{-n}(P); R\right)$ $= \bigcup_{n=1}^{\infty} H_*(g^{-n}(P); R) = \bigcup_{n=1}^{\infty} \varphi^{-n}(G)$. $\square$

Proposition 3.6 implies $H_*(X; R) \approx G \otimes_{\mathbb{Z}[\varphi]} \mathbb{Z}[\varphi, \varphi^{-1}]$ and limits the class of manifolds that can contain global attractors.

Question: We consider two manifolds of equal homotopy type. If one of them admits a self-homeomorphism with global attractor, does the other one necessarily share this property?

4. PROOF OF MAIN THEOREM AND CONCLUSIONS

Proof of Theorem 1.1: We observe $(X, f^n) = \overrightarrow{F}(f^n_{|P})$, $(Y, g^n) = \overrightarrow{F}(g^n_{|Q})$, $(A, f^n_{|A})$ $= \overleftarrow{F}(f^n_{|P})$ and $\left(B, g^n_{|B}\right) = \overleftarrow{F}\left(g^n_{|Q}\right)$. Therefore (i) implies (iii) by Proposition 3.4. (iii) implies $g^n_{|Q}\varphi \simeq \varphi f^n_{|P}$ and $\psi g^n_{|Q} \simeq f^n_{|P}\psi$. This means that $\left(P, f^n_{|P}\right)$ and $\left(Q, g^n_{|Q}\right)$ are equivalent in **Sem**, and now Lemma 3.5 applies. Similarly, the equivalence (ii) $\Leftrightarrow$ (iii) follows from Proposition 2.5 and Lemma 2.4. Diagram (1) is commutative with $\Phi_1 = \overleftarrow{F}(\varphi)$ and $\Phi_2 = \overrightarrow{F}(\varphi)$. $\square$

As illustration we want to prove the relations pictured in Figure 1, where $f : X \approx X$ is a homeomorphism on a manifold with global attractor A. The four remaining implications between the three boxes are false.

Lemma 4.1. *Let P be an open domain of attraction for A with $f\left(\overline{P}\right) \subseteq P$. If there exist a (not necessarily compact) polyhedron Q, an integer $n \geq 1$ and maps $u : P \to Q$, $v : Q \to P$ with $vu \simeq f^n_{|P}$ in P, such that $g := uv : Q \to Q$ is a homotopy equivalence, then the inclusion map $A \hookrightarrow X$ is a shape equivalence.*

Proof. We may assume that the polyhedron Q is σ-compact: If this condition is not satisfied, we replace u by $uf^n_{|P}$ to achieve that the new u has relatively compact image in Q. Then we construct an increasing sequence of compact subpolyhedra Q_n with $Q_1 \supseteq u(P)$ and such that the homotopies $uv \simeq g$, $gg^{-1} \simeq \text{id}_Q$, $g^{-1}g \simeq \text{id}_Q$ map $Q_n \times I$ to Q_{n+1}. Finally we replace Q by $\bigcup Q_n$.

We have: $f^n jvg^{-1}ui = jf^n_{|P}vg^{-1}ui \simeq jvuvg^{-1}ui \simeq jvgg^{-1}ui \simeq jvui \simeq jf^n_{|P}i = f^n ji$. Since f is a homeomorphism this means that following diagram is homotopy

$$\boxed{\begin{array}{l} \exists n \geq 1 : f^n \simeq \mathrm{id}_X \Leftrightarrow \\ \exists n \geq 1 : \eta\left(f_{|A}^n\right) = \mathrm{id}_A \end{array}}$$

$$\Downarrow$$

$$\boxed{\begin{array}{l} \text{The inclusion map } A \hookrightarrow X \text{ is a shape} \\ \text{equivalence } \Leftrightarrow A \text{ has the shape of a} \\ \text{polyhedron } \Leftrightarrow X \text{ is homotopy domi-} \\ \text{nated by a compact polyhedron} \end{array}}$$

$$\Uparrow$$

$$\boxed{\begin{array}{l} X \text{ has the homotopy type of} \\ \text{a compact polyhedron} \\ \Leftrightarrow A \text{ has the shape of a com-} \\ \text{pact polyhedron} \end{array}}$$

FIGURE 1

commutative, where $i : A \hookrightarrow P$ and $j : Q \hookrightarrow X$ are the inclusion maps:

$$(30) \qquad \begin{array}{ccccc} A & \xrightarrow{\ i\ } & P & \xrightarrow{\ u\ } & Q \\ {\scriptstyle ji}\Big\downarrow & & & & {\scriptstyle g^{-1}}\Big\downarrow \\ X & \xleftarrow{\ j\ } & P & \xleftarrow{\ v\ } & Q \end{array}$$

We are going to show that $ui : A \to Q$ is a shape equivalence and $jv : Q \to X$ is a pre-homotopy equivalence. It is known that every pre-homotopy equivalence between nice spaces (that's where we need σ-compactness) is pre-homotopic to a genuine homotopy equivalence. If $\varphi : Q \to X$ is a genuine homotopy equivalence with $\varphi \simeq_w jv$, then $ji \simeq jvg^{-1}ui \simeq \varphi g^{-1}ui$ because A is compact, therefore $ji : A \hookrightarrow X$ must be a shape equivalence.

To prove these properties of ui and jv we may assume $(X, f^n) = \overrightarrow{F}\left(f_{|P}^n\right)$ and $\left(A, f_{|A}^n\right) = \overleftarrow{F}\left(f_{|P}^n\right)$; with this identification the natural embedding $P \rightarrowtail \overrightarrow{P}$ corresponds to $j : P \hookrightarrow X$ and the natural projection $\overleftarrow{P} \to P$ to the inclusion map $i : A \hookrightarrow P$. We observe $f_{|P}^n v \simeq vg$, $uf_{|P}^n \simeq gu$ and apply Lemmata 2.3 and 3.3 to construct a shape morphism $\alpha \in \mathbf{sh}(Q, A)$ and a map $\varphi : X \to Q$ with $\eta(i)\alpha = \eta(v)$, $\eta\left(f_{|A}^n\right) = \alpha\eta(g)$, $\varphi j \simeq u$ and $g\varphi \simeq_w \varphi f^n$. This implies $\eta(ui)\alpha = \eta(uv) = \eta(g)$ and $\varphi jv \simeq uv \simeq g$. We set $\beta := \alpha\eta\left(g^{-1}ui\right) \in \mathbf{sh}(A, A)$ and $\psi := jvg^{-1}\alpha : X \to X$. Evidently we have: $\eta\left(f_{|A}^n\right)\beta = \alpha\eta(ui) = \beta\eta\left(f_{|A}^n\right)$, $f^n\psi \simeq jva \simeq_w \psi f^n$, $\eta(i)\beta\eta\left(f_{|A}^n\right) = \eta\left(if_{|A}^n\right) \Rightarrow \eta(i)\beta = \eta(i)$ and $f^n\psi j \simeq f^n j \Rightarrow \psi j \simeq j$. Now the uniqueness statement contained in Lemmata 2.3 and 3.3 implies $\beta = \mathrm{id}_A$ and

$\psi \simeq_w \mathrm{id}_X.$ □

Example 4.2. A self-map $f : P \to P$ of a compact polyhedron is a homotopy idempotent if $f^2 \simeq f$; f is said to split over a polyhedron Q if there are maps $u : P \to Q$, $v : Q \to P$ with $f \simeq vu$ and $uv \simeq \mathrm{id}_Q$. We turn $f : P \to P$ into a small embedding using Lemma 2.2, then the associated dynamical system $\left(\overrightarrow{P}, \overrightarrow{f}\right)$ satisfies the assumptions of Lemma 4.1 with $g := \mathrm{id}_Q$. Hence the inclusion map of the attractor $\overleftarrow{P}$ into the manifold $\overrightarrow{P}$ is a shape equivalence. The condition $f^2 \simeq f$ implies that the associated shift map $\overleftarrow{f} : \overleftarrow{P} \approx \overleftarrow{P}$ is an idempotent up to shape: $\eta\left(\overleftarrow{f}^2\right) = \eta\left(\overleftarrow{f}\right)$, and therefore $\eta\left(\overleftarrow{f}\right) = \mathrm{id}$. We observe that the proof of Lemma 4.1 also shows that $\overleftarrow{P}$ is shape equivalent to Q.

It is known [7, Ch.II,§9.5,Ex.2] that the data above may be chosen such that Q does not have the homotopy type of a *compact* polyhedron. In this case we obtain an attractor $\overleftarrow{P}$ that has the shape of a polyhedron, but not of a compact polyhedron.

Remark 4.3. A tower of the form $\mathbf{P} = \left\{ P \xleftarrow{f} P \xleftarrow{f} P \xleftarrow{f} \cdots \right\} \in$ pro–$\mathcal{C}$ generated by a morphism $f \in \mathcal{C}(P, P)$ in an arbitrary category $\mathcal{C}$ is stable [7, p.160], i.e., isomorphic to an object $Q \in \mathcal{C}$, if and only if there exist morphisms $u : P \to Q$, $v : Q \to P$ and an integer $n \geq 1$ with $vu = f^n$ and such that $uv : Q \to Q$ is an isomorphism. Hence the sufficient condition given in Lemma 4.1 is also necessary. It should be observed that it implies $f^{2n}g = f^n$ with $g := v(uv)^{-2}u$.

We claim that in the category of groups the tower created by an endomorphism $f : G \to G$ is stable if and only if there exists an integer $n \geq 1$ such that $\mathrm{im}\, f^m = \mathrm{im}\, f^n$ (Mittag-Leffler property) and $\ker f^m = \ker f^n$ for all $m \geq n$. This condition is necessary because for $f^n = vu$ as above we get $\ker f^{nk} = \ker u$ and $\mathrm{im}\, f^{nk} = \mathrm{im}\, v$ for all $k \geq 1$. To prove sufficiency we set $H := \mathrm{im}\, f^n$, define $u : G \twoheadrightarrow H$ by $u(x) := f^n(x)$ elementwise and denote by $v : H \hookrightarrow G$ the inclusion homomorphism. This means $vu = f^n$. We have $\mathrm{im}\, uv = \mathrm{im}\, f^{2n} = \mathrm{im}\, f^n = H$ and $\ker uv = \ker f^n \cap \mathrm{im}\, f^n = f^n\left(\ker f^{2n}\right) = f^n\left(\ker f^n\right) = 0$ and hence uv is an isomorphism.

We return to general categories, replace f by f^n to ensure $f^2g = f$ and consider f as "generalized idempotent", which "splits" if it is of the form $f = vu$ as described above. There exist unsplit generalized pointed homotopy idempotents on finite dimensional polyhedra: We take $\mathfrak{F}$ as free group with a sequence of free generators $x_1, x_2, x_3, \ldots$ and set P equal to the wedge of countably many circles, so that $P = K(\mathfrak{F}, 1)$. If $f, g : P \to P$ are pointed maps with $f_\#(x_1) = 1$, $f_\#(x_n) = x_{n-1}$ for $n > 1$ and $g_\#(x_n) = x_{n+1}$, then $fg \simeq \mathrm{id}$ and therefore $f^2g \simeq f$ in the pointed sense. f does not split, because $x_{n+1} \in \ker f_\#^{n+1} \setminus \ker f_\#^n$ for all n and therefore the fundamental pro-group of our tower is not stable.

We *don't* know if every "generalized (pointed) homotopy idempotent" on a *finite* complex splits. This would imply that every (pointed) movable attractor has the shape of a possibly infinite polyhedron.

Proposition 4.4. *The following conditions are equivalent:*

(i) *A has the shape of a (not necessarily compact) polyhedron.*

(ii) *X is homotopy dominated by a compact polyhedron.*

(iii) *The inclusion map $i : A \hookrightarrow X$ is a shape equivalence.*

Proof. (i) $\Rightarrow$ (iii): We suppose we are given a shape equivalence $\varphi : A \to Q$ into a polyhedron Q. Since polyhedra are absolute neighborhood extensors for compact spaces there is a map $\psi : W_1 \to Q$ on a neighborhood W_1 of A in X with $\psi_{|A} = \varphi$. We take an open domain of attraction P with $f(P) \subseteq P$; observing that P is an ANR-space and φ a shape equivalence we can find a map $v : Q \to P$ with $v\varphi : A \to P$ homotopic to the inclusion map $i : A \hookrightarrow P$. Then the map $v\psi : W_1 \to P$ is homotopic to i on A and this homotopy can be extended over a neighborhood $W_2 \subseteq W_1$. We choose $n \geq 1$ such that $f^n(P) \subseteq W_2$ and set $u := \psi f^n_{|P} : P \to Q$. Clearly $vu \simeq f^n_{|P}$ in P and $uv\varphi \simeq ui = \varphi f^n_{|A}$, and since φ is a shape equivalence and $f_{|A}$ a homeomorphism this means that $g := uv : Q \to Q$ is a homotopy equivalence. Now Lemma 4.1 tells us that $i : A \hookrightarrow X$ is a shape equivalence.

(iii) $\Rightarrow$ (i): X is an ANR-space and therefore has the homotopy type of a polyhedron [7, App.1,§2.2,Thm.5].

(ii) $\Rightarrow$ (iii): We suppose we are given a compact polyhedron K and maps $c : X \to K$, $b : K \to X$ with $bc \simeq \mathrm{id}_X$. Let P be an open domain of attraction with $f(\overline{P}) \subseteq P$ and choose $m \geq 0$ such that $b(K) \subseteq f^{-m}(P)$ and $bc_{|P} : P \to f^{-m}(P)$ is homotopic to the inclusion map in $f^{-m}(P)$. We consider the inclusion map $u : P \hookrightarrow X$ and set $v := f^m bc : X \to P$. Then $uv \simeq f^m$ and $vu \simeq f^m_{|P}$ in P; since X has the homotopy type of a polyhedron Lemma 4.1 implies that the inclusion map $i : A \hookrightarrow X$ is a shape equivalence.

(iii) $\Rightarrow$ (ii): Since X is an ANR-space and A is compact there exists a compact polyhedron K and maps $a : A \to K$, $b : K \to X$ with $ba \simeq i$. If i is a shape equivalence we can find a map $c : X \to K$ with $ci \simeq a$; then $bci \simeq i$ and therefore $bc \simeq \mathrm{id}_X$, i.e. X is homotopy dominated by K. $\qquad\square$

Remark 4.5. Two maps defined on a space, which is homotopy dominated by a compact space, are homotopic if and only if they are pre-homotopic. Hence the proof of Theorem 1.1 shows that the theorem can be strengthened, if the manifolds in question have this property: conditions (i) and (ii) remain equivalent if we impose the additional restriction $n = 1$.

Corollary 4.6. *We consider a polyhedron Q. There exists an attractor shape equivalent to Q if and only if Q is homotopy dominated by a compact polyhedron.*

Proof. Proposition 4.4 implies that the condition is necessary; sufficiency follows from Example 4.2. □

Corollary 4.7. *A has the shape of a compact polyhedron if and only if X has the homotopy type of a compact polyhedron. If this condition is satisfied, then the inclusion map $A \hookrightarrow X$ is a shape equivalence.*

Corollary 4.8. *A has trivial shape if and only if X is contractible.*

Example 4.9. In [1] Barge and Martin showed that for every self-map $f : I \to I$ of the unit segment the associated inverse limit $\overleftarrow{I}$ can be embedded in the plane in such a way, that the shift map $\overleftarrow{f} : \overleftarrow{I} \approx \overleftarrow{I}$ extends to an self-homeomorphism of the plane with $\overleftarrow{I}$ as global attractor. We want to show that this property still holds if $f : I \to I$ is replaced by a small self-embedding $f : D^2 \rightarrowtail D^2$ of the open unit disc, and that in this way we obtain all global attractors on $\mathbb{R}^2$ up to homeomorphism. To prove the first assertion it suffices to observe that $\overrightarrow{D^2}$ is a contractible 2-manifold and therefore homeomorphic to the plane (cf. [13, Footnote 26,p.320] and [11]). To prove the converse we consider a homeomorphism $g : \mathbb{R}^2 \approx \mathbb{R}^2$ containing a global attractor and choose an open domain of attraction P with $g\left(\overline{P}\right) \subseteq P$. If we can show that P may be chosen $\approx D^2$ we are done. Let K be a finite subcomplex of $\mathbb{R}^2$ with $\overline{g(P)} \subseteq K \subseteq P$; after replacing K by a regular neighborhood if necessary we may assume that K is a compact submanifold of $\mathbb{R}^2$ and hence a perforated disc, $K \approx D^2 \setminus \bigcup_{i=1}^{n} U_i$ with suitable open discs $U_i \subseteq \overline{U_i} \subseteq D^2$. Our proof is by induction on n. We set $\varphi := H_1\left(g_{|K}\right) : H_1(K; \mathbb{Z}) \to H_1(K, \mathbb{Z})$ and observe that the limit of the direct sequence with terms $H_1(K; \mathbb{Z})$ and bonding homomorphisms φ equals $H_1\left(\mathbb{R}^2; \mathbb{Z}\right) = 0$; this means $\varphi^m = 0$ for suitable $m \geq 1$. By abuse of notation we denote the homeomorphic images of the open discs U_i filling the holes of K again by U_i, orient their boundaries counterclockwise and observe that the homology classes x_i of these boundaries freely generate $H_1(K; \mathbb{Z})$. Let $M_i \subseteq \{1, \dots n\}$ be the set of indices j with $U_j \subseteq g\left(U_i\right)$. The sets M_i are pairwise disjoint and $\varphi\left(x_i\right) = \pm \sum \{x_j | j \in M_i\}$. We can't have $M_j \neq \emptyset$ for each j, because then φ would be an isomorphism in contradiction to $\varphi^m = 0$. If j is an index with $M_j = \emptyset$, then $g\left(K \cup U_j\right) \subseteq K \subseteq K \cup U_i =: K'$. K' has only $n - 1$ holes, and the induction hypothesis applies.

Proposition 4.10. *Every dynamical system $(X, f) \in \mathbf{Dyn}$ is similar to a real analytic dynamical system. This means in particular that the shape theoretic phenomena occurring on attractors are not limited by the assumption of real analyticity.*

Proof. Let P be an open domain of attraction in X with $f\left(\overline{P}\right) \subseteq P$. Since $\overline{P}$ is compact and P is an ANR-space there exist a compact polyhedron Q' and maps

$a : \overline{P} \to Q'$ and $b : Q' \to P$ with $ba \simeq f_{|\overline{P}}$. Observing Lemma 2.2 we can find a real analytic manifold Q, an analytic small self-embedding $g : Q \rightarrowtail Q$ and a homotopy equivalence $\varphi : Q' \to Q$ with $g\varphi \simeq \varphi ab$. Then $\overrightarrow{F}(g) \in \mathbf{Dyn}$ is a real analytic dynamical system similar to (X, f). $\qquad\square$

We know nothing about the complex analytic case.

Corollary 4.11. *For each number $n \geq 1$ we have $f^n \simeq \mathrm{id}_X \Leftrightarrow \eta\left(f_{|A}^n\right) = \mathrm{id}_A$, and these conditions imply that the inclusion map $A \hookrightarrow X$ is a shape equivalence.*

Proof. We may assume that the manifold X and the homeomorphism f are differentiable. Let $P \subseteq X$ be a compact domain of attraction with $f(P) \subseteq P^\circ$, and take a differentiable Urysohn function $\varphi : X \to I$ with $\varphi \equiv 0$ on $f(P)$ and $\varphi \equiv 1$ outside P. If $0 < \vartheta < 1$ is a regular value for φ, then $P' := \varphi^{-1}[0, \vartheta]$ is a compact domain of attraction with $f(P') \subseteq P'$, which is a bounded submanifold of X and in particular a compact ANR-space.

We set $f_\diamond := f_{|P'} : P' \to P'$. Both of the conditions stated above imply the existence of numbers $a \geq 0$, $b \geq 1$ with $f_\diamond^{a+b} \simeq f_\diamond^a$. We choose $r \geq 1$ such that $rb \geq a$; then $f_\diamond^{(r+1)b} = f_\diamond^{b+a} f_\diamond^{rb-a} \simeq f_\diamond^a f_\diamond^{rb-a} = f_\diamond^{rb}$ and by induction $f_\diamond^{(r+k)b} \simeq f_\diamond^{rb}$. For $k = r$ this means that $f_\diamond^{br} : P' \to P'$ is a homotopy idempotent, and since P' has the homotopy type of a compact polyhedron [7, App.I,§2.2,Thm.7] $f_\diamond^{br}$ splits [7, Ch.II,§9.5,Thm.20]. This means that there exist a polyhedron K and maps $\varphi : P' \to K$, $\psi : K \to P'$ with $f_\diamond^{br} \simeq \psi\varphi$ and $\varphi\psi \simeq \mathrm{id}_K$. As in Example 4.2 one shows that the inclusion map $j : A \hookrightarrow X$ is a shape equivalence, and then the equivalence of the conditions above follows from $fj = jf_{|A}$. $\qquad\square$

Remark 4.12. The converse of Corollary 4.11 does not hold: In example 4.17 we will construct a homeomorphism $f : X \approx X$ with X equal to the plane $\mathbb{R}^2$ punctured at three points, and with global attractor A homeomorphic to a disc with three holes (hence the inclusion map $A \hookrightarrow X$ is a shape equivalence), but $\eta\left(f_{|A}^n\right) \neq \mathrm{id}_A$ for $n \neq 0$.

Remark 4.13. Since a global attractor A "essentially" determines its encompassing manifold X one might conjecture that it also determines the complement $X \setminus A$ in the sense of duality theory. This turns out to be wrong: neither A nor $X \setminus A$ determines the other set, not even up to stable homotopy or stable shape. For proof we consider the standard representation of the torus T as a square identifying opposite edges, and choose a self-homeomorphism leaving center c and boundary fixed and moving all other points radially outward (cf. Figure 2). Then $A := a \cup b \approx S^1 \vee S^1$ is a global attractor of $X := T \setminus \{c\}$ with $X \setminus A \approx S^1 \times \mathbb{R} \simeq S^1$. On the other hand, taking the standard embedding of A in the plane one can construct a self-homeomorphism on $Y := \mathbb{R}^2 \setminus \{\text{two points}\}$ with A as global attractor. $Y \setminus A$

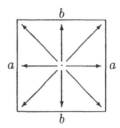

FIGURE 2. Flow on the torus used in Remark 4.13.

is the disjoint union of three copies of $S^1 \times \mathbb{R}$ and therefore $X \setminus A$ and $Y \setminus A$ are not stably equivalent. Also, the origin is a global attractor of the plane with $\mathbb{R}^2 \setminus \{0\} \approx S^1 \times \mathbb{R} \approx X \setminus A$, but of course A is not stably equivalent to a one point set.

Attractors in the plane can be classified up to shape using braids. Every continuum in the plane has the shape of a closed disc $D_n \subset \mathbb{R}^2$ with n holes, $0 \leq n < \infty$, or of the Hawaiian earring Y [7, Ch.II,§8.1,Rmk.1]. The Hawaiian earring cannot be an attractor by Proposition 2.7, because $\check{H}^1(Y;\mathbb{Q})$ has infinite dimension. The fundamental group $\pi_1(D_n) = \mathfrak{F}_n$ is the free group with n natural generators $\omega_1, \ldots \omega_n$; in addition we set $\omega_0^{-1} := \omega_1 \omega_2 \cdots \omega_n$. Every homeomorphism $h : D_n \approx D_n$ defines an automorphism $\pi_1(h) : \mathfrak{F}_n \approx \mathfrak{F}_n$, which is well defined up to an inner automorphism. An automorphism $\varphi : \mathfrak{F}_n \approx \mathfrak{F}_n$ is induced by an *orientation preserving* homeomorphism $D_n \approx D_n$ if and only if there is a permutation $\tau \in \mathfrak{S}_{n+1}$, such that $\varphi(\omega_k)$ is conjugate to $\omega_{\tau(k)}$ for $0 \leq k \leq n$ [8]. The subgroup $\mathfrak{H}_n \subseteq \mathrm{Aut}(\mathfrak{F}_n) / \mathrm{InnAut}(\mathfrak{F}_n)$ of all automorphisms obtained this way is generated by the automorphisms

$$(31) \qquad \sigma_k(\omega_j) := \begin{cases} \omega_{k+1} & \text{for } j = k \\ \omega_{k+1}^{-1} \omega_k \omega_{k+1} & \text{for } j = k+1 \\ \omega_j & \text{else} \end{cases}$$

with $0 \leq k < n$, and hence $\mathfrak{H}_n$ is a homomorphic (not isomorphic) image of the n+1-st braid group. For $h \in \mathfrak{H}_n$ we take a homeomorphism $\tilde{h} : D_n \approx D_n$ with $\pi_1(\tilde{h}) = h$ and form the object $A_h = (D_n, \tilde{h}) \in \mathbf{Atr}$. We add an open collar to the boundary of D_n and consider the resulting space U_n as $\mathbb{R}^2$ punctured at n points. Evidently $\tilde{h} : D_n \approx D_n$ can be extended to a homeomorphism $U_n \approx U_n$ with D_n as global attractor.

Proposition 4.14. *Every connected attractor of an orientation preserving homeomorphism $f : U \approx U$ of an open subset $U \subseteq \mathbb{R}^2$ is similar to A_h for suitable $h \in \mathfrak{H}_n$, $0 \leq n < \infty$. Two attractors A_h and A_g are similar if and only if there is $\varphi \in \mathrm{Aut}(\mathfrak{F}_n)$, such that h and $\varphi^{-1} g \varphi$ differ only by an inner automorphism.*

Proof. We may suppose that the attractor A in question is a *global* attractor of $f : U \approx U$; then U is necessarily connected and has finitely generated homology by Proposition 3.6. If n is the number of generators, then U is homeomorphic to $U_n = S^2 \setminus \{x_0, \ldots x_n\}$. We place small open discs B_k around the holes x_k and alter f isotopically such that $D_n := U_n \setminus \bigcup_{k=0}^{n} B_k$ becomes a global attractor of the new homeomorphism $f' : U_n \approx U_n$. Now Remark 4.5 implies that $(A, f_{|A})$ is similar to A_h, where $h : D_n \approx D_n$ is the restriction of f'. This proves the first statement, the second one follows immediately from the definitions. $\square$

Remark 4.15. We emphasize that the proof above shows that the open subset $U \subseteq \mathbb{R}^2$ cannot be esoteric when the attractor A is global. It is homeomorphic to the sphere S^2 punctured at finitely many points, and with this particular embedding $U \hookrightarrow S^2$ the homeomorphism f extends over the whole sphere. This means that every attractor of a self-homeomorphism of a planar open set occurs as an attractor of a homeomorphism $S^2 \approx S^2$ of the entire 2-sphere.

Example 4.16. Although the Plykin attractor [3, p.210] has the shape of D_3, it does not occur as attractor of a *homeomorphism* of a planar open set. If we denote the generators of $\pi_1(D_3)$ by α, β, γ, then such a homeomorphism would imply the existence of an automorphism $\varphi \in \mathfrak{H}_3$ with $\varphi(\alpha) := \beta$, $\varphi(\beta) := \beta\gamma\beta^{-1}$ and $\varphi(\gamma) := \alpha$. This is a contradiction, because $\varphi\left(\gamma^{-1}\beta^{-1}\alpha^{-1}\right) = \alpha^{-1}\beta\gamma^{-1}\beta^{-2}$ is not conjugate to $\gamma^{-1}\beta^{-1}\alpha^{-1}$.

Example 4.17. We consider the automorphism $\varphi := \sigma_1^2 \in \mathfrak{H}_3$. By induction one shows:

$$(32) \qquad \varphi^k(\omega_j) = \begin{cases} \gamma^{-k}\omega_j\gamma^k & \text{for } j \le 2 \\ \omega_j & \text{for } j = 3 \end{cases}$$

with $\gamma := \omega_1\omega_2$. If $f : D_3 \approx D_3$ is a homeomorphism with $\pi_1(f) = \varphi$, then f^n is not homotopic to the identity for $n \ne 0$ because φ^n is not an inner automorphism. This furnishes the counterexample mentioned in Remark 4.12.

Let us count the number of different dynamical structures on D_n, i.e., the number of similarity types of attractors in **Atr** with underlying compactum shape equivalent to D_n. Since D_0 is of trivial shape it carries only one. On D_1 there are two: they are defined by the automorphisms $\varphi_\pm : \mathfrak{F}_1 \approx \mathfrak{F}_1$, $\varphi_\pm(\omega_1) = \omega_1^{\pm 1}$ (now we also take into account orientation reversing maps).

There is an orientation reversing homeomorphism $D_2 \approx D_2$, which induces the automorphism $\chi : \mathfrak{F}_2 \approx \mathfrak{F}_2$, $\chi(\omega_i) = \omega_i^{-1}$.

Proposition 4.18. *a) D_2 carries six non similar dynamical structures determined by the automorphisms 1, σ_1, $\sigma_0\sigma_1$, χ, $\chi\sigma_1$ and $\chi\sigma_0\sigma_1$.*
b) For $n \geq 3$ D_n carries infinitely many non similar dynamical structures.

Proof. a) The relations [8, (14)] and [8, (19)] imply $\mathfrak{H}_2 = \mathfrak{S}_3$. Our automorphisms 1, σ_1 and $\sigma_0\sigma_1$ correspond to the permutations id, $(1,2)$ and $(0,1,2)$. One checks $\chi^2 = \mathrm{id}$ and $\chi\sigma_i\chi = \sigma_i$ modulo an inner automorphism, therefore the subgroup of $\Gamma := \mathrm{Aut}\,(\mathfrak{F}_n)\,/\,\mathrm{InnAut}\,(\mathfrak{F}_n)$ spanned by $\mathfrak{H}_2$ and χ is isomorphic to $\mathfrak{S}_3 \times \mathbb{Z}_2$ and contains six conjugacy classes; the automorphisms listed above form a complete system of representatives. However, we must make sure that no two of them determine the same conjugacy class in the encompassing group Γ. Since the automorphisms are of the order 1, 2, 3, 2, 2 and 4, respectively, and since orientation reversing automorphisms cannot be turned into orientation preserving ones by conjugation, it remains to show that χ and $\chi\sigma_1$ are not similar. The group Γ has a representation in $GL(2,\mathbb{Z})$ given by induced automorphisms of the abelianized group $\mathfrak{F}_2/\mathfrak{F}_2' \approx \mathbb{Z}^2$; the images of χ and $\chi\sigma_1$ under this representation are non conjugate matrices.

b) We recall the automorphism $\varphi := \sigma_1^2 : \mathfrak{F}_n \approx \mathfrak{F}_n$ for $n \geq 3$ from example 4.17, and denote by $A_k \in \mathbf{Atr}$ the attractor A_ϑ for $\vartheta = \varphi^k$, $k > 0$. We claim that A_k and A_m are not similar unless $k = m$ and have to show that for $k \neq m$ there does not exist $\psi \in \mathrm{Aut}\,(\mathfrak{F}_n)$, such that $\varphi^k\psi\varphi^{-m}\psi^{-1}$ is an inner automorphism.

In the sequel we consider each element of $\mathfrak{F}_n$ as word in the basic elements $\omega_1, \gamma, \omega_3, \omega_4, \ldots \omega_n$; again we have set $\gamma := \omega_1\omega_2$. For $\xi \in \mathfrak{F}_n$ and $k > 0$ we consider the subgroup $\Gamma_k(\xi) \subseteq \mathfrak{F}_n$ and the subset $\Gamma_k \subseteq \mathfrak{F}_n$ defined by

$$(33) \qquad \Gamma_k(\xi) \quad := \quad \left\{ \zeta \in \mathfrak{F}_n \,\middle|\, \varphi^k(\zeta) = \xi^{-1}\zeta\xi \right\}$$

$$(34) \qquad \Gamma_k \quad := \quad \left\{ \xi \in \mathfrak{F}_k \,\middle|\, \Gamma_k(\xi) \neq 1 \right\}.$$

We claim that the only elements $\zeta \in \mathfrak{F}_n$, such that $\varphi^k(\zeta)$ and ζ are conjugate, are those conjugate to elements of $\Gamma_k(1)$ or $\Gamma_k\left(\gamma^k\right)$.

For proof we observe $\gamma, \omega_3, \omega_4, \ldots \omega_n \in \Gamma_k(1)$ and $\omega_1, \gamma \in \Gamma_k\left(\gamma^k\right)$. Now suppose ζ is an element as above, which is not conjugate to an element of $\Gamma_k(1)$ or to ω_1^ℓ. Then there exist non zero integers $\ell_1, \ldots \ell_r$, $r \geq 1$, and non trivial words $\alpha_1, \ldots \alpha_r$ in the elements $\gamma, \omega_3, \omega_4, \ldots \omega_n$ such that

$$(35) \qquad \zeta' = \omega_1^{\ell_1}\alpha_1 \cdots \omega_1^{\ell_r}\alpha_r$$

is conjugate to ζ. (35) is a cyclically reduced word. We get

$$(36) \qquad \gamma^k\varphi^k\left(\zeta'\right)\gamma^{-k} = \omega_1^{\ell_1}\alpha_1' \cdots \omega_1^{\ell_r}\alpha_r'$$

with $\alpha_i' := \gamma^k\alpha_i\gamma^{-k}$. By assumption (35) and (36) are conjugate cyclically reduced words, hence there must be an integer s with

$$(37) \qquad \gamma^k\alpha_i\gamma^{-k} = \alpha_{i+s},$$

where $i + s$ is taken modulo r. (37) implies $\gamma^{rk} \alpha_i \gamma^{-rk} = \alpha_i$ and therefore $\alpha_i = \gamma^{\mu_i}$ with suitable integers μ_i. This means $\zeta' \in \Gamma_k\left(\gamma^k\right)$ and proves our claim. It also implies:

$$(38) \qquad \Gamma_k = \left\{ \alpha^{-1} \varphi^k(\alpha) \big| \alpha \in \mathfrak{F}_n \right\} \cup \left\{ \alpha^{-1} \gamma^k \varphi^k(\alpha) \big| \alpha \in \mathfrak{F}_n \right\}.$$

The image of Γ_k in the abelianized group $\mathfrak{A}_n = \mathfrak{F}_n / \mathfrak{F}'_n$ is

$$(39) \qquad \qquad \tilde{\Gamma}_k = \{0, k\gamma\},$$

because $\varphi^k = \text{id}$ in $\mathfrak{A}_n$. If φ^k and φ^m are related by an equation of the form

$$(40) \qquad \forall \zeta \in \mathfrak{F}_n : \varphi^k(\zeta) = \nu^{-1} \cdot \psi \varphi^m \psi^{-1}(\zeta) \cdot \nu$$

with $\psi \in \text{Aut}\left(\mathfrak{F}_n\right)$ and $\nu \in \mathfrak{F}_n$, then we must have $\psi\left(\Gamma_m\right) \cdot \nu = \Gamma_k$ and therefore, by (39), $\{\nu, \nu + m\psi(\gamma)\} = \{0, k\gamma\}$. This is possible only for $k = m$. $\qquad \square$

REFERENCES

[1] M. Barge, J. Martin: *The construction of global attractors;* Proc. Amer. Math. Soc. 110 (1990) 523–525.

[2] N. Bourbaki: *General Topology;* Hermann and Addison-Wesley, Elements of Mathematics (1966).

[3] R.L. Devaney: *Introduction to Chaotic Dynamical Systems;* Addison-Wesley, (1989) 2nd Edition.

[4] B. Günther, J. Segal: *Every attractor of a flow on a manifold has the shape of a finite polyhedron;* Proc. Amer. Math. Soc. 119 (1993) 321–329.

[5] B. Günther: *A compactum, that cannot be an attractor of a selfmap on a manifold;* To appear in Proc. Amer. Math. Soc.

[6] S.T. Hu: *Theory of Retracts;* Wayne State University Press, (1965).

[7] S. Mardešić, J. Segal: *Shape Theory;* North Holland, Mathematical Library 26 (1982).

[8] W. Magnus: *Über Automorphismen von Fundamentalgruppen berandeter Flächen;* Math. Ann. 109 (1934) 617–646.

[9] S. MacLane: *Kategorien;* Springer, Hochschultext (1972).

[10] R.F. Williams: *Classification of one-dimensional attractors;* Proc. Sympos. Pure Math. Vol. 14, AMS, (1970) 341–361.

[11] T. Rado: *Über den Begriff der Riemann'schen Fläche;* Acta Sci. Math. (Szeged) 2 (1925) 101–121.

[12] J.W. Robbin, D. Salamon: *Dynamical Systems, Shape Theory and the Conley Index;* Ergod. Th. & Dynam. Syst. 8* (1988) 375–393.

[13] H. Seifert, W. Threlfall: *Lehrbuch der Topologie;* Chelsea, (1980).

[14] E.H. Spanier: *Algebraic Topology;* McGraw-Hill, (1966).

[15] R.M. Switzer: *Algebraic Topology - Homotopy and Homology;* Springer, Grundlehren 212 (1975).

[16] C. Tezer: *Shift equivalence in homotopy;* Math. Z. 210 (1992) 197–201.

[17] R.M. Vogt: *A Note on Homotopy Equivalences;* Proc. Amer. Math. Soc. 32 (1972) 627–629.

E-mail address: `Guenther@mathematik.uni-frankfurt.d400.de` ''Bernd Günther''

16
SEMI-APOSYNDESIS AND CONTINUUM CHAINABILITY

CHARLES L. HAGOPIAN Department of Mathematics, California State University at Sacramento, Sacramento, California 95819, USA
LEX G. OVERSTEEGEN Department of Mathematics, University of Alabama at Birmingham, Birmingham, Alabama 35294, USA

ABSTRACT. Recently, Hagopian proved that every simply-connected plane continuum has the fixed-point property. We show that the arcwise-connectivity condition in this theorem cannot be replaced by semi-aposyndesis or continuum chainability. This is accomplished by constructing a nonarcwise-connected semi-aposyndetic continuum-chainable plane continuum that does not contain a simple closed curve and admits a period 2 fixed-point-free homeomorphism.

1. INTRODUCTION

A continuum M is *capped* if every simple closed curve in M bounds a disk in M. A plane continuum is simply connected if and only if it is capped and arcwise connected. Note that every plane continuum with the fixed-point property is capped.

Each outer boundary of each bounded complementary domain of an aposyndetic plane continuum is a simple closed curve [9, Theorem 43, p. 193] [8, 11, 13]. Thus no capped aposyndetic plane continuum separates the plane. In 1941, F. B. Jones [8] proved that every aposyndetic plane continuum that does not have infinitely many complementary domains is locally connected. Hence every capped aposyndetic plane continuum is a locally connected non-separating plane continuum and, by a theorem

1991 *Mathematics Subject Classification.* Primary 54F20.

Key words and phrases. fixed-point property, plane continuum, semi-aposyndesis, arcwise connected, continuum-chainable.

C. L. Hagopian was partially supported by the CSUS Research, Scholarship, and Creative Activities Program.

L. G. Oversteegen was partially supported by NSF–EPSCoR in Alabama.

of K. Borsuk [1], must have the fixed-point property. Recently, Hagopian [6] generalized this result by proving that every capped arcwise-connected plane continuum has the fixed-point property.

Since semi-aposyndesis and continuum chainability are natural generalizations of aposyndesis and arcwise connectivity, it is reasonable to ask the following:

Question 1.1. *If a capped plane continuum is either semi-aposyndetic or continuum chainable, then must it have the fixed-point property?*

All semi-aposyndetic E-continua [4] and all continuum-chainable plane continua with only finitely many complementary domains [5] are arcwise connected. Therefore, by [6], a capped continuum in either of these two classes must have the fixed-point property. Nevertheless, the answer to Question 1.1 is no. We construct a semi-aposyndetic continuum-chainable plane continuum that does not contain a simple closed curve and admits a period 2 fixed-point-free homeomorphism onto itself. This example is derived from Oversteegen's aposyndetic continuum-chainable plane continuum [10] that is not arcwise connected.

2. DEFINITIONS

A space S has the *fixed-point property* if for each map f of S into S there is a point p of S such that $f(p) = p$.

A space is *simply connected* if it is arcwise connected and its fundamental group is trivial.

A *continuum* is a nondegenerate compact connected metric space.

Suppose M is a continuum with distinct points p and q. Then M is *aposyndetic at p with respect to q* if an open subset G of M and a continuum H exist such that

$$p \in G \subset H \subset M \setminus \{q\}.$$

A continuum M is *aposyndetic* if M is aposyndetic at every point p of M with respect to every point q of $M \setminus \{p\}$.

A continuum M is *semi-aposyndetic* if for each pair p, q of distinct points of M, either M is aposyndetic at p with respect to q or M is aposyndetic at q with respect to p.

A finite collection $\{A_1, A_2, \ldots, A_n\}$ of sets is a *chain* from p to q provided $p \in A_1$, $q \in A_n$, and for integers i, j such that $1 \leq i \leq j \leq n$, $A_i \cap A_j \neq \emptyset$ if and only if $j - i \leq 1$. An *ϵ-continuum chain* from p to q in M is a chain from p to q whose elements are subcontinua of M each having diameter less than ϵ. A continuum M is *continuum chainable* if for each $\epsilon > 0$ and for each pair of points p, q of M, there is an ϵ-continuum chain in M from p to q [7, 2].

A plane continuum M is an *E-continuum* if for each $\epsilon > 0$ there are at most finitely many complementary domains of M of diameter greater than ϵ [12, p. 112].

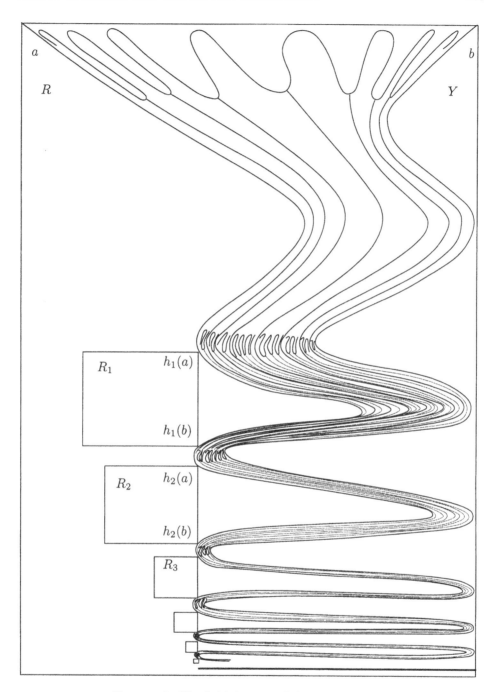

FIGURE 1. The initial stage of the construction

3. THE EXAMPLE

Define M_1 to be the union of the continuum Y and the closed rectangular disks $R_1, R_2, \ldots$ drawn in Figure 1. For each positive integer i, let h_i be a homeomorphism of the closed rectangular disk R onto R_i such that $h_i(a)$ and $h_i(b)$ are the points indicated in the figure. Define M_2 to be the union of Y and $h_1(M_1)$, $h_2(M_1)$, $h_3(M_1), \dots$. Then send R onto each $h_i(R_j)$ with appropriate homeomorphisms and define M_3 accordingly. Continuing this process, we define a nested sequence of continua $M_1, M_2, M_3, \dots$ whose intersection M is a semi-aposyndetic continuum-chainable plane continuum that does not contain a simple closed curve and is not arcwise connected.

Let h be a homeomorphism of R into the plane such that $h(a) = b$ and $h(b) = a$, and $h(R) \cap R$ is the interval from a to b. According to the figure, $M \cap h(M) = \{a, b\}$. Note that $M \cup h(M)$ is a semi-aposyndetic continuum-chainable plane continuum that does not contain a simple closed curve and is not arcwise connected. The continuum $M \cup h(M)$ admits a fixed-point-free homeomorphism that extends to a period 2 homeomorphism of the plane. This extension can be defined so that its only fixed point is the center point of the interval from a to b.

ADDENDUM. Every semi-aposyndetic plane continuum contains an arc [3, Theorem 1] [8]. In a subsequent paper, we will construct an arcless continuum-chainable plane continuum.

REFERENCES

1. K. Borsuk, *Einige Satze über stetige Streckenbilder,* Fund. Math. 18 (1932), 198–213.
2. W. D. Collins and E. J. Vought, *Continuum chainability and monotone decompositions in certain classes of unicoherent continua,* Houston J. of Math. 10 (1984), 457–465.
3. C. L. Hagopian, *Arcwise connectedness of semiaposyndetic plane continua,* Trans. Amer. Math. Soc. 158 (1971), 161–165.
4. _____, *Arcwise connectivity of semiaposyndetic plane continua,* Pacific J. of Math 37 (1971), 683–686.
5. _____, *Arcwise connectivity of continuum-chainable plane continua,* Houston J. of Math. 8 (1982), 69–74.
6. _____, *The fixed-point property for simply connected plane continua,* preprint.
7. C. L. Hagopian and L. E. Rogers, *Arcwise connectivity and continuum chainability,* Houston J. of Math. 7 (1981), 249–259.
8. F. B. Jones, *Aposyndetic continua and certain boundary problems,* Amer. J. Math 63 (1941), 545–553.
9. R. L. Moore, *Foundations of Point Set Theory,* rev. ed., Amer. Math. Soc. Colloq. Publ., Vol 13, Amer. Math. Soc., Providence, Rhode Island, 1962.
10. L. G. Oversteegen, *A continuum chainable aposyndetic plane continuum,* Houston J. of Math. 7 (1981), 271–274.
11. G. T. Whyburn, *Semi-locally-connected sets,* Amer. J. Math 61 (1939), 733–739.
12. _____, *Analytic Topology,* rev. ed., Amer. Math. Soc. Colloq. Publ., Vol 28, Amer. Math. Soc., Providence, Rhode Island, 1963.
13. R. L. Wilder, *Property S_n,* Amer. J. Math. 61 (1939), 823–832.

E-mail address: hagopian@csus.edu "Charles L. Hagopian"
E-mail address: overstee@math.uab.edu "Lex G. Oversteegen"

17
INVERSE LIMITS ON $[0,1]$ USING TENT MAPS AND CERTAIN OTHER PIECEWISE LINEAR BONDING MAPS

W. T. INGRAM Department of Mathematics and Statistics, University of Missouri - Rolla, Rolla, Missouri 65401, USA

ABSTRACT. In this paper, we show that inverse limits on $[0,1]$ using single bonding maps from the tent family, $T_\lambda(x) = 2\lambda x$ if $0 \le x \le \frac{1}{2}$ and $T_\lambda(x) = 2 - 2\lambda x$ if $\frac{1}{2} \le x \le 1$, where $0 \le \lambda \le 1$ contain indecomposable continua for $\lambda > \frac{1}{2}$. Inverse limits on intervals using single bonding maps chosen from one other parameterized family of piecewise linear mappings are also shown to exhibit similar behavior under suitable restrictions on the parameter.

0. INTRODUCTION

Suppose one constructs a mapping f_m of $[0,1]$ into itself by the following scheme: *Start out from the origin on a line of slope $m > 0$. If this line hits the top of $[0,1] \times [0,1]$, then start down from that point on a line of slope $-m$. If this line hits the bottom then start up again from that point on a line of slope m until this line hits the top again or exits on the side. Continue in this manner until a mapping from $[0,1]$ into itself is obtained.* Now, if one considers inverse limits on $[0,1]$ using a single bonding mapping so constructed, then when $m < 1$ the inverse limit is a single point and when $m = 1$ the inverse limit is an arc. It is well known that when $m \ge 2$ the inverse limit is an indecomposable continuum. In fact, these continua have been studied extensively when m is an integer greater than one, see [2, 7]. In this paper, we investigate these inverse limits for $1 < m < 2$. In particular, we show that for m in this range, the inverse limit contains an indecomposable continuum.

1991 *Mathematics Subject Classification.* primary 54H20, secondary 54F15, 58F03, 58F08.
Key words and phrases. inverse limit, periodic point, indecomposable continuum, families of mappings.

The mappings f_m for $1 < m < 2$ are topologically conjugate to members of a well-known family of mappings called the *tent family*. The tent family $\mathcal{T}$ of mappings on $[0, 1]$ is defined by

$$T_\lambda(x) = \begin{cases} 2\lambda x, & \text{if } 0 \leq x \leq \frac{1}{2} \\ 2\lambda(1 - x) & \text{if } \frac{1}{2} \leq x \leq 1 \end{cases} \text{ for } \frac{1}{2} \leq \lambda \leq 1;$$

Also, in this paper we investigate the presence of indecomposable subcontinua in inverse limits using a single bonding map chosen from one other family of mappings. This family is a collection of mappings on $[0, 1]$ mentioned to the author in conversation by William S. Mahavier several years ago. We denote this collection by $\mathcal{F}$ whose members are defined by

$$f_t(x) = \begin{cases} 2x, & \text{if } 0 \leq x \leq \frac{1}{2} \\ 2(t - 1)(x - 1) + t & \text{if } \frac{1}{2} \leq x \leq 1 \end{cases} \text{ for } 0 \leq t \leq 1$$

Dynamically, in contrast with the logistic family, these two families fail to exhibit period doubling bifurcations. Topologically, as the parameter values vary, the inverse limits exhibit a drastic change from possessing a simple structure (being an arc) to complicated structure (containing an indecomposable continuum).

We employ two different techniques to show these inverse limits contain indecomposable continua. For the tent family, we exhibit periodic point behavior which implies the existence of an indecomposable continuum in the inverse limit. Namely, we show that for $\lambda > \frac{1}{2}$, T_λ has a periodic point whose period has an odd factor greater than one. This is sufficient to show that $\varprojlim \{[0, 1], T_\lambda\}$ contains an indecomposable continuum [4, Theorem 4, p. 648]. For the Mahavier collection, $\mathcal{F}$, we show that for $t < \frac{1}{2}$ there is an interval J containing two subintervals J_1 and J_2 such that $f^4[J] = J$ and $f^4[J_1] = f^4[J_2] = J$. This is sufficient for the inverse limit to contain an indecomposable continuum.

By a *continuum* we mean a compact, connected subset of a metric space. A continuum is called *indecomposable* if it is not the union of two of its proper subcontinua. By a *mapping* we mean continuous function. If $X_1, X_2, X_3, \cdots$ is a sequence of metric spaces and, for each i, f_i is a mapping of X_{i+1} into X_i, by the *inverse limit* of the inverse limit sequence $\{X_i, f_i\}$ is meant the subset of the product $\prod_{i>0} X_i$ to which the point $(x_1, x_2, x_3, \cdots)$ belongs if and only if $f_i(x_{i+1}) = x_i$. The inverse limit is denoted $\varprojlim \{X_i, f_i\}$. It is well known that, in case each factor space X_i is a continuum, the inverse limit is a continuum. For additional information on inverse limits, see [6]. For some relationships between the nature of the bonding mappings in inverse limits and indecomposability see [1, 4].

Remark. The reader should be cautioned that, in this paper, where we are concerned with inverse limits using single bonding mappings from parameterized families, the usual superscript/subscript notation f_i^j prevalent in the literature on inverse limits to represent the mapping of the *jth* factor space to the *ith* factor space is not employed. Instead, here one will find this notation to represent the j-fold composition of the mapping f_i with itself.

1. The Tent Family

In this section we present the results on the tent family. However, instead of working directly with the tent family, we shift our attention to a family of mappings of $[0, 1]$ onto itself. A subfamily of this family yields mappings conjugate to the members of the tent family for $\lambda \geq \frac{1}{2}$. Specifically, we consider the family $\mathcal{C}$ whose members are defined by

$$f_m(x) = \begin{cases} mx, & \text{if } 0 \leq x \leq \frac{1}{m} \\ 2 - mx & \text{if } \frac{1}{m} \leq x \leq 1 \end{cases} \quad \text{for } 1 \leq m$$

For $\frac{1}{2} \leq \lambda \leq 1, T_\lambda$ is topologically conjugate to f_m for $m = 2\lambda$ via the homeomorphism, $h_\lambda(x) = \frac{1}{\lambda}x$, i.e., $T_\lambda = h_\lambda^{-1} \circ f_{2\lambda} \circ h_\lambda$. Thus, for our purposes, we can study inverse limits using single bonding maps chosen from the family $\mathcal{C}$ with $1 \leq m \leq 2$.

Remark. It is known that the inverse limit using a single bonding map from $\mathcal{C}$ with $m \geq 2$ is an indecomposable continuum, so the results of this paper show these inverse limits contain an indecomposable continuum for every $m > 1$. Indeed, the subcollection of $\mathcal{C}$ resulting from restricting the values of m to the integers greater than 1 has received considerable attention in the literature [2, 7].

Proofs of the theorems in this section involve some tedious calculations. However, they are almost worth wading through simply for the interesting numbers which occur in these calculations. Specifically, the golden mean, $\frac{1+\sqrt{5}}{2}$, and its $2^n th$ roots, along with the $2^n th$ roots of 2 occur. The first lemma yields ranges of values of m where the form of $f_m^{2^n}$ can be written explicitly enough to calculate (in the second lemma) the range of values of m where [5, Theorem 8] can be applied.

Denote by $\mathcal{P}$ the set of powers of 2, i.e., $\mathcal{P} = \{2^n \mid n = 0, 1, 2, \cdots\}$. For each j in $\mathcal{P}$, define a polynomial $P_j(m)$ by $P_1(m) = 2$ and $P_{2j}(m) = 2 - P_j(m) + m^j P_j(m)$ for $j \geq 1$. For each j in $\mathcal{P}$, define $\phi_j(x) = P_j(m) - m^j x$.

In the first lemma, we establish that, if j is in $\mathcal{P}$ and the parameter m is within the proper range of values, the graph of f_m^{2j} "ends" with a piece of negative slope. To be precise, we determine the range of values of m where f_m^{2j} is unimodal on $[f_m^{2j}(1), 1]$. We determine this by calculating the restriction on m required to keep $f_m^j(1)$ greater than the x-coordinate of the last relative low point on the graph of f_m^j. (It should be noted that $f_m^j(1)$ is always less than the first coordinate of the last point on the graph of f_m^j having second coordinate 1 since the slope of the ending piece of f_m^j is $-m^j$ and $m^j > 1$).

Lemma 1. *Suppose j is in $\mathcal{P}$, $1 < m < 2$ and, in case $j > 1, 1 < m < \sqrt[j]{\frac{1+\sqrt{5}}{2}}$. Then,*

$$f_m^{2j}(x) = \begin{cases} \cdots \\ 2 - \phi_{2j}(x) & \frac{P_j(m)-1}{m^j} \leq x \leq \frac{m^j P_j(m)-P_j(m)+1}{m^{2j}} \\ \phi_{2j}(x) & \frac{m^j P_j(m)-P_j(m)+1}{m^{2j}} \leq x \leq 1. \end{cases}$$

Proof. Note that f_m^2 has the desired form for all m in the range $1 < m < 2$. Suppose, inductively, that

$$f_m^{2k}(x) = \begin{cases} \cdots \\ 2 - \phi_{2k}(x) & \frac{P_k(m)-1}{m^k} \le x \le \frac{m^k P_k(m) - P_k(m) + 1}{m^{2k}} \\ \phi_{2k}(x) & \frac{m^k P_k(m) - P_k(m) + 1}{m^{2k}} \le x \le 1 \end{cases}$$

provided $m \le \sqrt[k]{\frac{1+\sqrt5}{2}}$ ($m \le \frac{1+\sqrt5}{2}$ in case $k = 2$). Then, $f_m^{2k}(1) = 2 - P_k(m) + m^k P_k(m) - m^{2k}$ and is not less than $\frac{P_k(m)-1}{m^k}$ if $m^{3k} - m^{2k} P_k(m) + m^k P_k(m) - 2m^k + P_k(m) - 1 \le 0$. Factoring yields $(m^{2k} - m^k - 1)(m^k - P_k(m) + 1) \le 0$. Since $\frac{P_k(m)-1}{m^k} < 1$ the second factor is positive, it follows that $\frac{P_k(m)-1}{m^k} \le f_m^{2k}(1)$ provided $m^{2k} - m^k - 1 \le 0$ which occurs when $m \le \sqrt[k]{\frac{1+\sqrt5}{2}}$. Thus for $m \le \sqrt[k]{\frac{1+\sqrt5}{2}}$,

$$f_m^{4k}(x) = \begin{cases} \cdots \\ \phi_{2k}(2 - \phi_{2k}(x)) \\ \phi_{2k}(\phi_{2k}(x)) \\ 2 - \phi_{2k}(\phi_{2k}(x)). \end{cases}$$

Note that $2 - \phi_{2k}(\phi_{2k}(x)) = 2 - P_{2k}(m) + m^{2k} P_{2k}(m) - m^{4k} x = \phi_{4k}(x)$. So the last two "pieces" we have listed for f_m^{4k} have the proper form. It remains to see that the last two "pieces" apply for the proper values for x. To this end, we first determine the value of x for which the third-from-the-last and the next-to-the-last "pieces" for f_m^{4k} agree, i.e., when $\phi_{2k}(\phi_{2k}(x)) = \phi_{2k}(2 - \phi_{2k}(x))$. This occurs when $P_{2k}(m) - m^{2k}(P_{2k}(m) - m^{2k} x) = P_{2k}(m) - m^{2k}(2 - P_{2k}(m) + m^{2k} x)$. Solving for x yields that $x = \frac{P_{2k}(m)-1}{m^{2k}}$. Further, $\phi_{2k}(\phi_{2k}(x)) = 1$ occurs when $x = \frac{m^{2k} P_{2k}(m) - P_{2k}(m) + 1}{m^{4k}}$. Note that $\frac{P_{2k}(m)-1}{m^{2k}} = \frac{2 - P_k(m) + m^k P_k(m) - 1}{m^{2k}} = \frac{P_k(m)-1}{m^k}(1 - \frac{1}{m^k}) + \frac{1}{m^k}$. Thus, since by the inductive hypothesis, $\frac{P_k(m)-1}{m^k} < 1$, the quantity $\frac{P_{2k}(m)-1}{m^{2k}} < 1$. Further, since $P_{2k}(m) - 1 < m^{2k}$ it follows easily that $\frac{P_{2k}(m)-1}{m^{2k}} < \frac{m^{2k} P_{2k}(m) - P_{2k}(m) + 1}{m^{4k}} < 1$. This completes the proof.

Lemma 2. *Suppose k is in $\mathcal{P}, 1 < m < 2$ and, in case $k > 1, 1 < m < \sqrt[k]{\frac{1+\sqrt5}{2}}$. If p_k denotes the last fixed point in $[0,1]$ for f_m^k, then $f_m^{2k}(1) \le p_k$ when $m \ge \sqrt[2k]{2}$.*

Proof. Suppose k is in $\mathcal{P}$. Then, by Lemma 1,

$$f_m^k(x) = \begin{cases} \cdots \\ 2 - \phi_k(x) \\ \phi_k(x) \end{cases}$$

where $\phi_k(x) = P_k(m) - m^k x$. Since $\phi_k(1) = P_k(m) - m^k$, $\phi_k(x) = m^k + \phi_k(1) - m^k x$. Thus, the last fixed point in $[0,1]$ for f_m^k occurs when $\phi_k(x) = x$. Solving for x yields $p_k = \frac{m^k + \phi_k(1)}{m^k + 1}$. For $m < \sqrt[k]{\frac{1+\sqrt5}{2}}$, $f_m^{2k}(1) = 2 - \phi_k(\phi_k(1))$. But, $\phi_k(\phi_k(1)) = P_k(m) - m^k \phi_k(1) = m^k + \phi_k(1) - m^k \phi_k(1)$. Since $f_m^{2k}(1) = 2 - m^k - \phi_k(1) + m^k \phi_k(1)$, $f_m^{2k}(1) < p_k$ when $(\phi_k(1) - 1)(m^{2k} - 2) < 0$. Since $\phi_k(1) < 1$, it follows that $m \ge \sqrt[2k]{2}$.

Theorem 1. *If k is in $\mathcal{P}$ and $m > \sqrt[2k]{2}$, f_m has a periodic point whose period is an odd multiple of k.*

Theorem 2. *If $m > 1$, $\varprojlim \{[0,1], f_m\}$ contains an indecomposable continuum.*

Proof. As we remarked earlier, it is well known that if $m \geq 2$ then the inverse limit is indecomposable. Choose m such that $1 < m < 2$ and k in $\mathcal{P}$ such that $\sqrt[2k]{2} \leq m < \sqrt[k]{2}$. Note that in case $k > 1$, $\sqrt[k]{2} < \sqrt[k]{\frac{1+\sqrt{5}}{2}}$. Since $m < \sqrt[k]{\frac{1+\sqrt{5}}{2}}$, f_m^{2k} has the proper form (i.e., the form in the conclusion of Lemma 2). If c_k denotes the last point of $[0,1]$ where f_m^k has value 1 and p_k denotes the last fixed point of f_m^k in $[0,1]$, f_m^k has no period two point between c_k and p_k. Since $m \geq \sqrt[2k]{2}$, $f_m^{2k}(1) \leq p_k$, by [5, Theorem 8], $\varprojlim \{[0,1], f_m\}$ contains an indecomposable continuum. $\blacksquare$

Corollary. *If $\frac{1}{2} < \lambda \leq 1$, $\varprojlim \{[0,1], T_\lambda\}$ contains an indecomposable continuum.*

The following question was asked of the author by Stu Baldwin at the special session on continuum theory at the summer meeting of the American Mathematical Society at Orono, Maine, in 1991.

Question 1. *If $\frac{1}{2} < \lambda_1 < \lambda_2 \leq 1$, is $\varprojlim \{[0,1], T_{\lambda_1}\}$ topologically different from $\varprojlim \{[0,1], T_{\lambda_2}\}$?*

There is a related question which the author has considered to be of interest for several years. He posed it at a problem session at the 1992 Spring Topology Conference in Charlotte for the special case of $n = 5$. Marcy Barge and Beverly Diamond have solved the problem as stated in Charlotte. See their paper in this volume for their solution.

Question 2. *Suppose $n \geq 4$ is a positive integer and H and K are chainable continua having the property that each is an indecomposable continuum with only n end points and every non-degenerate proper subcontinuum is an arc. Are H and K homeomorphic?*

[Question 2 has been solved by Barge and Diamond for $n = 5$.]

Continua having only five end points and every non-degenerate proper subcontinuum an arc arise as subcontinua in inverse limits on $[0,1]$ using members of the tent family. They occur as $\varprojlim \{[T_\lambda(\lambda), \lambda], T_\lambda\}$ for λ approximately 0.7564382, 0.8610419 and 0.96378099. Others with n end points occur for every $n \geq 3$ in the tent family.

2. THE FAMILY $\mathcal{F}$

In this section we examine inverse limits on $[0,1]$ using a single bonding map chosen from the family $\mathcal{F}$. It is well known and easily shown that for $t = \frac{1}{2}$, the inverse limit is the union of a topological ray R and an arc α such that $\alpha = \overline{R} - R$, i.e., a $\sin \frac{1}{x}$-curve. For $t > \frac{1}{2}$ the inverse limit is an arc. The main purpose of this section is to show that if $t < \frac{1}{2}$ and f_t is in $\mathcal{F}$ then $\varprojlim \{[0,1], f_t\}$ contains an indecomposable continuum. We cannot use the theorem employed in the previous section. For $t = \frac{7-\sqrt{17}}{8}$, $f_t^2(1) = \frac{t-2}{2t-3}$, ($\frac{t-2}{2t-3}$ is the positive fixed point for f_t).

Further, f_t has no period 2 point between $\frac{1}{2}$ and $\frac{t-2}{2t-3}$ Thus, we can conclude that for $t \leq \frac{7-\sqrt{17}}{8}$, $\varprojlim \{[0,1], f_t\}$ contains an indecomposable continuum (and f_t has a periodic point of odd period greater than 1). However, this phenomenon does not reoccur for f_t^2 as it did for the family $\mathcal{C}$. Specifically, $f_t^4(1)$ is not a fixed point for f_t^2 for any value for t in $(\frac{7-\sqrt{17}}{8}, \frac{1}{2})$. As a consequence, we use a different theorem to see that the inverse limit produces an indecomposable subcontinuum. As it turns out this technique makes for simpler calculations since it only requires looking at f_t^4.

Theorem 3. *For $t < \frac{1}{2}$ and f_t in $\mathcal{F}$, $\varprojlim \{[0,1], f_t\}$ contains an indecomposable continuum.*

Proof. As we observed in the remarks just preceding the statement of Theorem 3, if $0 \leq t \leq \frac{7-\sqrt{17}}{8}$, $\varprojlim \{[0,1], f_t\}$ contains an indecomposable continuum. Consequently, we will consider only values of the parameter between $\frac{7-\sqrt{17}}{8}$ and $\frac{1}{2}$. Note that for t in this range, $f_t^2(x) = 1$ for $x = \frac{2t-3}{4t-4}$ and $2t < \frac{2t-3}{4t-4} < 1$. Under f_t^4 each of the intervals $[2t, \frac{2t-3}{4t-4}]$ and $[\frac{2t-3}{4t-4}, 1]$ is thrown onto $[2t, 1]$. To see this requires checking that $f_t^2(2t) < \frac{2t-3}{4t-4}$ so f_t^2 throws $[2t, \frac{2t-3}{4t-4}]$ onto $[2t, 1]$ and noting that $f_t^2(2t) \geq 2t$ when $t \geq \frac{7-\sqrt{17}}{8}$ so f_t^2 throws $[2t, 1]$ onto itself. Consequently, $\varprojlim \{[2t, 1], f_t^4 | [2t, 1]\}$ is an indecomposable inverse sequence and thus the inverse limit is indecomposable [6, Theorem 2.7, p. 21].

Remarks. We close this section with some additional remarks. For $t = \frac{1}{4}$, $\frac{1}{2}$ is a periodic point of period 3 for f_t. At $t = \frac{1}{4}$, the inverse limit is the union of a ray and an indecomposable continuum C such that $C = \overline{R} - R$ and C has only three end points and every proper subcontinuum of C is an arc, see [3, p. 141]. At $t = 0.0625$ and for t approximately 0.2034649 and 0.3009195, C has only five end points and every proper subcontinuum of C is an arc.

REFERENCES

1. Marcy Barge and Joe Martin, *Chaos, periodicity, and snakelike continua*, Trans. Amer. Math. Soc. **289** (1985), 355–365.

2. W. Dębski, *On the simplest indecomposable continua*, Colloq. Math. **49** (1985), 13–24.

3. John G. Hocking and Gail W. Young, *Topology*, Addison-Wesley Pub. Co., Inc., Reading, MA, 1961.

4. W. T. Ingram, *Concerning periodic points in mappings of continua*, Proc. Amer. Math. Soc. **104** (1988), 643–649.

5. W. T. Ingram, *Periodicity and indecomposablility*, Proc. Amer. Math. Soc., to appear.

6. Sam B. Nadler, *An Introduction to Continuum Theory*, Marcel Dekker, New York, NY, 1992.

7. William Thomas Watkins, *Homeomorphic classification of certain inverse limit spaces with open bonding maps*, Pacific J. Math. **103** (1982), 589–601.

E-mail address: ingram@umr.edu "Tom Ingram"

18
ON COMPOSANTS OF INDECOMPOSABLE SUBCONTINUA OF SURFACES

ZBIGNIEW KARNO Institute of Mathematics, Warsaw University at Białystok, 15-267 Białystok 1, Poland, and Institute of Mathematics, Polish Academy of Sciences, 00-950 Warsaw, Poland

ABSTRACT. Let X be an indecomposable subcontinuum of a surface M. A composant C of X is called strongly external if there exists a subcontinuum L of M such that $L \cap C \neq \emptyset$, $L \setminus X \neq \emptyset$, and $L \cap X$ is a proper subcontinuum of X. It is proved that the union of all strongly external composants of X is a F_σ-set of the first category in X. This is an affirmative answer to a problem posed by J. Krasinkiewicz, and a generalization of some results of S. Mazurkiewicz and K. Kuratowski.

1. INTRODUCTION

A continuum (compact, connected metric space) is called *indecomposable* if it is not the union of two proper subcontinua. Let X be such continuum. By a *composant* C of a point x in X we understand the union of all proper subcontinua of X containing x. It is known that the composants of an indecomposable continuum X are pairwise disjoint, dense F_σ-sets of the first category in X (e.g., [6, p. 212]).

Let M be a surface, i.e., a compact connected 2-dimensional manifold without boundary. Let X be an indecomposable subcontinuum of M and let C be a composant of X. C is called *external* if there exists a subcontinuum L of M such that $L \cap C \neq \emptyset$, $L \setminus X \neq \emptyset$, and L does not intersect all composants of X; otherwise, C is called *internal* ([2], [4]).

The following theorem was proved by J. Krasinkiewicz in the case $M = S^2$ ([2], [4]). It was generalized to all surfaces by the author in [1].

1991 *Mathematics Subject Classification.* primary 54F15, secondary 54E52.
Key words and phrases. indecomposable continuum, composant, surface.

Theorem 1.1 ([4], [1]). *Let X be an indecomposable subcontinuum of a surface M. Then the union of all external composants of X is an F_σ-set of the first category in X; consequently, the union of all internal composants of X is a G_δ-set dense in X.*

The purpose is to prove an analogous theorem for the class of composants defined below.

A composant C of X is called *strongly external* (continuum accessible [7], K-composant [3]) if there exists a subcontinuum L of M such that $L \cap C \neq \emptyset$, $L \backslash X \neq \emptyset$, and $L \cap X$ is a proper subcontinuum of X (hence $L \cap X \subset C$); otherwise, C is called *weakly internal*. Such composants can be characterized in the following way: C is strongly external if and only if there is a ray (topological half-line) P in M disjoint from X with the remainder $Cl(P) \backslash P$ being a subcontinuum of X contained in C (see the next section). Every strongly external composant is external but not conversely. There are indecomposable plane continua with uncountable collections of external and weakly internal composants (see [3]).

In [7] K. Kuratowski proved that the union of all strongly external composants of an indecomposable plane continuum X is a first category subset of X. It was a generalization of an analogous results of S. Mazurkiewicz [8] for accessible composants. In [4] J. Krasinkiewicz posed the following problem:

Is the union of all strongly external composants of every indecomposable plane continuum X an F_σ-subset of X ?

Here the affirmative answer is given to this problem by proving the following theorem.

Theorem 1.2. *Let X be an indecomposable subcontinuum of a surface M. Then the union of all strongly external composants of X is an F_σ-set of the first category in X; consequently, the union of all weakly internal composants of X is a G_δ-set dense in X.*

2. CHARACTERIZATIONS OF STRONGLY EXTERNAL COMPOSANTS.

Let C be a composant of an indecomposable subcontinuum X of a surface M. Let G be a subset of $M \backslash X$. C is called *strongly external with respect to G* if there exists a subcontinuum L of M such that $L \cap C \neq \emptyset$, $L \backslash X \neq \emptyset$, $L \cap X$ is a proper subcontinuum of X, and $L \subset X \cup G$. Clearly, C is strongly external if and only if it is strongly external with respect to $M \backslash X$. Moreover, we have the following characterization.

Proposition 2.1. *Let C be a composant of X. Then C is strongly external if and only if there is a component G of $M \backslash X$ such that C is strongly external with respect to G.*

Proof. Let L_1 be a subcontinuum of M such that $L_1 \backslash X \neq \emptyset$ and $L \cap X \neq \emptyset$ is a subcontinuum of X contained in C. There is a component G of $M \backslash X$ which meets L_1. Let L_2 be the closure of some component of $G \cap L_1$. It follows that $L_2 \setminus G \subset L_1 \cap X$ and $L_2 \cap (L_1 \cap X) \neq \emptyset$. Hence the continuum $L = L_2 \cup (L_1 \cap X)$ has all required properties.

By a *ray* in M we mean here a subspace P of M which can be obtained as the homeomorphic image of the half-line $[0, \infty)$. Note that if P is a ray in M, then $Cl(P)$ and $Cl(P) \setminus P$ are both subcontinua of M.

Theorem 2.2. *Let C be a composant of X and let G be a component of $M \setminus X$. Then C is strongly external with respect to G if and only if there is a ray P in M such that $P \subset G$ and $Cl(P) \setminus P \subset C$.*

To prove this, first we shall show a few lemmas.

Lemma 2.3 ([5]). *Let U be an open connected and locally connected subset of a continuum Y and let A be a component of the boundary $Bd\, U$. Then there is a ray P in Y such that $P \subset U$ and $Cl(P) \setminus P \subset A$.*

Lemma 2.4. *Let D be a disk in M such that $X \cap Int\, D \neq \emptyset$ and let U be a component of $M \setminus (X \cup Int\, D)$. If A is a component of $X \cap Bd\, U$, then there is a ray P in M such that $P \subset U$ and $Cl(P) \setminus P \subset A$. In particular, if C is a composant of X meeting A, then C is strongly external with respect to U.*

Proof. Since $Y = M \setminus Int\, D$ is a locally connected continuum, it follows that the component U of $Y \setminus X$ is a locally connected and open (in Y) subset of Y. Since U is a component of $Y \setminus X$, it follows that $Bd_Y(U) = X \cap Bd\, U$. Then A is a component of $Bd_Y(U)$. Applying 2.3 to Y, U and A we get the conclusion.

Lemma 2.5. *Let D be a disk in M such that $X \cap Int\, D \neq \emptyset$ and let U be a component of $M \setminus (X \cup Int\, D)$. Let L be a subcontinuum of M such that $L \setminus X \neq \emptyset$, $L \cap D = \emptyset$ and $L \subset U \cup X$. If C is a composant of X meeting L, then there is a ray P in M such that $P \subset U$ and $Cl(P) \setminus P \subset C$. In particular, C is strongly external with respect to U.*

Proof. We first prove that if $L \cap C \neq \emptyset$, then $C \cap Bd\, U \neq \emptyset$. Since $L \setminus X \neq \emptyset$ and $L \subset U \cup X$, it follows that $L \cap U \neq \emptyset$. Let F be a component of $L \setminus U$ meeting C. Therefore by [6, p. 172]

$$\emptyset \neq F \cap Bd_L(L \setminus U) = F \cap Bd_L(L \cap U) \subset F \cap Bd\, U.$$

But F is a proper subcontinuum of X since F is a component of $L \setminus U = X \cap L$, $X \cap Int\, D \neq \emptyset$ and $L \cap D = \emptyset$. Moreover, F meets C. Hence $F \subset C$, and consequently $C \cap Bd\, U \neq \emptyset$.

Now let A be a component of $X \cap Bd\, U$ meeting C. Then A is a proper subcontinuum of X, because $A \subset X \setminus Int\, D$. Thus $A \subset C$. Applying 2.4 to D, U and A we get the conclusion.

Proof of Theorem 2.2. C is strongly external with respect to G. Then, by 2.1, there is a subcontinuum L of M such that

(1) $\emptyset \neq L \setminus X \subset G$, and

(2) $\emptyset \neq L \cap X \subset C$ is a continuum.

Let D be a disk in M such that $L \cap D = \emptyset$ and $X \cap Int\ D \neq \emptyset$. According to (1), let us consider a component U of $G \setminus Int\ D$ meeting L. Let L_0 be the closure of some component of $U \cap L$. Then L_0 is a subcontinuum of L such that $L_0 \setminus X \neq \emptyset$, $L_0 \cap D = \emptyset$, $L_0 \subset X \cup U$, and by (2), $L_0 \cap C \neq \emptyset$. Applying 2.5 to D, U, L_0 and C we infer that there exists a ray P in M such that $P \subset U \subset G$ and $Cl(P) \setminus P \subset C$. The reverse implication is obvious.

Immediately from 2.1 and 2.2 we get the following

Corollary 2.6. *Let C be a composant of X. Then C is strongly external if and only if there is a ray P in M such that $P \subset M \setminus X$ and $Cl(P) \setminus P \subset C$.*

3.   THE UNION OF STRONGLY EXTERNAL COMPOSANTS.

Throughout this section X will be an indecomposable subcontinuum of a surface M.

Theorem 3.1. *Let G be a component of $M \setminus X$. Then the union of all composants of X strongly external with respect to G is an F_σ-set of the first category in X.*

For the proof of this theorem we need the following two lemmas.

Lemma 3.2. *Let D be a disk in M such that $X \cap Int\ D \neq \emptyset$. Let G be a component of $M \setminus X$ and let U be a component of $G \setminus Int\ D$. If C is a composant of X such that $C \cap Cl\ U \neq \emptyset$, then C is strongly external with respect to G.*

Proof. Let D_0 be a disk in M such that $D_0 \subset Int\ D$ and $X \cap Int\ D_0 \neq \emptyset$. Consider the component V of $G \setminus Int\ D_0$ which contains U. Then the continuum $L = Cl\ U$ has the following properties: $L \setminus X \neq \emptyset$, $L \cap D_0 = \emptyset$, $L \subset V \cup X$ and $C \cap L \neq \emptyset$. By applying 2.5 to D_0, V, L and C we infer that there exists a ray P in M such that $P \subset V \subset G$ and $Cl(P) \setminus P \subset C$. Hence C is strongly external with respect G.

Lemma 3.3.   *Let D be a disk in M such that $X \cap Int\ D \neq \emptyset$. Let G be a component of $M \setminus X$ and let U be a component of $G \setminus Int\ D$. If F is the union of all components of $X \setminus Int\ D$ intersecting the continuum $Cl\ U$, then F is a closed subset of X contained in the union of all composants of X strongly external with respect to G.*

Proof. By 3.2, F is contained in the union of all composants of X strongly external with respect to G. We prove that F is closed in X. Let $x_1, x_2, \ldots$ be a sequence of points of F converging to the point x. It suffices to show that $x \in F$. For every n there is a component L_n of $X \setminus Int\ D$ such that

(1) $x_n \in L_n$, and

(2) $L_n \cap Cl\ U \neq \emptyset$.

Without loss of generality we may assume that the sequence $L_1, L_2, \ldots$ is topologically convergent with the limit continuum L contained in a component of $X \setminus Int\ D$.

Then $x \in L$ by (1), and $L \cap Cl\ U \neq \emptyset$ by (2). Hence $x \in F$.

Proof of Theorem 3.1. Let $D_1, D_2, \ldots$ be a sequence of disks in M with diameters converging to zero such that the sets $X \cap Int\ D_n \neq \emptyset$ constitute a base for X. Let $G \setminus Int\ D_n = G_{n,1} \cup G_{n,2} \cup \ldots$, where $G_{n,i}$, $i = 1, 2, \ldots$, are components of $G \setminus Int\ D_n$. For each pair of integers n, i let $F_{n,i}$ be the union of all components of $X \setminus Int\ D_n$ intersecting $Cl\ G_{n,i}$. Let F be the union of all composants of X strongly external with respect to G. By 3.3, $F_{n,i}$ is a closed subset of X and $F_{n,i} \subset F$. Hence

(1) $\bigcup_{n,i} F_{n,i}$ is an F_σ-set in X and $\bigcup_{n,i} F_{n,i} \subset F$.

We prove that

(2) $F \subset \bigcup_{n,i} F_{n,i}$.

Let $x \in F$. Then there is a composant C of X strongly external with respect to G such that $x \in C$. By 2.2, there is ray P in M such that $P \subset G$ and $Cl(P) \setminus P \subset C$. Let A be a proper subcontinuum of X intersecting $Cl(P) \setminus P$ such that $x \in A$. Then $A \cup Cl\ P$ is a continuum with the property

(3) $A \cup (Cl(P) \setminus P) \subset C$.

It follows that there is an index n such that

(4) $A \cup Cl\ P$ misses D_n.

Since $P \subset G$ is connected and it misses D_n, there is an index i such that $P \subset G_{n,i}$. Then from (4) and (3) we get that $A \cup (Cl(P) \setminus P) \subset F_{n,i}$. Hence $x \in F_{n,i}$, and (2) is proved.

Now from (1) and (2), we have that F is F_σ-subset of X. That F is of the first category in X follows from 1.1. This completes the proof.

Corollary 3.4. *Let G be the union of some components of $M \setminus X$. Then the union of all composants of X strongly external with respect to G is an F_σ-set of the first category in X.*

Theorem 1.2 we get as an immediate consequence of 3.4.

To complete, we now formulate the following problem.

PROBLEM. Let M be a closed surface and let X be an indecomposable subcontinuum of M. Is it true that there exist an indecomposable subcontinuum Y of M and a continuous mapping f from Y onto X such that every external composant of Y is strongly external and f maps the union of all weakly internal composants of Y homeomorphicaly onto the union of all internal composants of X ?

REFERENCES

1. Z. Karno, *On indecomposable subcontinua of surfaces*, Fund. Math. **123** (1984), 117–122.

2. J. Krasinkiewicz, *On the composants of indecomposable plane continua*, Bull. Acad. Polon. Sci., Sér. Sci. Math. Astron. Phys. **20** (1972), 935–940.

3. — *Concerning the accessibility of composants of indecomposable plane continua*, Bull. Acad. Polon. Sci., Sér. Sci. Math. Astron. Phys. **21** (1973), 621–628.

4. — *On internal composants of indecomposable plane continua*, Fund. Math. **84** (1974), 255–263.

5. J. Krasinkiewicz and P. Minc, *Continua with countable number of arc-components*, Fund. Math. **102** (1979), 119-127.

6. K. Kuratowski, *Topology*, vol. 2, Warsaw–New York, 1968.

7. — *Sur une condition qui caractérise les continus indécomposables*, Fund. Math. **14** (1929), 116–117.

8. S. Mazurkiewicz *Sur les points accessibles les continus indécomposables*, Fund. Math. **14** (1929), 107–115.

E-mail address: zkarnoimpan.impan.gov.pl "Zbigniew Karno"

19

MINIMAL SETS AND CHAOS IN THE SENSE OF DEVANEY
ON CONTINUUM-WISE EXPANSIVE HOMEOMORPHISMS

HISAO KATO Institute of Mathematics, University of Tsukuba, Tsukuba-City, Ibaraki 305, Japan

ABSTRACT. In [15], we proved that if a homeomorphism $f : X \to X$ of a compactum X is continuum-wise expansive and $\dim X > 0$, then there is a chaotic continuum Z of f and either f or f^{-1} is chaotic on almost all Cantor sets of Z in the sense of Li-Yorke. In [16], for a map $f : X \to X$, we defined the family $\mathcal{D}(f) = (\mathcal{M}^+(f))$ consisting of all minimal elements of not zero-dimensional, f-invariant closed subsets of X, and we showed that if $f : X \to X$ is a continuum-wise expansive homeomorphism of a compactum X with $\dim X > 0$, then $\mathcal{D}(f) \neq \emptyset$ and if $Y \in \mathcal{D}(f)$, both $f|Y : Y \to Y$ and $f^{-1}|Y : Y \to Y$ are sensitive and topologically transitive. In this paper, we study the minimal sets of continuum-wise expansive homeomorphisms. In particular, we show that if a homeomorphism $f : X \to X$ of a compactum X is continuum-wise expansive and $\dim X > 0$, then both $f|Y : Y \to Y$ and $f^{-1}|Y \to Y$ are *weakly chaotic* in the sense of Devaney for each $Y \in \mathcal{D}(f)$. We know that several chaotic properties are concentrated on such sets $Y \in \mathcal{D}(f)$.

1. INTRODUCTION.

All spaces under consideration are separable metric spaces. By a *compactum*, we mean a compact metric space. A *continuum* is a connected compactum. By $\dim X$, we mean the *topological dimension* of X (see [7]). Note that for a compactum X, $\dim X > 0$ if and only if there is a nondegenerate component of X.

Let Z be the set of all integers. A homeomorphism $f : X \to X$ of a compactum X with metric d is *expansive* (see [21] and [1]) if there is a positive number $c > 0$ such that if $x, y \in X$ and $x \neq y$, then there is an integer $n = n(x, y) \in Z$ such that

1991 *Mathematics Subject Classification.* primary 54H20, 54F50, secondary 54E50, 54B20.

Key words and phrases. (continuum-wise) expansive homeomorphism, chaos in the sense of Devaney, continuum, minimal set.

$$d(f^n(x), f^n(y)) \geq c.$$

A homeomorphism $f : X \to X$ of a compactum X is *continuum-wise expansive* [13] if there is $c > 0$ such that if A is nondegenerate subcontinuum of X, then there is an integer $n \in Z$ such that diam $f^n(A) \geq c$, where diam B denotes the diameter of a set B. Such a number $c > 0$ is called an *expansive constant* for f. Note that each expansive homeomorphism is continuum-wise expansive, but the converse assertion is not true.

Suppose that $f : X \to X$ is a map of a compactum X. Then f is said to be *sensitive* if there is $c > 0$ such that if $x \in X$ and U is any neighborhood of x in X, there is $y \in U$ and a natural number $n \geq 0$ such that $d(f^n(x), f^n(y)) \geq c$. Also, f is *topologically transitive* if there is a point x of X such that the set $O^s(x) = \{f^n(x) | n = 0, 1, 2, \ldots\}$ is dense in X. The map f is said to be *chaotic in the sense of Devaney* [4] if f is sensitive, topologically transitive and the set $P(f)$ of all periodic points of f is dense in X. A closed subset M of X is said to be a *minimal set* of $f : X \to X$ if M is f-invariant, i.e., $f(M) = M$, and any proper closed subset T of X is not f-invariant. Note that (1) a closed subset M of X is a minimal set of $f : X \to X$ if and only if $O^s(x)$ is dense in M for each $x \in M$, i.e., $Cl(O^s(x)) = M$, (2) if $x \in P(f)$, then $O^s(x)$ is a minimal set of f, (3) if M is a minimal set of f such that M is 0-dimensional and the cardinality $|M|$ of M is infinite, then M is a Cantor set.

Let $f : X \to X$ be a map of a compactum X. Consider the following inverse limit space

$$(X, f) = \{(x_n)_{n=1}^{\infty} | x_n \in X \text{ and } f(x_{n+1}) = x_n \text{ for each } n\}$$

topologized with the relativized product topology. Then the *shift map* $\tilde{f} : (X, f) \to (X, f)$ of f is defined by

$$\tilde{f}(x_1, x_2, \ldots) = (f(x_1), x_1, \ldots).$$

Note that $\tilde{f} : (X, f) \to (X, f)$ is a homeomorphism. The notion of shift map is convenient for dynamical systems.

Here, we give examples of homeomorphisms demonstrating some of these properties we are studying.

(1) Let S be the unit circle and let $f : S \to S$ be the natural covering map with degree 2. Then $X = (S, f)$ is the 2-adic solenoid, the shift map $\tilde{f} : (S, f) \to (S, f)$ of f is expansive and it is chaotic in the sense of Devaney.

(2) Let $f : T^2 \to T^2$ be an *Anosov diffeomorphism*, say

$$\begin{pmatrix} 2 & 1 \\ 1 & 1 \end{pmatrix}$$

on the 2-dimensional torus T^2. Then f is expansive and it is chaotic in the sense of Devaney (see [4]).

(3) Let $f : I \to I$ be a map of the unit interval $I = [0, 1]$ defined by

$$f(x) = \begin{cases} 2x & \text{if } 0 \leq x \leq 1/2, \\ -2x + 2 & \text{if } 1/2 \leq x \leq 1. \end{cases}$$

Then f is sensitive, the shift map $\tilde{f} : (I, f) \to (I, f)$ of f is not expansive, but it is continuum-wise expansive. Note that $\tilde{f}$ is chaotic in the sense of Devaney.

(4) Let $f : I \to I$ be a map defined by

$$f(x) = \begin{cases} -2x + 1 & \text{if } 0 \leq x \leq 1/6, \\ 2x + 1/3 & \text{if } 1/6 \leq x \leq 1/3, \\ -3x + 2 & \text{if } 1/3 \leq x \leq 2/3, \\ 2x - 4/3 & \text{if } 2/3 \leq x \leq 5/6, \\ -2x + 2 & \text{if } 5/6 \leq x \leq 1. \end{cases}$$

Then f is sensitive, the shift map $\tilde{f} : (I, f) \to (I, f)$ of f is not expansive, but it is continuum-wise expansive. Note that $\tilde{f}$ is not chaotic in the sense of Devaney, but the restriction $\tilde{f}|Y : Y \to Y$ is chaotic in the sense of Devaney, where $Y = ([0, 1/3] \cup [2/3, 1], f|[0, 1/3] \cup [2/3, 1]) \subset (I, f)$. Note that Y is not connected and $\dim Y > 0$.

In general, (continuum-wise) expansive homeomorphism do not always induce chaos in the sense of Devaney (see (3.5)). In this paper, we introduce a slightly weaker version of chaos in the sense of Devaney as follows: A map $f : X \to X$ of a compactum X is *weakly chaotic in the sense of Devaney* if f is sensitive, f is topologically transitive and $\cup\{M \,|\, M$ is a minimal set of $f\}$ is dense in X. We show that if $f : X \to X$ is a continuum-wise expansive homeomorphism of a compactum X with $\dim X > 0$, then f induces weak chaos in the sense of Devaney (see (2.7)).

2. CONTINUUM-WISE EXPANSIVE HOMEOMORPHISMS INDUCE WEAK CHAOS IN THE SENSE OF DEVANEY.

Let X be a compactum with metric d. By the *hyperspace* of X, we mean the set $2^X = \{A \,|\, A$ is a nonempty closed subset of $X\}$ with the *Hausdorff metric* d_H, i.e.,

$$d_H(A, B) = inf\{\epsilon > 0 \,|\, U_\epsilon(A) \supset B, U_\epsilon(B) \supset A\},$$

where $U_\epsilon(A)$ denotes the ϵ-neighborhood of A in X. Note that 2^X is a compact metric space with metric d_H.

Let $f : X \to X$ be a map of a compactum X. Consider the following sets:

(1) $\mathcal{I}(f) = \{A \in 2^X \,|\, A$ is f-invariant$\}$.

(2) $\mathcal{M}(f) = \{M \,|\, M$ is a minimal set of $f\}$.

(3) $\mathcal{P}(f) = \{O^s(p) \,|\, p$ is a periodic point of $f\}$.

(4) $\mathcal{M}_\infty(f) = \{M \in \mathcal{M}(f) | \, |M| \text{ is infinite}\} = \mathcal{M}(f) - \mathcal{P}(f)$.

(5) $\mathcal{I}^+(f) = \{A \in \mathcal{I}(f) | \dim A > 0\}$.

(6) $\mathcal{D}(f) = \{A \in \mathcal{I}^+(f) | \text{ if } B \in \mathcal{I}(f) \text{ and } B \text{ is a proper subset of } A,$
    then B is 0-dimensional$\}$,
        i.e., $\mathcal{D}(f)$ is the set of all minimal elements of $\mathcal{I}^+(f)$.

Then we see the following.

Proposition 2.1. *Let* $f : X \to X$ *be a map of a compactum* X. *Then*

(1) $\mathcal{I}(f)$ *is a closed set in* 2^X,
(2) $\mathcal{P}(f)$ *is an* F_σ-*set in* 2^X,
(3) $\mathcal{M}(f)$ *is a* G_δ-*set in* 2^X,
(4) $\mathcal{M}_\infty(f)$ *is a* G_δ-*set in* 2^X.

(1) and (2) are trivial. We shall show (3). Let $U = \{U_i\}_{i=1}^\infty$ be a countable open base of the compactum X. Put

$$W = \{(U, V) | \, U, V \in U \text{ and } Cl(U) \subset V\} (= \{(U_i, V_i)\}_{i=1}^\infty).$$

For each $(U_i, V_i) \in W$, consider the set

$$\mathcal{R}(U_i, V_i) = \{A \in \mathcal{I}(f) | A \cap Cl(U_i) \neq \emptyset \text{ and there is a closed}$$
$$\text{subset } B \text{ of } A \text{ such that } X - V_i \supset B \text{ and } B \in \mathcal{I}(f)\}.$$

Then we can easily see that $\mathcal{R}(U_i, V_i)$ is a closed subset of $\mathcal{I}(f)$. Now, we shall show that $\mathcal{M}(f) = \mathcal{I}(f) - \cup_{i=1}^\infty \mathcal{R}(U_i, V_i)$. Let $A \in \mathcal{I}(f) - \cup_{i=1}^\infty \mathcal{R}(U_i, V_i)$. Suppose, on the contrary that $A \notin \mathcal{M}(f)$. Then we can choose a proper closed subset B of A such that $B \in \mathcal{I}(f)$. Choose $(U_i, V_i) \in W$ such that $Cl(U_i) \cap A \neq \emptyset$ and $B \subset X - V_i$. Then $A \in \mathcal{R}(U_i, V_i)$. This is a contradiction. The converse inclusion is trivial. (4) follows from (2) and (3).

By the proof of [13, (2.3)], we have

Lemma 2.2. *Let* $f : X \to X$ *be a continuum-wise expansive homeomorphism of a compactum* X *with an expansive constant* $c > 0$, *and let* $0 < \epsilon < c/2$. *Then there is* $\epsilon > \delta > 0$ *such that if* A *is a subcontinuum of* X *with* $\operatorname{diam} A \leq \delta$ *and* $\operatorname{diam} f^m(A) \geq \epsilon$ *for some* $m \in Z$, *then one of the two following conditions holds:*

(1) *If* $m \geq 0$, *for each* $n \geq m$ *and* $x \in f^n(A)$, *there is a subcontinuum* B *of* A *such that* $x \in f^n(B)$, $\operatorname{diam} f^j(B) \leq \epsilon$ *for* $0 \leq j \leq n$ *and* $\operatorname{diam} f^n(B) = \delta$.
(2) *If* $m < 0$, *for each* $n \geq -m$ *and* $x \in f^{-n}(A)$, *there is a subcontinuum* B *of* A *such that* $x \in f^{-n}(B)$, $\operatorname{diam} f^{-j}(B) \leq \epsilon$ *for* $0 \leq j \leq n$, *and* $\operatorname{diam} f^{-n}(B) = \delta$.

Lemma 2.3. [13, (2.4)] *Let $f : X \to X$ be a continuum-wise expansive homeomorphism of a compactum X with $\dim X > 0$. If $\delta > 0$ is as in (2.2), then for each $\gamma > 0$ there is a natural number $N(\gamma)$ such that if A is a subcontinuum of X with $\operatorname{diam} A \geq \gamma$, then either $\operatorname{diam} f^n(A) \geq \delta$ for each $n \geq N(\gamma)$ or $\operatorname{diam} f^{-n}(A) \geq \delta$ for each $n \geq N(\gamma)$ holds.*

Proposition 2.4. *If $f : X \to X$ is a continuum-wise expansive homeomorphism of a compactum X with $\dim X > 0$, then*

(1) $\mathcal{I}^+(f)$ *is a closed set in 2^X,*
(2) $\mathcal{D}(f) \neq \emptyset$ *is a G_δ-set in 2^X.*

Proof. By (2.3), we see that if $A \in \mathcal{I}^+(f)$, then $\operatorname{diam} A \geq \delta$, where δ is a positive number as in (2.3). Then (1) follows from this fact. By [16, (3.1)], $\mathcal{D}(f) \neq \emptyset$. Also, by using the similar proof to one of (2.1), we see that $\mathcal{D}(f)$ is a G_δ-set in 2^X.

In [18], Mañé proved that minimal sets of expansive homeomorphisms of compacta are 0-dimensional. Moreover, in [13] we proved that the result is also true for the case of continuum-wise expansive homeomorphisms. Moreover, we show the following.

Proposition 2.5. *Let $f : X \to X$ be a continuum-wise expansive homeomorphism of a compactum X with $\dim X > 0$. Then*

(1) $\dim(\cup\{A \mid A \in \mathcal{I}(f) \text{ and } \dim A = 0\}) = 0$, *in particular,* $\dim(\cup\{M \mid M \in \mathcal{M}(f)\}) = 0$, *and*
(2) *if each component of X is nondegenerate, then $\cup\{A \in \mathcal{I}(f) \mid \dim A = 0\}$ is of the first category in X, and hence $\cup\{M \mid M \in \mathcal{M}(f)\}$ is of the first category in X.*

Proof. Put $E = \cup\{A \in \mathcal{I}(f) \mid \dim A = 0\}$. Let $\delta > 0$ be a positive number as in (2.3). Consider a countable open base U of X such that if $U \in U$, then $\operatorname{diam} U \leq \delta/3$. For each natural number $n \geq 1$, put

$$W_n = \{(U_1, U_2, \ldots, U_n) \mid U_i \in U \text{ for each } i, \text{ and } Cl(U_i) \cap Cl(U_j) = \emptyset \text{ if } i \neq j\},$$

and $W = \cup_{n=1}^\infty W_n$. Note that W is a countable set. For each $(U_1, \ldots, U_n) \in W$, consider the set

$$W(U_1, \ldots, U_n) = \{x \in X \mid O(x) = \{f^n(x) \mid n \in Z\} \subset \cup_{i=1}^n Cl(U_i)\}.$$

Then we can easily see that $W(U_1, \ldots, U_n)$ is a closed set of X and it is f-invariant. By (2.3), we see that $W(U_1, \ldots, U_n)$ contains no nondegenerate subcontinuum, which implies that it is 0-dimensional. Hence $W(U_1, \ldots, U_n) \subset E$. Next, we shall show that

$$\bigcup_{(U_1, \ldots, U_n) \in W} W(U_1, \ldots, U_n) \supset E.$$

If $A \in \mathcal{I}(f)$ and $\dim A = 0$, we can choose $(U_1, \ldots, U_n) \in W$ such that $A \subset \cup_{i=1}^n U_i$. Then $A \subset W(U_1, \ldots, U_n)$. Hence $E = \cup_{(U_1, \ldots, U_n) \in W} W(U_1, \ldots, U_n)$.

Since $\dim W(U_1, \ldots, U_n) = 0$ for each $(U_1, \ldots, U_n) \in W$, by the sum theorem of dimension theory (see [7]), we see that $\dim E = 0$. By the subset theorem of dimension theory (see [7]), we see that $\dim(\cup\{M \,|\, M \in \mathcal{M}(f)\}) = 0$. Moreover, if each component of X is nondegenerate, we see that $W(U_1, \ldots, U_n)$ is nowhere dense, which implies that E is an F_σ-set of the first category.

Lemma 2.6. *Let* $f : X \to X$ *be a continuum-wise expansive homeomorphism of a compactum* X *with* $\dim X > 0$. *If* $A \in \mathcal{I}(f)$ *and* $\dim A = 0$, *then there is* $M \in \mathcal{M}(f)$ *such that* $M \cap A = \emptyset$.

Proof. Let $\delta > 0$ be as in (2.3). Let W be the same set as in the proof of (2.5). Since $\dim A = 0$, we can choose $(U_1, \ldots, U_n) \in W$ such that $\cup_{i=1}^n U_i \supset A$. Choose a positive number $\lambda > 0$ such that $\lambda < \delta$ and if F is a subcontinuum of X with $\operatorname{diam} F \geq \delta$, then there is a subcontinuum H of F such that $H \cap Cl(\cup_{i=1}^n U_i) = \emptyset$ and $\operatorname{diam} H \geq \lambda$. By (2.3), we can choose a natural number $N(\lambda)$ such that if B is a subcontinuum of X with $\operatorname{diam} B \geq \lambda$, then either $\operatorname{diam} f^n(B) \geq \delta$ for each $n \geq N(\lambda)$ or $\operatorname{diam} f^{-n}(B) \geq \delta$ for each $n \geq N(\lambda)$. Since $\dim X > 0$, we can choose a subcontinuum B of X with $\operatorname{diam} B = \lambda$. Without loss of generality, we may assume that $\operatorname{diam} f^n(B) \geq \delta$ for each $n \geq N(\lambda)$. Since $\operatorname{diam} f^{N(\lambda)}(B) \geq \delta$, we can choose a subcontinuum B_1 of $f^{N(\lambda)}(B)$ such that $\operatorname{diam} B_1 = \lambda$ and $B_1 \cap Cl(\cup_{i=1}^n U_i) = \emptyset$. By (2.3), we see that $\operatorname{diam} f^{N(\lambda)}(B_1) \geq \delta$. Choose a subcontinuum B_2 of $f^{N(\lambda)}(B_1)$ such that $\operatorname{diam} B_2 = \lambda$ and $B_2 \cap Cl(\cup_{i=1}^n U_i) = \emptyset$. By induction, we can obtain a sequence $B_1, B_2, \ldots,$ of subcontinua of X such that $B_{j+1} \subset f^{N(\lambda)}(B_j)$, $\operatorname{diam} B_j = \lambda$ and $B_j \cap Cl(\cup_{i=1}^n U_i) = \emptyset$ for each $j = 1, 2, \ldots$. Choose a point $x \in \cap_{j=1}^\infty f^{-j \cdot N(\lambda)}(B_j)$. Note that $f^{j \cdot N(\lambda)}(x) \notin \cup_{i=1}^n U_i$ for each $j = 1, 2, \ldots$. Then we can easily see that there is $\eta > 0$ such that $d(f^j(x), A) > \eta$ for each $j = 1, 2, \ldots$. Since $Cl(O^s(x)) \cap A = \emptyset$, and $f(Cl(O^s(x))) \subset Cl(O^s(x))$, we can choose $M \in \mathcal{M}(f)$ such that $M \subset Cl(O^s(x))$. Then $M \cap A = \emptyset$. This completes the proof.

By using (2.6), we obtain the following.

Theorem 2.7. *Let* $f : X \to X$ *be a continuum-wise expansive homeomorphism of a compactum* X *with* $\dim X > 0$. *If* $Y \in \mathcal{D}(f)$, *then both* $f|Y : Y \to Y$ *and* $f^{-1}|Y : Y \to Y$ *are weakly chaotic in the sense of Devaney.*

Proof. Let $Y \in \mathcal{D}(f)$. By [16, (3.1)], we see that $f|Y$ and $f^{-1}|Y$ are sensitive and topologically transitive. We must show that the set $\cup\{M \,|\, M \in \mathcal{M}(f|Y)\}$ is dense in Y. Put $Y_1 = Cl(\cup\{M \,|\, M \in \mathcal{M}(f|Y)\}) \subset Y$. Note that $\dim Y > 0$ and $Y_1 \in \mathcal{I}(f)$. If $\dim Y_1 = 0$, by (2.6) we can choose $M \in \mathcal{M}(f|Y)$ such that $M \cap Y_1 = \emptyset$. This is a contradiction. Hence $\dim Y_1 > 0$. Since $Y \in \mathcal{D}(f)$, $Y_1 \subset Y$ and $\dim Y_1 > 0$, we see that $Y = Y_1$. This completes the proof.

3. TRANSITIVE ORBITS OF $f|Y : Y \to Y$ FOR $Y \in \mathcal{D}(f)$.

In this section, we study the set of points of $Y \in \mathcal{D}(f)$ at which $f|Y$ is topologically transitive.

Let $f : X \to X$ be a homeomorphism of a compactum X and let $x \in X$. Consider the following sets:

(1) $O(x) = \{f^n(x) | n \in Z\}$,

(2) $O^s(x) = \{f^n(x) | n = 0, 1, \ldots, \}$,

(3) $O^u(x) = \{f^{-n}(x) | n = 0, 1, \ldots, \}$,

(4) $T(f) = \{x \in X | O(x) \text{ is dense in } X\}$,

(5) $T^s(f) = \{x \in X | O^s(x) \text{ is dense in } X\}$,

(6) $T^u(f) = \{x \in X | O^u(x) \text{ is dense in } X\}$.

For a homeomorphism $f : X \to X$, define sets of stable and unstable nondegenerate subcontinua of X as follows (see [14]):

$$V^s = \{A | A \text{ is a nondegenerate subcontinuum of } X$$
$$\text{such that } \lim_{n \to \infty} \text{diam } f^n(A) = 0\}.$$

$$V^u = \{A | A \text{ is a nondegenerate subcontinuum of } X$$
$$\text{such that } \lim_{n \to \infty} \text{diam } f^{-n}(A) = 0\}.$$

Lemma 3.1. *Let $f : X \to X$ be a continuum-wise expansive homeomorphism of a compactum X with $\dim X > 0$. Then $V^s \neq \emptyset$ or $V^u \neq \emptyset$. Moreover, if $X \in \mathcal{D}(f)$ and $V^\sigma \neq \emptyset$ ($\sigma = s$ or u), then for each point $x \in X$, there is $x \in A_x \in V^\sigma$.*

Proof. By [13, (2.5)], we see that $V^s \neq \emptyset$ or $V^u \neq \emptyset$. Suppose that $X \in \mathcal{D}(f)$ and $V^s \neq \emptyset$. Choose $A \in V^s$. Note that if $O^u(A) = \cup\{f^{-n}(A) | n = 0, 1, \ldots, \}$, then $f^{-1}(Cl(O^u(A)) \subset Cl(O^u(A))$ and $\cap_{n=1}^{\infty} f^{-n}(Cl(O^u(A))$ contains a subcontinuum whose diameter is $\geq \delta$, where ϵ and δ are positive numbers as in (2.2). Since $B = \cap_{n=1}^{\infty} f^{-n}(Cl(O^u(A)))$ is f-invariant, we see that $\cup_{n=1}^{\infty} f^{-n}(A)$ is dense in X. Let $x \in X$. By (2.2), we see that there are sequences $\{x_n\}$ of points of X and $\{B_n\}$ of subcontinua of A such that $\lim_{n \to \infty} x_n = x, x_n \in f^{-n}(B_n)$, diam $f^{-n}(B_n) = \delta$, and diam $f^{-j}(B_n) \leq \epsilon$ for $0 \leq j \leq n$. We may assume that $\lim_{n \to \infty} f^{-n}(B_n) = A_x$. Then $x \in A_x \in V^s$ (see [13, (2.1)]).

Theorem 3.2. *Let $f : X \to X$ be a continuum-wise expansive homeomorphism of a compactum X with $\dim X > 0$. Suppose that $X \in \mathcal{D}(f)$. Then the following are satisfied:*

(1) *$\cup\{M | M \in \mathcal{M}(f)\}$ is of the first category in X.*

(2) *For each natural number n, there is a minimal set $M \in \mathcal{M}(f)$ such that $|M| > n$.*

(3) *$\dim(X - T(f)) = 0$, and $\dim T(f) \geq \max\{\dim X - 1, 1\}$.*

(4) *$T(f) = T^s(f) \cup T^u(f), T^s(f) \cap T^u(f)$ is a dense G_δ-set in X, and both $T^s(f) - T^u(f)$ and $T^u(f) - T^s(f)$ are dense in X.*

(5) If $V^s(f) \neq \emptyset$, then $\dim(X - T^s(f)) > 0, \dim T^s(f) > 0$, and also if $V^u(f) \neq \emptyset$, then $\dim(X - T^u(f)) > 0, \dim T^u(f) > 0$.

Proof. By (3.1), each component of X is nondegenerate. hence (1) follows from (2.5). Note that for each $n \geq 1, P_n(f) = \{p \in P(f)| f^n(p) = p\}$ is a 0-dimensional closed subset of X. Since $\cup\{M| M \in \mathcal{M}(f)\}$ is dense in X, we see that there is $M \in \mathcal{M}(f)$ with $|M| > 0$. This implies (2). We shall show (3). Let A be a subcontinuum of X. Then $O(A) = \cup\{f^n(A)| n \in Z\}$ is f-invariant, and hence $O(A)$ is dense in X, because $X \in \mathcal{D}(f)$. Let $U = \{U_i\}$ be a countable open base of X. By induction, we can obtain a sequence $A = A_1 \supset A_2 \supset \ldots$, of nondegenerate subcontinua of A and a sequence $N(1), N(2), \ldots$, of integers such that $f^{N(i)}(A_i) \subset U_i$ for each i. Take a point $a \in \cap_{i=1}^{\infty} A_i$. Then $O(a)$ is dense in X, hence we see that each nondegenerate subcontinuum of X contains a point of $T(f)$. Since $T(f)$ is a G_δ-dense set in X, $T(f) = \cap_{j=1}^{\infty} G_j$, where G_j is an open set of X. Since the closed set $X - G_j$ contains no nondegenerate subcontinuum, we see that $\dim(X - G_j) = 0$ for each j. By the sum theorem of dimension theory, we see that $\dim(X - T(f)) = \dim(\cup_{j=1}^{\infty}(X - G_j)) = 0$. By the addition theorem of dimension theory, we see that $\dim T(f) \geq \dim X - \dim(X - T(f)) - 1 = \dim X - 1$. Also, the fact $\dim T(f) \geq 1$ follows from (5) below. Next, we shall show (4). Let $x \in T(f)$. Then $X = Cl(O(x)) = Cl(O^s(x) \cup Cl(O^u(x))$. Since $\dim X > 0$, we see that $\dim Cl(O^s(x)) > 0$, or $\dim Cl(O^u(x)) > 0$. We assume that $\dim Cl(O^s(x)) > 0$. If $\omega(x)$ is the ω-limit set of x, then $\dim \omega(x) > 0$ and $\omega(x)$ is f-invariant. Hence $\omega(x) = X = Cl(O^s(x))$. This implies that $x \in T^s(x)$. Hence $T(f) = T^s(f) \cup T^u(f)$. Since $T^\sigma(f)(\sigma = s$ and $u)$ is a dense G_δ-set of X, we see that $T^s(f) \cap T^u(f)$ is also a dense G_δ-set of X. By (3.1), we may assume that $V^s \neq \emptyset$. For each $m \in M \in \mathcal{M}(f)$, choose a small subcontinuum A_m such that $m \in A_m \in V^s$. Note that if $A \in V^s$, then $\cup_{n=0}^{\infty} f^{-n}(A)$ is dense in X. By using this fact and the above argument, we can choose a point $y \in A_m$ such that $y \in T^u(f)$. Since $\lim_{n\to\infty} d(f^n(m), f^n(y)) = 0, m \in M$ and M is a proper closed subset of X, we see that $y \notin T^s(f)$. Since $\cup\{M| M \in \mathcal{M}(f)\}$ is dense in X, we see that $T^u(f) - T^s(f)$ is dense in X. Next, we shall show that $T^s(f) - T^u(x)$ is dense in X. Let $x \in T^s(f)$. By (3.1), we can choose a small subcontinuum A_x such that $x \in A_x \in V^s$. Choose a minimal set $M \in \mathcal{M}(f)$. By the same proof of (2.6), we can choose a point $y \in A_x$ such that $Cl(O^u(y)) \cap M = \emptyset$, which implies that $y \notin T^u(f)$. Since $\lim_{n\to\infty} d(f^n(x), f^n(y)) = 0$ and $x \in T^s(f), y \in T^s(f)$. Since $T^s(f)$ is dense in X, we can see that $T^s(f) - T^u(f)$ is dense in X. The case $V^u \neq \emptyset$ is similarly proved. Finally, we shall show (5). Suppose that $V^s \neq \emptyset$. If $x \in T^s(f)$, choose a subcontinuum A_x such that $x \in A_x \in V^s$. Then $A_x \subset T^s(f)$, which implies that $\dim T^s(f) > 0$. If $x \in M \in \mathcal{M}(f), A_x \subset X - T^s(f)$. Hence $\dim(X - T^s(f)) > 0$. The case $V^u \neq \emptyset$ is similarly proved.

By [13, (3.2)], [16, (5.3)], and (3.2), we obtain the following.

Corollary 3.3. *Let $f : X \to X$ be a sensitive map of a finite graph G (= compact 1-dimensional polyhedronset). Then $V^s = \emptyset$ and $V^u \neq \emptyset$, and $\mathcal{D}(f)$ is a finite family of subgraphs of G such that for each $K \in \mathcal{D}(f), f|K : K \to K$ is chaotic in the sense of Devaney and the shift map $\tilde{f}_K : (K, f_K) \to (K, f_K)$ satisfies the properties (1)-(5) as in (3.2), where $f_K = f|K : K \to K$.*

Remark 3.4. *Several complicated dynamical properties of (continuum-wise) expansive homeomorphisms f are concentrated in the sets $Y \in \mathcal{D}(f)$. Note that there is a σ-chaotic continuum $Z \subset Y$ of f ($\sigma = s$ or u) such that $\cup_{i=k}^{\infty} f^i(Z)$ and $\cup_{i=k}^{\infty} f^{-i}(Z)$ are dense in Y for each $k = 0, 1, 2, \ldots$, and either f or f^{-1} is chaotic on almost all Cantor sets of $f^n(Z)$ ($n \in Z$) in the sense of Li-Yorke (see [15, (4.1)]).*

Example 3.5. *There is an expansive homeomorphism $f : X \to X$ of a compactum X with $\dim X > 0$ such that f has no periodic point. Let $g_1 : C \to C$ be an expansive homeomorphism of a Cantor set C such that C is itself a minimal set of g_1 (see [6]). Let $g_2 : Y \to Y$ be any expansive homeomorphism of a compactum Y with $\dim Y > 0$. Consider the product $f = g_1 \times g_2 : C \times Y \to C \times Y$. Then f is a desired expansive homeomorphism. Hence in (2.7) we can not conclude that f induces chaos in the sense of Devaney.*

Question 3.6. *If $f : X \to X$ is a continuum-wise expansive homeomorphism of a compactum X with $\dim X > 0$, then does there exist a Cantor set C in X which is a minimal set of f ?*

REFERENCES

1. N, Aoki, Topological dynamics, in : Topics in general topology (eds, K. Morita and J. Nagata), *Elsevier Science Publishers B. V.*, (1989), 625-740.

2. J. Banks, J. Brooks, G. Cairns, G. Davis and P. Stacey, On Devaney's definition of chaos, *Amer. Math. Monthly*, 99 (1992), 332-334.

3. B. F. Bryant, Unstable self-homeomorphisms of a compact space, *Vanderbilt University*, Thesis, 1954.

4. R. Devaney, An Introduction to Chaotic Dynamical Systems, second edition, *Addison-Wesley*, 13046 (1989).

5. W. Gottschalk, Minimal sets; an introduction to topological dynamics, *Bull. Amer. Math. Soc.*, 64 (1958), 336-351.

6. W. Gottschalk and G. Hedlund, Topological dynamics, *Amer. Math. Soc. Colloq.*, 34 (1955).

7. W. Hurewicz and H. Wallman, Dimension Theory, *Princeton University Press*, Princeton, NJ 1948.

8. J. F. Jacobson and W. R. Uts, The nonexistence of expansive homeomorphisms of a closed 2-cell, *Pacific J. Math.*, 10 (1960), 1319-1321.

9. H. Kato, The nonexistence of expansive homeomorphisms of Peano continua in the plane, *Topology and its appl.*, 34 (1990), 161-165.

10. H. Kato, On expansiveness of shift homeomorphisms of inverse limits of graphs, *Fund. Math.*, 137 (1991), 201-210.

11. H. Kato, Expansive homeomorphisms in contiinuum theory, *Topology and its appl.*, 45 (1992), 223-243.

12. H. Kato, Expansive homeomorphisms and indecomposability, *Fund. Math.*, 139 (1991), 49-57.

13. H. Kato, Continuum-wise expansive homeomorphisms, *Canad. J. Math.*, 45 (1993), 576-598.

14. H. Kato, Striped structures of stable and unstable sets of expansive homeomorphisms and a theorem of K. Kuratowski on independent sets, *Fund. Math.*, 143 (1993), 153-165.

15. H. Kato, Chaotic continua of (continuum-wise) expansive homeomorphisms and chaos in the sense of Li-Yorke, *Fund. Math.*, to appear.

16. H. Kato, Chaos of continuum-wise expansive homeomorphisms and dynamical properties of sensitive maps of graphs, preprint.

17. T. Y. Li and J. A. Yorke, Period three implies chaos, *Amer. Math. Monthly*, 82 (1975), 985-992.

18. R. Mañé, Expansive homeomorphisms and topological dimension, *Trans. Amer. Math. Soc.*, 252 (1979), 313-319.

19. S. B. Nadler, Jr., Hyperspaces of sets, *Pure and Appl. Math.*, 49, Dekker, New York, 1978.

20. W. Reddy, The existence of expansive homeomorphisms of manifolds, *Duke Math. J.*, 32 (1965), 627-632.

21. W. Utz, Unstable homeomorphisms, *Proc. Amer. Math. Soc.*, 1 (1950), 769-774.

20
CHARACTERIZATIONS OF MENGER MANIFOLDS AND HILBERT CUBE MANIFOLDS IN TERMS OF PARTITIONS

KAZUHIRO KAWAMURA Institute of Mathematics, University of Tsukuba, Tsukuba-City, Ibaraki 305, Japan

ABSTRACT. We characterize compact Menger manifolds and Hilbert cube manifolds, in the same spirit, as compacta which have partitions of appropriate connectivities and certain kinds of universality. The proofs are based on characterization theorems due to Bestvina and Toruńczyk.

1. INTRODUCTION

When Bestvina [Be] established his characterization theorem of Menger manifolds, it became apparent that there is a deep analogy between Hilbert cube manifold (= I^∞ -manifold) theory and Menger manifold (= μ^n -manifold) theory. To some extent, we can say that μ^n -manifold theory is an n-dimensional analogue of I^∞-manifold theory. Since the partition technique, originally due to Anderson and Bing, plays an essential role in Menger manifold theory, it is natural to ask whether we can detect the properties of partitions of spaces which characterize Menger manifolds. Further, in light of the above analogy, we may ask the same question for I^∞-manifolds. Similar attempts have been made for topological manifolds (see, for example, [Bi2], and [H]).

The purpose of this note to give an answer to the above questions. It turns out that partitions with appropriate connectivity properties in the sense of homotopy

1991 *Mathematics Subject Classification.* 54C55, 57A20.
Key words and phrases. Menger manifold, Hilbert cube manifold, partition.
The author is supported by an NSERC International Fellowship.
The paper was written while the author was visiting the University of Saskatchewan.

and a certain kind of universality characterize Menger manifolds and Hilbert cube manifolds in the same spirit. Although it is difficult to verify these conditions in practice to detect these manifolds, it give us some insight to the roles of partitions. The proof is, of course, based on characterizations due to Toruńczyk [To] and Bestvina [Be] of these types of manifolds.

The author would like to express his thanks to the referee for his detailed comments.

2. Preliminaries

Definition 2.1. (1) *A partition P of a compactum X is a finite collection of subcontinua of X satisfying the following conditions.*

(1.1) *Each member of P is regularly closed and $\{int(p) \mid p \in P\}$ forms a mutually disjoint collection.*

(1.2) *$\cup P = X$.*

When X is a compact polyhedron, we call P a polyhedral partition, if each intersection of members of P is a subpolyhedron of X. For a member p of a partition P, $\mathcal{ST}(p, P)$ denotes the collection of all members of P which meet p and let $st(p, P) = \cup \mathcal{ST}(p, P)$. For a partition P, stP denotes the cover $\{st(p, P) \mid p \in P\}$.

(2) Suppose that P and Q are partitions of X. Q is called a refinement of P, denoted by $Q < P$, if each element of Q is contained in an element of P. If $stQ < P$, then Q is called a star refinement of P, denoted by $Q <^* P$.

Definition 2.2. *Let P be a partition of a compactum X.*

(1) P is called a C^k-partition if it satisfies:

(C^k) for each subcollection $\{p_1, ..., p_t\} \subset P$, $\cap p_i$ is LC^{k-t+1} and C^{k-t+1} (if $t > k+1$, then the condition is vacuous).

If $\cap p_i$ is LC^k and C^k for all $\{p_1, ..., p_t\} \subset P$ with non-empty intersection, it is called a strong C^k-partiton.

(2) P is called an AR-partition if it satisifies:

(AR) for each subcollection $\{p_1, ..., p_t\} \subset P$, $\cap p_i$ is an AR compactum if it is not empty.

Definition 2.3.

(1) A function $T : P \to Q$ between partitions of compacta X and Y is called a k-bijection, if

(1.1) it is a bijection, and

(1.2) for each $\{p_1, ..., p_t\} \subset P$, where $t \leq k+1$, $\cap T(p_i) \neq \emptyset$ if and only if $\cap p_i \neq \emptyset$.

(2) If a function $T : P \to Q$ between partitions is a k-bijection for each k, it is called an one-to-one correspondence.

(3) *Suppose that $T : P \to Q$ and $S : U \to V$ are k-bijections or one-to-one correspondences between partitions, and U is a refinement of P, V is a refinement of Q. Then S is said to be compatible with T, if for each pair $(p, u) \in P \times U$,*

(1) $u \subset p$ *if and only if* $S(u) \subset T(p)$, *and*
(2) *if* $u \cap p \neq \emptyset$, *then* $S(u) \cap T(p) \neq \emptyset$.

Examples 2.4.

(1) Let M be a I^∞-manifold. By the Triangulation Theorem [Chap], there is a compact polyhedron K such that M is homeomorphic to $K \times I^\infty$. Let $\pi : M \to K$ be the natural projection (under the above identification) and let P_K be a polyhedral AR partition of K. Such a partition is easily constructed using a triangulation of K. Then it is clear that $P = \pi^{-1}(P_K)$ is an AR-partition of M.

(2) Let M be a μ^k-manifold. By the Triangulation Theorem for μ^k-manifolds [Dr], there is a UV^{k-1} map $f : M \to K$ onto a compact polyhedron K. Embed K into a higher dimensional Euclidean space R^N ($N \geq 2k + 1$) and take a regular neighbourhood $N(K)$ of K in R^N. There is a CE retraction $r : N(K) \to K$ onto K. By [Dr], there is a UV^{k-1} map $f^* : X^* \to N(K)$ of a μ^k-manifold X^* onto $N(K)$ such that, for each LC^{k-1} compactum A in $N(K)$, $f^{*-1}(A)$ is a LC^{k-1} compactum. Notice that $r \circ f^*$ is a UV^{k-1} map. By the uniqueness of resolution [Be], 5.1.1, X and X^* are homeomorphic. In what follows, we identify X with X^* and let $p = f^* \circ r : X \to N(K)$. It has the following property:

for each subpolyhedron A of $N(K)$, $p^{-1}(A)$ is a LC^{k-1} compactum. Take a polyhedral AR partition H of $N(K)$ and let $Q = f^{*-1}(H)$. Then Q is a strong C^{k-1} partiton by the above property and the UV^{k-1} condition.

(3) Let M be a μ^k-manifold and let $\{M_i, L_i\}$ be a defining sequence (See [Be]) for M. Each M_i is a compact PL manifold of dimension $\geq 2k + 1$ with a triangulation L_i defined inductively as follows.

M_0 is a compact manifold with $\dim M = m \geq 2k + 1$, and L_0 is a triangulation of M_0. If (M_i, L_i) is defined, then $M_{i+1} = \cup\{st(b_\sigma, \beta^2 L_i) \mid b_\sigma$ is the barycenter of $\sigma \in L_i$ with $\dim \sigma \geq m - k\}$ and $L_{i+1} = \beta^2 L_i \mid M_{i+1}$. Let P_i be the handlebody decomposition of M_i defined by $\{st(b_\sigma, \beta^2 L_i) \mid b_\sigma$ is the barycenter of $\sigma \in L_i$ with $\dim \sigma \geq m - k\}$, then $M = \cap M_i$ and $P = P_i \mid M$ is a C^{k-1} partition.

Remark. We do not know a characterization of a compact ANR (a LC^k-compactum resp.) which admits an AR-partition (a C^k-partition resp.) with arbitrarily small mesh. A classical example of Borsuk and Mazurkiewicz (See [Bor], p.152-153) shows that not every compact ANR admits such a partition (in [Bo], an ANR admitting an AR-partition is said to satisfy the condition (Γ)). Such an example exists even in the class of generalized manifolds ([DW]). The author is grateful to Professor R.J. Daverman for this information.

3. Main results

Our main results are as follows:

Theorem 3.1. *For a compactum X, the following conditions are equivalent:*

(A) *X is a I^∞-manifold.*

(B) *For each $\epsilon > 0$ and for each integer $k > 0$, there is an AR partition P of X satisfying the following properties:*

  (a) *$\mathrm{mesh}\,P < \epsilon$,*

  (b) *there are a compact PL manifold M_P, a sequence of polyhedral AR partitions $\{U_i \mid i \geq 0\}$ of M_P, a sequence of AR partitions $\{P_i \mid i \geq 0\}$ of X, and a sequence of one-to-one correspondences $\{T_i : P_i \to U_i\}$ such that*

   (b.1) *$\dim M_P \geq 2k + 1$, and*
   (b.2) *$P_0 = P$, $P_{i+1} <^* P_i$, $U_{i+1} <^* U_i$ and $\mathrm{mesh}\,U_i \to \infty$ as $i \to \infty$.*
   (b.3) *T_{i+1} is compatible with T_i.*

For μ^k-manifolds, we give two types of characterizations in terms of partitions. Roughly speaking, they correspond to the partition induced by Menger construction [M] and by Bestvina construction [Be] respectively.

Theorem 3.2. *For a k-dimensional compactum X, the following conditions are equivalent:*

(A) *X is a μ^k-manifold.*

(B) *For each $\epsilon > 0$, there is a strong C^{k-1} partition P of X satisfying the following properties:*

  (a) *$\mathrm{mesh}\,P < \epsilon$.*

  (b) *there are a compact PL manifold M_P, a sequence of polyhedral AR partitions $\{U_i \mid i \geq 0\}$ of M_P, a sequence of strong C^{k-1} partitions $\{P_i \mid i \geq 0\}$ of X, and a sequence of one-to-one correspondences $\{T_i :: P \to U_i \mid i \geq 0\}$ such that*

   (b.1) *$\dim M_P \geq 2k + 1$,*
   (b.2) *$P_0 = P$, $P_{i+1} <^* P_i$, $U_{i+1} <^* U_i$ and $\mathrm{mesh}\,U_i \to \infty$ as $i \to \infty$, and*
   (b.3) *T_{i+1} is compatible with T_i.*

Theorem 3.3. *For a compactum X, the following conditions are equivalent:*

(A) *X is a μ^k-manifold.*

(B) *For each $\epsilon > 0$, there is a C^{k-1}-partition P of X satisfying the following properties:*

  (a) *$\mathrm{mesh}\,P < \epsilon$ and $\mathrm{ord}\,P \leq k + 1$.*

(b) *there are a compact PL manifold M_P, a sequence of polyhedral AR partitions $\{U_i \mid i \geq 0\}$ of M_P, a sequence of C^{k-1}-partitions $\{P_i \mid i \geq 0\}$ of X, and a sequence of k-bijections $\{T_i : P_i \to U_i\}$ such that*

(b.1) *$dim M_P \geq 2k + 1$,*

(b.2) *$P_0 = P$, $P_{i+1} <^* P_i$, $U_{i+1} <^* U_i$ and $mesh U_i \to 0$ as $i \to \infty$, and*

(b.3) *T_{i+1} is compatible with T_i.*

(c) *for each $p \in P$, there corresponds an open C^{k-1} neighbourhood $N(p)$ of p such that $diam N(p) < \epsilon$.*

Remark. (1) The condition (b) of (B) in all theorems state that if we have a refinement U_{i+1} of U_i, we may "realize the pattern of U_{i+1} in U_i as the one of a refinement P_{i+1} of P_i." This kind of "universality" has appeared in several situations, for example, such as in the proof of the uniquness of the pseudo-arc. For one of the formulations. see [OTy]. For a related notion, see [K], where the corresponding property is stated as the property (∗).

(2) The condition (c) of Theorem 3.3 is of technical nature to guarantee that the space under consideration is LC^{k-1}.

The proofs of implications (A) $\Rightarrow$ (B) in all theorems are verifications of the properties of partitions of Examples 2.4. The proofs of the reverse implications consist of verifying the appropriate general position properties.

Proof of Theorem 3.1.

(A)$\Rightarrow$ (B). By the Triangulation Theorem of I^∞-manifolds, there is a compact polyhedron K such that X is homeomorphic to $K \times I^\infty$. Since any polyhedron which is simple homotopy equivalent to K gives a homeomorphic I^∞-manifold by taking the product with I^∞ ([Chap]), we may assume that K is a compact PL manifold (with boundary) with $dim K \geq 2k+1$. Let $\pi : X \to K$ be the natural projection. By replacing K by $K \times I^n$ for large n, if necessary, we may assume that π is an $\epsilon/4$ map. Take a handlebody decomposition U_0 of K with $mesh < \epsilon/4$, and let $P = \pi^{-1}(U_0)$. The projection π induces an one-to-one correspondence $T_0 : P \to U_0$. If V is an AR refinement of U_0, define an AR refinement Q by $Q = \pi^{-1}(V)$. Again the projection induces an one-to-one correspondence $S : Q \to V$ which is compatible with T. From this observation, it is easy to construct the sequences of partitons and one-to-one correspondences required in (2). This completes the proof of (A)$\Rightarrow$(B).

(B) $\Rightarrow$(A). Let X be a compactum satisfying the condition (B). Applying the condition (b) for a small ϵ, we see that $X = \cup X_i$, where each $X_i \cap X_j$ is a compact AR ($i, j = 1, ..., m$). It follows from the Sum Theorem of ANR's [Bor], Chap.4, 6.1, we see that X is an ANR. Therefore, it suffices to prove that X satisfies the $DD^n P$ for each $n \geq 1$ ([To]).

Fix an integer $n > 0$ and take any $\epsilon > 0$. We take a partition P of X applying condition (b) for $\epsilon/3^3$ and $k = n$. Using the condition (2), we may construct a map $\alpha : X \to M_P$ such that

(1) for each $p \in P$, $\alpha(p) \subset st(T_0(p), U_0)$.

Proof of (1). This can be proved by a standard method (see, for example, [Be], 1.3.2). We sketch the outline. Take a point $x \in X$ and let $\{p_i(x)\}$ be a sequence of members of partitions such that $x \in p_i(x) \in P_i$. We show that $st(T_{i+1}(p_{i+1}((x)), U_{i+1}) \subset st(T_i(p_i(x)), U_i)$.

By the compatibility of T_i and T_{i+1}, we have that $T_{i+1}(p_{i+1}(x)) \cap T_i(p_i(x)) \neq \emptyset$. Since $U_{i+1} <^* U_i$, there is an element $u_i \in U_i$ such that $st(T_{i+1}(p_{i+1}(x)), U_{i+1}) \subset u_i$. Then clearly $u_i \in \mathcal{ST}(T_i(p_i(x)), U_i)$. Then $st(T_{i+1}(p_{i+1}(x)), U_{i+1}) \subset u_i \subset st(T_i(p_i(x)), U_i)$, we have the required inclusion.

Then we define $\alpha(x) = \cap st(T_i(p_i(x)), U_i)$, and it can be seen that it is a continuous map satisfying the desired property (1).

For any pair $f, g : I^n \to X$ of maps of the n-cell, we consider $f* = \alpha \circ f$ and $g* = \alpha \circ g$. Since $dim M_P \geq 2n + 1$, there are PL maps $F, G : I^n \to M_P$ such that

(2) $F*$ is stU_0-close to $f*$, $G*$ is stU_0-close to $g*$, and $imF^* \cap imG^* = \emptyset$.

Take a sufficiently large i so that $st^2(imF^*, U_i) \cap st^2(imG^*, U_i) = \emptyset$. We consider $T_0 : P \to U_0$ and $T_i : P_i \to U_i$, and construct maps $F, G : I^n \to X$ as follows. For simplicity, let $P_i = Q$, $U_0 = U$, $U_i = V$ and let $T = T_0 : P \to U$, and $S = T_i : Q \to V$.

$$
\begin{array}{ccc}
P & \xrightarrow{\ T = T_0\ } & U = U_0 \\[4pt]
\vee & & \vee \\[4pt]
P_i = Q & \xrightarrow[\ S = T_i\ ]{} & V = U_i
\end{array}
$$

Consider the set $W = \cup\{v_1 \cap v_2 \mid v_i \in V, v_1 \neq v_2\}$. Since W is a polyhedron, there is a neighbourhood retraction $r : N \to W$. We may change r slightly into $r' : N \to W$ so that there is a $\delta > 0$ such that, for each subset A in N with $diam A < \delta$, $r'(A)$ is contained in an element of V (see [B], 2.4.6).

Take a sufficiently fine triangulation Σ of I^n so that $mesh F^*(\Sigma) < \delta$ and $mesh G^*(\Sigma) < \delta$. Then $F^{*\prime} = r' \circ F^*$ and $G^{*\prime} = r' \circ G^*$ satisfy

(3) for each $\sigma \in \Sigma$, $F^{*\prime}(\sigma) \subset v(\sigma, F)$ for some $v(\sigma, F) \in V$, and $G^{*\prime}(\sigma) \subset v(\sigma, G)$ for some $v(\sigma, G) \in V$.

For notational simplicity, $F^{*\prime}$ and $G^{*\prime}$ are simply denoted by F^* and G^* respectively. Let $q(\sigma, F) = S^{-1}(v(\sigma, F))$ and $q(\sigma, G) = S^{-1}(v(\sigma, G))$. For any simplex τ of Σ, $\Sigma(\tau)$ denotes the set of all simplexes containing τ.

For each vertex a of Σ, $F^*(a) \in \cap\{v(\sigma, F) \mid \sigma \in \Sigma(a)\}$. Choose a point $F(a)$ in $\cap\{q(\sigma, F) \mid \sigma \in \Sigma(a)\}$ ($\neq \emptyset$, because S is an one-to-one correspondence). This defines $F : \Sigma^{(0)} \to X$. Next take an 1-simplex $\tau \in \Sigma$ so that $\partial \tau = \{a, b\}$. Clearly $\Sigma(\tau) \subset \Sigma(a) \cap \Sigma(b)$. We also have that

(4) $F^*(\tau) \subset \{v(\sigma, F) \mid \sigma \in \Sigma\tau)\}$, and

$\{F(a), F(b)\} \subset \cap\{q(\sigma, F) \mid \sigma \in \Sigma(a)\} \cap \{q(\sigma, F)) \mid \sigma \in \Sigma(b)\} \subset \cap\{q(\sigma, F) \mid \sigma \in \Sigma(\tau)\}$.

By the AR property of $\cap\{q(\sigma, F) \mid \sigma \in \Sigma(\tau)\}$, we can extend $F|\partial\tau$ to $F \mid \tau : \tau \to \cap\{q(\sigma, F) \mid \sigma \in \Sigma(\tau)\}$. Repeating this procedure for each 1-simplex of Σ, we obtain a map $F : \Sigma^{(1)} \to X$.

Continuing this process, we obtain a map $F : I^n \to X$ such that

(5-F) for each $\tau \in \Sigma$, $F(\tau) \subset \cap\{q(\sigma, F) \mid \sigma \in \Sigma(\tau)\} = \cap\{S^{-1}(v(\sigma, F) \mid \sigma \in \Sigma(\tau)\}$

Similarly, we obtain a map $G : I^n \to X$ such that

(5-G) for each $\tau \in \Sigma$, $G(\tau) \subset \cap\{q(\sigma, G) \mid \sigma \in \Sigma(\tau)\} = \cap\{S^{-1}(v(\sigma, G) \mid \sigma \in \Sigma(\tau)\}$.

Claim 1. $d(F, f) < \epsilon$ and $d(G, g) < \epsilon$.

Suppose that $x \in \tau \in \Sigma$ and $f(x) \in p \in P$. By the condition (1), $f^*(x) \in st(T(p), U)$. The condition (2) implies that $F^*(x) \in st^2(T(p), U)$. On the other hand, $F^*(x) \in F^*(\tau) \subset v(\tau, F) \in V$. Take an elememt $u \in U$ such that $v(\tau, F) \subset u$. Then we see that $u \in \mathcal{ST}^3(T(p), U)$. By the compatibility of S and T, we have that

$$S^{-1}(v(\tau, F)) \subset T^{-1}(u) \subset \mathcal{ST}^3(p, U).$$

By the condition (4), $F(x) \in S^{-1}(v(\tau, F))$. Thus, $d(F(x), f(x)) < mesh\ st^3 P < 3^3 \cdot \epsilon/3^3 = \epsilon$.

The same argument applies to G and g.

Claim 2. $im F \cap im G = \emptyset$.

We prove that $\mathcal{ST}(im F, Q) \subset S^{-1}(\mathcal{ST}^2(im F*, V))$. Since S is an one-to-one correspondence and $st^2(im F^*, V) \cap st^2(im G^*, V) = \emptyset$, this is enough for the proof of Claim 2.

Take a $q \in \mathcal{ST}(im F, Q)$ and take a point $x \in I^n$ such that $F(x) \in q$ and let $x \in int\tau$, $\tau \in \Sigma$. Then by (5-F), we have that

$$F(x) \in F(\tau) \subset \cap\{S^{-1}(v(\sigma, F)) \mid \sigma \in \Sigma(\tau)\}.$$

This implies that $q \in \mathcal{ST}(S^{-1}(v(\sigma, F), Q)$, and thus, $S(q) \in st(v(\sigma, F), V)$ for each $\sigma \in \Sigma(\tau)$. On the other hand, $F^*(x) \in \cap\{v(\sigma, F) \mid \sigma \in \Sigma(\tau)\}$ by (4). Thus $S(q) \in \mathcal{ST}^2(im F^{(}*), Q)$ as desired.

These two claims complete the proof. □

Proof of Theorem 3.2.

(A) $\Rightarrow$(B). Let X be a μ^k-manifold and take any $\epsilon > 0$. As in Example 2.4 (2), there is a compact PL manifold M with $dim M \geq 2k+1$ and a UV^{k-1} map $f : X \to M$ such that

(1) for each LC^{k-1} compactum A of M, $f^{-1}(A)$ is LC^{k-1} ([Dr], Theorem 2).

Further, from the construction of the map f in [Dr], we may assume that

(2) f is an $\epsilon/2$ map.

Let δ be a positive number such that, for any subset S of M with $diam S < \delta$, we have that $diam f^{-1}(S) < \epsilon$. Take a polyhedral AR partition U_0 of M and let $P = f^{-1}(U_0)$. Then P is a strong C^{k-1} partiton of X with $mesh P < \epsilon$ by the property (1) and the UV^{k-1} property of f. Clearly f induces an one-to-one correspondence between P and U_0. Let $M_P = M$. The rest of the proof proceeds as in Theorem 3.1. This completes the proof of (A)$\Rightarrow$ (B).

(B) $\Rightarrow$ (A). By the Sum Theorem of $LC^{k-1} \cap C^{k-1}$ compacta, it follows that X is LC^{k-1}. By the characterization theorem of Bestvina ([Be]), it suffices to verify the $DD^k P$. Recalling that any LC^{k-1} and C^{k-1} compactum is an absolute extensor for k-dimensional compacta, the proof proceeds as in Theorem 3.1 again. $\square$

Proof of Theorem 3.3.

(A) $\Rightarrow$(B). Let $\{(M_i, L_i)\}$ be a defining sequence of a μ^k-manifold X and let $m = dim M_i \geq 2k+1$. Recall that

$$M_{i+1} = \cup\{st(b_\sigma, \beta^2 L_i) \mid b_\sigma \text{ is the barycenter of } \sigma \in L_i \text{ with } dim\sigma \geq m-k\}$$
$$= \text{a regular neighbourhood of the dual } k\text{-skeleton of } L_i.$$

And $L_{i+1} = \beta^2 L_i \mid M_{i+1}$. The above construction determines a polyhedral AR partition

$$H_{i+1} = \{st(b_\sigma, \beta^2 L_i) \mid b_\sigma \text{ is the barycenter of } \sigma \in L_i \text{ with } dim\sigma \geq m-k\}$$

of M_{i+1} such that $ord H_{i+1} \leq k+1$.

Let $\epsilon > 0$. Take a large i such that $mesh H_i < \epsilon$ and define a partition P of X by $P = H_i \mid X$. The natural "inclusion" $T_0 : P \to H_i$ defines an one-to-one correspondence. Let $U_0 = H_i$.

Take the second barycentric subdivision $\beta^2 L_i$ and let

$$V = \{st(b_\sigma, \beta^2 L_i) \mid b_\sigma \text{ is the barycenter of } \sigma \in L_i\},$$

be a polyhedral AR partition of M_i which is a refinement of U_0. From the definition, $H_{i+1} = V|M_{i+1}$. Let $I : H_{i+1} \to V$ be the "inclusion" defined by $I(h_{i+1}) = h_{i+1}$ for each $h_{i+1} \in H_{i+1}$. Now we proceed as in [Be], 2.4-2.7 and Appendix.

Step 1. Following the method of [Be], 2.4.8 (p.40-41), we define a partition H^* of M_{i+1} and a k-bijection $J : H^* \to V$ such that $H^* < H_{i+1}$ and J is compatible with T_0.

Take an elelment $v \in V - I(H_{i+1})$ such that $v \cap I(H_{i+1}) \neq \emptyset$ and suppose that $v \subset h_i \in H_i$. There is an element $h_{i+1} \in H_{i+1}$ such that $I(h_{i+1}) \cap v \neq \emptyset$, and

$h_{i+1} \subset h_i$. Take an m-ball B in $int(h_{i+1})$, and let $h'_{i+1} = h_{i+1} - int(B)$. A new partition $H' < H_{i+1}$ of M_{i+1} is defined by $H' = (H_{i+1} \cup \{h_{i+1} \cup \{B, h'_{i+1}\}$. A function $I' : H' \to V$, defined by

$$I' \mid H_{i+1} - \{h_{i+1} = I \mid H_{i+1} - \{h_{i+1}\},$$

$$I'(h'_{i+1}) = I(h_{i+1}), and \ I'(B) = v,$$

has $I(H_{i+1}) \cup \{v\}$ as its image.

Repeating this process, we may construct a partition H'' and a sujection $I'' : H'' \to V$. Next, we change I'' into a k-bijection $J : H^* \to V$ which is compatible with T_0 using the method in [Be], p.41 (See also [Be] Appendix). Notice that elements of H^* need not have "nice intersections" with X.

Step 2. By applying Isotopy Moves (ibid. p.31-33), we may change H^* into a partition Q of M_j for some large $j > i + 1$ such that each element Q underlies a subcomplex L_j. By a property of Isotopy Moves, we can define a k-bijection $S : Q \to V$ which is compatible with T_0. Observe that Q may not be a C^{k-1} partition.

Step 3. Next we modify Q into a C^{k-1}-partition.

Since V is an AR-partition, we may perform all of the arguments of [Be], 2.5-2.7 to construct a C^{k-1}-partiton P_1^* of M_s for some large $s > j$ and a k-bijection $R : P_1^* \to V$ which is compatible with T_0. Applying the Isotopy Move again, we may assume that each member of P_1^* underlies a subcomplex of L_t for some large ts. Then $P_1^* \mid X$ forms a C^{k-1}-partiton of X. Define $U_1 = V$, $P_1 = P \mid X$, and $T_1 : P_1 \to U_1$ by $T_1(p_1 \cap X) = R(p_1)$ for each $p_1 \in P_1^*$. Then T_1 is compatible with T_0.

Continuing this process, we get the desired sequence.

$(B) \Rightarrow (A)$. Assume that a compactum X satisfies the condition (B). By the condition (1), we see that $dim X \le k$. Take a parition P required in (B) with a small mesh. By the condition (3), eacn $N(p)$ is $(k-1)$-connected for each $p \in P$. Since the mesh of P can be arbitrarily small, X is LC^{k-1}. It remains to verify the DD^kP.

Except for some technical details, the proof proceeds as in Theorem 3.1. Take any $\epsilon > 0$ and take a partition P of X which is required in the condition (B) such that $mesh P < \epsilon/4$. Since $T : P \to U_0$ is a k-bijection, $k \ge 1$, the method of Theorem 3.1 can be used to construct a map $\alpha : X \to M_P$ such that

(1) for each $p \in P$, $\alpha(p) \subset st(T_0(p), U_0)$.

For any pair $f, g : I^k \to X$ of maps of the k-cell, we consider $f^* = \alpha \circ f$ and $g^* = \alpha \circ g$ and by general position, find PL maps $F^*, G^* : I^k \to M$ such that

(2) F* is stU_0-close to f^*, G^* is stU_0-close to g^*, and $im F^* \cap im G^* = \emptyset$.

Take a sufficiently large i so that $st^2(imF^*, U_i) \cap st^2(imG^*, U_i) = \emptyset$, and let $T = T_0 : P \to U = U_0$ and $S = T_i : Q = P_i \to V = U_i$.

For a triangulation Σ of I^k, we define a polyhedral AR partition Ω of I^k by $\Omega = \{st(v, \beta\Sigma) \mid v \in \Sigma^{(0)}\}$.

It has $order \le k+1$ and for each mutually distinct collection $\{\omega_1, ..., \omega_t\} \subset \Omega$, we have that $dim \cap \{\omega_i \mid 1 \le i \le t\} = k - t + 1$ if $\cap\{\omega_i \mid 1 \le i \le t\} \ne \emptyset$. Let

$$S^j(\Omega) = \{\cap\omega_s \mid \omega_s \in \Omega \; for \; 1 \le s \le k - j + 1\},$$

and

$$S^j(\Omega) = \cup \mathcal{S}^j(\Omega),$$

Then each element of $\mathcal{S}^j(\Omega)$ is a j-dimensional cell, and hence, $dim S^j(\Omega) = j$.

As in (3) of Theorem 3.1, taking a sufficiently fine triangulation Σ if necessary, we may adjust F^* and G^* slightly so that they satisfy

(3) for each $\omega \in \Omega$, $F^*(\omega) \subset v(\omega, F)$ for some $v(\omega, F) \in V$, and $G^*(\omega) \subset v(\omega, G)$ for some $v(\omega, G) \in V$.

Using $S^j(\Omega)$'s instead of skeletons, we define maps $F, G : I^k \to X$ as in Theorem 3.1. For any element ω of Ω, $\Omega(\omega)$ denotes the set of all elements of Ω containing ω. Notice that $\#\Omega(\omega) \le k + 1$.

Let $q(\omega, F) = S^{-1}(v(\omega, F))$. For each 0-cell a of $S^{(0)}(\Omega)$, $F^*(a) \in \cap\{v(\omega, F) \mid \omega \in \Omega(a)\}$. Choose a point $F(a)$ in $\cap\{q(\omega, F) \mid \omega \in \Omega(a)\}$. This defines $F : S^{(0)}(\Omega) \to X$. Next take an 1-cell $\tau = \cap\{\omega_i \mid 1 \le i \le k\}$ and let $\partial\tau = \{\tau \cap \omega \mid \omega \in \Omega - \{\omega_1, ..., \omega_k\}\} \subset S^{(0)}(\Omega)$. We have further that

(4) $F^*(\tau) \subset \cap\{v(\omega, F) \mid \omega \in \Omega(\tau)\}$, and

$F(\partial\tau) \subset \cup\{q(\omega, F) \mid \omega \in \Omega(a), a \in \partial\tau\} \subset \cap\{q(\omega, F) \mid \omega \in \Omega(\tau)\}$. By the 0-connectedness of $\cap\{q(\omega, F) \mid \omega \in \Sigma(\tau)\}$, we can extend $F \mid \partial\tau$ to $F \mid \tau : \tau \to \cap\{q(\omega, F) \mid \omega \in \Sigma(\tau)\}$. Repeating this procedure, we obtain a map $F : S^{(1)} \to X$.

In general, we use the fact that $\cap\{p_t \mid 1 \le t \le k-j+1\}$ is $(k-(k-j+1))(= j-1)$-connected and each element of $S^j(\Omega)$ is a j-dimensional cell to extend $F \mid S^{(j-1)}$ to a map $F : S^{(j)} \to X$. Continuing this process, we obtain a map $F : I^k \to X$ such that

(5-F) for each $\omega \in \Omega$, $F(\omega) \subset \cap\{q(\chi, F) \mid \chi \in \Omega(\omega)\} = \cap\{S^{-1}(v(\chi, F) \mid \chi \in \Omega(\omega)\}$.

Similarly, we obtain a map $G : I^k \to X$ such that

(5-G) for each $\omega \in \Omega$, $G(\omega) \subset \cap\{q(\chi, G) \mid \chi \in \Omega(\omega)\} = \{S^{-1}(v(\chi, G)) \mid \chi \in \Omega(\omega)\}$.

Since only the intersections of at most $(k + 1)$ elements of Ω are involved, k-bijection plays the same role as one-to-one correspondence does in Theorem 3.1.

Hence, as in Theorem 3.1, we can prove that the maps F and G are the desired approximations of f and g.

This completes the proof. □

REFERENCES

[Be]   M. Bestvina, *Characterizing k-dimensional universal Menger compacta*, Mem. Amer. Math. Soc. vol. 71, no. 380 (1988).

[Bi1]   R.H. Bing, *Partitioning a set*, Bull. Amer. Math. Soc., 58 (1952), 536-556.

[Bi2]   _____ , *A characterization of 3-space by partitionings*, Trans. Amer. Math. Soc., 70 (1951), 15-27.

[Bor]   K. Borsuk, *Theory of Retracts*, PWN Warszawa, 1966.

[Chap] T.A. Chapman, *Lectures on Hilbert cube manifolds*, CBMS Regional conf. Series in Math. 28 (1976), Amer. Math. Soc.

[Chig] A. Chigogidze, *Theory of n-shapes*, Russian Math. Surveys, 44;5 (1989), 145-174.

[DaW]   R.J. Daverman and J.J. Walsh, *A ghastly generalized n-manifold*, Illinois J. Math. 25 (1981), 555-576.

[Dr]   A.N. Dranishnikov, *Universal Menger compacta and universal mappings*, Math. USSR Sbornik 57 (1987), 131-149.

[H]   O.G. Harrold, *A characterization of locally Euclidean spaces*, Trans. Amer. Math. Soc. 118 (1965), 1-16.

[K]   K. Kawamura, *Near-homeomorphisms of hereditarily indecomposable circle-like continua*, Tsukuba J. Math. 13 (1989), 165-173..

[L]   R.C. Lacher, *Cell-like mappings and their generalizations*, Bull. Amer. Math. Soc., 83 (1977), 495-552.

[M]   K. Menger, *Allgemeine Raume und Cartesian Raume, Zweite Mitteilung: "Uber umfassendste n-dimensionale Mengen"*, Proc. Akad.Amsterdam 29 (1926), 1125-1128.

[OTy]   L.G. Oversteegen and E.D. Tymchatyn, *On hereditarily indecomposable compacta*, Geometric and Algebraic Topology, ed. by H. Toruńczyk, S. Jackowski, and S. Spiez, Banach Center Publ. 18 PWN, 1986, 407-417..

[To]   H. Toruńczyk, *On CE images of Hilbert cube and a characterization of Q-manifolds*, Fund. Math. 106 (1980), 31-40.

21
HOMOLOGY SEPARATION AND 2-HOMOGENEITY

KRYSTYNA M. KUPERBERG Department of Mathematics, Auburn University, Auburn, Alabama 36849, USA

WŁODZIMIERZ KUPERBERG Department of Mathematics, Auburn University, Auburn, Alabama 36849, USA

WILLIAM R. R. TRANSUE Department of Mathematics, Auburn University, Auburn, Alabama 36849, USA

ABSTRACT. The homology separation axiom is introduced to investigate homogeneity properties of Cartesian products of continua. We prove that under certain conditions of a homological nature the products are factorwise rigid. Also, if the product $X \times Y$ of an n-dimensional representable continuum X and a non-degenerate continuum Y is 2-homogeneous, then X is locally n-acyclic. In particular, the product $B \times Y$ of a μ^n-manifold B (based on the n-dimensional Menger universal compactum μ^n) with any non-degenerate continuum Y is not 2-homogeneous. These theorems generalize a previous result of the same authors and some results of J. Kennedy Phelps.

INTRODUCTION

In 1975 G. S. Ungar [17] proved that every 2-homogenous continuum is locally connected, thereby giving rise to a variety of questions concerning the relationship between n-homogeneity and local connectedness in homogeneous continua. It seemed then reasonable to expect that locally connected homogeneous continua must have some stronger homogeneity properties. However, in [13] the authors showed that a homogeneous locally connected continuum need not be 2-homogeneous. One example is the Cartesian product $M \times S^1$ of the Menger universal

1991 *Mathematics Subject Classification.* primary 54F35, secondary 54F15.

Key words and phrases. n-homogeneous, representable, factorwise rigid, locally n-acyclic, semi-locally n-acyclic, Menger universal curve, Menger manifold, homology separation.

curve M and the circle S^1 (for a proof of homogeneity of M see [1]), another one is $M \times M$. J. Kennedy Phelps [9] proved that under certain conditions 2-homogeneity and representability are equivalent. As a corollary she obtained that $M \times X$, where X is an arbitrary continuum, is not 2-homogeneous.

The proof in [13] of non-2-homogeneity of $M \times S^1$ includes a stronger feature of the product space, namely every homeomorphism $h : M \times S^1 \to M \times S^1$ sends each S^1-fiber onto an S^1-fiber. Similarly, every homeomorphism $h : M \times M \to M \times M$ is of the form $h(x_1, x_2) = (h_1(x_1), h_2(x_2))$ or $h(x_1, x_2) = (h_2(x_2), h_1(x_1))$ with $h_i : M \to M$. This provided ground for examples of continua with peculiar homogeneity properties [10], [11], and especially for the example of a locally connected, homogeneous, non-bihomogeneous continuum [12].

Each of the above examples is locally a Cartesian product with at least one of the factors being the Menger curve M. The "preservation of fibers in auto-homeomorphisms" property of the product space $X \times Y$ is called factorwise rigidity and is a useful tool in solving problems involving homogeneity and related properties of product spaces. For example, such rigidity exists in the Cartesian product of several copies of the Sierpiński curve (as well as the Menger curve), and since the Sierpiński curve is not homogeneous, the product is not homogeneous either [8].

Following are the main results of this paper:

1. *If X is n-dimensional and not locally n-acyclic at any point, and if Y is locally k-acyclic for $k = 0, 1, \ldots, n$, then $X \times Y$ is factorwise rigid with respect to Y* (Theorem 3.3 in Section 3);

2. *If X is n-dimensional, representable, and not locally n-acyclic, then $X \times Y$ is not 2-homogeneous* (Theorem 4.2 in Section 4).

1. PRELIMINARIES

All spaces considered in this paper are metric non-degenerate continua. By a neighborhood of a point x we understand a closed set containing x in its interior. To indicate that a set is a neighborhood of a point x, we often use the subscript x, e.g., U_x, for brevity. The projections of $X \times Y$ onto X and Y are denoted by p_X and p_Y, respectively. Inclusion maps are denoted by i and j (possibly with subscripts) and the symbol $\hookrightarrow$.

A space is *n-homogeneous* if for any pair of n-point sets the space admits an auto-homeomorphism sending one of the sets onto the other. A *homogeneous* space is a 1-homogeneous space. An auto-homeomorphism is *primitively stable* if it is the identity on some non-empty open set. A space is *primitively stable* if it admits a primitively stable auto-homeomorphism different from the identity. A space X is *representable* if for every point $x \in X$ and every neighborhood U_x there is a neighborhood $V_x \subset U_x$ such that for every $y \in V_x$ there is a homeomorphism $h : X \to X$ with $h(x) = y$ and $h(z) = z$ for $z \in X - U_x$. M denotes the Menger universal curve.

The Cartesian product $X = \Pi_{\lambda \in \Lambda} X_\lambda$ is *factorwise rigid* if every homeomorphism $h : X \to X$ is of the form $h(x_\lambda) = (h_\lambda(x_{\alpha(\lambda)}))$, where $\alpha : \Lambda \to \Lambda$ is a one-to-one and onto correspondence, and $h_\lambda : X_\lambda \to X_{\alpha(\lambda)}$ is a homeomorphism. The product $X_1 \times X_2$ is *factorwise rigid with respect to X_2 [X_1]* if every homeomorphism $h : X_1 \times X_2 \to X_1 \times X_2$ has the property that for every $x \in X_1$ [$\in X_2$], there is a $y \in X_1$ [$\in X_2$] such that $h(\{x\} \times X_2) = \{y\} \times X_2$ [$h(\{x\} \times X_1) = \{y\} \times X_1$]. The authors of this paper have proved [13] that $M \times X$, where X is a manifold, is factorwise rigid with respect to X, and $M \times M$ is factorwise rigid. Kennedy Phelps [8] generalized the latter result to a countable product of the Menger curves. Bellamy and Łysko [4] proved that the product of two pseudo-arcs is factorwise rigid, and, using the homogeneity of the pseudo-arc, Bellamy and Kennedy Phelps [3] generalized this theorem to an arbitrary product of pseudo-arcs. The results mentioned above deal with spaces of a complicated local structure: the pseudo-arc is not locally connected and the Menger curve M is not locally simply connected. Section 3 contains spaces with homological properties more general than local connectedness and local simple connectedness.

It is known, and it is not difficult to prove, that each representable space is n-homogeneous for every $n \geq 1$. Following is a surprising result of Kennedy Phelps [9]:

Theorem KP. *Each 2-homogeneous primitively stable space is representable.*

Using this theorem Kennedy Phelps proved that the Cartesian product of M with an arbitrary continuum is not 2-homogeneous. Section 4 generalizes this conclusion.

Also, we should mention here some results of H. Patkowska [15], closely related to the topic of this paper: 1) If a product of finitely many Peano curves is 2-homogeneous, then each factor is S^1; 2) A product of finitely many 1- or 2-dimensional ANR's is locally homogeneous if and only if each factor is a manifold. (A space X is *locally homogeneous*, also called *micro-homogeneous* in [19] and [10], if for every pair of points $x, y \in X$ there are neighborhoods U_x and U_y and a homeomorphism h of U_x onto U_y with $h(x) = y$.)

The collection of continua with complicated but succintly described local structure includes the universal k-dimensional Menger compacta μ^k (generalizing the Menger curve) and μ^k-manifolds. An impressive topological characterization and a proof of homogeneity of μ^k was given by M. Bestvina in [5]. (For a thorough treatment of this subject see *Menger Manifolds* by A. Chigogidze, K. Kawamura and E. Tymchatyn [6], in this volume.) Other examples of homogeneous continua with complicated local structure are described by W. Jakobsche [7] and P. R. Stallings [16]. The results of Sections 3 and 4 can be applied to the types of spaces mentioned above.

As a tool in our investigations we use the Vietoris-Čech reduced homology groups with coefficients in the group of rationals modulo 1, denoted by $\check{H}$. The dimension used here is the covering dimension or, equivalent for the spaces considered, the small or large inductive dimension. The following characterization of dimension of finite-dimensional spaces, due to Alexandroff, can be found in most standard dimension theory texts, for instance in [14].

Theorem A. *For a finite-dimensional space X and an integer $n \geq 0$, the inequality* $\dim X \leq n$ *holds if and only if the homomorphism* $i_* : \check{H}_n(A) \to \check{H}_n(X)$, *induced by* $i : A \hookrightarrow X$, *is a monomorphism for every closed subset A of X.*

A closed subset A of X is a *support* for an element $a \in \check{H}_n(X)$ if $a \in i_*(\check{H}_n(A))$, where $i : A \hookrightarrow X$.

Corollary A. *Let* $\dim X = n$. *If* $a \in \check{H}_n(X)$ *has two disjoint supports, then $a = 0$.*

Proof. Suppose that $A_1 \cap A_2 = \emptyset$ and each of A_1 and A_2 is a support of a, say $a = i_{1*}(a_1) = i_{2*}(a_2)$, where $a_k \in \check{H}_n(A_k)$ and $i_k : A_k \hookrightarrow X$. Then $a_1 \oplus (-a_2) \in \check{H}_n(A_1 \cup A_2)$, since $\check{H}_n(A_1 \cup A_2) = \check{H}_n(A_1) \oplus \check{H}_n(A_2)$. But $i_*(a_1 \oplus (-a_2)) = a - a = 0$, where $i : A_1 \cup A_2 \hookrightarrow X$. Since $\dim X = n$, we get that $a_1 \oplus (-a_2) = 0$ (in $\check{H}_n(A_1 \cup A_2)$), from which the conclusion follows.

2. THE HOMOLOGY SEPARATION AXIOM AND 2-HOMOGENEITY

A space X is *n-acyclic* if $\check{H}_n(X) = 0$. A space X is *locally n-acyclic* at a point $x \in X$ if for every neighborhood U_x there is a neighborhood $V_x \subset U_x$ such that $\check{H}_n(V_x) = 0$. A space X is *semi-locally n-acyclic* at a point $x \in X$ if for every neighborhood U_x there is a neighborhood $V_x \subset U_x$ such that the homomorphism $i_* : \check{H}_n(V_x) \to \check{H}_n(X)$ induced by $i : V_x \hookrightarrow X$ is trivial. A space is *locally n-acyclic* [*semi-locally n-acyclic*] if it is locally n-acyclic [semi-locally n-acyclic] at each of its points.

If the points $x, y \in X$ have respective neighborhoods U_x and U_y such that

$$i_*(\check{H}_n(U_x)) \cap j_*(\check{H}_n(U_y)) = 0,$$

where $i : U_x \hookrightarrow X$ and $j : U_y \hookrightarrow X$, then x and y are *homologically separated in dimension n*, and U_x and U_y are *separating neighborhoods* (in dimension n). Observe that if U_x and U_y are separating neighborhoods, and if $V_x \subset U_x$ and $V_y \subset U_y$, then V_x and V_y are separating neighborhoods as well. In the spirit of the Hausdorff separation axiom (in the point-set topology), we introduce the following:

Homology Separation Axiom. A space X is *n-$T_{\check{H}}$* if every two distinct points x, y in X are homologically separated in dimension n.

The following Lemma 2.1 and 2.2 are immediate consequences of Theorem A and Corollary A.

Lemma 2.1. *In an n-dimensional space the notions of local and semi-local n-acyclicity at a point x are equivalent.*

Lemma 2.2. *If* $\dim X = n$, *then X is n-$T_{\check{H}}$.*

Proposition 2.3. *A locally n-acyclic space is n-$T_{\check{H}}$.*

Theorem 2.4. *If $X \times Y$ is n-$T_{\check{H}}$, then X is semi-locally n-acyclic.*

Proof. Let x be a point in X, and let y_1 and y_2 be distinct points in Y. By hypothesis, there exists a neighborhood W_k of (x, y_k) in $X \times Y$ ($k = 1, 2$) such that $(*)\ i_{1*}(\check{H}_n(W_1)) \cap i_{2*}(\check{H}_n(W_2)) = 0$, where $i_k : W_k \hookrightarrow X \times Y$. Let U neighborhood of x in X such that $U_k = U \times \{y_k\} \subset W_k$ for $k = 1, 2$. Let $i : U \hookrightarrow X$. We will show that $i_*(\check{H}_n(U)) = 0$. Choose an arbitrary $a \in \check{H}_n(U)$. Let $a_k = a \times \{y_k\} \in \check{H}_n(U_k)$, $j_k : U_k \hookrightarrow W_k$, and $p_1 = p_X|_{U_1}$ (see diagram).

$$
\begin{array}{ccccc}
U_1 & \underset{j_1}{\hookrightarrow} & W_1 & \underset{i_1}{\hookrightarrow} & X \times Y \\[2mm]
p_1 \downarrow & & & & \downarrow p_X \\[2mm]
U & & \underset{i}{\hookrightarrow} & & X
\end{array}
$$

Since Y is connected, we get $(i_1 j_1)_*(a_1) = (i_2 j_2)_*(a_2)$. Thus by $(*)$, $(i_1 j_1)_*(a_1) = 0$. Since the above diagram commutes, it follows that

$$i_*(a) = (i p_1)_*(a_1) = (p_X i_1 j_1)_*(a_1) = 0,$$

which completes the proof.

Define a relation between points of a space X as follows. Say that $x \asymp_n y$ ($x, y \in X$) if for every pair of neighborhoods U_x, U_y, there exist neighborhoods $V_x \subset U_x$, $V_y \subset U_y$ such that $\check{H}_n(V_x)$ and $\check{H}_n(V_y)$ are isomorphic. The relation is reflexive and symmetric, but perhaps not transitive. Observe that if there exists a homeomorphism $h : X \to X$ with $h(x) = y$, then $x \asymp_n y$. Therefore, we get:

Proposition 2.5. *If X is homogeneous, then $x \asymp_n y$ for every x and y in X.*

Another relation, similarly defined, is relevant to 2-homogeneity. Say that $x \bowtie_n y$ ($x, y \in X$) if for every pair of neighborhoods U_x, U_y, there exist neighborhoods $V_x \subset U_x$, $V_y \subset U_y$ such that $i_*(\check{H}_n(V_x)) = j_*(\check{H}_n(V_y))$, where $i : V_x \hookrightarrow X$ and $j : V_y \hookrightarrow X$. Again, this relation is reflexive and symmetric, but perhaps not transitive. Observe that if $h : X \to X$ is a homeomorphism and $x \bowtie_n y$, then $h(x) \bowtie_n h(y)$. Therefore, we get:

Proposition 2.6. *If X is 2-homogeneous, then either no two points are in relation $\bowtie_n$ to each other or $x \bowtie_n y$ for all x and y in X.*

Remark. Each of the two possibilities named in the above Proposition can occur, as the examples of the Menger curve (no two points are in relation $\bowtie_1$ to each other) and of any n-manifold ($x \bowtie_n y$ for all x and y) show.

Lemma 2.7. *If Y is locally k-acyclic for $k = 0, 1, \ldots, n$, then $(x, y_1) \bowtie_n (x, y_2)$ in $X \times Y$ for every $x \in X$ and $y_1, y_2 \in Y$.*

Proof. Let U_1 and U_2 be neighborhoods of (x, y_1) and (x, y_2), respectively. Then there exist a neighborhood U_x (in X), and k-acyclic ($k = 0, 1, \ldots, n$) neighborhoods V_{y_1} and V_{y_2} (in Y) such that $V_m = U_x \times V_{y_m} \subset U_m$ ($m = 1, 2$). Let $p_m = p_X|_{V_m}$: $V_m \to U_x$ and $i_m : V_m \hookrightarrow X \times Y$. Since V_{y_m} is k-acyclic ($k = 0, 1, \ldots, n$), the homomorphism $p_{m*} : \check{H}_n(V_m) \to \check{H}_n(U_x)$ is, by the Vietoris-Begle theorem [2], an isomorphism, thus the correspondence between $\check{H}_n(V_1)$ and $\check{H}_n(V_2)$ defined by $p_{1*}^{-1}(a) \leftrightarrow p_{2*}^{-1}(a)$ ($a \in \check{H}_n(U_x)$) is an isomorphism. On the other hand, since Y is connected, we get $i_{1*}p_{1*}^{-1}(a) = i_{2*}p_{2*}^{-1}(a)$ for every $a \in \check{H}_n(U_x)$, which completes the proof.

Lemma 2.8. *If X is n-$T_{\check{H}}$ and X is not semi-locally n-acyclic at x_1, then the relation $(x_1, y_1) \bowtie_n (x_2, y_2)$ does not hold in $X \times Y$ for any $x_2 \in X$, $x_2 \neq x_1$, and $y_1, y_2 \in Y$.*

Proof. Let U_{x_1} and U_{x_2} be disjoint, separating neighborhoods in X. For every $V_{x_1} \subset U_{x_1}$ there exists an $a \in \check{H}_n(X)$, $a \neq 0$, with support V_{x_1}. Note that $i_*(a) \neq 0$ in $\check{H}_n(X \times Y)$, where $i : X \hookrightarrow X \times Y$, $i(x) = (x, y_1)$. There is no $b \in \check{H}_n(X \times Y)$ with support $U_{x_2} \times Y$ such that $i_*(a) = b$ in $\check{H}_n(X \times Y)$; otherwise $a = p_{X*}(b)$ contradicting Corollary A.

Proposition 2.6, Lemma 2.7 and Lemma 2.8 yield:

Theorem 2.9. *If X is n-$T_{\check{H}}$, X is not semi-locally n-acyclic, and Y is locally k-acyclic for $k = 0, 1, \ldots, n$, then $X \times Y$ is not 2-homogeneous.*

As a corollary, by Lemmas 2.1 and 2.2, we get:

Theorem 2.10. *If $\dim X = n$, X is not locally n-acyclic, and Y is locally k-acyclic for $k = 0, 1, \ldots, n$, then $X \times Y$ is not 2-homogeneous.*

The following example shows that the assumption "$\dim X = n$" is indispensable in Theorem 2.10.

Example. Let X be the product of countably many circles S^1 and let $Y = S^1$. X is not locally n-acyclic (not even semi-locally n-acyclic) for every $n \geq 1$, and Y is locally contractible. However, by [18], $X \times Y$, which is the countable product of circles S^1, is n-homogeneous for every n.

Problem P$_1$. Suppose $\dim X < \infty$, Y is locally contractible, and X is not locally n-acyclic [semi-locally n-acyclic] for some n. Can $X \times Y$ be 2-homogeneous?

The following examples show that the assumptions of Theorem 2.10 are weaker than the assumptions of Theorem 2.9.

Examples. Here we describe examples of 2-dimensional spaces which are not semi-locally 1-acyclic at any point but are 1-$T_{\check{H}}$. Let $X_0 = [0, 1] \times [0, 1]$, and, for $n \geq 1$, X_n is obtained from X_{n-1} by attaching a circle at each point of a finite ε_n-dense subset of X_{n-1}, where $\varepsilon_n > 0$, $\varepsilon_n \to 0$. Let $f_n : X_n \to X_{n-1}$ be the identity on X_{n-1} while collapsing each attached circle to the corresponding point of attachment. Let $X = \varprojlim(X_n, f_n)$. Clearly, X is is not semi-locally n-acyclic at any point. It is also easy to see that X is 1-$T_{\check{H}}$. A very similar example is obtained by attaching a facet of the unit cube in $\mathbb{R}^3$ to the Menger universal curve M in its standard construction as a subset of the cube. Many of the homogeneous (in fact, representable) continua constructed by Jakobsche [7] and Stallings [16] can serve here as examples as well.

3. FACTORWISE RIGIDITY

Lemma 3.1. *Suppose that X is n-$T_{\check{H}}$, X is not semi-locally n-acyclic at x_1, and Y is locally k-acyclic for $k = 0, 1, \dots, n$. If $h : X \times Y \to X \times Y$ is a homeomorphism, then there is an $\widetilde{x}_1 \in X$ such that $h(x_1 \times \{Y\}) \subset \widetilde{x}_1 \times \{Y\}$.*

Proof. Suppose that there are points (x_1, y_1), (x_1, y_2) in the fiber $x_1 \times \{Y\}$ such that $h(x_1, y_1) = (\widetilde{x}_1, \widetilde{y}_1)$, $h(x_1, y_2) = (\widetilde{x}_2, \widetilde{y}_2)$, and $\widetilde{x}_1 \neq \widetilde{x}_2$. By Lemma 2.7, $(x_1, y_1) \bowtie_n (x_1, y_2)$, but by Lemma 2.8, $(\widetilde{x}_1, \widetilde{y}_1)$, and $(\widetilde{x}_2, \widetilde{y}_2)$ are not $\bowtie_n$ related; a contradiction.

As a corollary we get:

Theorem 3.2. *If X is n-$T_{\check{H}}$, X is not semi-locally n-acyclic at any point, and Y is locally k-acyclic for $k = 0, 1, \dots, n$, then $X \times Y$ is factorwise rigid with respect to Y.*

By Lemmas 2.1 and 2.2, the above theorem yields the following:

Theorem 3.3. *If $\dim X = n$, X is not locally n-acyclic at any point, and Y is locally k-acyclic for $k = 0, 1, \dots, n$, then $X \times Y$ is factorwise rigid with respect to Y.*

It is easy to see that if $X \times Y$ is factorwise rigid with respect to Y, then $X \times Y$ is not 2-homogeneous.

Corollary 3.4. *Under the assumptions of Theorem 2.2 or Theorem 2.3, $X \times Y$ is not 2-homogeneous.*

Following T. Yagasaki, the authors of [6] mention that the proof of [8] can be generalized to show that the Cartesian product $X = \Pi_{\lambda \in \Lambda} X_\lambda$ of copies a μ^n-manifold is factorwise rigid.

Problem P_2. Are Cartesian products of Menger manifolds (of possibly different dimensions) [or μ_k^n compacta], factorwise rigid?

Problem P_3. Is the Cartesian product $X = \Pi_{\lambda \in \Lambda} X_\lambda$, where each X_λ is one of the homogeneous continua constructed in [7] and [16], factorwise rigid?

4. REPRESENTABILITY AND 2-HOMOGENEITY

Theorem 4.1. *If X is n-$T_{\breve{H}}$ but not semi-locally n-acyclic, then $X \times Y$ is not representable for any Y.*

Proof. Suppose to the contrary that $X \times Y$ is representable. Let x_1 be a point at which X is not semi-locally n-acyclic, and $y_1 \in Y$. Since the product space is representable, it admits a auto-homeomorphism that "bends" the Y-fiber over x_1, *i.e.*, a homeomorphism $h : X \times Y \to X \times Y$ such that $h(x,y) = (x,y)$ in some neighborhood U of (x_1, y_1), and $h(x_1, \widetilde{y}_1) = (x_2, y_2) \notin \{x_1\} \times Y$ for some $\widetilde{y}_1$.

Let V_{x_1} and V_{x_2} be separating, disjoint neighborhoods in X, such that $V_{x_1} \times \{y_1\} \subset U$. There is an $a \in \breve{H}_n(V_{x_1} \times \{y_1\})$ such that $p_{X*}(a) \neq 0$ in $\breve{H}_n(X)$; hence $a \neq 0$ in $\breve{H}_n(X \times Y)$. Since Y is connected, there is a corresponding $\widetilde{a} \in \breve{H}_n(V_{x_1} \times \{\widetilde{y}_1\})$; $a = \widetilde{a}$ in $\breve{H}_n(X \times Y)$. We have $a = h_*(a)$. It follows that $(p_X h)_*(a) = (p_X h)_*(\widetilde{a})$ in $\breve{H}_n(X)$ (see Figure), which contradicts Corollary A. Consequently, $X \times Y$ is not representable.

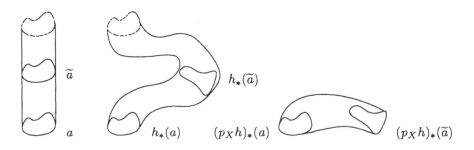

FIGURE. Bending a fiber.

Theorem 4.2. *If X is n-dimensional, representable, but not locally n-acyclic, then $X \times Y$ is not 2-homogeneous.*

Proof. Since X is n-dimensional, by Lemma 2.2 X is n-$T_{\breve{H}}$. Since X is representable, X is primitively stable, and so is $X \times Y$. Thus, if $X \times Y$ were 2-homogeneous, by Theorem KP $X \times Y$ would be representable, contradicting Theorem 4.1.

Since every μ^n-manifold (based on the n-dimensional universal Menger compactum μ^n) is representable but not locally n-acyclic, where n is the dimension of the manifold, Theorem 4.2 implies:

Corollary 4.3. *If B is a μ^n-manifold, then $B \times X$ is not 2-homogeneous for any X.*

Problem P_4. Can the assumption "$\dim X = n$" in Theorem 4.2 be replaced with "$\dim X < \infty$"?

Problem P₅. Do statements similar to the theorems and lemmas in this paper hold if one replaces Čech-Vietoris homology with singular homology, cohomology, or homotopy?

REFERENCES

1. R. D. Anderson, *A characterization of the universal curve and the proof of its homogeneity*, Ann. of Math. **67** (1958), 313–324.

2. E. G. Begle, *The Vietoris mapping theorem for bicompact spaces*, Ann. of Math. **51** (1950), 534-550.

3. D. Bellamy and J. A. Kennedy, *Factorwise rigidity of products of pseudo-arcs*, Topology Appl. **24** (1986), 197–205.

4. D. Bellamy and J. Łysko, *Factorwise rigidity of products of two pseudo-arcs*, Topology Proc. **8** (1983), 21–27.

5. M. Bestvina, *Characterizing k-dimensional universal Menger compacta*, Mem. Amer. Math. Soc. **71** (1988).

6. A. Chigogidze, K. Kawamura and E. Tymchatyn, *Menger manifolds*, Marcel Dekker, this volume.

7. W. Jakobsche, *Homogeneous cohomology manifolds which are inverse limits*, Fund. Math. **137** (1991), 81–95.

8. J. Kennedy Phelps, *Homeomorphisms of products of universal curves*, Houston J. Math. **6** (1980), 127–143.

9. J. Kennedy Phelps, *A condition under which 2-homogeneity and representability are the same in continua*, Fund. Math. **121** (1984), 89–98.

10. K. Kuperberg, *A locally connected micro-homogeneous nonhomogeneous continuum*, Bull. Acad. Pol. Sci. **28** (1980), 627–630.

11. K. Kuperberg, *Homogeneity and twisted products*, Topology Proc. **13** (1988), 237–248.

12. K. Kuperberg, *On the bihomogeneity problem of Knaster*, Trans. Amer. Math. Soc. **321** (1990), 129–143.

13. K. Kuperberg, W. Kuperberg, W. R. R. Transue, *On the 2-homogeneity of Cartesian products*, Fund. Math. **110** (1980), 131–134.

14. J. Nagata, *Modern dimension theory*, Bibl. Mat. **6** (1965).

15. H. Patkowska, *On the homogeneity of Cartesian products of Peano continua*, Bull. Acad. Pol. Sci. **32** (1984), 343–350.

16. P. R. Stallings, *On Jakobsche's construction of n-homogeneous continua*, Marcel Dekker, this volume.

17. G. S. Ungar, *On all kinds of homogeneous spaces*, Trans. Amer. Math. Soc. **212** (1975), 393–400.

18. G. S. Ungar, *Products and n-homogeneity*, preprint.

19. D. van Dantzig, *Ueber topologish homogene Kontinua*, Fund. Math. **15** (1930), 102–125.

E-mail address: kuperkm@mail.auburn.edu "Krystyna Kuperberg"
E-mail address: kuperwl@mail.auburn.edu "Wlodek Kuperberg"
E-mail address: transwr@mail.auburn.edu "Bill Transue"

22
SOLENOIDS AND BIHOMOGENEITY

PIOTR MINC Department of Mathematics, Auburn University, Auburn, Alabama 36849, USA

ABSTRACT. Let Σ_N denote a solenoid. We observe that for 'most' pairs a and b, each map $g : \Sigma_N \to \Sigma_N$, switching a and b, must be homotopic to the map $h_{a,b}$ given by $h_{a,b}(x) = ax^{-1}b$. This observation yields a relatively simple construction of a homogeneous but not bihomogeneous continuum.

1. INTRODUCTION

All spaces considered in this paper are metric. A space X is *homogeneous* [*bihomogeneous*] provided that for each two points a and b in X there is a homeomorphism $h : X \to X$ such that $h(a) = b$ [and $h(b) = a$]. A *continuum* is a compact and connected space.

Around 1921, B. Knaster asked whether every homogeneous space is bihomogeneous. In 1922, K. Kuratowski [6] constructed an example of (non-locally-compact) homogeneous but not bihomogeneous space. This result was improved by H. Cook (1986, a locally-compact example [1]) and by K. Kuperberg (1990, a locally connected continuum [5]). In 1930, D. van Dantzig [2] defined solenoids and noticed that the map $h_{a,b}$, defined by $h_{a,b}(x) = ax^{-1}b$, is a homeomorphism switching points a and b. It turns out that for 'most' pairs a and b of points of a solenoid Σ_N, each map $g : \Sigma_N \to \Sigma_N$, switching a and b, must be homotopic to $h_{a,b}$. We use this observation to construct another example of a homogeneous but not bihomogeneous continuum. Roughly speaking, we get a continuum Δ_N by replacing arc components in a solenoid by 'stacks' of some mapping cylinders and thus eliminating all

1991 *Mathematics Subject Classification.* Primary 54F15, Secondary 54H20.
Key words and phrases. solenoid, homogeneity, bihomogeneity.
This paper was partially supported by a travel grant from NSF EPSCOR in Alabama.

297

homeomorphisms that 'reverse the orientation'. To ensure homogeneity we take the Cartesian product of Δ_N and the Hilbert cube. The construction presented in this paper is simpler, but does not have as strong properties as the spectacular construction by K. Kuperberg. In particular, the example described in Section 3 is not locally connected.

2. Maps switching pairs of points in solenoids

Let S^1 denote the unit circle in the complex plane. If n is an integer, let f_n denote the mapping of S^1 into itself given by the formula $f_n(z) = z^n$. If N is a sequence of integers $n_1, n_2, \ldots$ each of which is greater than 1, we will denote by Σ_N the solenoid which is the inverse limit of the system $S^1 \xleftarrow{f_{n_1}} S^1 \xleftarrow{f_{n_2}} S^1 \xleftarrow{f_{n_3}} \ldots$. Denote by p_i the projection of Σ_N onto its i-th component. If $x = (x_0, x_1, x_2, \ldots)$ is a point of Σ_N then $p_i(x) = x_i$ and $f_{n_{i+1}}(x_{i+1}) = x_i$ for $i = 0, 1, \ldots$. For $j > i$, let $n[i,j]$ denote the product $n_{i+1} \times n_{i+2} \times \cdots \times n_j$. By $[i,i]$ we will denote 1. Observe that $f_{n[i,j]}(x_j) = x_i$. The solenoid Σ_N is a topological group with the multiplication induced by the multiplication on coordinates (see [2]). Observe that $e = (1, 1, \ldots,)$ is the identity element for the multiplication. Denote by E the arc component of e.

In this section we examine maps $g : \Sigma_N \to \Sigma_N$ with the property that $g(g(e)) \in E$. We show (Theorem 1) that if $g(e)$ does not belong to one of some countable set of arc components, then g is homotopic to the map assigning $x^{-1}g(e)$ to each $x \in \Sigma_N$. To prove Theorem 1, we need Propositions 1 through 5. The first four of them are widely known. They can be for example inferred from [3, Lemma 3.8], [4, Theorem 1.1] and [8, Corollary 2]. For the sake of completeness, we include here elementary proofs of Propositions 1 to 4 using only the classic technique of covering maps.

Let $g : \Sigma_N \to \Sigma_N$ be a map and let r be a rational number. We will say that g is the power r map provided that for every $i = 0, 1, 2, \ldots$ there is an index $j \geq i$ such that $m = r \times n[i,j]$ is an integer and $p_i(g(x)) = (p_j(x))^m$ for each $x \in \Sigma_N$. Note that the map assigning x^{-1} to each $x \in \Sigma_N$ is the power -1 map. It is easy to prove that if $r \neq 0$, then the power r map restricted to any arc component of Σ_N is a one to one map onto an arc component of Σ_N.

Proposition 1. *Let k and m be two integers, and let $g : \Sigma_N \to \Sigma_N$ be a map such that $g(e) = e$ and $(p_k(x))^m = p_0(g(x))$ for every $x \in \Sigma_N$. Then g is the power $m/n[0,k]$ map.*

Proof. Let i be a positive integer. There is a positive number ϵ such that for every two different points s and z of S^1, if the distance between s and z is less than ϵ then $s^{n[0,i]} \neq z^{n[0,i]}$. There is an integer j greater than both i and k such that the diameter of $p_i\left(g\left(p_j^{-1}(z)\right)\right)$ is less than ϵ for each $z \in S^1$. Since $(p_i(g(x)))^{n[0,i]} = p_0(g(x)) = (p_k(x))^m = z^{m \times n[k,j]}$ for every $x \in p_j^{-1}(z)$, it follows that $p_i \left(g \left(p_j^{-1} \left(z \right) \right) \right)$ is a single point. Denote this point by $\varphi \left(z \right)$. Observe that φ is a continuous function of S^1 into itself. Observe also that φ is a lifting of $f_{m \times n [k,j]}$ through $f_{n[0,i]}$. It follows from [9, Th5§4Ch.2] that there is an integer m_i such that $m \times n \, [k,j] = m_i \times n \, [0,i]$. Since $\varphi \left(1 \right) = 1$, it follows from [9,Th2§2Ch.2] that $\varphi = f_{m_i}$. So $p_i \left(g \left(x \right) \right) = \varphi \left(p_j \left(x \right) \right) = \left(p_j \left(x \right) \right)^{m_i}$ for every $x \in \Sigma_N$. Since $n \, [k,j] = n \, [0,j] / n \, [0,k]$ and $n \, [0,j] = n \, [0,i] \times n \, [i,j]$, we have that $m_i = \left(m/n \, [0,k] \right) \times n \, [i,j]$.

Proposition 2. *Two different power maps of Σ_N are not homotopic in Σ_N.*

Proof. Let q_0 be the map of the set of real numbers $\mathbf{R}$ defined by the formula $q_0 \left(t \right) = \exp \left(2\pi t \mathbf{i} \right)$. Let $q_j : \mathbf{R} \to S^1$ be defined by the formula $q_j \left(t \right) = q_0 \left(t/n \, [0,j] \right)$. Observe that $q_j = f_{n[j,k]} \circ q_k$ for $k > j$. Let $q : \mathbf{R} \to \Sigma_N$ be defined by the formula $q \left(t \right) = \left(q_0 \left(t \right), q_1 \left(t \right), q_2 \left(t \right), \dots \right)$. Note that q is one-to-one map onto E. Suppose that there is a homotopy $H : \Sigma_N \times [0,1] \to \Sigma_N$ such that $H \left(\cdot, 0 \right)$ and $H \left(\cdot, 1 \right)$ are two power maps of Σ_N, with different powers r and s, respectively. Clearly, $q^{-1} \left(H \left(q \left(t \right), 0 \right) \right) = r \times t$ and $q^{-1} \left(H \left(q \left(t \right), 1 \right) \right) = s \times t$ for each real number t. Observe that the homotopy H', defined by the formula $H' \left(x, z \right) = H \left(x, z \right) / H \left(e, z \right)$, keeps e constant. So, without loss of generality, we can assume that $H \left(e, z \right) = e$ for each $z \in [0,1]$. There is a positive number ϵ such that for each point $x \in \Sigma_N$, if the distance between x and e is less than ϵ, then $p_0 \left(H \left(x, z \right) \right) \neq -1$ for each $z \in [0,1]$. There is a positive integer k such that $|r \times n \, [0,k] - s \times n \, [0,k]| > 1$ and the distance between $q \left(n \, [0,k] \right)$ and e is less than ϵ. Denote $n \, [0,k]$ by t. Since $q^{-1} \left(H \left(q \left(t \right), 0 \right) \right) = r \times t$, $q^{-1} \left(H \left(q \left(t \right), 1 \right) \right) = s \times t$ and $|r \times t - s \times t| > 1$, there is a number $z \in [0,1]$ and there is an integer m such that $q^{-1} \left(H \left(q \left(t \right) \right), z \right) = m + \frac{1}{2}$. This contradicts the choice of k, because $p_0 \left(H \left(q \left(t \right), z \right) \right) = q_0 \left(q^{-1} \left(H \left(q \left(t \right), z \right) \right) \right) = -1$.

The following proposition is very closely related to [4, Theorem 1.1] and [8, Corollary 2].

Proposition 3. *Suppose $g : \Sigma_N \to \Sigma_N$ is a map such that $g \left(e \right) = e$. Then there is exactly one rational number r such that g is homotopic to the power r map.*

Proof. There is an index k and a map $\psi : S^1 \to S^1$ such that $\psi \left(1 \right) = 1$ and $|\psi \left(p_k \left(x \right) \right) - p_0 \left(g \left(x \right) \right)| < 2$ for every $x \in \Sigma_N$. There is a homotopy $H_1 : \Sigma_N \times [0,1] \to S^1$ such that $H_1 \left(x, 0 \right) = p_0 \left(g \left(x \right) \right)$ and $H_1 \left(x, 1 \right) = \psi \left(p_k \left(x \right) \right)$ for every $x \in \Sigma_N$ and $H_1 \left(e, t \right) = 1$ for every $t \in [0,1]$. Let m be an integer such that ψ is homotopic to f_m. There is a homotopy $H_2 : \Sigma_N \times [0,1] \to S^1$ such that $H_2 \left(x, 0 \right) = \psi \left(p_k \left(x \right) \right)$ and $H_2 \left(x, 1 \right) = \left(p_k \left(x \right) \right)^m$ for every $x \in \Sigma_N$, and $H_2 \left(e, t \right) = 1$ for every $t \in [0,1]$. Using H_1 and H_2 we construct a homotopy $H : \Sigma_N \times [0,1] \to S^1$ such that $H \left(x, 0 \right) = p_0 \left(g \left(x \right) \right)$ and $H \left(x, 1 \right) = \left(p_k \left(x \right) \right)^m$ for every $x \in \Sigma_N$ and $H \left(e, t \right) = 1$ for every $t \in [0,1]$. Since f_m is a covering map the homotopy H can be lifted (see [9, Th3§2Ch.2]) step by step to each S^1 of the inverse system defining Σ_N. The sequence of such defined homotopies induces a homotopy $\tilde{H} : \Sigma_N \times [0,1] \to \Sigma_N$ such that $\tilde{H}(x,0) = g(x)$ and $p_0((x,1)) = (p_k(x))^m$ for every $x \in \Sigma_N$ and $\tilde{H}(e,t) = e$ for every $t \in [0,1]$. By Proposition 1, $\tilde{H}(\cdot,1)$ is the power $m/n\,[0,k]$ map. The uniqueness follows from Proposition 2.

Proposition 4. *For each map $g : \Sigma_N \to \Sigma_N$ there is exactly one rational number $\varrho(g)$ such that g is homotopic to the power $\varrho(g)$ map multiplied by $g(e)$ (the multiplication is in Σ_N).*

Proof. By the Proposition 3, g multiplied by $(g(e))^{-1}$ is homotopic to a power map. To prove the proposition multiply the homotopy by $g(e)$

If $g : \Sigma_N \to \Sigma_N$ is a map, the number $\varrho(g)$ satisfying Proposition 4 will be called the order of g.

Proposition 5. *Let $r \neq -1$ be a rational number. There is a finite collection $\mathcal{K}(r)$ of arc components of Σ_N with the property that if $g : \Sigma_N \to \Sigma_N$ is a map of order r such that $g(g(e)) \in E$, then $g(e)$ belongs to one of the elements of $\mathcal{K}(r)$.*

Proof. Let u and v be two positive and relatively prime integers such that $|r+1| = u/v$. Observe that the equation $x^u = e$ has at most u solutions in Σ_N. Let $\mathcal{K}(r)$ be the collection of arc components of these solutions.

Let h denote the power r map and let $g : \Sigma_N \to \Sigma_N$ be a map such that $g(g(e)) \in E$ and g is homotopic to $g(e)h$. Denote $g(e)$ by a and let A be the arc component of a. Observe that $ah(a) \in E$. Consider the map $f : \Sigma_N \to \Sigma_N$ defined by the formula $f(x) = xh(x)$ for each $x \in \Sigma_N$. Observe that f is the power $r+1$ map. Since A is mapped by f onto E, there is a point $b \in A$ such that $f(b) = e$. For every index i, there is an integer $j \geq i$ such that $(p_j(b))^{(r+1)\times n[i,j]} = p_i(f(b)) = 1$. Thus $(p_i(b))^u = (p_j(b))^{u\times n[i,j]} = \left((p_j(b))^{(r+1)\times n[i,j]}\right)^v = 1$ for every $i = 0, 1, \dots$. It follows that $b^u = e$ and therefore $A \in \mathcal{K}(r)$.

The following theorem is a simple consequence of Proposition 5.

Theorem 1. *There is a countable collection $\mathcal{K}$ of arc components of Σ_N with the property that if $g : \Sigma_N \to \Sigma_N$ is a map of order different than -1 such that $g(g(e)) \in E$, then $g(e)$ belongs to one of the elements of $\mathcal{K}$.*

3. HOMOGENEOUS BUT NOT BIHOMOGENEOUS CONTINUA MODELED ON SOLENOIDS

Let $m \geq 2$ be an integer. Denote by C the mapping cylinder of f_m. The cylinder C is $S^1 \times [0,1]$ with points $(z,1)$ and $(s,1)$ identified if and only if $z^m = s^m$. Let τ be the projection of $S^1 \times [0,1]$ onto C. Let $\gamma : \tau(S^1 \times \{1\}) \to \tau(S^1 \times \{0\})$ be the map such that $\gamma(\tau(z,1)) = \tau(z^m,0)$. Observe that γ is a homeomorphism.

Let Z denote the set of integers. Let K denote the quotient space of $C \times Z$ with each point of the form $(\tau(z,1),i)$ identified with the point $(\gamma(\tau(z,1)),i+1)$, where $z \in S^1$ and $i \in Z$. Let κ denote the projection of $S^1 \times [0,1] \times Z$ onto K.

Let C_i denote the set $\kappa \left(S^1 \times [0,1] \times \{i\} \right)$ and let T_i be the simple closed curve $\kappa \left(S^1 \times \{1\} \times \{i\} \right)$.

In this section we show how to use Theorem 1 to construct another example of a homogeneous but not bihomogeneous continuum. First we observe that the 'direction' of K cannot be reversed by non-trivial maps (Proposition 6). Then we construct a space Δ_N by replacing each arc component of Σ_N by an one-to-one image of K. The product $\Delta_N \times Q$, where Q is the Hilbert cube, is homogeneous. The proof that $\Delta_N \times Q$ is not bihomogeneous will follow from Theorem 1 and Proposition 6.

Proposition 6. *Suppose $g : K \to K$ is a map such that its restriction to T_0 essential in K. Then for each integer i there is an integer l such that for each integer $j \geq l$, the set $g(T_j)$ is not contained in $\bigcup_{k=-\infty}^{i} C_k$.*

Proof. For each integer n, there is a retraction r_n of $\bigcup_{k=-\infty}^{n} C_k$ to T_n. Observe that there is a homotopy $H_n : \bigcup_{k=-\infty}^{n} C_k \times [0,1] \to \bigcup_{k=-\infty}^{n} C_k$ such that $H_n(x,0) = x$ and $H_n(x,1) = r_n(x)$ for each $x \in \bigcup_{k=-\infty}^{n} C_k$. Observe also that if $k < n$, then r_n restricted to T_k is a map of degree m^{n-k}. Let i be an arbitrary integer. Without loss of generality we may assume that $g(T_0) \subset \bigcup_{k=-\infty}^{i} C_k$. Let d denote the degree of $r_i \circ g$ restricted to T_0. Clearly, $d \neq 0$. Let l be an integer such that $m^l > |d|$.

Suppose that there is an integer $j \geq l$ such that $g(T_j) \subset \bigcup_{k=-\infty}^{i} C_k$. Let d_j denote the degree of $r_i \circ g$ restricted to T_j. Let $n \geq i$ be an integer such that $g(H_j(T_0 \times [0,1])) \subset \bigcup_{k=-\infty}^{n} C_k$. Note that the degree of $r_n \circ g$ restricted to T_0 is dm^{n-i}. On the other hand, this degree is equal to $d_j m^j m^{n-i}$. It follows that $d = d_j m^j$. Since d_j is an integer and $d \neq 0$, $|d| \geq m^j$ and consequently $|d| \geq m^l$, a contradiction with the choice of l.

For each positive integer n, let Z_n denote the set $\{0, 1, \ldots, n-1\}$. Let K_n denote the set $C \times Z_n$ in which each point of the form $(\tau(z,1), i)$ is identified with the point $(\gamma(\tau(z,1)), j)$, where $z \in S^1$, $i, j \in Z_n$ and $i + 1 = j \mod n$. Let κ_n denote the projection of $S^1 \times [0,1] \times Z_n$ onto K_n.

Let N be the sequence from the previous section. For each positive integer i, let $\delta_i : K_{n[0,i]} \to K_{n[0,i-1]}$ be the map defined by $\delta_i \left(\kappa_{n[0,i]}(z,t,j) \right) = \kappa_{n[0,i-1]}(z,t,k)$, where $z \in S^1$, $t \in [0,1]$, $j \in Z_{n[0,i]}$, $k \in Z_{n[0,i-1]}$ and $j = k \mod n[0,i-1]$. Let Δ_N denote the inverse limit of the system $K_{n[0,0]} \xleftarrow{\delta_1} K_{n[0,1]} \xleftarrow{\delta_2} K_{n[0,2]} \xleftarrow{\delta_3} \ldots$.

Let S_i denote $\bigcup_{k \in Z_{n[0,i]}} \kappa_{n[0,i]}(\{1\} \times [0,1] \times \{k\})$. Observe that S_i is a simple closed curve and $\delta_i(S_i) = S_{i-1}$ for each positive integer i. Let $\sigma_i : S_i \to S^1$ be defined by $\sigma_i \left(\kappa_{n[0,i]}((1,t,k)) \right) = \exp(2\pi i (t+k)/n[0,i])$. Observe that σ_i is a homeomorphism such that $\sigma_{i-1} \circ \delta_i = f_{n_i} \circ \sigma_i$. So, the sequence $\sigma_1^{-1}, \sigma_2^{-1}, \ldots$ induces an embedding of Σ_N into Δ_N. We will identify Σ_N with its image under this embedding.

Let $q_i : K_{n[0,i]} \to S_i$ be defined by $q_i \left(\kappa_{n[0,i]}((z,t,k)) \right) = \kappa_{n[0,i]}((1,t,k))$ for each $z \in S^1$, $t \in [0,1]$ and $k \in Z_{n[0,i]}$. Observe that $\delta_i \circ q_i = q_{i-1} \circ \delta_i$. Let $q : \Delta_N \to \Sigma_N$

be the map induced by the sequence $q_0, q_1, \ldots$. Observe that $q^{-1}(x)$ is a simple closed curve for each $x \in \Sigma_N$.

Proposition 7. *Suppose $h : \Delta_N \to \Delta_N$ is a continuous map such that $h(h(e)) = e$ and $q \circ h(e)$ does not belong to any of the components in $\mathcal{K}$ (see Theorem 1). Then h restricted to $q^{-1}(e)$ is not essential.*

Proof. For each positive integer n, let α_n denote the map of K onto K_n given by $\alpha_n(\kappa(z,t,i)) = \kappa_n(z,t,j)$ where $z \in S^1$, $t \in [0,1]$, $i \in Z$, $j \in Z_n$ and $j = i \mod n$. For each integer m, let $\tilde{\delta}_m : K \to K$ denote the map defined by $\tilde{\delta}_m(\kappa(z,t,i)) = \kappa(z,t,i+m)$ where $z \in S^1$, $t \in [0,1]$, $i \in Z$. Observe that if, for each positive integer j, m_j is an integer multiple of $n[0,j-1]$, then the diagram

$$(*) \qquad \begin{array}{ccccccc} K & \xleftarrow{\;\tilde{\delta}_{m_1}\;} & K & \xleftarrow{\;\tilde{\delta}_{m_2}\;} & K & \xleftarrow{\;\tilde{\delta}_{m_3}\;} & \cdots \\ \Big\downarrow{\scriptstyle \alpha_{n[0,0]}} & & \Big\downarrow{\scriptstyle \alpha_{n[0,1]}} & & \Big\downarrow{\scriptstyle \alpha_{n[0,2]}} & & \\ K_{n[0,0]} & \xleftarrow{\;\delta_1\;} & K_{n[0,1]} & \xleftarrow{\;\delta_2\;} & K_{n[0,2]} & \xleftarrow{\;\delta_3\;} & \cdots \end{array}$$

is commutative. Since the inverse limit of the system $K \xleftarrow{\;\tilde{\delta}_{m_1}\;} K \xleftarrow{\;\tilde{\delta}_{m_2}\;} K \xleftarrow{\;\tilde{\delta}_{m_3}\;} \cdots$ is homeomorphic to K, the diagram $(*)$ induces a one-to-one map of K onto an arc component of Δ_N. Let $v_e : K \to \Delta_N$ denote the map induced by $(*)$ in the case where $m_j = 0$ for each positive j. Observe that E (the arc component of e) is the image of K under v_e. There is a sequence $m_1, m_2, \ldots$ of integer multiples of $n[0,0], n[0,1], \ldots$, respectively, such that the map induced by the diagram $(*)$ maps K onto the arc component of $h(e)$. We will denote this induced map by v. Let $g : K \to K$ denote the composition $v^{-1} \circ h \circ v_e$. Observe that g is continuous.

Let $\tilde{g}$ denote the restriction of $q \circ h$ to Σ_N. By Theorem 1, $\tilde{g}$ is a map of order -1. Let $H : \Sigma_N \times [0,1] \to \Sigma_N$ be a homotopy such that $H(\cdot,0)$ is $\tilde{g}$ and $H(\cdot,1)$ is the power -1 map multiplied by $\tilde{g}(e)$.

Claim 1. *There is an integer m' such that $p_{m'} \circ H(\{x\} \times [0,1])$ is a proper subset of S^1 for each $x \in \Sigma_N$.*

Proof of Claim 1. Suppose that for each integer $i \geq 0$, there is a point $x_i \in \Sigma_N$ such that $p_i \circ H(\{x_i\} \times [0,1]) = S^1$. Without loss of generality we may assume that the sequence $x_0, x_1, \ldots$ converges to a point $x \in \Sigma_N$. It follows that $H(\{x\} \times [0,1]) = \Sigma_N$, which is not possible because Σ_N is not arc-wise connected.

Claim 2. *There is an integer m'' such that $\tilde{g}(x) \in v\left(\bigcup_{k=-j-m''}^{-j+m''} C_k\right)$ for each integer j and each $x \in \Sigma_N \cap v_e(C_j)$.*

Proof of Claim 2. Let ν be such that $h(e) \in v(C_\nu)$. (If $h(e) \in v(C_n \cap C_{n-1})$, let $\nu = n$.) Let $m'' = 1 + |\nu| + n[0,m']$, where m' satisfies Claim 1. Let j be an arbitrary integer and let x be an arbitrary point of $\Sigma_N \cap v_e(C_j)$. Observe that $H(x,1) \in v(C_{-j-1+\nu})$. Since $\tilde{g}(x) = H(x,0)$, the claim follows from Claim 1.

Claim 3. *Then there is an integer μ such that $p_\mu \circ q \circ h \left(q^{-1} \left(x \right) \right)$ is a proper subset of S^1 for each $x \in \Sigma_N$.*

Proof Claim 3. Suppose that for each integer $i \geq 0$, there is a point $x_i \in \Sigma_N$ such that $p_i \circ q \circ h \left(q^{-1} \left(x_i \right) \right) = S^1$. Without loss of generality we may assume that the sequence $x_0, x_1, \ldots$ converges to a point $x \in \Sigma_N$. It follows that $q \circ h \left(q^{-1} \left(x \right) \right) = \Sigma_N$, which is not possible because $q^{-1} \left(x \right)$ is a simple closed curve and Σ_N is not arc-wise connected.

Let $m = m'' + n\left[0, \mu\right]$, where m'' and μ satisfy Claims 2 and 3, respectively. Observe that $g\left(T_j\right) \subset \bigcup_{k=-j-m}^{-j+m} C_k$ for each integer j. In particular,

$$(**) \qquad g\left(T_j\right) \subset \bigcup_{k=-\infty}^{0} C_k \qquad \text{for each integer } j \geq m.$$

Suppose the proposition is false. Then h restricted to $q^{-1}\left(e\right)$ is essential. Since $q^{-1}\left(e\right) = v_e\left(T_0\right)$, the restriction of g to T_0 is essential in K. Now $(**)$ contradicts Proposition 6. Thus Proposition 7 is true.

Let Q denote the Hilbert cube.

Theorem 2. $\Delta_N \times Q$ *is a homogeneous but not bihomogeneous continuum.*

Proof. It follows from Proposition 7 that $\Delta_N \times Q$ is not bihomogeneous. We will show that it is homogeneous. Let $a = \left(\tilde{a}, z_a\right)$ and $b = \left(\tilde{b}, z_b\right)$ be two arbitrary points of $\Delta_N \times Q$. We have to show that there is a homeomorphism $h : \Delta_N \times Q \to \Delta_N \times Q$ such that $h\left(a\right) = b$. Since there is a homeomorphism of Δ_N onto itself taking the arc component of $\tilde{a}$ in Δ_N onto the arc component of $\tilde{b}$, we may assume that $\tilde{a}$ and $\tilde{b}$ are in one arc component of Δ_N. Let $\tilde{L} \subset \Delta_N$ be an arc joining $\tilde{a}$ and $\tilde{b}$. Recall that q is the projection of Δ_N onto Σ_N and $p_i : \Sigma_N \to S^1$ is the projection onto the i-th component of the inverse system defining Σ_N. There is an integer i such that $p_i \circ q \left(\tilde{L}\right)$ is a proper subset of S^1. Let $M \subset S^1$ be a closed arc containing $p_i \circ q \left(\tilde{L}\right)$ in its interior. Let G denote $\left(p_i \circ q\right)^{-1}\left(M\right)$ and let P be the component of G containing $\tilde{L}$. Observe that P is a polyhedron and G is homeomorphic to the product of P and the Cantor set. Let r_1 and r_2 denote the projections of G onto P and the Cantor set, respectively. Let P_0 denote the intersection of P and the interior of G.

Let L be an arc in $\tilde{L} \times Q$ joining a and b. By [7, Th. 7.8.1], $P \times Q$ is a Q-manifold. Let w_1 and w_2 denote the projections of $P \times Q$ onto P and Q, respectively. There is a open covering $\mathcal{U}\left\{U_1, U_2, \ldots, U_k\right\}$ of L in $P_0 \times Q$ such that U_i is homeomorphic to a connected open subset of the Hilbert cube for each $i = 1, \ldots, k$. Without loss of generality we may assume that $a \in U_1$, $b \in U_k$ and, for each $i = 1, \ldots, k-1$, there is a point $t_i \in U_i \cap U_{i+1}$. Set $t_0 = a$ and $t_k = b$. Using [7, Exercise 3, p. 261] one can get a set $F_i \subset U_i$ and a homeomorphism $\tilde{h}_i$ of U_i onto itself such that

$\tilde{h}_i\left(t_{i-1}\right) = t_i$, F_i is closed in $\Delta_N \times Q$ and $\tilde{h}_i\left(x\right) = x$ for each $x \in U_i \backslash F_i$. For an arbitrary point $y = \left(s, z\right) \in \Delta_n \times Q$, we will define $h_i\left(y\right)$ in the following way. If $s \notin G$, then set $h_i\left(y\right) = y$. Now, suppose $s \in G$. Let u denote the point $\left(r_1\left(s\right), z\right)$. If $u \notin U_i$, then, like before, set $h_i\left(y\right) = y$. Finally, suppose $u \in U_i$. In this case let $\tilde{s} = \left(w_1\left(\tilde{h}_i\left(u\right)\right), r_2\left(s\right)\right)$ and $\tilde{z} = w_2\left(\tilde{h}_i\left(u\right)\right)$. Set $h_i\left(y\right) = \left(\tilde{s}, \tilde{z}\right)$. Observe that h_i is a homeomorphism of $\Delta_N \times Q$ onto itself such that $h_i\left(t_{i-1}\right) = t_i$. Since $t_0 = a$ and $t_k = b$, the composition $h = h_k \circ h_{k-1} \circ \ldots h_1$ is a homeomorphism of $\Delta_N \times Q$ onto itself such that $h\left(a\right) = b$.

REFERENCES

1. H. Cook, *A locally compact, homogeneous metric space which is not bihomogeneous*, Topology Proc. **11** (1986), 25–27.

2. D. van Dantzig, *Ueber topologisch homogene Kontinua*, Fund. Math. **15** (1930), 102–125.

3. J. Keesling, *The group of homeomorphisms of a solenoid*, Trans. A. M. S. **172** (1972), 119–131.

4. J. Keesling, *Shape theory and compact connected Abelian topological groups*, Trans. A. M. S. **194** (1974), 349–358.

5. K. Kuperberg, *On the bihomogeneity problem of Knaster*, Trans. A. M. S. **321** (1990), 129–143.

6. C. Kuratowski, *Un probleme sur les ensambles homogènes*, Fund. Math. **3** (1923), 14–19.

7. J. van Mill, *Infinite-Dimensional Topology, Prerequisites and Introduction*, North-Holland, 1989.

8. W. Scheffer, *Maps between topological groups that are homotopic to homomorphisms*, Proc. Amer. Math. Soc. **33** (1972), 562–567.

9. E. Spanier, *Algebraic Topology*, Springer-Verlag, 1966.

E-mail address: mincpio@mail.auburn.edu "Piotr Minc"

23
OPENLY HOMOGENEOUS CONTINUA IN 2-MANIFOLDS: A GENERALIZATION OF A THEOREM OF BING

JANUSZ R. PRAJS Institute of Mathematics, Opole University, ul. Oleska 48, 45-951 Opole, Poland

ABSTRACT. Among other things it is shown that a plane continuum which contains an arc, and which is homogeneous with respect to light open mappings, is a simple closed curve. This generalizes a classical result of Bing for the usual homogeneity.

This paper gives a complete classification of continua in 2-manifolds containing an arc, which are homogeneous with respect to light open mappings. It is proved that they are exactly those continua, which are homogeneous (with respect to homeomorphisms), i.e. a simple closed curve and a 2-manifold without boundary. For the usual homogeneity this theorem was proved in 1924 by Mazurkiewicz [5] in the locally connected case, and in the general case by Bing [1] in 1960. In the papers [6] and [7] the author has generalized these theorems to higher dimensional cases. However, he has no idea to do such generalization for the homogeneity with respect light open mappings.

It seems to be quite possible to answer the question below in the affirmative. This would be another natural generalization of Bing's theorem [1] and of the main result of this paper.

Question. Let X be a plane continuum containing an arc. If X is homogeneous with respect to light confluent mappings, must it be a simple closed curve ?

In fact, since light confluent and light open mappings coincide on locally connected continua, the locally connected case is already proved (Theorem 5.8 below).

1991 *Mathematics Subject Classification.* 54F15.
Key words and phrases. atriodic continuum, confluent mapping, generalized homogeneity, light mapping, local separating point, open mapping, S-curve, solenoid.

Nevertheless, some essential tools are missing for the general case.

As the author has shown in [9], Bing's theorem cannot be extended to homogeneity with respect to all open mappings. Indeed, the 2-cell, as well as some other planar examples happen to be openly homogeneous. Recently, the author has also shown that the Sierpiński universal plane curve is homogeneous with respect to monotone open mappings.

Except the main theorem, the paper contains a number of results concerning continua homogeneous with respect to: open, light open and light confluent mappings. Though they are used as steps to prove the main result, some of them might be interesting by their own right.

The main tools employed in the paper are a number of results of Whyburn: concerning light open mappings [11], concerning local separating points [12], and concerning the characterization of the Sierpiński universal plane curve [13]. We also essentially use an earlier result of the author, which characterizes solenoids with the help of open homogeneity [8]. A weak version of the Effros theorem proved for open homogeneity by Charatonik and Maćkowiak [2] is also applied.

1. Definitions and main tools

Throughout the paper spaces are assumed to be metric and mappings are assumed to be continuous. A mapping $f : X \to Y$ is said to be:

- *open* if for any open set U in X the set $f(U)$ is open in Y,

- *confluent* if for any continuum K in Y and any component C of $f^{-1}(K)$ we have $f(C) = K$, or

- *light* if for any point $p \in Y$ the set $f^{-1}(p)$ is totally disconnected.

If $f, g : X \to Y$ are mappings and d is a metric on a compact space Y, then we put $\tilde{d}(f,g) = \sup\{d(f(x),g(x)) : x \in X\}$, and, if moreover $X \subset Y$, $d(f) = \sup\{d(f(x),x) : x \in X\}$.

The following fact is well-known.

1.1.Fact. Each open surjection of a compact space is confluent.

The following lemma is a particular case of [[4], Theorem 5.1].

1.2.Lemma. If $f : X \to Y$ is a light confluent surjection of a compact space X onto a locally connected space Y, then f is open.

The next lemma is a consequence of [[11], Theorem (2.4), p. 188] and of Lemma 1.2. It was essentially used (in a weaker version) in the proof of [[8], (3), p. 136].

1.3.Lemma. Let $f : X \to Y$ be a light confluent surjection between compact spaces X and Y, and let p be a point of a dendrite D in Y. Then for each point $q \in f^{-1}(p)$ there is a dendrite $D' \subset f^{-1}(D)$ containing q such that the map $f|D' : D' \to D$ is a homeomorphism.

Proof. Putting $L = f^{-1}(D)$ observe that the map $f|L : L \to D$ is open by Lemma 1.2. Applying [[11], Theorem (2.4), p. 188] the lemma follows.

A space X is said to be *homogeneous with respect to a class M of mappings* provided for all $x, y \in X$ there is a surjection $f : X \to X$ belonging to M such that $f(x) = y$. If M is the class of all homeomorphisms, this notion coincides with the usual notion of homogeneity. We say that a space is *openly homogeneous* instead of saying that it is homogeneous with respect to the class of all open mappings. In short we write X is *h.l.o.m.* to express that X is homogeneous with respect to the class of all light open mappings.

Though the well known Effros theorem is not valid for generalized homogeneity, we have a weaker version of this theorem for open homogeneity due to Charatonik and Maćkowiak [[2], Theorem 5.9] (see also [[8], Remark, p. 138]). To formulate this theorem we need the following definition. Given a set A in a space X, the union of all open sets U such that $U - A$ is of the first category we call the *quasi-interior* of A and denote it by A^*.

1.4.Theorem(Charatonik, Maćkowiak). Let X be a compact space homogeneous with respect to the class M of all open (light open) autosurjections of X. For each point $a \in X$ define a map $T_a : M \to X$ by letting $T_a(f) = f(a)$. Then for all points $a, b \in X$ there is a map f in M such that $b = f(a) \in (T_a(H))^*$ for each neighborhood H of f in M.

A continuum is said to be *indecomposable*, if it is not the union of two of its proper subcontinua.

A continuum T is called a *triod* provided it is the union of three subcontinua A, B, C such that the intersection $S = A \cap B \cap C$ is a continuum, $A \cap B = B \cap C = C \cap A = S$, and the sets $T - (A \cup B)$, $T - (B \cup C)$, $T - (C \cup A)$ are nonempty. If A, B, C are arcs with S consisting of a common end point of them, T is called a *simple triod*. If a continuum contains a triod, it is said to be *triodic*. Otherwise it is *atriodic*. The following result of Sorgenfrey [[10], Theorem 1.8, p.443] concerning triodic continua will be used.

1.5.Theorem(Sorgenfrey). If a continuum X is the union of subcontinua A, B, C with nonempty intersection, such that $A \cup B \neq X \neq B \cup C$ and $C \cup A \neq X$, then X is triodic.

A point p of a continuum X is called a *local separating point* of X provided there is a compact neighborhood N of p such that if C is a component of N containing p, then $N - \{p\}$ is separated between some two points of $C - \{p\}$. Observe that if X is locally connected, then p is a local separating point of if and only if the set $N - \{p\}$ is not connected for each sufficiently small neighborhood N of p.

A point p of a continuum X has *order* n provided n is the least integer such that p has a basis of neighborhoods with boundaries consisting of n points. We will apply the following theorem of Whyburn [[12], Theorem 9, p.309].

1.6. Theorem (Whyburn). All save possibly a countable number of the local separating points of any continuum are points of order 2.

The term 2-manifold means here a connected manifold of dimension 2 (not necessarily compact, with or without boundary).

Any homeomorphic copy of the Sierpiński universal plane curve is called an S-curve. A continuum which has a basis of compact neighborhoods composed of S-curves will be called an S-manifold.

Whyburn characterized S-curves [[13], Theorem 4, p.323] as planar, locally connected curves containing no local separating point. Using this characterization, one can easily verify the following theorem.

1.7. Theorem. A continuum X is an S-manifold if and only if X is a locally connected curve with no local separating point and X has a basis of planar neighborhoods.

2. CONTINUA CONTAINING ARCS AND CONTAINING NO SIMPLE TRIOD

Denote by $\mathcal{S}$ the class of all continua X such that there is a $\tau > 0$ such that each nondegenerate subcontinuum of X with diameter less than τ is an arc. First, we observe that a number of results from [8] proved for continua in $\mathcal{S}$ remain valid for larger classes of continua. This chapter should be read together with [8]. Since the arguments for the results 2.1-2.5 below are the same as those for corresponding results in [8], we present them without proofs. Observe that the class of all continua as in the title of the chapter is essentially larger than $\mathcal{S}$, even for homogeneous continua. In fact, the Cartesian product of the pseudo-arc and any solenoid is homogeneous, contains an arc, contains no simple triod, and does not belong to $\mathcal{S}$.

We begin with the following lemma, which is a generalization of Lemma 3 from [8], where $X, Y \in \mathcal{S}$.

2.1. Lemma. Let continua X and Y contain no simple triod, and let $f : X \to Y$ be a light confluent surjection. If an arc component B of Y is a one-to-one image of the real line, then for any arc component A of $f^{-1}(B)$ the following statements hold:

(1) A is an arc component of X.
(2) $f(A) = B$.
(3) A is a one-to-one image of the real line,
(4) The map $f|A : A \to B$ is one-to-one.

In Propositions 2.2-2.8 the letter K denotes an arbitrary continuum with a metric d, containing an arc and containing no simple triod. Similarly as in [[8], Proposition 5, Lemma 6, p.137] we have the following two propositions by Lemmas 1.3 and 2.1.

2.2.Proposition. If K is homogeneous with respect to light confluent mappings, then either each of its arc components is a simple closed curve, or each of its arc components is a one-to-one image of the real line.

2.3.Proposition. If A and B are simple closed curves in K, then for each light confluent surjection $f : X \to X$ such that $f(A) \subset B$, the map $f|A : A \to B$ is an open surjection onto B.

Given a continuum X with a metric r and a point $x \in X$, denote by A_x the arc component of X containing x. Further, define the set $F(x)$ of all surjections f from the real line $\mathbb{R}$ onto $A - x$ such that $f(0) = x$ and either f is a universal covering mapping (if A_x is a simple closed curve), or a one-to-one mapping. Define a function $\hat{\rho} : X^2 \to \mathbb{R}$ as follows:
$$\rho(\hat{x}, y) = \begin{cases} 0, & \text{if } x = y \\ \inf\{\tilde{\rho}(f, g) : f \in F(x), g \in F(y)\}, & \text{if } F(x) \neq \emptyset \neq F(y) \\ \text{diam } (X, \rho), & \text{if } x \neq y \text{ and either } F(x) = \emptyset, \text{ or } F(y) = \emptyset. \end{cases}$$
It is observed in [[8], p.137] that $\hat{\rho}$ is a metric on X not necessarily equivalent to ρ. Using the same arguments as those for Lemmas 7 and 9 from [[8], pp.138, 139], we obtain two further propositions. In Propositions 2.4-2.8 $\hat{d}$ denotes the metric defined similarly the above metric $\hat{\rho}$ for the continuum (K, d).

2.4.Proposition. If K is h.l.o.m., then the metrics d and $\hat{d}$ are equivalent.

2.5.Proposition. If the metrics d and $\hat{d}$ are equivalent , then the family of all closures of arc components of K yields a continuous decomposition of K.

The next proposition is a consequence of the definition of $\hat{d}$.

2.6.Proposition. If the metrics d and $\hat{d}$ are equivalent, then for each $p \in K$, for each arc $ab \subset K$ with $p \in ab$, and for each sequence $\{p_n\} \subset K$ converging to p, there are homeomorphisms $h_n : ab \to h_n(ab) \subset K$ such that $h_n(p) = p$ and $\lim d(h_n) = 0$.

Proof. Since d and $\hat{d}$ are equivalent, there are mappings $f_n \in F(p_n)$ and $g_n \in F(p)$ such that $\lim \tilde{d}(f_n, g_n) = 0$ and there are intervals $I_n = [a_n, b_n] \subset \mathbb{R}$ containing 0 such that $g_n|I_n : I_n \to ab$ are homeomorphisms. Then for sufficiently large n the maps $f_n|I_n : I_n \to f_n(I_n)$ are also homeomorphisms. The required maps are $(f_n|I_n)(g_n|I_n)^{-1}$.

Using Proposition 2.6 the following result is easily obtainable by compactness of K. We omit the proof.

2.7.Proposition. If the metrics d and $\hat{d}$ are equivalent, then there is a $\xi > 0$ such that for any $p \in K$ there are arcs $ap, pb \subset K$ with $ap \cap pb = \{p\}$, and $\min\{\text{diam } ap, \text{diam } pb\} > \xi$.

2.8.Proposition. If the metrics d and $\hat{d}$ are equivalent, and K does not contain arbitrarily small triods, then K belongs to $\mathcal{S}$.

Proof. Supposing $K \notin \mathcal{S}$, for any ϵ with $0 < \epsilon < \xi$ (ξ is as in Proposition 2.7) there is a continuum $C \subset K$ contained in no arc with diam $C < \epsilon/3$. Take any point $p \in C$. By Proposition 2.7 there are arcs ap and pb in K such that $ap \cap pb = \{p\}$ and diam $ap = $ diam $pb = \epsilon/3$. The union $ap \cup pb \cup C$ is a triod of diameter less than ϵ . This completes the proof.

Applying Theorem 15 from [[8], p.146] we obtain another characterization of solenoids. This characterization will be employed in the next section.

2.9. Theorem. For any continuum (X, ρ) the following statements are equivalent:

(a) X is a solenoid,

(b) X contains an arc and does not contain arbitrarily small triods, and X is h.l.o.m.,

(c) X does not contain arbitrarily small triods, and the metrics ρ and $\hat{\rho}$ are equivalent.

Proof. (a) $\rightarrow$ (b) is well known, (b) $\rightarrow$ (c) is proved in Proposition 2.4. If (c) is satisfied, X contains arcs by the equivalence of ρ and $\hat{\rho}$. Thus X is in $\mathcal{S}$ by Proposition 2.8. Hence X is a solenoid by [[8], Theorem 15, implication (g) $\rightarrow$ (a), p. 146].

3. ATRIODIC PROPERTIES OF OPENLY HOMOGENEOUS CONTINUA IN 2-MANIFOLDS

The crucial theorem of this chapter is the following.

3.1. Theorem. Let X be an openly homogeneous continuum in a 2-manifold. If U is an open set in X such that each connected subset of U has empty interior, then U contains no triod.

Proof. Suppose $T = A \cup B \cup C$ is a triod in U with continua A, B, C satisfying $A \cap B \cap C \neq \emptyset \neq A - (B \cup C)$ and $B - (A \cup C) \neq \emptyset \neq C - (A \cup B)$. Take a point $p \in A \cap B \cap C$, any point $q \in X$, and an open surjection $f : X \rightarrow X$ with $f(q) = p$ guaranteed by Theorem 1.4. Let A_1, B_1, C_1 be the components of $f^{-1}(A)$, $f^{-1}(B)$, $f^{-1}(C)$, respectively, containing q. Since f is confluent, we have $f(A_1) = A$, $f(B_1) = B$ and $f(C_1) = C$. Furthermore, for some neighborhood N of f in the space M of all open mappings of X onto X, for each $g \in N$ the union $g(A_1) \cup g(B_1) \cup g(C_1)$ contains a triod by Theorem 1.5. Let $H \subset N$ be a neighborhood of f in M such that $g(A_1 \cup B_1 \cup C_1) \subset U$ for any $g \in H$. Hence for each $g \in H$ the point $g(q)$ lies in a triodic continuum in U. Note that the set $T_q(H) = \{g(q) : g \in H\}$ is not of the first category, for the set $(T_q(H))^*$ is nonempty by theorem 1.4. Since each countable union of components of U is of the first category, the set $T_q(H)$ meets uncountably many components of U. Therefore the 2-manifold contains uncountably many mutually exclusive triods, a contradiction to the generalized version [3] of the well known Moore triodic theorem. The proof is complete.

Since proper subcontinua of indecomposable continua have empty interiors, we have the following result by the previous theorem.

3.2.Theorem. Indecomposable, openly homogeneous continua in 2-manifolds are atriodic.

To obtain another application of Theorem 3.1 we prove the following proposition.

3.3.Proposition. If an openly homogeneous continuum X is not locally connected, then there is a $\tau > 0$ such that int $C = \emptyset$ for each subcontinuum C of X with diam $C < \tau$.

Proof. Suppose there is a sequence $\{C_n\}$ of subcontinua of X with nonempty interiors, such that $\lim \operatorname{diam} C = 0$. This sequence has an accumulation point, and thus each point of X is a limit point of such a sequence by the open homogeneity of X. Therefore the set

$$U_n = \cup\{\text{int } C : C \text{ is a subcontinuum of } X \text{ with diam } C < 1/n\}$$

is a dense open subset of X for each positive integer n. It follows that the set $U_1 \cap U_2 \cap ...$ of all points of local connectedness of X is nonempty by the Baire theorem. Hence X is locally connected by the open homogeneity of X. This completes the proof.

Combining Proposition 3.3 and Theorem 3.1 we obtain another atriodic property of openly homogeneous continua in 2-manifolds.

3.4.Theorem. If an openly homogeneous continuum X in a 2-manifold is not locally connected, then there is a $\tau > 0$ such that each subcontinuum C of X with diam $C < \tau$ is atriodic.

4. LOCAL SEPARATING POINTS IN OPENLY HOMOGENEOUS CONTINUA

Now we present a nice observation on local separating points in openly homogeneous continua. It is a consequence of Whyburn's theorem (Theorem 1.6).

4.1.Theorem. An openly homogeneous continuum is either a simple closed curve, or it contains at most countably many local separating points.

Proof. If an openly homogeneous continuum contains uncountably many local separating points , it contains a point of order 2 by theorem 1.6. The order of a point cannot increase by open mappings [[11], Corollary (7.31), p.147], so each point of X has order 2, and thus X is a simple closed curve.

5. LOCALLY CONNECTED CONTINUA IN 2-MANIFOLDS, WHICH ARE H.L.O.M.

Combining Theorems 3.4 and 2.9 the following result is obtained.

5.1.Theorem. Let X be a continuum in a 2-manifold, which is h.l.o.m. and contains an arc. Then X is locally connected.

Proof. Suppose X is not locally connected. Then X does not contain arbitrarily small triods by Theorem 3.4. So it is a solenoid by Theorem 2.9. As is is well known, only the trivial solenoid, i.e. the simple closed curve, is embeddable in a 2-manifold. Hence X is locally connected, a contradiction.

5.2.Lemma. Let X be a locally connected continuum in a 2-manifold. If X is h.l.o.m., then either X is a simple closed curve or X contains no local separating points.

Proof. Assume X is nondegenerate, and is not a simple closed curve. Then X contains a point p which does not locally separate X (Theorem 4.1). Given $q \in X$, let $f : X \to X$ be a light open mapping such that $f(p) = q$. Since all points of X must have order greater than 2 (for an open mapping cannot increase the order of a point), the set $f^{-1}(q)$ is finite [[11], Corollary (3.21), p. 190]. Therefore there are neighborhoods U_n of p such that each $U_n - \{p\}$ is connected, $\lim \operatorname{diam} U_n = 0$, and $U_n \cap f^{-1}(q) = \{p\}$. The sets $f(U_n)$ form a basis of neighborhoods of q and each $f(U_n) - \{q\} = f(U_n - \{p\})$ is connected. Hence q does not locally separate X. The proof is complete.

The following theorem is a consequence of two results of Whyburn from [11].

5.3.Theorem. Let X be a continuum h.l.o.m. in a 2-manifold M with $\dim X = 2$. Then $X = M$ and M is a manifold without boundary.

Proof. If $\dim X = 2$, X contains an open set U homeomorphic to the plane. For any light open map $f : X \to X$ the map $f|U : U \to f(U)$ remains open and light. Since light open mappings preserve 2-manifolds [[11], Theorem (4.4), p. 197], the continuum X is a 2-manifold (for X is h.l.o.m.). Light open mappings on 2-manifolds preserve boundary points [[11], (iii'), p. 195], so X is a manifold without boundary, and thus the theorem follows.

Now we are going to prove that no S-manifold is h.l.o.m.. Any S-curve in the plane is composed of so called rational points, i.e., all points in the boundaries of the complementary domains in the plane, and of points which are not rational (called irrational). It is known that the notion of a rational point in an S-curve is a topological invariant. Similarly, we say that a point p is a rational (irrational) point of an S-manifold N provided there is an S-curve C in N such that $p \in \operatorname{int} C$ and p is a rational (irrational) point of C. Denote by $R(N)$ and $I(N)$ the sets of all rational and irrational points of N, respectively. It is not difficult to see that $R(N) \cap I(N) = \emptyset$, $R(N) \cup I(N) = N$ and $R(N) \neq \emptyset \neq I(N)$ (these are consequences of the topological invariance of the notion of a rational point in an S-curve). In the following theorem we will prove that the notion of a rational point in an S-manifold is also invariant with respect to light open mappings.

5.4.Theorem. Each light open mapping between S-manifolds preserves rational points.

Recall the well known observation on the irrational points of an S-curve P. Identifying to a point each component of $R(P)$, a topological 2-sphere is obtained from P. The quotient map is a homeomorphism on $I(P)$ and the number of components of $R(P)$ is countable. Therefore each point of $I(P)$ has a neighborhood in $I(P)$ homeomorphic to some set $\mathbb{R}^2 - A$, where $\mathbb{R}^2$ is the plane and A is countable. The last statement holds true also for any S-manifold substituted for P. In view of this observation we will apply Facts 5.6 and 5.7 below, concerning the plane. To see that they are very easy and actually known, notice an obvious fact (Fact 5.5 below). The detailed proofs of these facts are omitted. In Facts 5.5.-5.7 the letter A denotes any countable set.

5.5.Fact. For any point $x \in \mathbb{R}^2 - A$ and any sequence $\{U_n\}$ of nonempty open subsets of $\mathbb{R}^2$, there are straight-line segments $x_0 x_n$ in $\mathbb{R}^2 - A$ such that $x_n \in U$ and $x_0 x_i \cap x_0 x_j = \{x_0\}$ for $i \neq j$.

5.6.Fact. For any sequence $\{x_n\} \subset \mathbb{R}^2 - A$ converging to a point $x_0 \in \mathbb{R}^2 - A$ such that $x_i \neq x_j$ for $i \neq j$, $i, j = 0, 1, 2, ...$, there are arcs $x_0 x_n \subset \mathbb{R}^2 - A$ satisfying $x_0 x_i \cap x_0 x_j = \{x_0\}$ for $i \neq j$, and $\lim \operatorname{diam} x_0 x_n = 0$.

5.7.Fact. No subset of an arc in $\mathbb{R}^2 - A$ separates $\mathbb{R}^2 - A$.

Proof of Theorem 5.4. Let $f : M \to N$ be a light open surjection between S-manifolds M and N. Suppose there are $p \in R(M)$, $q \in I(N)$ such that $f(p) = q$. Let U, V be S-curves in M and N, respectively, such that $p \in R(U) \cap \operatorname{int} U$ and $q \in I(V) \cap \operatorname{int} V$. Since arc components of $R(V)$ are closed, for any arc L in $R(U)$ containing p we have $f(L - \{p\}) \cap I(V) \neq \emptyset$. The set $f^{-1}(q)$ is finite by [[11], Corollary (3.21), p. 190], and thus there is an arc L in $R(U)$ containing p, and a sequence $\{p_n\} \subset L$ converging to p such that $q_n = f(p_n) \in I(V)$ and $q_i \neq q_j \neq q$ for $i \neq j$. Moreover, there are arcs $q_n q \subset I(V)$ with $q_i q \cap q_j q = \{q\}$ and $\lim \operatorname{diam} q_n q = 0$ by Fact 5.6. Furthermore, there are arcs $p_n p'_n \subset N$ such that the maps $f | p_n p'_n : p_n p'_n \to q_n q$ are homeomorphisms by Lemma 1.3. We have $p'_n \in f^{-1}(q)$, $\lim p_n = p$, $\lim \operatorname{diam} p_n p'_n = 0$ (for f is light), and p is an isolated point of $f^{-1}(q)$. Hence it follows that $p'_n = p$ and $p_n p'_n \subset \operatorname{int} U$ for almost all n. Without loss of generality assume this holds true for all n, and denote the arcs $p_n p'_n$ by $p_n p$. Observe that almost all arcs $p_n p$ are not contained in $R(U)$, and thus, for these arcs, there are open nonempty sets W_n in M such that $\operatorname{bd} W_n \subset p_n p$ and $\lim \operatorname{diam} W_n = 0$. Since f is open, each $f(W_n)$ is open and $\operatorname{bd} f(W_n) \subset f(\operatorname{bd} W_n) \subset f(p_n p) = q_n q$. Moreover, $\lim \operatorname{diam} f(W_n) = 0$ by the continuity of f. But the arcs $q_n q$ cannot contain boundaries of small open sets in M by Fact 5.7, a contradiction. The proof is complete.

Now we are ready to formulate the main result for locally connected curves in 2-manifolds.

5.8.Theorem. For any locally connected curve in 2-manifold the following statements are equivalent:

(a) X is homogeneous with respect to light open mappings,

(b) X is homogeneous with respect to light confluent mappings,

(c) X is a simple closed curve.

Proof. (a) and (b) are equivalent by Fact 1.1 and Lemma 1.2. (c) $\rightarrow$ (a) is obvious. Assume X is h.l.o.m. and suppose it is not a simple closed curve. Then X contains no local separating point by Theorem 5.2 and X is locally planar. Therefore X is an S-manifold by Theorem 1.7. Hence X is not h.l.o.m. by Theorem 5.4, a contradiction.

We end the paper with the following summarizing theorem.

5.9. Theorem. Let X be a continuum in a 2-manifold M containing an arc. If X is homogeneous with respect to light open mappings, then either

(a) $X = M$ and M is a compact 2-manifold without boundary, or

(b) X is a simple closed curve.

Proof. If $\dim X = 2$, the statement (a) holds true by Theorem 5.3. Assume $\dim X = 1$. Since X is locally connected by Theorem 5.1, (b) follows by Theorem 5.8.

REFERENCES

1. R. H. Bing, *A simple closed is the only homogeneous bounded plane continuum that contains an arc*, Canad. J. Math. 12(1960), 209-230.
2. J. J. Charatonik and T. Maćkowiak, *Around Effros' theorem*, Trans. Amer. Math. Soc. 298(1986), 579-602.
3. A. Lelek, *On the Moore triodic theorem*, Bull. Acad. Polon. Sci. Math. Astronom. Phys. 8(1960), 271-276.
4. A. Lelek and D. R. Read, *Compositions of confluent mappings and some other classes of functions*, Colloq. Math. 29(1974), 101-112.
5. S. Mazurkiewicz, *Sur les continus homogenes*, Fund. Math. 5(1924), 137-146.
6. J. R. Prajs, *Homogeneous continua in Euclidean (n+1)-space which contain an n-cube are locally connected*, Trans. Amer. Math. Soc. 307(1988), 383-394.
7. J. R. Prajs, *Homogeneous continua in Euclidean (n+1)-space which contain an n-cube are n-manifolds*, Trans. Amer. Math. Soc. 318(1990), 143-148.
8. J. R. Prajs, *Openly homogeneous continua having only arcs for proper subcontinua*, Topology Appl. 31(1989), 133-147.
9. J. R. Prajs, *On open homogeneity of closed balls in the Euclidean spaces*, preprint.
10. R. H. Sorgenfrey, *Concerning triodic continua*, Amer. J. Math. 66(1944), 439-460.
11. G. T. Whyburn, *Analytic Topology*, Amer. Math. Soc. Colloq. Publ., vol. 28, Amer. Math. Soc., Providence, R.I., 1963.
12. G. T. Whyburn, *Local separating points of continua*, Monats. fur Mathematik und Physik 36(1929), 305-314.
13. G. T. Whyburn, *Topological characterization of the Sierpiński curve*, Fund. Math. 45(1958), 320-324.

E-mail address: jrprajs@sparc-1.wsp.opole.pl "Janusz Prajs"

24
A CONTINUOUS DECOMPOSITION OF THE SIERPIŃSKI CURVE

CARL R. SEAQUIST Department of Mathematics, Auburn University, Auburn, Alabama 36849, USA

ABSTRACT. This paper describes a construction of a continuous decomposition of the Sierpiński curve into non-degenerate cellular continua. The decomposition space under the quotient topology is then shown to be homeomorphic to the Sierpiński curve. Thus the paper shows the existence of a monotone open map from the Sierpiński curve onto the Sierpiński curve that is not a homeomorphism.

1. INTRODUCTION

The main result of this paper is a continuous decomposition of the Sierpiński curve so that each member of the decomposition space is a non-degenerate cellular continuum, and so that the decomposition space is homeomorphic to the Sierpiński curve. By the Sierpiński curve we mean the plane universal curve; see for instance [3]. R. D. Anderson in [1] shows how to create a continuous decomposition of a planar one dimensional curve so that the decomposition space is homeomorphic to the plane and thus shows the existence of a monotone open map which raises dimension. Our construction shows the existence of a monotone open map from the Sierpiński curve onto the Sierpiński curve which is not a homeomorphism. R. D. Anderson achieves his result by carefully closing all the holes of the planar curve; i.e., by mapping boundaries of bounded complementary regions of the planar curve to single points. We avoid closing any holes in our construction. Besides being of intrinsic interest, if this construction were modified to selectively close holes, then the Sierpiński curve would be shown to be open monotone homogeneous [2, 8].

1991 *Mathematics Subject Classification.* primary 54B15, secondary 54C10, 54E45, 54F50.
Key words and phrases. Continuous decompositions, monotone open map.
The author was supported by a GAANN fellowship.

315

Our construction is based on extensions found in [9] to the construction described by W. Lewis and J. J. Walsh in [7] where they show in detail how the plane can be continuously decomposed into pseudo-arcs. In [9] we modify their construction to create a continuous cellular decomposition of the disk into non-degenerate elements. In this paper we further modify their construction to accommodate the holes in the Sierpiński curve. The construction of the decomposition occurs concurrently with the construction of the Sierpiński curve.

2. PRELIMINARIES AND OVERVIEW OF CONSTRUCTION

Before giving an overview of our construction we introduce the following notation. If P is a collection of sets, then P^* denotes the union of members of P. If p is a set, then $\mathrm{st}^1(p, P) = \{p' \in P : p' \cap p \neq \emptyset\}$ and inductively $\mathrm{st}^i(p, P) = \mathrm{st}^1(\mathrm{st}^{i-1}(p, P)^*, P)$. We abbreviate $\mathrm{st}^1(p, P)$ by $\mathrm{st}(p, P)$. By $\mathrm{Cl}_B(A)$ we mean the closure of A with respect to B and by $\mathrm{Int}_B(A)$ we mean the interior of A with respect to B. By $\mathrm{Cl}(A)$ and $\mathrm{Int}(A)$ we mean closure and interior respectively relative to E^2 unless otherwise stated. By A^c we mean the complement of A with respect to E^2.

We will construct a sequence $\{Y_n\}_{n=1}^{\infty}$ of continua so that $Y_{n+1} \subset Y_n$ and so that $Y = \cap_{n=1}^{\infty} Y_n$ is the Sierpiński curve. Simultaneously while constructing Y_n, we will describe a partition P_n of Y_n into cells with non-overlapping interiors so that $G = \{\cap_{n=0}^{\infty} \mathrm{st}(p_n, P_n)^* : \cap_{n=0}^{\infty} p_n \neq \emptyset$ where $p_n \in P_n\}$ is a continuous decomposition of Y. Rather than describe the cells of P_n directly, we proceed as in [7, 9] by first describing a partition Q_n of a continuum X_n into fairly simple cells with non-overlapping interiors and then taking Y_n to be the image of X_n under a homeomorphism H_n. The collection P_n is obtained by applying the same homeomorphism H_n to each of the cells $q_n \in Q_n$. In the construction, X_n will be a continuum with finitely many complementary regions with disjoint boundaries that are simple closed curves.

The construction starts with $Y_1 = X_1 = D = [0, 1] \times [0, 1]$ and proceeds inductively. Assuming we are at stage n, we are given the continuum X_n and $\widehat{R}_n$, a division of D into either congruent vertical or congruent horizontal strips.

Definition 1. *A vertical (respectively horizontal) division of D is the collection $R = \{[(i-1)a, ia] \times [0, 1] : i \in 1, ..., 1/a\}$ (respectively $\{[0, 1] \times [(i-1)a, ia] : i \in 1, ..., 1/a\}$) where $(1/a) \in \mathbb{Z}^+$. The mesh of R denoted by $\mathrm{mesh}(R)$ is a. Each member of R is called a vertical strip (respectively horizontal strip).*

Given the vertical (respectively horizontal) division $\widehat{R}$ of D, a division R of D is a refinement of $\widehat{R}$ if for every strip $X \in R$ there is a strip $Y \in \widehat{R}$ so that $X \subset Y$.

We will call the bounded complementary regions of X_n *holes*. We insure that the unbounded complementary region of X_n will be precisely the complement of D and that the boundary of X_n will be a finite number of disjoint simple closed curves. We will further constrain the holes of X_n to be open squares with sides parallel to either the x-axis or the y-axis. For the continua X_n that arise in the construction we define vertical boundary and horizontal boundary.

Definition 2. *Given a continuum $X_n \subset D$, the vertical boundary of X_n is the left and right edges of D unioned with the vertical line segments that make up the left and right edges of the holes of X_n. Horizontal boundary is similarly defined.*

Given the vertical (horizontal) division $\widehat{R}_n$ we take and refine it to obtain R_n. The common part of each strip of R_n and X_n is then partitioned into cells with non-over lapping interiors to obtain the collection of cells Q_n. Once Q_n is defined, a homeomorphism $h_n : D \to D$ is defined so that $\{h_n^{-1}(q_n) : q_n \in Q_n\}$ is a collection of identical rectangles with non-over lapping interiors whose union is X_n. We will define h_n so that it will leave the boundary of X_n invariant. Now we set $Y_n = h_1 \circ \cdots \circ h_{n-1}(X_n)$ and the collection P_n is defined to be $\{h_1 \circ \cdots \circ h_{n-1}(q_n) : q_n \in Q_n\}$. The homeomorphism $h_1 \circ \cdots \circ h_{n-1}$ will be denoted by H_n. To continue on to stage $n+1$ we use $\{h_n^{-1}(q_n) : q_n \in Q_n\}$ to define $\widehat{R}_{n+1}$, a horizontal (resp. vertical) division of D. Thus the construction alternates between working with horizontal and vertical divisions of D. Arbitrarily, we let $\widehat{R}_n$ be a vertical division when n is odd and a horizontal division when n is even. Throughout the paper we will only describe the construction when n is odd unless it is necessary to discuss two stages at once. The description of the construction when n is even is entirely analogous. To continue on to stage $n+1$ we must also define X_{n+1}. For each $q_n \in Q_n$ we define a small open s_n by s_n square hole, referred to as w_n, which will be centered in the rectangle $h_n^{-1}(q_n)$. The parameter s_n is a rational number which helps control the construction at stage n. We set $W_n = \{w_n : \exists q_n \in Q_n$ and w_n is an open s_n by s_n square centered in $h_n^{-1}(q_n)\}$. We will define X_{n+1} to be $X_n \backslash W_n^*$. Note that $X_{n+1} = D \backslash (\cup_{i=1}^n W_n^*)$. See Figures 1A and 1B.

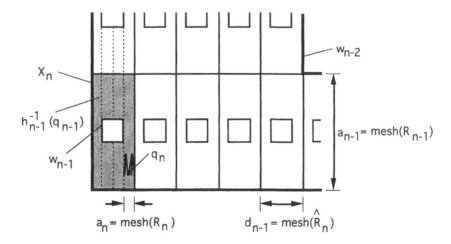

FIGURE 1A. Shows the lower left corner of X_n.

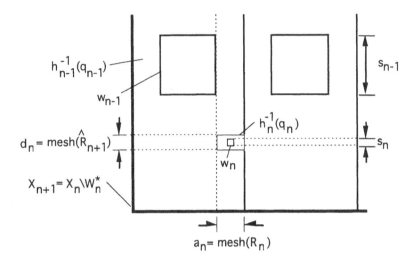

FIGURE 1B. The lower left corner of X_{n+1} showing lower part of $h_{n-1}^{-1}(q_{n-1})$.

The construction is constrained at stage n by five positive rational constants, a_n, a_n', b_n, c_n, s_n, and a positive integer k_n. To facilitate our discussion of the application of these parameters we give an informal description of the cells $q_n \in Q_n$. First, however, we introduce the following definitions when n is odd; i.e., when we are decomposing vertical strips. (Of course similar definitions exist when n is even.)

Definition 3. *Given a cell q, the top boundary of q is $\{(x,y) \in q : \forall (x,y') \in q,\ y' \leq y\}$. The bottom boundary of q is $\{(x,y) \in q : \forall (x,y') \in q,\ y' \geq y\}$. The left boundary of q is the left most vertical line segment contained in q and the right boundary of q is the right most vertical line segment contained in q. The union of the left boundary and right boundary of a cell is its vertical boundary.*

There are two cell types: Those that do not lie along side of holes $w_{n-1} \in W_{n-1}$ which were created during the previous stage and those that do. These are referred to as *Type* 1 and *Type* 2 cells respectively. See Figure 2. A cell q_n will consist of $k_n/2$ congruent pieces on the left joined by a rectangle of width s_n to $k_n/2$ congruent pieces on the right. The cell will be symmetrical about a vertical line running through the center of the rectangle. Thus the pieces to the left of the rectangle will be reflections of the pieces to the right. Each of the k_n pieces, which we call a *cell-piece*, will consist of two symmetrical parts called *cell-points*. Note that the width of a cell-piece is $(a_n - s_n)/k_n$. In Type 1 cells the two cell-points of a cell-piece are congruent and "point" in the same direction. For a typical cell $q_n \in Q_n$ of Type 1 see Figure 3. Note that a_n defines the width of q_n. The cell has a height of at least c_n but less than $b_n + c_n$. The thickness; i.e., vertical transverse thickness, of the cell is limited by b_n. Thus the Type 1 cells are very similar to those used in the construction described in [9] except for the small rectangular piece in the middle of the cell. This modification is needed to accommodate the image under h_n of the hole w_n in a relatively unstretched fashion. The image under h_n of the hole w_n will be centered horizontally in the rectangular piece.

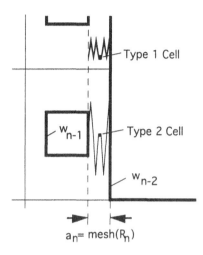

FIGURE 2. There are two cell types. Type 2 Cells lie along side holes in W_{n-1}.

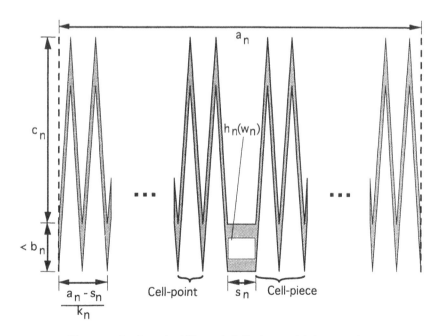

FIGURE 3. A typical Type 1 Cell along with image of w_n.

For a typical cell $q_n \in Q_n$ of Type 2 see Figure 4. Like Type 1 cells, a_n defines the width of q_n and the thickness of the cell is limited by b_n. In Type 2 cells, however, the two cell-points of a cell-piece will not in general be congruent nor will they "point" in the same direction. In addition the height of Type 2 cells can be greater than $c_n + b_n$ and in fact can have height greater than $c_n + s_{n-1}$ where s_{n-1} is the length of a side of the square holes in W_{n-1}. The cell-pieces of a cell of Type 2 are forced to extend at least $c_n/2$, but no more than $c_n + b_n$, beyond (either above or below) the hole in W_{n-1} that has a vertical edge containing a component of the vertical boundary of the cell. The use of Type 2 cells makes it possible to go around the most recently introduced holes in the construction. Note that we will always choose s_n; i.e., the length of the side of the holes in W_n to be less than b_n, the transverse thickness of a cell in Q_n. The constant b_n will in turn always be less than a_n, the width of a cell in Q_n.

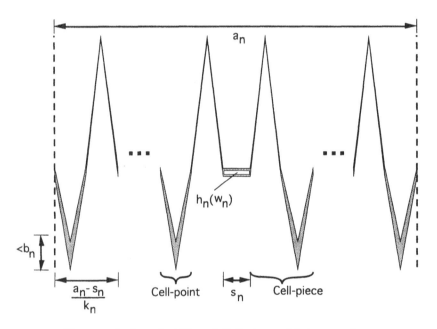

FIGURE 4. A typical Type 2 Cell along with image of w_n.

3. THE GENERAL CONSTRUCTION

We now describe the general construction in more detail. We assume we are at stage n where $n > 1$ is odd and that we are given a vertical division of D, $\widehat{R}_n$, and a continuum, $X_n \subset D$, which is essentially a disk minus a finite number of open squares with disjoint boundaries. The smallest holes; i.e., those removed during the last stage, will be those in W_{n-1}. To start we refine $\widehat{R}_n$ to create R_n letting $a_n = \text{mesh}(\widehat{R}_n)/4$ be the mesh of R_n. The size and position of the holes of X_n will have been previously chosen carefully so that the vertical edges of the holes will lie on the vertical edges of the strips of R_n.

3.1. The Creation of Q_n. Our strategy in defining the cells of Q_n will be to define a sequence of disjoint polygonal lines $\{L_n^j\}_{j=-1}^{m_n+1}$ which run horizontally across the vertical strips of R_n. Note that by $L_n^j(x)$ we mean the y such that $(x, y) \in L_n^j$. For each $i \in 1, ..., 1/a_n$ and $j \in 0, ..., (m_n + 1)$ we will define the open cell $\widehat{q}_{i,j}$ as follows:

$$\widehat{q}_{i,j} = \{(x, y) \in D : (i - 1)a_n < x < ia_n \text{ and } L_n^{j-1}(x) < y < L_n^j(x)\}.$$

We define Q_n to be the closure of the non-empty cells:

$$Q_n = \{\text{Cl}(\widehat{q}_{i,j}) : X_n \cap \widehat{q}_{i,j} \neq \emptyset \text{ and } i \in 1, ..., \frac{1}{a_n} \text{ and } j \in 0, ..., (m_n + 1)\}.$$

To carry out this strategy we define a polygonal arc ℓ_n which will establish the underlying pattern of the cells. To define ℓ_n we first define the set of points which determine the pattern for the left half of the cell which will consist of $k_n/2$ identical pieces:

$$A_n'' = \bigcup_{i=1}^{k_n/2} \left\{ \left((i - 1)t_n, 0\right), \left((i - 1)t_n + \frac{t_n}{4}, c_n\right), \left((i - 1)t_n + \frac{t_n}{2}, 0\right), \right.$$
$$\left. \left((i - 1)t_n + \frac{3t_n}{4}, c_n\right), \left((i - 1)t_n + t_n, 0\right) \right\}$$

where $t_n = (a_n - s_n)/k_n$. Then we define the points of the right half of the pattern for the cell and combine these with the points that determine the pattern of the left half of the cell:

$$A_n' = \left\{ \left(x + \frac{a_n + s_n}{2}, y\right) : (x, y) \in A_n'' \right\} \cup A_n''$$

Note that the right most point of the left half of the pattern and the left most point of the right half of the pattern determine a horizontal line segment of length s_n. Finally we repeat this cell pattern $1/a_n$ times across D:

$$A_n = \bigcup_{i=1}^{1/a_n} \left\{ (x + a_n(i - 1), y) : (x, y) \in A_n' \right\}.$$

Order the points of A_n in ascending order by abscissa and connect pairs of consecutive points under this ordering by a line segment. The union of these line segments is the polygonal arc ℓ_n.

We will let O_n be the ordinates of the top and bottom edges of the holes in X_n. To simplify our discussion we will say that the bottom edge of X_n is the top edge of a hole and that the top edge of X_n is the bottom edge of a hole. Thus $0 \in O_n$ and $1 \in O_n$. In other words, O_n is the projection of the horizontal boundary of X_n onto the y-axis. For each $y \in O_n$ we define the polygonal arc N_n^y as follows:

$$N_n^y(x) = \begin{cases} y + \ell_n(x) & \text{if } y \text{ is the ordinate of the top edge of a hole in } X_n; \\ y - \ell_n(x) & \text{if } y \text{ is the ordinate of the bottom edge of a hole in } X_n; \end{cases}$$

In our construction if y and y' are the ordinates of the bottom and top edges of a hole and $\widehat{y}$ and $\widehat{y}'$ are the ordinates of the bottom and top edges of another hole, then if either $y \leq \widehat{y} \leq y'$ of $y \leq \widehat{y}' \leq y'$ we will have that $y \leq \widehat{y} < \widehat{y}' \leq y'$. Thus there is no ambiguity in the above definition of N_n^y. For each $y \in O_n$ define $\underline{M}_n^y(x)$, $M_n^y(x)$, and $\overline{M}_n^y(x)$ as follows:

$$\underline{M}_n^y(x) = \begin{cases} y - b_n & \text{if } (x, y - b_n) \in \text{Cl}(X_n^c); \\ N_n^y(x) - b_n & \text{otherwise}; \end{cases}$$

$$M_n^y(x) = \begin{cases} y & \text{if } (x, y) \in \text{Cl}(X_n^c); \\ N_n^y(x) & \text{otherwise}; \end{cases}$$

$$\overline{M}_n^y(x) = \begin{cases} y + b_n & \text{if } (x, y + b_n) \in \text{Cl}(X_n^c); \\ N_n^y(x) + b_n & \text{otherwise}; \end{cases}$$

We assume now but will show later that c_n and b_n can be chosen small enough so that members of $\cup_{y \in O_n} \{\underline{M}_n^y, M_n^y, \overline{M}_n^y\}$ are pairwise disjoint. If $y \in O_n \backslash \{1\}$ we denote by $\text{Nxt}(y)$ the least element of O_n larger than y. See Figure 5.

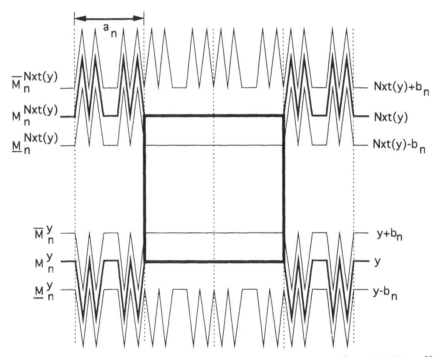

FIGURE 5. Shows the definition of polygonal arcs $\underline{M}_n^y$, M_n^y, $\overline{M}_n^y$, $\underline{M}_n^{\text{Nxt}(y)}$, $M_n^{\text{Nxt}(y)}$, and $\overline{M}_n^{\text{Nxt}(y)}$.

If $y \in O_n \backslash \{1\}$, then one of the following cases hold:

(1) y and $\text{Nxt}(y)$ are both ordinates of top edges of holes or are both ordinates of bottom edges of holes.

(2) y is the ordinate of the top edge of a hole and $\text{Nxt}(y)$ is the ordinate of the bottom edge of a hole.

(3) y is the ordinate of the bottom edge of a hole and $\text{Nxt}(y)$ is the ordinate of the top edge of a hole.

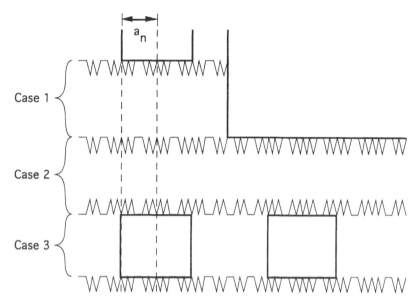

FIGURE 6. Three different cases arise in constructing polygonal arcs.
The first two result in Type 1 cells while the last results in Type 2 cells.

See Figure 6. Note that Case (3) will occur only when y and $\mathrm{Nxt}(y)$ are ordinates of the bottom and top edges respectively of a hole $w_{n-1} \in W_{n-1}$; i.e., one of the smallest holes of X_n. In Case (3) then we will be creating Type 2 cells between the polygonal arcs $\overline{M}_n^y$ and $\underline{M}_n^{\mathrm{Nxt}(y)}$ and in Cases (1) and (2) Type 1 cells. In any case we will want to create a sequence of polygonal arcs $\{L_y^j\}_{j=0}^{m_y}$ so that $L_y^0 = \overline{M}_n^y$ and $L_y^{m_y} = \underline{M}_n^{\mathrm{Nxt}(y)}$ and so that

(i) $b_n \geq \max\{|L_y^j(x) - L_y^{j-1}(x)| : x \in [0,1]\}$ for every $j \in 1,...,m_y$. Thus b_n will control the thickness of cells.

(ii) For each $j \in 1,...,m_y$ and for every $x \in [0, 1-a_n]$, $\quad L_y^j(x) = L_y^j(x+a_n)$. Thus all cells between the two polygonal arcs L_y^j and L_y^{j+1} will be congruent.

(iii) For each $j \in 1,...,m_y$ the polygonal sub-arc $\{L_y^j : x \in [0, a_n]\}$

    (a) is symmetrical about the vertical line passing through $a_n/2$, and

    (b) is made up of $k_n/2$ polygonal sub-arcs, called pieces, each of which is the horizontal displacement of another. These pieces are then followed by a horizontal line segment of length s_n, followed by $k_n/2$ more congruent polygonal sub-arcs.

This condition insures that the cells are vertically symmetrical and have the general shape described above in the overview.

In Cases (1) and (2) we will want additionally that

(iv) $c_n = \max\{|L_y^j(x) - L_y^j(x')| : x, x' \in [0,1]\}$ for every $j \in 0, 1,...,m_y$. Thus c_n will control the height of Type 1 cells to be between c_n and $c_n + b_n$.

In case (3) we will want to insure instead that each piece of a cell extends either above Nxt(y) or below y by at least $c_n/2$. Thus the following condition will be required:

(v) Either $\min\{L_y^j(x) : x \in [0,1]\} \leq y - c_n/2$ or $\max\{L_y^j(x) : x \in [0,1]\} \geq$ Nxt(y) $+ c_n/2$ for every $j \in 0, 1, ..., m_y$.

The fact that we can create the sequence of disjoint polygonal arcs $\{L_y^j\}_{j=0}^{m_y}$ satisfying conditions (i)-(iv) in Case (1) is immediate since the polygonal arcs can simply be translations of L_y^0. The fact that we can create the polygonal arcs in Case (2) when $5c_n/2 + 2b_n <$ Nxt(y) $- y$ follows from the following lemma taken from [9].

Lemma 4. *Let* $c > b > 0$. *Let* M^0 *be the polygonal arc connecting the points,* $\{(0,0), (1/2, c), (1,0)\}$ *in order and let* M^1 *be the polygonal arc connecting the points,* $\{(0, r), (1/2, r - c), (1, r)\}$ *in order. If* $5c/2 < r$ *and* $\widehat{m} \in \mathbb{Z}^+$, *then there exists an* $m \in \mathbb{Z}^+$, $m > \widehat{m}$ *and a sequence of polygonal lines* $\{L^j\}_{j=0}^m$ *so that* $L^0 = M^0$ *and* $L^m = M^1$ *such that*

(i) $b \geq \max\{|L^j(x) - L^{j-1}(x)| : x \in [0,1]\}$ *for every* $j \in 1, ..., m$.

(ii) $c = \max\{|L^j(x) - L^j(x')| : x, x' \in [0,1]\}$ *for every* $j \in 0, 1, ..., m$.

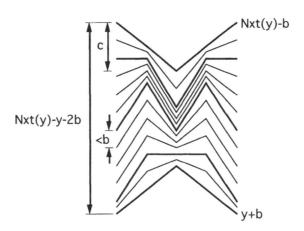

FIGURE 7. Example of the shapes of cell-points in Case 2.

Figure 7 shows how a part of the intervening polygonal arcs between $\overline{M}_n^y$ and $\underline{M}_n^{\text{Nxt}(y)}$ might look in Case (2). In Case (3) the polygonal arcs $\{L_y^j\}_{j=0}^{m_y}$ must satisfy conditions (i)-(iii) and (v) rather than (iv). That this is possible when $b_n < c_n/2$ and $2b_n < (\text{Nxt}(y) - y) = s_{n-1}$ follows from the following lemma where $s = s_{n-1}$, $b = b_n$, and $c = c_n$.

Lemma 5. *Let* $c > b > 0$ *and* $s > 0$. *Let* M^0 *be the polygonal arc connecting the points,* $\{(0, b), (1/4, b - c), (1/2, b), (3/4, b - c), (1, b)\}$ *in order and let* M^1 *be the polygonal arc connecting the points,* $\{(0, s - b), (1/4, s + c - b), (1/2, s - b), (3/4, s + c - b), (1, s - b)\}$ *in order. If* $2b < s$ *and* $b < c/2$ *and* $\widehat{m} \in \mathbb{Z}^+$, *then* *there exists an $m \in \mathbb{Z}^+$, $m > \widehat{m}$ and a sequence of polygonal arcs $\{L^j\}_{j=0}^m$ so that $L^0 = M^0$ and $L^m = M^1$ such that*

(i) $b \geq \max\{|L^j(x) - L^{j-1}(x)| : x \in [0,1]\}$ *for every* $j \in 1, ..., m$.

(ii) *Either* $s + c/2 < \max\{L^j(x) : x \in [0,1]\}$ *or* $-c/2 > \min\{L^j(x) : x \in [0,1]\}$ *for every* $j \in 0, 1, ..., m$.

Proof. Let $N \in \mathbb{Z}^+$ so that $2N > \widehat{m}$ and

$$b' = \frac{2(s-b) + 3c}{2N} < b.$$

For $j \in 0, 1, ..., N$ let L^j be the horizontal polygonal arc connecting the following points:

$$\left\{ \left(0, b + \frac{j(s-2b)}{2N}\right), \left(\frac{1}{4}, b - c + jb'\right), \left(\frac{1}{2}, b + \frac{j(s-2b)}{2N}\right), \right.$$
$$\left. \left(\frac{3}{4}, b - c + \frac{j(c-2b)}{2N}\right), \left(1, b + \frac{j(s-2b)}{2N}\right)\right\}.$$

See Figure 8. Define $\{L^j\}_{j=N+1}^{2N}$ so the configuration has symmetry with respect to the point $(1/2, s/2)$. $\square$

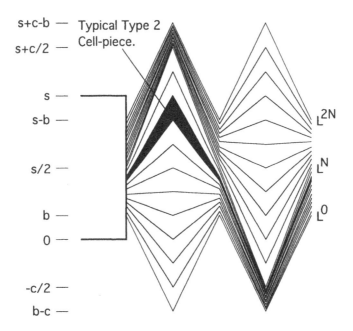

FIGURE 8. Example of the shapes of cell-pieces in Case 3. See Lemma 5.

Thus for each $y \in O_n \setminus \{1\}$ we can define a sequence of polygonal arcs $\{L_y^j\}_{j=0}^{m_y}$ so that $L_y^0 = \overline{M}_n^y$ and $L_y^{m_y} = \underline{M}_n^{\mathrm{Nxt}(y)}$ and that satisfy the appropriate conditions, (i)-(iii) and either (iv) or (v) depending on whether or not $\overline{M}_n^y$ intersects W_{n-1}^*. It should be noted that the polygonal arcs $L_{m_y}^j$ are defined for all $0 \leq x \leq 1$ and that they run "straight" across the holes. For each $y \in O_n\backslash\{1\}$ we define an additional polygonal arc $L_y^{m_y+1} = M_n^{\text{Nxt}(y)}$ and define d_y to be $(\text{Nxt}(y) - y)/(m_y + 2)$. If O_n only contains rational numbers, we can choose m_y for each $y \in O_n\backslash\{1\}$ so that $d_y = (\text{Nxt}(y) - y)/(m_y + 2)$ is fixed. In other words, we can make $d_y = d_{\hat{y}}$ for all $y, \hat{y} \in X\backslash\{1\}$. This fact insures that the vertical boundary of X_n can be made to lie on the vertical edges of the strips of R_{n+1}. Let

$$m_n = \left(\sum_{y \in O_n\backslash\{1\}} (m_y + 2) \right) - 2$$

which is two less then the total number of polygonal arcs in $\{\{L_y^j\}_{j=0}^{m_y+1}\}_{y \in O_n\backslash\{1\}}$. We relabel all polygonal arcs $\{\{L_y^j\}_{j=0}^{m_y+1}\}_{y \in O_n\backslash\{1\}}$ in ascending order by $\{L_n^j\}_{j=0}^{m_n+1}$ and define L_n^{-1} to be M_n^0; i.e., $L_n^{-1} = 0$. Let $d_n = 1/(m_n + 2)$. Note that $d_n = d_y$ for all $y \in O_n\backslash\{1\}$. Also note that $d_n < b_n$.

3.2. Definition of the Homeomorphisms. We define $h_n : D \to D$ to be a homeomorphism which maps vertical lines onto themselves each in a piecewise linear manner so that the pre-image of the polygonal arcs $\{L_n^j\}_{j=-1}^{m_n+1}$ is a collection of horizontal straight lines evenly spaced apart at the distance d_n. Thus $h_n^{-1}(Q_n)$ is a collection of a_n by d_n rectangles with disjoint interiors. Note that because of the way the polygonal arcs $\{L_n^j\}_{j=-1}^{m_n+1}$ were defined h_n maps the boundary of each hole of X_n onto itself. Thus if $w_i \in W_i$, then $h_n(\text{Cl}(w_i)) = \text{Cl}(w_i)$ for all $i < n$. For cells which lie along either the top or bottom edge of a hole of X_n the map h_n can cause a great deal stretching. In an approach analogous to that described in [9] we control where this stretching can occur. We force it to occur at a distance of between $a_n'/128$ and $a_n'/64$ from the horizontal boundary of X_n. See Figure 9. Typically a_n' is some fraction of d_n and determines a_{n+1}. We define H_n to be $h_1 \circ \cdots \circ h_{n-1}$ when $n > 1$ and H_1 to be Id_D. Thus $Y_1 = H_1(X_1) = D$ and $P_1 = H_1(Q_1) = Q_1$.

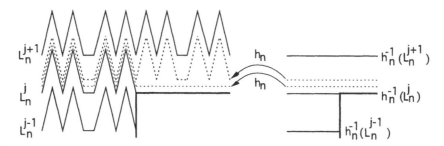

FIGURE 9. The large stretching by h_n along horizontal border is confined to between the dotted lines at distances of $a_n'/128$ and $a_n'/64$ from the horizontal border.

3.3. Preparation of the Next Stage. When $n > 1$ we set

$$P_n = H_n(Q_n) \quad \text{and} \quad Y_n = H_n(X_n).$$

To continue the construction we define

$$W_n = \left\{ \left(x - \frac{s_n}{2}, x + \frac{s_n}{2} \right) \times \left(y - \frac{s_n}{2}, y + \frac{s_n}{2} \right) : \right.$$
$$x = (i-1)a_n + a_n/2 \text{ and } y = jd_n + d_n/2$$
$$\left. \text{for } i = 1, ..., 1/a_n \text{ and } j = 0, ..., (m_n + 1) \right\}.$$

We set

$$X_{n+1} = X_n \backslash W_n^*$$

and

$$\widehat{R}_{n+1} = \{[0,1] \times [jd_n, (j+1)d_n] : j \in 0, ..., (m_n + 1)\}.$$

Thus $\widehat{R}_{n+1}$ is a horizontal division of D and the construction can continue to stage $(n+1)$. Note that $\mathrm{mesh}(\widehat{R}_{n+1}) = d_n = 1/(m_n + 2)$.

4. The Specific Construction

We will now apply the construction to build the continuum $Y = \cap_{n=1}^\infty Y_n$ and the collection $G = \{\cap_{n=1}^\infty \mathrm{st}(p_n, P_n)^* : \cap_{n=1}^\infty p_n \neq \emptyset$ where $p_n \in P_n\}$ of subsets of Y. We will show that at each stage n we can pick a_n, a_n', b_n, c_n, s_n, and k_n so that

(i) It is possible to apply lemmas 4 and 5 and to continue the construction to stage $n+1$.

(ii) The continuum $Y = \cap_{n=1}^\infty Y_n$ is the Sierpiński curve.

(iii) The collection G is a continuous decomposition of Y.

(iv) Each member of G is non-degenerate.

(v) The collection G under the quotient topology is homeomorphic to the Sierpiński curve.

4.1. The Constants are Chosen. Let $a_1 = 1/128$ so

$$R_1 = \{[(i-1)\frac{1}{128}, i\frac{1}{128}] \times [0,1] : i \in 1, ..., 128\}.$$

Let $c_1 = 1/4$ and $b_1 = 1/8192$. Since $5c_1/2 + 2b_1 = 5/8 + 1/4096 < 1$ we can use Lemma 4 to compute m_1 and d_1. Note that Lemma 5 is not needed here since there are no holes in X_1. Let $a_1' = d_1/4$ and $s_1 = d_1/2$. Let $k_1 \geq 64$ so that $4|k_1$ and $k_1 > a_1/a_1'$.

Assume we are at the beginning of stage n having just created $\widehat{R}_n$ and X_n. Thus the collections R_i, Q_i, W_i, and P_i; the continua X_i and Y_i; and the homeomorphism h_i have all been defined as described previously for $i \in 1, ..., (n-1)$, along with the function H_i for $i \in 1, ..., n$. In addition, the rational constants a_i, a_i', b_i, c_i, s_i, and k_i described above have been defined for each stage $i \in 1, ..., (n-1)$. The number of strips in the division $\widehat{R}_{i+1}$ is $m_i + 2$ and $\mathrm{mesh}(\widehat{R}_{i+1}) = d_i = 1/(m_i + 2)$ for $i \in 1, ..., (n-1)$. Let $\delta_i > 0$ so that $|x - x'| < \delta_i \Rightarrow |H_i(x) - H_i(x')| < 1/2^{i+7}$ for $i \in 1, ..., n$. We assume that these parameters in previous stages $i \in 1, ..., (n-1)$ were chosen as described below:

Let $a_i = a'_{i-1}$ and let $c_i = a_{i-1}/9$.

Let $b_i > 0$ be rational so that

   1a) $b_i < \delta_i/4$;

   1b) $b_i < a_i/(2k_{i-1})$;

   1c) $b_i < b_{i-1}(a_{i-1} - s_{i-1})/(4k_{i-1}(c_{i-1} + b_{i-1}))$.

Define m_i and d_i as described in 3.1.

Let $a'_i = d_i/4$ and $s_i = 2a'_i = d_i/2$.

Let k_i be an integer so that

   2a) $k_i \geq 64$;

   2b) $4 | k_i$;

   2c) $k_i > a_i/a'_i$.

Now we show we can proceed at stage $i = n$. We must be sure we can apply Lemmas 4 and 5 in order to define m_n and d_n. To see that Lemma 4 can be applied note that the smallest applicable distance here occurs between a hole in W_{n-1} and the horizontal boundary of X_{n-1}. This distance can be no smaller than $(a_{n-1} - s_{n-1})/2$. But $s_{n-1} < d_{n-1} < b_{n-1} < a_{n-1}/2k_{n-2} < a_{n-1}/128$ and so we have that $(a_{n-1} - s_{n-1})/2 > 127a_{n-1}/256 > a_{n-1}/3 = 3c_n > 5c_n/2 + b_n$ since $b_n < a_{n-1}/128 < a_{n-1}/18 = c_n/2$. Lemma 5 can be used because $b_n < a_n/128 = a'_{n-1}/128 = s_{n-1}/256$ and $b_n < c_n/2$.

4.2. The Sierpiński Curve is Constructed.
In this subsection we show that the resulting continuum that we decompose is indeed the Sierpiński curve.

Lemma 6. *The continuum $Y = \cap_{n=1}^{\infty} Y_n$ is the Sierpiński curve.*

Proof. Recall that $Y_n = H_n(X_n) = h_1 \circ \cdots \circ h_{n-1}(X_n)$. Since each h_i is defined on all of D, the holes of Y_n are the images of the holes of X_n under H_n. Thus the holes of Y_n are $H_n(\cup_{i=1}^{n-1} W_i)$. But if $i < n$, then $h_n(W_i) = W_i$. Therefore the collection of holes of Y is given by

$$W' = \cup_{i=1}^{\infty} H_{i+1}(W_i).$$

To show that Y is the Sierpiński curve it suffices by Toruńczyk's Lemma [5] or by [10] to show that

   (i) The boundaries of W' are disjoint simple closed curves.

   (ii) The diameters of the members of W' tend to zero.

   (iii) The members of W' are dense in D.

That (i) is satisfied is immediate since the boundaries of the members of $\cup_{i=1}^{\infty} W_i$ are disjoint simple closed curves. To see that (ii) is satisfied note that if $w_i \in W_i$ then

$$\text{diam}(h_i(w_i)) < (b_i + s_i) < 2b_i < \delta_i.$$

See Figures 3 and 4. Thus by our choice of δ_i we have that

$$\text{diam}(H_{i+1}(w_i)) = \text{diam}(H_i(h_i(w_i))) < \frac{1}{2^{i+7}}.$$

In order to show that (iii) holds we will show that for every point $p' \in Y_{i+1}$ there is a $w_i \in W_i$ so that $d(p', H_{i+1}(w_i)) < 1/2^{i+6}$. To do this we will actually show that for every $p \in h_{i-1} \circ h_i(X_{i+1})$ there is a $w_i \in W_i$ so that $d(p, h_{i-1} \circ h_i(w_i)) < \delta_{i-1}$. This is sufficient since we can choose p so that $H_{i-1}(p) = p'$ and since then

$$d(p', H_{i+1}(w_i)) = d(H_{i-1}(p), H_{i-1}(h_{i-1} \circ h_i(w_i))) < 1/2^{i+6}.$$

Without loss of generality we will assume that i is even. First observe that in $h_i(X_{i+1})$ there will be a columns of images under h_i of members of W_i spaced horizontally apart by a distance of no more than b_i. See Figure 10A. Because h_{i-1} maps vertical lines onto themselves, the horizontal spacing of these columns will be unchanged by h_{i-1}. We must, however, check the vertical spacing between the members of $h_{i-1} \circ h_i(W_i)$ where h_{i-1} has caused a great deal of stretching along the horizontal boundary of X_{i-1}. The stretching between members of $h_i(W_i)$ can be large since they are at a distance of $(a_i - s_i)/2$ from the horizontal boundary of X_{i-1} and since

$$\frac{a_i - s_i}{2} > \frac{a_i - b_i}{2} > \frac{a_i - (a_i/128)}{2} > \frac{127a_i}{256} > \frac{a'_{i-1}}{64}.$$

Thus members of $h_i(W_i)$ are actually pulled away from the horizontal boundary of X_{i-1} and from each other by h_{i-1}. Fortunately the cell-pieces of cells in Q_{i-1} where this pulling occurs are quite narrow. In fact the width of the cell pieces is less than b_{i-1} since

$$\frac{a_{i-1} - s_{i-1}}{k_{i-1}} < \frac{a_{i-1}}{k_{i-1}} < a'_{i-1} < d_{i-1} < b_{i-1}.$$

Also the columns of holes are within a horizontal distance of b_i from each other and so the maximum that two holes can be pulled apart in the vertical distance by h_{i-1} is

$$\frac{4b_i k_{i-1}(c_{i-1} + b_{i-1})}{a_{i-1} - s_{i-1}},$$

and so by 1c) we have that

$$\frac{4b_i k_{i-1}(c_{i-1} + b_{i-1})}{a_{i-1} - s_{i-1}} < b_{i-1}.$$

See Figure 10B. Therefore there is $w_i \in W_i$ so that

$$d(p, h_{i-1} \circ h_i(w_i)) < 4b_{i-1} < \delta_{i-1}$$

as desired. $\square$

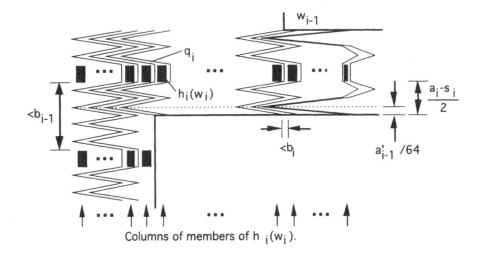

FIGURE 10A. Shows the images of some of the holes in X_{i+1} under h_i.

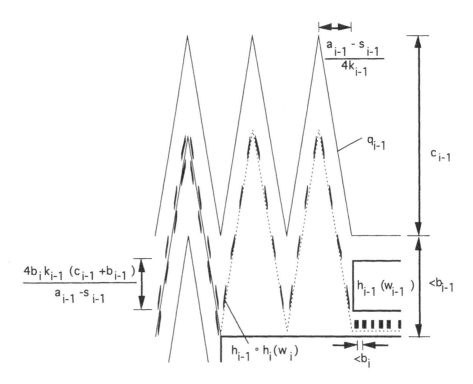

FIGURE 10B. Shows the images of some of the holes in X_{i+1} under $h_{i-1} \circ h_i$.

4.3. The Decomposition is Continuous. To prove that G is a continuous decomposition of Y we state and prove the following lemma. This lemma is a modification of Proposition 3.1 of Lewis and Walsh in [7] and our proof follows the outline of theirs. We will use this lemma by showing that the sequence $\{P_n\}_{n=1}^{\infty}$ in our construction satisfies the five conditions of the lemma.

Lemma 7. *Let $\{Y_n\}_{n=1}^{\infty}$ be a sequence of subspaces of the metric space Y_0 and let $\{P_n\}_{n=1}^{\infty}$ be a sequence satisfying the following five conditions:*

(1) *For each n,*
   (i) *the set P_n is a locally finite collection of non-empty compact subsets of Y_n with $P_n^* = Y_n$;*
   (ii) *the elements of P_n have pairwise disjoint interiors;*
   (iii) *for each $p_n \in P_n$ we have $\mathrm{Cl}(\mathrm{Int}(p_n)) = p_n$;*
   (iv) *the closure of each component of $Y_n \backslash Y_{n+1}$ is contained in the interior of some unique $p_n \in P_n$;*
   where the notions of closure and interior above are all relative to Y_0.
(2) *For each $p_{n-1} \in P_{n-1}$ we have that $\mathrm{st}^3(p_{n-1}, P_n)^* \subset \mathrm{st}(p_{n-1}, P_{n-1})^*$.*
(3) *There is a positive number L such that for each pair $p_n, p_n' \in P_n$ such that $p_n \cap p_n' \neq \emptyset$, we have $p_n \subset \mathrm{N}_{L/2^n}(p_n')$.*
(4) *There is a positive number K such that for each $p_n \in P_n$, there is a $p_{n-1} \in P_{n-1}$ with $p_n \cap p_{n-1} \neq \emptyset$ and $p_{n-1} \subset \mathrm{N}_{K/2^n}(p_n)$.*
(5) *If $p_n \in P_n$ and $p_{n-1}, p_{n-1}' \in P_{n-1}$ and $p_n \cap p_{n-1} \neq \emptyset$ and $p_n \cap p_{n-1}' \neq \emptyset$, then $p_{n-1} \cap p_{n-1}' \neq \emptyset$. (This condition actually follows from conditions (1) and (2) but its inclusion simplifies the proof of the lemma.)*

Define G to be $\{\cap_{n=1}^{\infty} \mathrm{st}(p_n, P_n)^* : \cap_{n=1}^{\infty} p_n \neq \emptyset$ where $p_n \in P_n\}$. Then G is a continuous decomposition of $Y = \cap_{n=1}^{\infty} Y_n$.

Proof. The following observations will be useful in proving the lemma.

Observation 1. If $\cap_{n=1}^{\infty} p_n \neq \emptyset$, then for all $n \in \mathbb{Z}^+$ we have $\mathrm{st}^5(p_{n+2}, P_{n+2})^* \subset \mathrm{st}(p_n, P_n)^*$.

To verify this note that since $p_{n+1} \cap p_{n+2} \neq \emptyset$ we have

$$\mathrm{st}^5(p_{n+2}, P_{n+2}) \subset \mathrm{st}^6(p_{n+1}, P_{n+2}).$$

Thus from Condition (2) of the hypothesis of the lemma

$$
\begin{aligned}
\mathrm{st}^6(p_{n+1}, P_{n+2})^* &= \mathrm{st}^3(\mathrm{st}^3(p_{n+1}, P_{n+2})^*, P_{n+2})^* \\
&\subset \mathrm{st}^3(\mathrm{st}(p_{n+1}, P_{n+1})^*, P_{n+2})^* \\
&\subset \mathrm{st}(\mathrm{st}(p_{n+1}, P_{n+1})^*, P_{n+1})^* \\
&= \mathrm{st}^2(p_{n+1}, P_{n+1})^*.
\end{aligned}
$$

but $p_n \cap p_{n+1} \neq \emptyset$, so $\mathrm{st}^2(p_{n+1}, P_{n+1})^* \subset \mathrm{st}^3(p_n, P_{n+1})^* \subset \mathrm{st}(p_n, P_n)^*$.

Observation 2. For every $n \in \mathbb{Z}^+$ and $A \subset Y_n$, we have that

$$\mathrm{st}(A, P_n)^* \subset \mathrm{Int}_{Y_n}(\mathrm{st}^2(A, P_n)^*).$$

To verify this it will suffice to show that if $y \in Y_n$ then $y \in \mathrm{Int}_{Y_n}(\mathrm{st}(y, P_n)^*)$, for if $y \in \mathrm{st}(A, P_n)^*$ then $\mathrm{st}(y, P_n) \subset \mathrm{st}^2(A, P_n)$ and $y \in \mathrm{Int}_{Y_n}(\mathrm{st}(y, P_n)^*) \subset \mathrm{Int}_{Y_n}(\mathrm{st}^2(A, P_n)^*)$. Let $P = \{p_n \in P_n : y \notin p_n\}$ and let $V = U \backslash P^*$ where U is open in Y_n and $y \in U$. Now V is open since P_n is locally finite and $y \in V$. Also $V \subset \mathrm{st}(y, P_n)^*$ since if $x \in V$ there is, because $P_n^* = Y_n$, a $p_n \in P_n$ so $x \in p_n$. But $y \in p_n$ since if $y \notin p_n$ then x would be in P^* and not in V. Thus $x \in \mathrm{st}(y, P_n)^*$, and Observation 2 is verified.

Observation 3. *If $g = \cap_{n=1}^{\infty} \mathrm{st}(p_n, P_n)^*$ and $\cap_{n=1}^{\infty} p_n \neq \emptyset$, then for each $N \in \mathbb{Z}^+$ we have $g \subset \mathrm{Int}_{Y_N}(\mathrm{st}(p_N, P_N)^*)$.*

This is true because Observation 1 implies that $\mathrm{st}^5(p_{N+2}, P_{N+2})^* \subset \mathrm{st}(p_N, P_N)^*$ and so with Observation 2 we get

$$
\begin{aligned}
g &\subset \mathrm{st}(p_{N+2}, P_{N+2})^* \\
&\subset \mathrm{Int}_{Y_N}(\mathrm{st}^2(p_{N+2}, P_{N+2})^*) \\
&\subset \mathrm{Int}_{Y_N}(\mathrm{st}^5(p_{N+2}, P_{N+2})^*) \\
&\subset \mathrm{Int}_{Y_N}(\mathrm{st}(p_N, P_N)^*).
\end{aligned}
$$

We now show that G is indeed a decomposition of Y. Clearly $G^* = Y$. Let $g = \cap_{n=1}^{\infty} \mathrm{st}(p_n, P_n)^*$ where $\cap_{n=1}^{\infty} p_n \neq \emptyset$ and $g' = \cap_{n=1}^{\infty} \mathrm{st}(p'_n, P_n)^*$ where $\cap_{n=1}^{\infty} p'_n \neq \emptyset$. Assume that $g \cap g' \neq \emptyset$. We will show that $g = g'$. Since $g \cap g' \neq \emptyset$ we have that if $n \in \mathbb{Z}^+$, then $\mathrm{st}(p_{n+2}, P_{n+2})^* \cap \mathrm{st}(p'_{n+2}, P_{n+2})^* \neq \emptyset$. Thus $\mathrm{st}(p'_{n+2}, P_{n+2})^* \subset \mathrm{st}^4(p_{n+2}, P_{n+2})^*$. See Figure 11. So by Observation 1 $\mathrm{st}(p'_{n+2}, P_{n+2})^* \subset \mathrm{st}(p_n, P_n)^*$ and $g' \subset g$. Similarly $g \subset g'$.

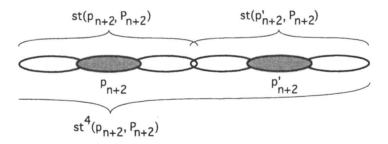

FIGURE 11. If $\mathrm{st}(p_{n+2}, P_{n+2})^* \cap \mathrm{st}(p'_{n+2}, P_{n+2})^* \neq \emptyset$, then $\mathrm{st}(p'_{n+2}, P_{n+2})^* \subset \mathrm{st}^4(p_{n+2}, P_{n+2})^*$.

To show that G is upper semi-continuous we will show that for any $g \in G$ and $U' \subset Y$ open in Y with $g \subset U'$ there is an open set V' in Y with $g \subset V'$ so that if $g' \in G$ and $g' \cap V' \neq \emptyset$, then $g' \subset U'$. Let $g = \cap_{n=1}^{\infty} \mathrm{st}(p_n, P_n)^*$ where $\cap_{n=1}^{\infty} p_n \neq \emptyset$ and let U' be open in Y with $g \subset U'$. Let U be open in Y_0 so that $U \cap Y = U'$. Since for every $n \in \mathbb{Z}^+$ each $p_n \in P_n$ is compact and P_n is locally finite we have that $\mathrm{st}(p_n, P_n)^*$ is compact. Thus there is an $N \in \mathbb{Z}^+$ so that $\mathrm{st}(p_N, P_N)^* \subset U$.

Let

$$V' = \text{Int}_{Y_{N+2}}(\text{st}(p_{N+2}, P_{N+2})^*) \cap Y.$$

From Observation 3 we have that $g \subset V'$. Now let $g' = \cap_{n=1}^{\infty} \text{st}(p'_n, P_n)^*$ where $\cap_{n=1}^{\infty} p'_n \neq \emptyset$ and $g' \cap V' \neq \emptyset$. Thus $\text{st}(p_{N+2}, P_{N+2})^* \cap \text{st}(p'_{N+2}, P_{N+2})^* \neq \emptyset$ but as before (recall Figure 11), this means that $g' \subset \text{st}(p'_{N+2}, P_{N+2})^* \subset \text{st}^4(p_{N+2}, P_{N+2})^*$ and so by Observation 1 $g' \subset \text{st}(p_N, P_N)^* \subset U$ and $g' \subset U'$.

We now will show that G is a lower semi-continuous decomposition of Y, and thus that G is continuous. Let $g \in G$. To show that G is lower semi-continuous at g it suffices to show that for every $\epsilon > 0$ there is an open set V' in Y containing g so that if $g' \in G$ and $g' \cap V' \neq \emptyset$, then $g \subset \text{N}_\epsilon(g')$. To accomplish this we state the following Claims taken from [7]. Because of the changes we have made to Condition (1) in [7] to obtain our Condition (1) we give an expanded proof of Claim A. The others follow easily from the preceding claims as in [7].

Claim A. *If $p_n \in P_n$ and $p_{n-1} \in P_{n-1}$ with $p_n \cap p_{n-1} \neq \emptyset$, then*

$$\text{st}(p_{n-1}, P_{n-1})^* \subset \text{N}_{(4L+K)/2^n}(p_n).$$

To verify this recall that Condition (4) guarantees a $p'_{n-1} \in P_{n-1}$ so that $p_n \cap p'_{n-1} \neq \emptyset$ and $p'_{n-1} \subset \text{N}_{K/2^n}(p_n)$. By condition (5), $p_{n-1} \cap p'_{n-1} \neq \emptyset$. Now Condition (3) implies that $\text{st}^2(p'_{n-1}, P_{n-1})^* \subset \text{N}_{2L/2^{n-1}}(p'_{n-1})$. Therefore $\text{st}(p_{n-1}, P_{n-1})^* \subset \text{st}^2(p'_{n-1}, P_{n-1})^* \subset \text{N}_{2L/2^{n-1}}(p'_{n-1}) \subset \text{N}_{2L/2^{n-1}}(\text{N}_{K/2^n}(p_n)) \subset \text{N}_{(4L+K)/2^n}(p_n)$.

Claim B. *If $\cap_{n=1}^{\infty} p_n \neq \emptyset$, then*

$$\text{st}(p_{n-1}, P_{n-1})^* \subset \text{N}_{(4L+K)(2-1/2^m)/2^n}(p_{n+m}).$$

Claim C. *Let $g = \cap_{n=1}^{\infty} \text{st}(p_n, P_n)^*$ where $\cap_{n=1}^{\infty} p_n \neq \emptyset$, then*

$$\text{st}(p_{n-1}, P_{n-1})^* \subset \text{N}_{(12L+3K)/2^n}(g).$$

Claim D. *Let $g = \cap_{n=1}^{\infty} \text{st}(p_n, P_n)^*$ where $\cap_{n=1}^{\infty} p_n \neq \emptyset$ and $g' = \cap_{n=1}^{\infty} \text{st}(p'_n, P_n)^*$ where $\cap_{n=1}^{\infty} p'_n \neq \emptyset$. If $\text{st}(p_n, P_n)^* \cap \text{st}(p'_n, P_n)^* \neq \emptyset$, then*

$$g \subset \text{N}_{(18L+3K)/2^{n+1}}(g').$$

Now we continue with our proof that G is lower semi-continuous. Let $\epsilon > 0$ and let $g = \cap_{n=1}^{\infty} \text{st}(p_n, P_n)^*$ where $\cap_{n=1}^{\infty} p_n \neq \emptyset$. Pick $N \in \mathbb{Z}^+$ so that $(18L + 3K)/2^{N+1} < \epsilon$. Let $V' = \text{Int}_{Y_N}(\text{st}(p_N, P_N)^*) \cap Y$. Now $g \subset V'$ and if $g' \cap V' \neq \emptyset$ where $g' = \cap_{n=1}^{\infty} \text{st}(p'_n, P_n)^*$ and $\cap_{n=1}^{\infty} p'_n \neq \emptyset$, then $\text{st}(p_N, P_N)^* \cap \text{st}(p'_N, P_N)^* \neq \emptyset$. But by Claim D this means that $g \subset \text{N}_\epsilon(g')$. $\square$

To show that our construction does indeed result in a continuous decomposition of Y it is only necessary to show that the first four conditions of Lemma 7 are satisfied by our construction.

4.3.1. *Condition* (1). We let Y_0 be E^2 and condition (1) is quickly verified.

4.3.2. *Condition* (2). To show that condition (2) holds it suffices to show that if $q_{n-1} \in Q_{n-1}$ then $\mathrm{st}^3(h_{n-1}^{-1}(q_{n-1}), Q_n)^* \subset \mathrm{st}(h_{n-1}^{-1}(q_{n-1}), h_{n-1}^{-1}(Q_{n-1})^*)$. See Figure 12. Note that for cell $q_n \in Q_n$ the width of q_n is a_n and the height of q_n is less than $2c_n + s_{n-1}$. But $3a_n < d_{n-1}$ and $2c_n + s_{n-1} + 3b_n < a_{n-1}$. Thus the desired inclusion holds.

4.3.3. *Condition* (3). We will show that for each pair $p_n, p_n' \in P_n$ with $p_n \cap p_n' \neq \emptyset$ we have $p_n \subset \mathrm{N}_{L/2^n}(p_n')$ where $L = 1/64$. Let $q_n = H_n^{-1}(p_n)$ and $q_n' = H_n^{-1}(p_n')$. Thus both q_n and q_n' are in Q_n. We consider two cases. See Figure 13.

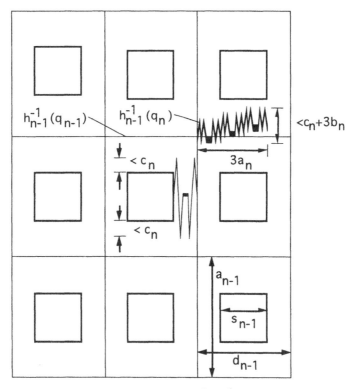

FIGURE 12. Shows that $\mathrm{st}^3(h_{n-1}^{-1}(q_{n-1}), Q_n) \subset$
$\mathrm{st}(h_{n-1}^{-1}(q_{n-1}), h_{n-1}^{-1}(Q_{n-1}))$.

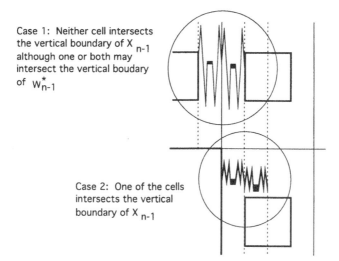

Case 1: Neither cell intersects the vertical boundary of X$_{n-1}$ although one or both may intersect the vertical boudary of w$_{n-1}^*$

Case 2: One of the cells intersects the vertical boundary of X$_{n-1}$

FIGURE 13. Shows the two Cases in proving Condition 3.

Case 1: (Assume neither q_n and q_n' intersect the vertical boundary of X_{n-1}). See Figure 14 showing this case. Consider $h_{n-1}(q_n)$ and $h_{n-1}(q_n')$. Our figure shows the situation when q_n and q_n' are in different strips of $\widehat{R}_n$. The other situation is similar. In either situation

$$h_{n-1}(q_n) \subset \mathrm{N}_{2(a_{n-1}-s_{n-1})/k_{n-1}+s_{n-1}+b_{n-1}}\big(h_{n-1}(q_n')\big).$$

This follows from the fact that $b_n < a_n/(2k_{n-1}) < (a_{n-1}-s_{n-1})/k_{n-1}$ and the fact that $c_n/2 > (a_{n-1}-s_{n-1})/k_{n-1}$. To see the first fact note that $a_n < b_{n-1}/4 < a_{n-1}/512$ and that $a_{n-1} - s_{n-1} > 127a_{n-1}/128$. The second fact follows from $c_n/2 = a_{n-1}/18 > (a_{n-1}-s_{n-1})/k_{n-1}$. Since

$$\frac{a_{n-1}-s_{n-1}}{k_{n-1}} < \frac{a_{n-1}}{k_{n-1}} < a_{n-1}' < d_{n-1} < b_{n-1}$$

and since $s_{n-1} < d_{n-1} < b_{n-1}$ we have that

$$h_{n-1}(q_n) \subset \mathrm{N}_{4b_{n-1}}\big(h_{n-1}(q_n')\big).$$

Thus

$$h_{n-1}(q_n) \subset \mathrm{N}_{\delta_{n-1}}\big(h_{n-1}(q_n')\big).$$

But by choice of δ_{n-1} we have that

$$p_n \subset \mathrm{N}_{1/2^{n+6}}\big(H_{n-1} \circ h_{n-1}(q_n')\big) = \mathrm{N}_{1/2^{n+6}}\big(p_n'\big) \subset \mathrm{N}_{L/2^n}\big(p_n'\big).$$

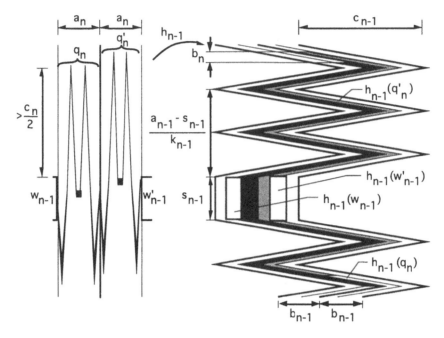

FIGURE 14. Shows that

$$h_{n-1}(q_n) \subset \mathrm{N}_{2(\frac{a_{n-1}-s_{n-1}}{k_{n-1}})+s_{n-1}+b_{n-1}}(h_{n-1}(q'_{n-1})).$$

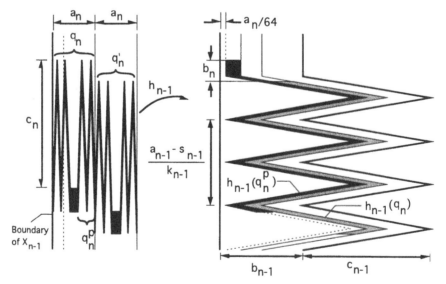

FIGURE 15. Shows that $h_{n-1}(q'_n) \subset \mathrm{N}_{4b_{n-1}}(h_{n-1}(q^p_{n-1}))$
and that $h_{n-1}(q_n) \subset \mathrm{N}_{4b_{n-1}}(h_{n-1}(q'_{n-1}))$

Case 2: (Assume q_n intersects the vertical boundary of X_{n-1}). See Figure 15.
It suffices to show $h_{n-1}(q_n) \subset \mathrm{N}_{\delta_{n-1}}(h_{n-1}(q'_n))$ and $h_{n-1}(q'_n) \subset \mathrm{N}_{\delta_{n-1}}(h_{n-1}(q_n))$.

Notice that both q_n and q'_n must be in the same strip of $\widehat{R}_n$ since q_n is assumed to intersect the vertical boundary of X_{n-1} and $a_n = (d_{n-1})/4$. Also $a'_{n-1}/64 = a_n/64$ since $a_n = a'_{n-1}$. Now since q_n consists of two identical halves at least one of which must contain a cell-piece, q_n^p, that is at a distance of greater than $a_n/64$ from the vertical boundary of X_{n-1}. Thus

$$h_{n-1}(q'_n) \subset \mathrm{N}_{4b_{n-1}}(h_{n-1}(q_n^p)) \subset \mathrm{N}_{\delta_{n-1}}(h_{n-1}(q_n)).$$

Since $(a_{n-1} - s_{n-1})/k_{n-1} < b_{n-1}$ we also have that $h_{n-1}(q_n) \subset \mathrm{N}_{4b_{n-1}}(h_{n-1}(q'_n))$. Thus

$$h_{n-1}(q_n) \subset \mathrm{N}_{4b_{n-1}}(h_{n-1}(q'_n)) \subset \mathrm{N}_{\delta_{n-1}}(h_{n-1}(q'_n)).$$

4.3.4. *Condition* (4). Let $p_n \in P_n$. We will show that there is a $p_{n-1} \in P_{n-1}$ with $p_n \cap p_{n-1} \neq \emptyset$ so that $p_{n-1} \subset \mathrm{N}_{K/2^n}(p_n)$ where $K = 1/16$. To accomplish this we will prove two lemmas. In order to state the lemmas we must introduce the following notation and terminology. For any $q_n \in Q_n$ let q_n^p be any cell-piece of q_n. Thus the width of q_n^p is $(a_n - s_n)/k_n$. When n is odd we say that $q_n \in Q_n$ *crosses* q_{n-1}^p if q_n intersects both top and bottom edges of $h_{n-1}^{-1}(q_{n-1}^p)$. (When n is even we say that $q_n \in Q_n$ *crosses* q_{n-1}^p if q_n intersects both left and right edges of $h_{n-1}^{-1}(q_{n-1}^p)$). Note that $h_{n-1}^{-1}(q_{n-1}^p)$ is a rectangle of width d_{n-1} and height $(a_{n-1} - s_{n-1})/k_{n-1}$.

We now state and prove the following lemmas:

Lemma 8. *If $q_n \in Q_n$ and q_n^p is further than $a_n/64$ from the vertical boundary of X_{n-1}, then*

$$h_{n-1}(q_n) \subset \mathrm{N}_{\delta_{n-1}}(h_{n-1}(q_n^p)).$$

Proof. Let $q_n \in Q_n$. We will consider two cases:

Case 1: (Assume q_n does not intersect the vertical boundary of X_{n-1}). Figure 16 illustrates this case. With arguments similar to those above it can be seen that $h_{n-1}(q_n) \subset \mathrm{N}_{b_{n-1}}(h_{n-1}(q_n^p)) \subset \mathrm{N}_{\delta_{n-1}}(h_{n-1}(q_n^p))$.

Case 2: (Assume q_n intersects the vertical boundary of X_{n-1}). Figure 17 illustrates this case. Again it can be shown that $h_{n-1}(q_n) \subset \mathrm{N}_{2b_{n-1}}(h_{n-1}(q_n^p))$. Therefore $h_{n-1}(q_n) \subset \mathrm{N}_{\delta_{n-1}}(h_{n-1}(q_n^p))$. $\square$

Lemma 9. *If $q_n \in Q_n$ crosses q_{n-1}^p where $q_{n-1} \in Q_{n-1}$, then*

$$q_{n-1}^p \subset \mathrm{N}_{\delta_{n-1}}(h_{n-1}(q_n)).$$

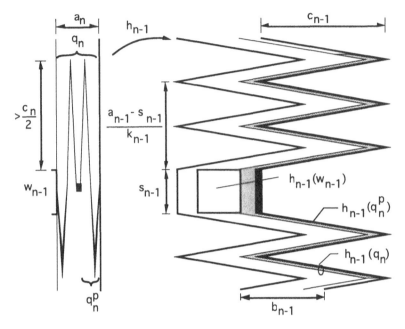

FIGURE 16. Shows that $h_{n-1}(q_n) \subset N_{b_{n-1}}(h_{n-1}(q_n^p))$ when q_n does not intersect vertical boundary of X_{n-1}.

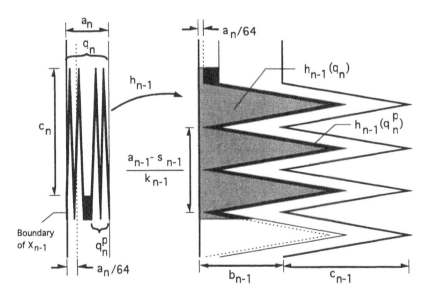

FIGURE 17. Shows that $h_{n-1}(q_n) \subset N_{2b_{n-1}}(h_{n-1}(q_n^p))$ when q_n does intersect vertical boundary of X_{n-1}.

Proof. Let $q_n \in Q_n$, and $q_{n-1} \in Q_{n-1}$ with cell-piece q_{n-1}^p so that q_n crosses q_{n-1}^p; that is, q_n contains points on both the top and bottom boundaries of the rectangle $h_{n-1}^{-1}(q_{n-1}^p)$. We will look at two cases:

Case 1: (Assume q_n does not intersect the vertical boundary of X_{n-1}). Figure 18 shows this case and that $q_{n-1}^p \subset N_{b_{n-1}}(h_{n-1}(q_n)) \subset N_{\delta_{n-1}}(h_{n-1}(q_n))$.

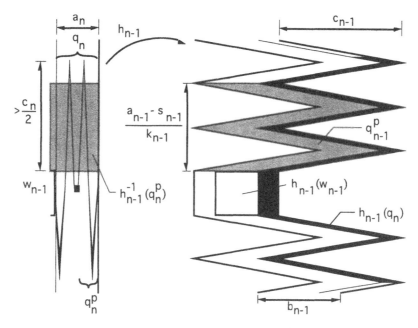

FIGURE 18. Shows that in Case 1 of Lemma 9 $\quad q_{n-1}^p \subset N_{b_{n-1}}(h_{n-1}(q_n))$.

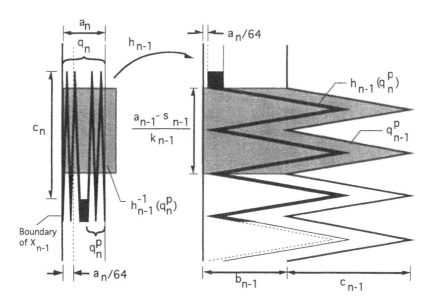

FIGURE 19. Shows that in Case 2 of Lemma 9 $\quad q_{n-1}^p \subset N_{4b_{n-1}}(h_{n-1}(q_n^p))$.

Case 2: (Assume q_n intersects the vertical boundary of X_{n-1}). Figure 19 illustrates this case. Let q_n^p be a piece of q_n which is at a distance of more than

$a_n/64$ from the vertical boundary of X_{n-1}. Thus $q_{n-1}^p \subset \mathrm{N}_{4b_{n-1}}(h_{n-1}(q_n^p)) \subset \mathrm{N}_{\delta_{n-1}}(h_{n-1}(q_n))$. $\square$

Now we can verify that condition (4) holds. Let $q_n = H_n^{-1}(p_n)$. Now there are two cases: Either q_n is of Type 1 or of Type 2.

If q_n is of Type 1 then the height of q_n is greater than $c_n = a_{n-1}/9$ and at least half of this height must lie in a strip of R_{n-1}. But

$$\frac{2(a_{n-1}-s_{n-1})}{k_{n-1}} + \frac{a_{n-1}}{64} < \frac{2a_{n-1}}{k_{n-1}} + \frac{a_{n-1}}{64} < 3a_{n-1}/64 < \frac{1}{2}\frac{a_{n-1}}{9} = \frac{c_n}{2}.$$

So there is a piece q_{n-1}^p of some q_{n-1} which is crossed by q_n and which is at a distance of more than $a_{n-1}/64$ from the horizontal boundary of X_{n-2}. Thus Lemma 8 will be applicable.

If q_n is of Type 2 then q_n lies along side of a hole $w_{n-1} \in W_{n-1}$. Let $q_{n-1} \in Q_{n-1}$ so that w_{n-1} is the hole corresponding to q_{n-1}. Now q_n extends beyond w_{n-1} (above or below) by at least $c_n/2$. But as seen before $c_n/2 > (a_{n-1}-s_{n-1})/k_{n-1}$. Thus there is a cell-piece of q_{n-1}, say q_{n-1}^p, which is crossed by q_n. Also note that q_{n-1}^p can be the cell-piece of q_{n-1} that is right next to the rectangular connecting piece of q_{n-1} and so its distance from the horizontal boundary of X_{n-2} is at least

$$\frac{a_{n-1}-s_{n-1}}{2} - \frac{a_{n-1}-s_{n-1}}{k_{n-1}} > \frac{127}{256}a_{n-1} - \frac{a_{n-1}}{64} > \frac{a_{n-1}}{64}.$$

Thus Lemma 8 will again be applicable.

In either case by Lemma 8 we have that $h_{n-2}(q_{n-1}) \subset \mathrm{N}_{\delta_{n-2}}(h_{n-2}(q_{n-1}^p)))$. Thus

$$p_{n-1} = H_{n-1}(q_{n-1}) = H_{n-2} \circ h_{n-2}(q_{n-1}) \subset \mathrm{N}_{1/2^{n+5}}(H_{n-2} \circ h_{n-2}(q_{n-1}^p)).$$

But by Lemma 9 we have that $q_{n-1}^p \subset \mathrm{N}_{\delta_{n-1}}(h_{n-1}(q_n))$ and

$$H_{n-1}(q_{n-1}^p) \subset \mathrm{N}_{1/2^{n+6}}(H_{n-1} \circ h_{n-1}(q_n)) = \mathrm{N}_{1/2^{n+6}}(p_n).$$

Combining this with the inclusion above we get

$$\begin{aligned}
p_{n-1} &\subset \mathrm{N}_{1/2^{n+5}}(H_{n-2} \circ h_{n-2}(q_{n-1}^p)) \\
&= \mathrm{N}_{1/2^{n+5}}(H_{n-1}(q_{n-1}^p)) \\
&\subset \mathrm{N}_{1/2^{n+5}}(\mathrm{N}_{1/2^{n+6}}(p_n)) \\
&\subset \mathrm{N}_{1/2^{n+4}}(p_n)) \\
&= \mathrm{N}_{K/2^n}(p_n)
\end{aligned}$$

where $K = 1/16$. Therefore the decomposition satisfies Condition (4).

Above we have shown that G is a continuous decomposition of Y. We can also show that each member of G is non-degenerate.

Lemma 10. *Each member of G is non-degenerate.*

Proof. Note that since the construction satisfies Claim C used in the proof of Lemma 7 when $K = 1/16$ and $L = 1/64$ we have that $p_1 \subset \mathrm{st}(p_{2-1}, P_{2-1})^* \subset \mathrm{N}_{(12L+3K)/2^2}(g) = \mathrm{N}_{3/32}(g)$. But the diameter of p_1 is at least $1/4$ and so the diameter of g is at least $1/16$. $\square$

4.4. The Decomposition is the Sierpiński Curve.

To show that the decomposition G is homeomorphic to the Sierpiński curve we first need the following lemma.

Lemma 11. *No member of G contains more than one point of the set $\{\mathrm{Bd}(U) : U$ is a component of $Y^c\}^*$.*

Proof. We let $\mathbb{Q}_G = \{\mathrm{Bd}(U) : U$ *is a component of* $Y^c\}^*$. By way of contradiction suppose there is a $g = \cap_{n=1}^{\infty} \mathrm{st}(p_n, P_n)^*$ where $\cap_{n=1}^{\infty} p_n \neq \emptyset$ and $g \cap \mathbb{Q}_G$ contains two points x and y. We consider two cases:

Case 1) There is a single hole w in Y with $x, y \in \mathrm{Bd}(w)$ where $w_k \in W_k$ and $H_{k+1}(w_k) = w$. Let $x_n = H_n^{-1}(x)$, $y_n = H_n^{-1}(y)$, and $q_n = H_n^{-1}(p_n)$ for all $n \in \mathbb{Z}^+$ where $n > k$. Because of the shape of the cells in $\mathrm{st}(q_n, Q_n)$ which are around the hole w_k when $n > k$, we can always have chosen $x, y \in g \cap \mathbb{Q}_G$ so than x_n and y_n are on the same edge of w_k. For example, if x_n and y_n are on different edges, then these edges must intersect at a common vertex. The image of this vertex under H_n will be in $g \cap \mathbb{Q}_G$ and could have been chosen instead of y. Without loss of generality we assume that x_n and y_n are both on the same edge of w_k and that this edge is vertical.

Let $N \in \mathbb{Z}^+$ so that N is odd, $N > k$, and so that $1/2^{N+7} < \mathrm{d}(x, y)$. Note that $x_N, y_N \in \mathrm{st}(q_N, Q_N)^*$. Now x_N and y_N both lie on the same vertical edge of w_k and so both lie on a vertical transversal of $\mathrm{st}(q_N, Q_N)^*$. Thus $\mathrm{d}(x_N, y_N) < 3b_N < \delta_N$ and so $\mathrm{d}(x, y) < 1/2^{N+7}$ contrary to our choice of N.

Case 2) There are two holes w and w' in Y with $x \in \mathrm{Bd}(w) \cap g$ and $y \in \mathrm{Bd}(w') \cap g$ where $w = H_{k+1}(w_k)$ and $w' = H_{k'+1}(w_{k'})$. We assume that $k \geq k'$, so $w' = H_{k+1}(w_{k'})$. Now $w_k \neq w_{k'}$ so without loss of generality assume the distance between the x-projection of w_k and $w_{k'}$ is non-zero, say t. Let $x_n = H_n^{-1}(x)$, $y_n = H_n^{-1}(y)$, and $q_n = H_n^{-1}(p_n)$ for all $n \in \mathbb{Z}^+$ where $n > k$. Now $\lim_{n \to \infty} a_n = 0$ so pick $N \in \mathbb{Z}^+$ so that N is odd, $N > k + 1$, and $a_N < t/3$. Because each cell of Q_N is confined to a strip of width a_N, we see that the x-projection of $\mathrm{st}(q_N, Q_N)^*$ has diameter of no more than $3a_N$ which is less than t. Thus $\mathrm{st}(q_N, Q_N)^*$ cannot possibly intersect both $\mathrm{Bd}(w_k)$ and $\mathrm{Bd}(w_{k'})$ and so g does not intersect both $\mathrm{Bd}(w)$ and $\mathrm{Bd}(w')$. $\square$

Lemma 12. *The set $G = \{\cap_{n=1}^{\infty} \mathrm{st}(p_n, P_n)^* : \cap_{n=1}^{\infty} p_n \neq \emptyset\}$ is the Sierpiński curve.*

Proof. Let $\pi : Y \to G$ be the natural projection. Now G is locally connected because Y is. Since G is upper semi-continuous we know that by adding the points of the holes of Y we get an upper semi-continuous decomposition of E^2 which we will call G'. Thus we extend π to $\hat{\pi} : E^2 \to G'$. Now no member of G' separates the plane so by R.L. Moore's Theorem [4] we know that G' is homeomorphic to the plane. Thus G is planar. By Lemma 11 we get that $(\widehat{\pi})|\operatorname{Cl}(U)$ is a homeomorphism for any U a component of Y^c. Thus the boundaries of the components of Y^c are simple closed curves, are dense in G, and have diameters that go to zero. Therefore G is homeomorphic to the Sierpiński curve by Toruńczyk's Lemma [5]. $\square$

Thus we have proved the following theorem.

Theorem 13. *There is a monotone open map f from the Sierpiński curve onto the Sierpiński curve so that for each y in the range of f we have that $f^{-1}(y)$ is non-degenerate.*

REFERENCES

1. R.D. Anderson, *On monotone interior mappings in the plane*, Trans. Amer. Math. Soc. **73** (1952), 211–222.
2. J.J. Charatonik, *On generalized homogeneity of locally connected plane continua*, Comment. Math. Univ. Carolin. **32** (1991), 769–774.
3. R. Engelking and K. Sieklucki, *Topology A Geometric Approach*, (Heldermann Verlag, Berlin 1992).
4. R.J. Daverman, *Decompositions of Manifolds*, (Academic Press, New York 1986).
5. W. Jakobsche, *The Bing-Borsuk conjecture is stronger then the Poincaré conjecture*, Fund. Math. **106** (1980), 127–134.
6. K. Kuratowski, *Topology*, (Academic Press, New York 1968).
7. W. Lewis and J.J. Walsh, *A continuous decomposition of the plane into pseudo-arcs*, Houston J. Math. **4** (1978), 209–222.
8. J.R. Prajs, *On open homogeneity of closed balls in the Euclidean spaces*, Preprint.
9. C.R. Seaquist, *A new continuous cellular decomposition of the disk into non-degenerate elements*, Top. Proc., to appear.
10. G.T. Whyburn, *Topological characterization of the Sierpiński curve*, Fund. Math. **45** (1958), 320–324.
11. G.T. Whyburn, *Analytical Topology*, Amer. Math. Soc. Colloq. Publ. **28**(1942).

E-mail address: seaqucr@mail.auburn.edu "Carl R. Seaquist"

25
A WILD k-DIMENSIONAL MENGER COMPACTUM IN R^{2k+1}, ALL CELL-LIKE SUBSETS OF WHICH ARE CELLULAR

RICHARD B. SHER Department of Mathematics, University of North Carolina at Greensboro, Greensboro, North Carolina 27412, USA

ABSTRACT. A wildly embedded Menger compactum is constructed in R^{2k+1}, $k \geq 2$, all of whose cell-like subsets are cellular.

In their recent survey of Menger manifold theory [1, Problem 4.1.12], Chigogidze, Kawamura and Tymchatyn ask if there is a wild embedding of the k-dimensional Menger compactum in R^{2k+1}, all cell-like subsets of which are cellular. We show that there is such an embedding if $k \geq 2$. (Note that the case $k = 0$ is trivial, while in the case $k = 1$ even the standard (tamely embedded) Menger manifold contains cell-like sets that are noncellular.) We begin by constructing such an embedding that is locally tame modulo an arc. This in turn suggests the construction in (2) in which the embedded Menger compactum is wild at each of its points. Thanks are extended to Bob Daverman for suggesting the use of the disjoint disks property to show that the decomposition used in the second construction is shrinkable.

1. Let μ^k denote the canonical (i.e. as constructed in [4]) universal k-dimensional Menger compactum in R^{2k+1}, where $k \geq 2$. It follows from the construction that dem $\mu^k = k$. Since $k \geq 2$, it also follows that every cell-like subset of μ^k satisfies the cellularity criterion and is therefore cellular in R^{2k+1}. By [5, Example 2.6.6] there is an embedding $h \colon [0,1] \times [0,1] \to R^{2k+1}$ such that

(a) $h([0,1] \times \{1\})$ is everywhere wild and cellular, and

---

1991 *Mathematics Subject Classification.* primary 57N45, secondary 57N60, 57N35, 57M30.
Key words and phrases. Menger compactum, wild embedding, cell-like set, cellular set, cellularity criterion, disjoint disks property.

(b) $h([0,1] \times [0,1])$ is locally tame modulo $h([0,1] \times \{1\})$.

We may assume that there is an arc B in μ^k such that

(c) $\mu^k \cap h([0,1] \times [0,1]) = h([0,1] \times \{0\}) = B$.

Now let G denote the upper semicontinuous decomposition of R^{2k+1} whose nondegenerate elements are the arcs $h(\{t\} \times [0,1])$, $t \in [0,1]$. It is easy to see by a direct shrinking argument (cf. [2, Chapter II]) that the decomposition space R^{2k+1}/G is topologically equivalent to R^{2k+1} and that the image of $h([0,1] \times [0,1])$ under the natural projection mapping $p \colon R^{2k+1} \to R^{2k+1}/G$ is an everywhere wild and cellular arc A. We shall identify R^{2k+1}/G with R^{2k+1} and shall denote $p(\mu^k) \subset R^{2k+1}$ by M. Then $M \cong \mu^k$, M is locally tame modulo A, and M is wild since A is wild. We claim that each cell-like subset of M is cellular.

Let X be a cell-like subset of M. Since cell-like subsets of $M - A$ are cellular, we may as well suppose that $X \cap A \neq \emptyset$. In order to show that X is cellular it suffices to show that X satisfies the cellularity criterion. Let U be a neighborhood of X, and let V be a neighborhood of X lying in U such that V is nullhomotopic in U. Let α be a loop in $V - X$. We may assume that α is nonsingular and bounds a piecewise linear disk D in U. By [3, Theorem 6], $X \cap A$ is definable by cells. We may use this fact to move D off of $X \cap A$ while staying inside U and keeping α fixed. Now α bounds a disk in $U - (X \cap A)$, and using the fact that M is locally tame modulo A we may make another slight adjustment, again staying inside U and keeping α fixed, to obtain a disk in $U - X$ bounded by α. This shows that X satisfies the cellularity criterion and completes the proof of the claim.

2. We now modify the above construction to obtain an example in which the embedded Menger compactum is wild at each of its points. Let $\{A_i\}$ be a null sequence of pairwise disjoint arcs in μ^k whose union is dense in μ^k. Along the arc A_i we attach a "fin" F_i just as in (c) above, making sure that the F_i's are pairwise disjoint and form a null sequence. We let G denote the upper semicontinuous decomposition of R^{2k+1} whose nondegenerate elements are the appropriate arcs of the fins. We now sketch a proof of the fact that the decomposition space R^{2k+1}/G satisfies the disjoint disks property. Begin by considering two maps of the 2-cell into R^{2k+1}/G. Using the fact that G is a cell-like decomposition, we may approximately lift these maps through the projection mapping $p \colon R^{2k+1} \to R^{2k+1}/G$. It will suffice to show that we can arbitrarily closely approximate the two lifted mappings in R^{2k+1} by ones whose compositions with p have disjoint images. To see that this is possible, first adjust the two mappings by a small amount so that they are tame embeddings with disjoint images. Then alter each slightly so as to miss μ^k. Since the fins used in constructing G form a null sequence, each image will now intersect only finitely many fins. The next adjustment keeps the intersection of the images with the wild arcs of these fins fixed, while moving the images off of the remaining portion of the fins. This results in two embeddings whose images are disjoint and intersect the union of μ^k and the fins only at points of the wild arcs of the fins.

Since each point of such a wild arc intersects precisely one nondegenerate element of G, it follows that the compositions of these embeddings with p are embeddings with disjoint images in R^{2k+1}/G.

Having now established that R^{2k+1}/G satisfies the disjoint disks property, it follows from [2, Theorem 3, pg. 181] that R^{2k+1}/G is topologically equivalent to R^{2k+1}. We denote $p(\mu^k) \subset R^{2k+1}/G$ by M. Then $M \cong \mu^k$, and M fails to be locally tame at each of its points. Let X be a cell-like subset of M. Then X is cellular in R^{2k+1}/G if $p^{-1}(X)$ is cellular in R^{2k+1}. We verify that this is the case by using the cellularity criterion. Let U be a neighborhood of $p^{-1}(X)$, and let V be a neighborhood of $p^{-1}(X)$ lying in U such that V is nullhomotopic in U. Let α be a loop in $V - X$. We may assume that α is nonsingular and bounds a piecewise linear disk D in U. We adjust D much as in the previous construction. To begin, move D slightly to obtain a disk D' bounded by α that misses μ^k. Since the fins used in constructing G form a null sequence, D' intersects at most finitely many of them. Let the union of the fins that intersect D' be denoted by A, and let the union of the noncellular arcs of these fins be denoted by B. There is a neighborhood of B that intersects neither μ^k nor any of the fins other than those that comprise A. The first adjustment to D' changes nothing outside of this neighborhood, and results in a disk D'' in U that intersects $p^{-1}(M)$ only in $A - B$. Now let $\delta > 0$ be the distance from D'' to the union of μ^k and all of the fins not intersected by D''. The proof is completed by moving D'' off of A while displacing no point by as much as δ, resulting in a disk in $U - p^{-1}(X)$ bounded by α.

REFERENCES

1. A. Chigogidze, K. Kawamura and E. D. Tymchatyn, *Menger manifolds*, Marcel Dekker, this volume.
2. Robert J. Daverman, *Decompositions of manifolds*, Academic Press, 1986.
3. D. R. McMillan, Jr., *A criterion for cellularity in a manifold*, Ann. of Math., **79** (1964), 327-337.
4. K. Menger, *Kurventheorie*, Teubner, Berlin-Leipzig, 1932.
5. T. Benny Rushing, *Topological embeddings*, Academic Press, 1973.

E-mail address: sherrb@iris.uncg.edu

26

AN EXTENSION OF JAKOBSCHE'S CONSTRUCTION OF
n-HOMOGENEOUS CONTINUA TO THE NONORIENTABLE
CASE

PAUL R. STALLINGS   Electronic Data Systems, 13736 Riverport Drive, Maryland
Heights, Missouri 63043, USA

ABSTRACT. In 1980, assuming the negation of Poincaré's conjecture, W. Jakobsche
constructed a homogeneous continuum that provided a counterexample to the Bing
- Borsuk conjecture. In 1991, he extended the work to include continua constructed
from other orientable manifolds. This paper contains a completion of the extension
to the nonorientable case, and a proof that the resulting continua are representable,
hence n-homogeneous.

1. INTRODUCTION

A space X is *n-homogeneous* [*countably dense homogeneous*] if for every pair A, B
of n-element subsets [countable dense subsets] of X, there exists a homeomorphism
of $X \to X$ which maps A onto B. A space is *homogeneous* if it is 1-homogeneous.
A space X is *representable* (or *strongly locally homogeneous*) if for every element p
of X and every open neighborhood U of p, there exists an open neighborhood V of
p such that if q is an element of V, then there exists a homeomorphism of X onto
X which maps p to q and is the identity on the complement of U.

A space is called LC^0 at the point p if it is locally arcwise connected at p. A
space is called *semi-LC^0* at the point p if given any neighborhood U of p, there is
a neighborhood V of p such that every point in V can be joined to p with an arc
in the space. If $n > 0$, then a space X is called LC^n (*semi-LC^n*) at the point p if
it is LC^{n-1} (semi-LC^{n-1}) at the point p, and if given any neighborhood U of p,
there is a neighborhood V of p such that the homomorphism $\pi_n(V, p) \to \pi_n(U, p)$

1991 *Mathematics Subject Classification.* primary 54F35, secondary 57P99.
Key words and phrases. n-homogeneous, representable, p.l.-manifold.

347

$[\pi_n(V, p) \to \pi_n(X, p)]$, induced by the inclusion, is trivial. A space that is LC^1 is also called *locally simply connected*.

R.B. Bennett [3] showed that a locally compact representable separable metric space is countably dense homogeneous, which, by a result of G.S. Ungar [12], implies that the space is n-homogeneous. In 1958, R.D. Anderson [2] proved that the Menger universal curve M is n-homogeneous for every n; however, the Menger universal curve is not locally simply connected. In 1975, Ungar [11] proved that every 2-homogeneous metric continuum is locally connected. It was shown by K. Kuperberg, W. Kuperberg, and W.R.R. Transue [8] that the Cartesian product of the circle S^1 and the Menger universal curve is not 2-homogeneous. The Cartesian product of the circle S^1 and the Menger universal curve is homogeneous, of dimension 2 and not locally simply connected. Hence, homogeneous 2-dimensional continua need not be locally simply connected. B. Fitzpatrick when working on [4] asked whether or not a 2-homogeneous continuum of dimension at least two must be locally simply connected. This paper answers Fitzpatrick's question by constructing a representable continuum of dimension n which is not locally simply connected, for every integer $n \geq 2$.

In 1980, Jakobsche [6] proved that if the Poincaré conjecture in dimension 3 is false, then there is a 3-dimensional compact ANR which is homogeneous but is not a manifold. This provided a possible counterexample to the Bing - Borsuk conjecture. The construction is similar to that of the Pontryagin disk; the final continuum is obtained by pasting in copies of another continuum in a dense fashion, which can also be expressed in terms of inverse limits [13]. In 1991, Jakobsche [7] generalized the construction to include continua constructed from other orientable manifolds (see also [1]), which vary over the construction.

The results of this paper were obtained in 1987/88 as part of the author's Ph. D. thesis and were presented at the 94th Annual Meeting of the AMS, January 1988, but were never published. Jakobsche's independent generalization does not include the nonorientable case which warrants this publication.

The natural question, related to Pontryagin's construction, whether by repeatedly and densely inserting Möbius strips into S^2 one can obtain a homogeneous continuum, was put forth in a seminar conducted by Borsuk in Warsaw in the 1960's. This paper answers the question by showing that such continuum is representable. The author would like to point out that it can also be shown that Jakobsche's continua [7] are representable.

2. Basic Definitions and Theorems

For most of the construction we will use the piecewise linear category. The reader may find the basic notions such as p.l.-cell, star of a vertex, combinatorial triangulation, p.l.-manifold, p.l.-homeomorphism, p.l.-isotopy, connected sum, pseudo-radial projection, etc., in [5] or [10].

We will need the following well-known facts:

If M is a p.l.-n-manifold, $\varepsilon > 0$, and k is a p.l.-n-cell in the interior M, then there exists a p.l.-n-cells U_k in M, which is a neighborhood of k, such that U_k is a subset of the ε-neighborhood of k.

The connected sum of two p.l.-n-manifolds is a p.l.-n-manifold.

If B and C are p.l.-n-cells with $B \subset Int(C)$, then the closure of $C - B$ is p.l.-homeomorphic to $\partial B \times [0, 1]$.

If M is a connected piecewise-linear n-manifold and h_1 and h_2 are p.l.-embeddings of $[-1, 1]^n$ into the interior of M, then h_1 is ambient isotopic to one of h_2 or $h_2 \circ r_n$, where $r_n\left(\langle x_1, \cdots, x_n \rangle\right) = \langle -x_1, x_2, \ldots, x_n \rangle$.

The last two are known as the *p.l.-Annulus* and *-Disk Theorems* and can be found in [10] page 36 and 44, respectively.

Definition 1. *A compact manifold (with or without boundary) will be called admissible if it is piecewise linear and connected.*

Definition 2. *If B and C are p.l.-n-cells, and f_1 and f_2 are p.l.-embeddings from ∂B into the interior of C, then f_1 and f_2 are compatible with respect to C if there exists a p.l.-homeomorphism $h : C \to C$ such that $h(x) = x$ for all $x \in \partial C$ and $h|f_1(\partial B) = f_2 \circ f_1^{-1}$.*

Denote the standard n-simplex by Δ^n.

3. THE CONSTRUCTION OF THE CONTINUA

If M and N are admissible n-manifolds, then $C(M, N)$ will be defined to be the intersection of a nested sequence of polyhedra, which will be denoted by $(B_1, B_2, B_3, \ldots)$. Polyhedron B_m will be constructed from an admissible n-manifold K_m.

Let K_1 be a polyhedral underlying space of a combinatorial triangulation of M in $\mathbb{R}^{2n+1}$. Next, pick a $(2n+1)$-simplex σ_α for each n-simplex α in the triangulation of K_1 so that α is a face of σ_α, and such that $\sigma_\alpha \cap \sigma_\beta = \alpha \cap \beta$ for each pair of n-simplices α and β in K_1. Note that σ_α may be formed by taking as its vertices the vertices of α along with a set S of $n + 1$ points that are sufficiently close to the barycenter of α, such that S along with the vertices of α form a linearly independent set. Let B_1 be the union of the set of all σ_α's where α is an n-simplex in the triangulation of K_1.

Assume that B_m and K_m have been constructed. Then construct B_{m+1} and K_{m+1} by induction as follows: For each n-simplex α in the triangulation of K_m, pick two n-simplices α_1 and α_2 in the third barycentric subdivision of the triangulation of K_m so that α_1 and α_2 are disjoint subsets of the interior of α. Next let β be an n-simplex in the triangulation of N and pick two homeomorphisms, $f_{\alpha_1} : N \to N_{\alpha_1}$ and $f_{\alpha_2} : N \to N_{\alpha_2}$ such that:

1. f_{α_i} is piecewise linear and $f_{\alpha_i}(\beta) = \alpha_i$ for $i \in \{1,2\}$;
2. $N_{\alpha_i} \subset \sigma_\alpha$;
3. the simplices of N_{α_i} have diameter no larger than the diameter of α_i;
4. $N_{\alpha_i} \cap \partial \sigma_\alpha = \alpha_i$ for $i \in \{1,2\}$;
5. $N_{\alpha_1} \cap N_{\alpha_2} = \emptyset$;
6. the function $f_{\alpha_1}|\beta$ is not compatible with the function $f_{\alpha_2}|\beta$ in the interior of α.

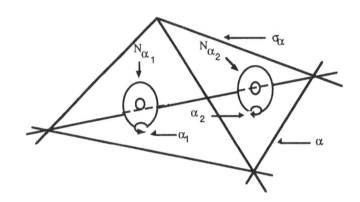

FIGURE 1.   The Placement of N_{α_i} in σ_α.

Let A_m be the set of all α_1 and α_2 such that α is in the triangulation of K_m. Let K_{m+1} be the closure of $K_m \cup (\cup\{N_{\alpha_i} : \alpha_i \in A_m \text{ and } i \in \{1,2\}\}) - \cup A_m$ with the combinatorial triangulation given to it by the third barycentric subdivision of the triangulation of K_m together with the given triangulations of the elements of $\{N_{\alpha_i} : \alpha_i \in A_m\}$. Next, pick a $(2n+1)$-simplex σ_α that is a subset of B_m for each n-simplex α in the triangulation of K_{m+1} such that α is the face of largest diameter of σ_α, and such that $\sigma_\alpha \cap \sigma_\beta = \alpha \cap \beta$ for each pair of n-simplices α and β in K_{m+1}. Let B_{m+1} be the union of the set of all σ_α's where α is an n-simplex in the triangulation of K_{m+1}. Moreover, one can construct the K_1, K_2, $K_3, \ldots$ such that for all $(2n+1)$-simplices β in B_m, and for all $(2n+1)$-simplices α in B_j we have that $K_{m+1} \cap \beta$ is the image of an affine transformation of $K_{j+1} \cap \alpha$, for all $m, j \in \mathbb{Z}^+$. After constructing the K_m's and B_m's for all integers $m \geq 1$, let $C(M, N) = \cap\{B_m : m \in \mathbb{Z}^+\}$.

4.   PROPERTIES OF $C(M, N)$

Theorem 3. *$C(M, N)$ is an n-dimensional continuum.*

Proof. Note that $C(M, N)$ is the intersection of a nested set of continua, hence $C(M, N)$ is a continuum. Similarly as in [7], but using $\mathbb{Z}_2$ coefficients, we can show that the dimension of $C(M, N)$ is n.

The following theorem will be proved in Section 6:

Theorem 4. *If M and N have empty boundaries, then $C(M, N)$ is representable.*

Local properties of $C(M, N)$ depend on the global properties of N. Specifically, if $\pi_k(N)$ is not trivial for some $0 < k \leq n$, then $C(M, N)$ is not semi-LC^n. Note that if M is a p.l.-n-dimensional sphere and N is the Cartesian product of a p.l.-m-dimensional sphere with a p.l.-$(n - m)$-dimensional sphere, then $C(M, N)$ is a continuum of dimension n that is not semi-LC^m, where $n > m > 1$. Hence, Fitzpatrick's question is answered by:

Theorem 5. *Let n be a positive integer. For each m with $1 < m < n$, there is a homogeneous continuum of dimension n that is not semi-LC^m.*

Also the question of Borsuk is answered by:

Theorem 6. *If M is the 2-dimensional sphere S^2 and N is the projective plane $\mathbb{P}^2$, then $C(M, N)$ is homogeneous.*

5. Extension of Toruńczyk's Lemma to the nonorientable case

Let $\mathcal{L}$ be a set of n-cells contained in the interior of a given admissible n-manifold M with boundary. $S(\mathcal{L})$ will be used to denote the union of the interiors of the elements of $\mathcal{L}$. Similarly as in Jakobsche's papers, define in a manifold a good set of cells:

Definition 7. *A set $\mathcal{L}$ of n-cells in the interior of a given n-manifold M will be called good if:*

1) *the elements of $\mathcal{L}$ are mutually disjoint;*
2) *if $\varepsilon > 0$, then only finitely many elements of $\mathcal{L}$ have diameter more than ε;*
3) *$S(\mathcal{L})$ is dense in M.*

And a null family:

Definition 8. *A collection $\mathcal{L}$ of subsets of a metric space is a null family if it satisfies Condition 2 of the definition of a good set of cells in a manifold.*

Note that the image of a good set of cells under an autohomeomorphism of a compact manifold is a good set of cells. The same may be said about a null family.

The notion of a good set of cells and Lemma 13 in this section stated for orientable manifolds are due to H. Toruńczyk. The lemma is proved for the orientable case in dimension 3 in [6]. The generalization to higher dimensional orientable manifolds is mentioned in [7]. For nonorientable manifolds the lemma requires a slight modification but the proof is similar. First, we need the following definition:

Definition 9. *Let M be an admissible n-manifold, $\mathcal{K}$ be a good set of p.l.-n-cells contained in the interior of M, and for each $k \in \mathcal{K}$, let f_k be a homeomorphism with domain $\partial \Delta^n$ and range ∂k. The set $\{f_k : k \in \mathcal{K}\}$ is called a dense orientation of $\mathcal{K}$ if for each $k \in \mathcal{K}$ and each p.l.-n-cell U_k that is a neighborhood of k, there exists a $j \in \mathcal{K}$ such that $j \subset U_k$ and such that the function f_k and f_j are not compatible in U_k.*

Since there are orientable manifolds not admitting an orientation reversing homeomorphism, the notion of a dense orientation of a good set of cells introduced here is also useful for the construction involving orientable manifolds only. If N is oriented and such that the two connected sums $N \# N$ and $N \#(-N)$ are not homeomorphic, the assumption of dense orientation is necessary in Section 6, even if both M and N are orientable.

For every positive integer n pick an n-simplex, $\sigma(n)$ in $\mathbb{R}^n$ such that its vertices all have distance 1 from the origin. Next, let $\sigma(n, r)$ be the image of $\sigma(n)$ under the linear map $f(x) = rx$, where $x \in \mathbb{R}^n$ and $r > 0$.

The following two lemmas 10 and 11, were proved for the topological case in dimension 3 by D.V. Meyer [9]. However, we will need them in the piecewise linear case for all positive finite dimensions. Hence, modified versions of Meyer's lemmas are provided below.

Lemma 10. *If $\mathcal{L}$ is a null family of compact subsets of $\sigma(n, 2) - \sigma(n, 1)$ such that for all $l \in \mathcal{L}$, $diam(l) < \varepsilon$, then there exists a p.l.-autohomeomorphism h of $\sigma(n, 2)$ such that $h(\sigma(n, 1)) \subset \sigma(n, \varepsilon/2)$, and $diam(h(l)) < \varepsilon$ for all $l \in \mathcal{L}$. Moreover, if $\delta > 1$, then h can be chosen so that $h(x) = x$ for all x in the closure of $\sigma(n, 2) - \sigma(n, \delta)$.*

Proof. First note that if $\varepsilon \geq 2$ then h could be the identity. Hence we will assume that $2 > \varepsilon > 0$. Let m be a positive integer such that $0 < (1 - (m\varepsilon)/3) < \varepsilon/2$. If $\varepsilon/2 < a < b < c < 2$, then let $G_{a,b,c}$ be the p.l.-autohomeomorphism of $\sigma(n, 2)$ that:

1. leaves $\sigma(n, 2) - \sigma(n, c)$ fixed;
2. takes $\sigma(n, c) - \sigma(n, b)$ to $\sigma(n, c) - \sigma(n, b - \varepsilon/3)$ with a pseudo-radial projection with respect to the origin;
3. takes $\sigma(n, b) - \sigma(n, a)$ to $\sigma(n, b - \varepsilon/3) - \sigma(n, a - \varepsilon/3)$ by a pseudo-radial projection with respect to the origin;
4. takes $\sigma(n, a)$ to $\sigma(n, a - \varepsilon/3)$ by a pseudo-radial projection with respect to the origin. Note that if $p \in \sigma(n, 2)$, then $dist(p, G_{a,b,c}(p)) \leq \varepsilon/3$. Hence, if p and q are points in $\sigma(n, 2)$ such that $dist(p, q) < \varepsilon/3$, then $dist(G_{a,b,c}(p), G_{a,b,c}(q)) \leq dist(G_{a,b,c}(p), p) + dist(p, q) + dist(q, G_{a,b,c}(q)) \leq \varepsilon$.

Hence, if $l \subset \sigma(n, 2)$ and $diam(l) < \varepsilon/3$, the $diam(G_{a,b,c}(l)) < \varepsilon$. Since $\mathcal{L}$ is a null family there exists a positive number c_1 such that $1 < c_1 < \delta$ and if $l \in \mathcal{L}$ and $l \cap \sigma(n, c_1) \neq \emptyset$, then $diam(l) < \varepsilon/3$. Let $a_1 = 1$ and $b_1 = (a_1 + c_1)/2$. Let $h_1 = G_{a_1 b_1 c_1}$ and note that since h_1 is uniformly continuous we have that $h_1(\mathcal{L})$ is a null family.

Let j be a positive integer, $1 \leq j < m$, and assume that $h_1, h_2, \ldots, h_j$ have been chosen. Let c_{j+1} be a positive number such that $(a_j - \varepsilon/3) < c_{j+1} < (b_j - \varepsilon/3)$ and if $l \in \mathcal{L}$ and $(h_j \circ h_{j-1} \circ \ldots \circ h_1(l)) \cap \sigma(n, c_1) \neq \emptyset$, then $diam(h_j \circ h_{j-1} \circ \ldots \circ h_1(l)) < \varepsilon/3$. Let $a_{j+1} = 1$ and $b_{j+1} = (a_{j+1} + c_{j+1})/2$. Let $h_{j+1} = G_{a_{j+1} b_{j+1} c_{j+1}}$.

Let $h = h_m \circ h_{m-1} \circ \ldots \circ h_1$. Note that since h is the finite composition of piecewise linear autohomeomorphisms of $\sigma(n, 2)$ that are all fixed on the closure of $\sigma(n, 2) - \sigma(n, \delta)$ we have that h is a piecewise linear autohomeomorphism of $\sigma(n, 2)$ that is fixed on the boundary of $\sigma(n, 2)$. Also, if $l \in \mathcal{L}$, then diam $(h_j \circ h_{j-1} \circ \ldots \circ h_1(l)) < \varepsilon$ because $h_j \circ h_{j-1} \circ \ldots \circ h_1$ only moves the elements of $h_{j-1} \circ \ldots \circ h_1(\mathcal{L})$ that have diameter less than $\varepsilon/3$. Hence, if $l \in \mathcal{L}$, then diam $(h(l)) < \varepsilon$. Moreover, notice that $h(\sigma(n, 1)) = \sigma(n, 1 - (m\varepsilon)/3) \subset \sigma(n, \varepsilon/2)$.

Lemma 11. *Suppose that M is an admissible n-manifold with boundary, $\varepsilon > 0$, $\mathcal{K}$ is a good collection of p.l.-n-cells in the interior of M, $k \in \mathcal{K}$, and U_k is a p.l.-n-cell in M that contains k such that U_k intersects no element of $\mathcal{K} - \{k\}$ with diameter greater than or equal to ε. Then there exists a p.l.-autohomeomorphism h of M such that $h(x) = x$ for all $x \in M - Int(U_k)$, and diam $(h(s)) < \varepsilon$ for all $s \in \mathcal{K}$ such that $s \cap U_k \neq \emptyset$.*

Proof. Let f_1 be a p.l.-homeomorphism with domain U_k and range $\sigma(n, 2)$. By the disk theorem there exists a p.l.-autohomeomorphism f_2 of $\sigma(n, 2)$ that is fixed on the boundary of $\sigma(n, 2)$ such that $f_2 \circ f_1(k) = \sigma(n, 1)$. Let $f = f_2 \circ f_1$.

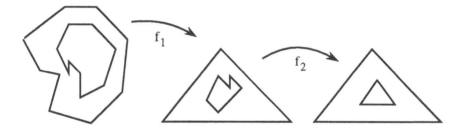

FIGURE 2. A function Diagram of f_1 and f_2.

Since f is uniformly continuous on U_k and f^{-1} is uniformly continuous on $\sigma(n, 2)$ there exists a number $d > 0$ such that if $l \subset \sigma(n, 2)$ and diam $(l) < d$, then diam $(f^{-1}(L)) < \varepsilon$; and there exists a $b > 0$ such that if $q \subset U_k$ and diam $(q) < b$, then diam $(f(q)) < d$. Also, there exists a number $c > 1$ such that if $s \in \mathcal{K} - \{k\}$ and $s \cap f^{-1}(\sigma(n, c)) \neq \emptyset$, the diam $(s) < b$ and $s \subset U_k$. Note that $\mathcal{L} = \{s : s \cap f^{-1}(\sigma(n, c)) \neq \emptyset \text{ and } s \in \mathcal{K}\}$ is a countable collection of compact subsets of $\sigma(n, 2) - \sigma(n, 1)$ that is a null family, and if $s \in \mathcal{L}$, the diam $(s) < d$. Hence by Lemma 10 there exists a p.l.-autohomeomorphism g of $\sigma(n, 2)$ such that:

1. $g(x) = x$ for all x in the closure of $\sigma(n, 2) - \sigma(n, c)$;
2. $g(\sigma(n, 1)) \subset \sigma(n, d/2)$;
3. if $t \in \mathcal{L}$, the diam $(g(t)) < d$.

Let $h(x) = x$ for all $x \in M - Int(U_k)$, and let $h(x) = f^{-1} \circ g \circ f(x)$. Notice that if $s \in \mathcal{K}$ and $s \cap f^{-1}(\sigma(n, c)) = \emptyset$, then $h(s) = s$. And, if $s \cap f^{-1}(\sigma(n, c)) \neq \emptyset$, then $\mathrm{diam}\,(s) < b$, hence $\mathrm{diam}\,(f(s)) < d$. Therefore $\mathrm{diam}\,(g \circ f(s)) < d$ and $\mathrm{diam}\,(f^{-1} \circ g \circ f(s)) < \varepsilon$. Moreover, $g \circ f(k) \subset \sigma(n, d/2)$, hence $\mathrm{diam}\,(g \circ f(k)) < d$. Therefore, $\mathrm{diam}\,(f^{-1} \circ g \circ f(k)) < \varepsilon$.

Lemma 12. *If M is an admissible n-manifold with boundary, $0 < \delta < \varepsilon$, $\mathcal{K}$ is a good collection of p.l.-n-cells in the interior of M, and $\mathcal{T}$ is a finite subset of $\mathcal{K}$ such that if $k \in \mathcal{K} - \mathcal{T}$, the $\mathrm{diam}\,(k) < \varepsilon$, then there exists a p.l.-autohomeomorphism g of M such that $g\,|\partial M = id|\,\partial M$, $g\,|k = id|\,k$ for all $k \in \mathcal{T}$, $\mathrm{diam}\,(g(k)) < \delta$ for all $k \in \mathcal{K} - \mathcal{T}$, and $\mathrm{dist}\,(g, id) < \varepsilon$, (and hence $\mathrm{dist}\,(g^{-1}, id) < \varepsilon$), where id is the identity on M.*

Proof. Let $\mathcal{B} = \{k \in \mathcal{K} - \mathcal{T} : \mathrm{diam}\,(k) \geq \delta\}$ and note that $\mathcal{B}$ is finite. Hence, if $k \in \mathcal{B}$, then there exists a p.l.-n-cell U_k in M, which is a neighborhood of k, such that $U_k \cap U_j = \emptyset$ if $k \neq j$ and so that $U_k \cap t = \emptyset$ for all $t \in \mathcal{T}$. By Lemma 11, if $k \in \mathcal{B}$, then there exists a p.l.-homeomorphism $g_k : M \to M$ such that $g_k(x) = x$ for all $x \in M - \mathrm{Int}\,(U_k)$ and $\mathrm{diam}\,(g_k(c)) < \delta$ for all $c \in \mathcal{K} - (\mathcal{T} \cup \mathcal{B})$. Let g be the compositions of the g_k's, where $k \in \mathcal{B}$, and note that g satisfies the conclusion of the Lemma.

Next, we prove Toruńczyk's Lemma extended to the nonorientable case:

Lemma 13. *If M and N are admissible n-manifolds with boundary, $H : M \to N$ is a p.l.-homeomorphism, $\mathcal{K}$ and $\mathcal{L}$ are two good sets of p.l.-n-cells contained in the interior of M and N respectively, and $\{f_k : k \in \mathcal{K}\}$ and $\{f_l : l \in \mathcal{L}\}$ are dense orientations of $\mathcal{K}$ and $\mathcal{L}$ respectively, then there exists a one-to-one and onto function $\mu : \mathcal{K} \to \mathcal{L}$ and a homeomorphism $G : M - S(\mathcal{K}) \to N - S(\mathcal{L})$ such that $G\,|\partial M = H|\,\partial M$ and $G\,|\partial k = f_{\mu(k)} \circ f_k^{-1}$ for all $k \in \mathcal{K}$.*

Proof. First for each $\alpha \in \mathcal{K} \cup \mathcal{L}$ extend the domain of f_α to obtain a new p.l.-homeomorphism $F_\alpha : \Delta^n \to \alpha$. Assume the hypothesis, and without loss of generality assume that the diameters of M and N are less than one. If $\mathcal{Y}$ is a good collection of p.l.-n-cells in a manifold, then let $\mathcal{Y}_m = \{k \in \mathcal{Y} : \mathrm{diam}\,(k) \geq 1/2^m\}$. Also, if $\mathcal{F}$ is a family of subsets of the domain of a function f, then we will use $f(\mathcal{F})$ to denote $\{f(t) : t \in \mathcal{F}\}$, where $f(t)$ will be used to denote $\{f(x) : x \in t\}$. It should be clear from the context which definition is being used.

Let P be a manifold with diameter less than one which is p.l.-homeomorphic to M, and for the sake of notation assume that P, M, and N are pairwise disjoint. We shall inductively construct a sequence of homeomorphisms $h_0, h_1, h_2 \ldots$ such that for all m, $h_m : M \to P$, and a sequence of homeomorphisms $g_0, g_1, g_2, \ldots$ such that for all m, $g_m : N \to P$, such that the following conditions are satisfied:

(a_m) if $k \in \mathcal{K}_m$, then there exists a $l \in \mathcal{L}$ such that $h_m(k) = g_m(l)$ and $g_m^{-1} \circ h_m|k = F_l \circ F_k^{-1}$;

$(a_m)'$ if $l \in \mathcal{L}_m$, then there exists a $k \in \mathcal{K}$ such that $h_m(k) = g_m(l)$ and $g_m^{-1} \circ h_m|k = F_l \circ F_k^{-1}$;

(b_m) $\operatorname{diam} h_m(k) < 1/2^m$ for every $k \in \mathcal{K} - \left(\mathcal{K}_m \cup h_m^{-1} \circ g_m(\mathcal{L}_m)\right)$;

$(b_m)'$ $\operatorname{diam} g_m(l) < 1/2^m$ for every $l \in \mathcal{L} - \left(\mathcal{L}_m \cup g_m^{-1} \circ h_m(\mathcal{K}_m)\right)$;

(c_m) $h_m |k = h_{m-1}| k$ for every $k \in \mathcal{K}_{m-1} \cup h_{m-1}^{-1} \circ g_{m-1}(\mathcal{L}_{m-1})$;

$(c_m)'$ $g_m |l = g_{m-1}| l$ for every $l \in \mathcal{L}_{m-1} \cup g_{m-1}^{-1} \circ h_{m-1}(\mathcal{K}_{m-1})$;

(d_m) $\operatorname{dist}(h_m, h_{m-1}) < 1/2^{m-2}$, $\operatorname{dist}\left(h_m^{-1}, h_{m-1}^{-1}\right) < 1/2^{m-2}$;

$(d_m)'$ $\operatorname{dist}(g_m, g_{m-1}) < 3/2^{m-1}$, $\operatorname{dist}\left(g_m^{-1}, g_{m-1}^{-1}\right) < 3/2^{m-1}$;

(e_m) $g_m^{-1} \circ h_m |\partial M = H|\partial M$.

Note that if $h = \lim\limits_{m \to \infty} h_m$ and $g = \lim\limits_{m \to \infty} g_m$, then (d_m), $(d_m)'$, and (e_m) give us that $g^{-1} \circ h$ is a homeomorphism from M to N with $g^{-1} \circ h |\partial M = H|\partial M$, and (a_m), $(a_m)'$, (c_m) and $(c_m)'$ give us that $g^{-1} \circ h(M - S(\mathcal{K})) = N - S(\mathcal{L})$, and $g^{-1} \circ h|k = F_l \circ F_k^{-1}$ for every $k \in \mathcal{K}$. The (b) conditions are used in the inductive construction. And, (d_m) and $(d_m)'$ will make $g^{-1} \circ h$ a homeomorphism from M to N. Hence, we will let $G = g^{-1} \circ h$ and let $\mu : \mathcal{K} \to \mathcal{L}$ be such that if $\mu(k) = l$, then $G(\partial k) = \partial(l)$.

To start the inductive construction: we will let h_0 and g_0 be any two p.l.-homeomorphisms such that $g_0^{-1} \circ h_0 = H$. Also, note that e_0 is true, and that because the diameters of M, N and P are assumed to be less than one, conditions (a_m), $(a_m)'$, (b_m), and $(b_m)'$ are vacuously true for $m = 0$ and conditions $(c_m)'$, $(c_m)'$, $(d_m)'$, and $(d_m)'$ do not apply for $m = 0$. Suppose that we have already constructed h_{m-1} and g_{m-1} for some $m \geq 1$. If $\mathcal{K}_m - \mathcal{K}_{m-1} \neq \emptyset$, then let $\delta = \min\{ \operatorname{diam}(h_{m-1}(k)) : k \in \mathcal{K}_m - \mathcal{K}_{m-1}\}$, otherwise let $\delta = 1$.

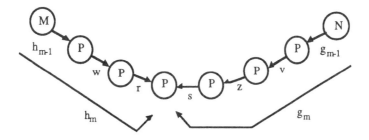

FIGURE 3.   The Complete Function Diagram for Lemma 13.

Remark. The diagram in Figure 3 illustrates the matching of cells of $\mathcal{K}$ and $\mathcal{L}$. (For this remark only, we use the phrases "cells of $\mathcal{K}$" and "cells of $\mathcal{L}$" for the images of such cells under appropriate homeomorphisms.) The matching preformed in the inductive step m does not alter the matching of step $m - 1$; homeomorphisms in the diagram influence all cells of $\mathcal{K} \cup \mathcal{L}$ unmatched in step $m - 1$. The action of the homeomorphisms in the diagram is as follows:

v shrinks the unmatched cells of $\mathcal{L}$;

w matches cells of $\mathcal{K}$ to cells of $\mathcal{L}$;

r shrinks the remaining cells of $\mathcal{K}$;

z matches cells of $\mathcal{L}$ to cells of $\mathcal{K}$;

s shrinks the remaining cells of $\mathcal{L}$.

Now, we return to the inductive procedure: Let $\mathcal{T} = \mathcal{L}_{m-1} \cup g_{m-1}^{-1} \circ h_{m-1}(\mathcal{K}_{m-1})$. Notice that if $t \in \mathcal{L} - \mathcal{T}$, then, by condition $(b_{m-1})'$, we have $\operatorname{diam}(g_{m-1}(t)) < 1/2^{m-1}$. Let j be a positive integer such that $1/2^j < \delta$, and for each element y of $(g_{m-1}(\mathcal{L} - \mathcal{T}))_j$ pick a cell N_y, that is a neighborhood of $g_{m-1}(y)$, such that $\operatorname{diam}(N_y) < 1/2^{m-1}$ and $\operatorname{diam}(g_{m-1}^{-1}(N_y)) < 1/2^{m-1}$. Moreover, pick the N_y's to be pairwise disjoint and disjoint from $h_{m-1}(\mathcal{K}_{m-1}) \cup g_{m-1}(\mathcal{L}_{m-1})$. Then, by the Lemma 12, there is a p.l.-homeomorphism $v : P \to P$ such that:

1a. $v\,|\partial P = id|\,\partial P$;

2a. $v \circ g_{m-1}\,|t = g_{m-1}|\,t$ for every $t \in \mathcal{T}$;

3a. $\operatorname{diam}(v \circ g_{m-1}(t)) < \delta$ for every $t \in \mathcal{L} - \mathcal{T}$;

4a. $\operatorname{dist}(v, id) < 1/2^{m-1}$;

5a. $v\,|(P - \cup\{N_y : y \in (g_{m-1}(\mathcal{L} - \mathcal{T}))_j\})$ is the identity;

6a. $v\,|\psi = id|\,\psi$ for all $\psi \in g_{m-1}(\mathcal{L})$ such that there exists a $y \in (g_{m-1}(\mathcal{L} - \mathcal{T}))_j$ so that $\psi \cap \partial N_y \neq \emptyset$.

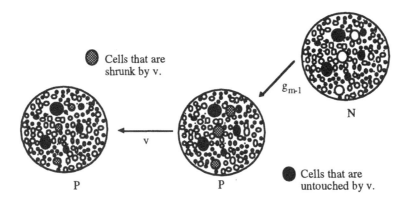

FIGURE 4.   A Function Diagram of g_{m-1} and v.

Next note that if $k \in \mathcal{K}_m - (\mathcal{K}_{m-1} \cup h_{m-1}^{-1} \circ g_{m-1}(\mathcal{L}_{m-1}))$, then by condition (b_{m-1}) $\operatorname{diam}(h_{m-1}(k)) < 1/2^{m-1}$. Hence there exists a cell U_k that is a neighborhood of $h_{m-1}(k)$ such that the $\operatorname{diam}(U_k) < 1/2^{m-1}$ and $\operatorname{diam}(h_{m-1}^{-1}(U_k)) < 1/2^{m-1}$. Moreover, the U_k's may be chosen to be p.l.-n-cells that are pairwise disjoint and disjoint from $h_{m-1}(\mathcal{K}_{m-1}) \cup g_{m-1}(\mathcal{L}_{m-1})$. Also, note that v was constructed so that $h_{m-1}(k)$ would not be contained in any element of $v \circ g_{m-1}(\mathcal{L} - \mathcal{T})$. Hence, there exists $l \in \mathcal{L} - \mathcal{T}$ such that $v \circ g_{m-1}(l) \subset h_{m-1}(k)$. Moreover, l may be chosen so that the function $v \circ g_{m-1} \circ f_l$ is compatible with the function $h_{m-1} \circ f_k$ with respect to U_k. Since $h_{m-1}(k)$ and $v \circ g_{m-1}(l)$ are p.l.-n-cells in P, and one is a subset of the other, it follows from the annulus theorem, that there is a p.l.-homeomorphism $w_k : P \to P$ such that:

1b. $w_k\,|\partial P = id|\,\partial P$;

2b. $w_k(x) = x$ if $x \notin \mathrm{Int}\,(U_k)$;

3b. $w_k\,|\psi = id|\,\psi$ for all $\psi \in h_{m-1}(\mathcal{K})$ such that $\psi \cap \partial U_k \neq \emptyset$;

4b. $w_k \circ h_{m-1}(k) = v \circ g_{m-1}(l)$;

5b. $g_{m-1}^{-1} \circ v^{-1} \circ w_k \circ h_{m-1}\,\big|\,k = F_l \circ F_k^{-1}$.

Let w be the composition of the identity and the w_k's, where

$$k \in \mathcal{K}_m - \left(\mathcal{K}_{m-1} \cup h_{m-1}^{-1} \circ g_{m-1}(\mathcal{L}_{m-1})\right).$$

Since, for all $k \in \mathcal{K}_m - \left(\mathcal{K}_{m-1} \cup h_{m-1}^{-1} \circ g_{m-1}(\mathcal{L}_{m-1})\right)$ the U_k's are pairwise disjoint, and $\mathrm{dist}\,(w_k, id) < 1/2^{m-1}$ we have that $\mathrm{dist}\,(w, id) < 1/2^{m-1}$.

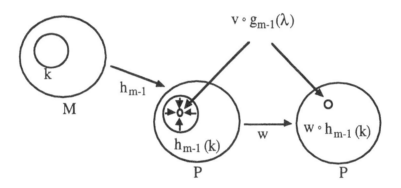

FIGURE 5. A function Diagram of h_{m-1} and w.

Let ρ be the minimum of $1/2^m$ and $\{\mathrm{diam}\,(v \circ g_{m-1}(l)) : l \in \mathcal{L}_m - \mathcal{L}_{m-1}\}$. Let $\mathcal{H} = \mathcal{K}_m \cup h_{m-1}^{-1} \circ g_{m-1}(\mathcal{L}_{m-1})$ and notice that if $\tau \in \mathcal{K} - \mathcal{H}$ then by condition (b_{m-1}) and the construction of w we have that $\mathrm{diam}\,(w \circ h_{m-1}(\tau)) < 1/2^{m-1}$ and $\mathrm{diam}\,\left(h_{m-1}^{-1} \circ w^{-1}(\tau)\right) < 1/2^{m-1}$. For each $\tau \in \mathcal{K} - \mathcal{H}$ such that $\mathrm{diam}\,(w \circ h_{m-1}(\tau)) \geq \rho$, pick a cell V_τ that is a neighborhood of $w \circ h_{m-1}(\tau)$ such that $\mathrm{diam}\,\left(h_{m-1}^{-1} \circ w^{-1}(V_\tau)\right) < 1/2^{m-1}$ and $\mathrm{diam}\,(V_\tau) < 1/2^{m-1}$. Moreover, choose the V_τ's so that they are pairwise disjoint and disjoint from $w \circ h_{m-1}(\mathcal{K}_m) \cup g_{m-1}(\mathcal{L}_{m-1})$. Hence, by Lemma 12, there is a p.l.-homeomorphism $r : P \to P$ such that:

1c. $r\,|\partial P = id|\,\partial P$;

2c. $r\,|\varphi = id|\,\varphi$ for all $\varphi \in w \circ h_{m-1}(\mathcal{K})$ such that there exists a $\tau \in \mathcal{K} - \mathcal{H}$ so that $\varphi \cap V_t \neq \emptyset$;

3c. $r\,|w \circ h_{m-1}(t) = id$ for every $\tau \in \mathcal{H}$;

4c. $\mathrm{diam}\,(r \circ w \circ h_{m-1}(\tau)) < \rho$ for every $\tau \in \mathcal{K} - \mathcal{H}$;

5c. $\mathrm{dist}\,(r, id) < 1/2^{m-1}$, which implies $\mathrm{dist}\,(r^{-1}, id) < 1/2^{m-1}$;

6c. $r|(P - \cup\{V_t : t \in \mathcal{K} - \mathcal{H}\})$ is the identity.

Let $h_m = r \circ w \circ h_{m-1}$. Notice that condition 3c. gives us that $\mathrm{diam}\,(r \circ w \circ h_{m-1}(\tau)) < \rho \leq 1/2^m$ for every $\tau \in \mathcal{K} - \mathcal{H}$, and hence condition (b_m)

is satisfied. To show that the first part of (d_m) is satisfied notice that $\text{dist}\,(r \circ w \circ h_{m-1}, h_{m-1}) < 1/2^{m-1} + 1/2^{m-1} = 1/2^{m-2}$. To show that the second part of (d_m) is satisfied first notice that if $p \in P$, then $\text{dist}\,\big(p, w^{-1} \circ r^{-1}(p)\big) < 1/2^{m-1} + 1/2^{m-1} = 1/2^{m-2}$.

To show $\text{dist}\,\big(h_{m-1}^{-1}(p),\ h_{m-1}^{-1} \circ w^{-1} \circ r^{-1}(p)\big) < 1/2^{m-1}$ one needs only to observe that with the exception of finitely many cells $w^{-1} \circ r^{-1}$ is the identity, and h_{m-1}^{-1} sends each of these cells to a cells in M of diameter less than $1/2^{m-1}$. Hence, $h_{m-1}^{-1}(p)$ and $h_{m-1}^{-1} \circ w^{-1} \circ r^{-1}(p)$ either lie in the same cell of diameter less than $1/2^{m-1}$ or in intersecting cells each having diameter less than $1/2^{m-1}$.

Given $\beta \in \mathcal{L}_m - \big(\mathcal{L}_{m-1} \cup g_{m-1}^{-1} \circ v^{-1} \circ h_m(\mathcal{K}_m)\big)$, the set $v(\beta)$ is not a subset of an element of $h_m(\mathcal{K})$ and hence there is a $k \in \mathcal{K}$ such that $h_m(k) \subset v(\beta)$. Moreover, k may be chosen so that the function $v \circ g_{m-1} \circ f_\beta$ is compatible with the function $h_m \circ f_k$ with respect to a sufficiently small cell that contains $h_m(k)$ in its interior. Hence, we can construct a new p.l.-homeomorphism $z : P \to P$ in an analogous way as we constructed the following w. Notice that $\text{dist}\,(z, id) < 1/2^{m-1}$, $z \circ v \circ g_{m-1}|\beta = v \circ g_{m-1}|\beta$ for all $\beta \in \mathcal{L}_{m-1} \cup g_{m-1}^{-1} \circ v^{-1} \circ h_m(\mathcal{K}_m)$, and for every $l \in \mathcal{L}_m$ there exists a $k \in \mathcal{K}$ such that $h_m(k) = z \circ v \circ g_{m-1}(l)$ and $g_{m-1}^{-1} \circ v^{-1} \circ z^{-1} \circ h_m\,|k = F_l \circ F_k^{-1}$.

Let $\mathcal{W} = \{l \in \mathcal{L} - (\mathcal{L}_m \cup g_m^{-1} \circ v^{-1} \circ z^{-1} \circ h_m(\mathcal{K}_m)) : \text{diam}\,(z \circ v \circ g_{m-1}(l)) \geq 1/2^m\}$. In an analogous way as we constructed the function v, construct a new function $s : P \to P$ such that $s(x) = x$ for all x in an element of $\mathcal{L}_m \cup g_m^{-1} \circ v^{-1} \circ z^{-1} \circ h_m(\mathcal{K}_m)$, and so that $\text{diam}\,(s \circ z \circ v \circ g_{m-1}(l)) < 1/2^m$ for all $l \in \mathcal{L} - (\mathcal{L}_m \cup g_m^{-1} \circ v^{-1} \circ z^{-1} \circ h_m(\mathcal{K}_m))$. Moreover, since $\text{diam}\,(z \circ v \circ g_{m-1}(l)) < 1/2^{m-1}$ for all $l \in \mathcal{L} - (\mathcal{L}_m \cup g_m^{-1} \circ v^{-1} \circ z^{-1} \circ h_m(\mathcal{K}_m))$ we have $\text{dist}\,(z, id) < 1/2^{m-1}$.

Now, if we let $g_m = s \circ z \circ v \circ g_{m-1}$, then $\text{dist}\,(s \circ z \circ v \circ g_{m-1}, g_{m-1}) < 1/2^{m-1} + 1/2^{m-1} + 1/2^{m-1} = 3/2^{m-1}$. Similarly, $\text{dist}\,(g_{m-1}^{-1} \circ v^{-1} \circ s^{-1}, g_{m_1}^{-1}) < 3/2^m$. Hence, conditions (a_m), $(a_m)'$, (b_m), $(b_m)'$, (c_m), $(c_m)'$, (d_m), $(d_m)'$, and (e_m) are satisfied. This completes the inductive construction and the proof.

6. Proof that C(M,N) is Representable

Definition 14. *If P is a subpolyhedron of K_m, then by $\mathcal{B}_m(P)$ we mean the union of all $(2n+1)$-simplices in B_m that intersect P at an n-simplex.*

Definition 15. *Let φ be an n-simplex in K_1, let $F(N) = \mathcal{B}_1(\varphi) \cap C(M,N)$. Since $\partial\varphi \subset F(N)$ we will let $\partial F(N)$ denote $\partial\varphi$.*

Definition 16. *If M and N are subset of $\mathbb{R}^{2n+1}$ that are admissible n-manifolds with boundary, then by $(M,N)^*$ we mean the set of all subsets X of $\mathbb{R}^{2n+1}$ which satisfy the following conditions:*

1. there exists a good family $\mathcal{L}$ of p.l.-n-cells in M;

2. *for each $l \in \mathcal{L}$ there exist an unique $F(N)_l \subset \mathbb{R}^{2n+1}$ and a homeomorphism*
   $g_l : F(N) \to F(N)_l$ such that $g_l|\partial F(N)$ p.l.-maps $\partial F(N)$ to ∂l and so that
   $\{g_l|\partial F(N) : l \in \mathcal{L}\}$ is a dense orientation of $\mathcal{L}$;

3. *for each l and ξ in $\mathcal{L}$, $F(N)_l$ and $F(N)_\xi$ are disjoint if $l \neq \xi$;*

4. *$\partial l = (M - S(\mathcal{L})) \cap F(N)_l$;*

5. *if $\varepsilon > 0$, then only finitely many of the teh sets $F(N)_l$ have diameter larger*
   than ε;

6. *$X = M - S(\mathcal{L}) \cup (\cup\{F(N)_l : l \in \mathcal{L}\})$.*

By ∂X we will mean ∂M.

The following is known as the Pasting Lemma and the proof is omitted.

Lemma 17. *Let M be a compact metric space, and let $\mathcal{Y}$ be a null family of pairwise disjoint closed subsets of M. Let μ be a permutation of $\mathcal{Y}$, and let f be an autohomeomorphism of $\overline{M - \cup \mathcal{Y}}$. Suppose that for each $l \in \mathcal{Y}$, there is a homeomorphism $f_l : l \to \mu(l)$ such that $f(p) = f_l(p)$ for all $p \in l \cap \overline{M - \cup \mathcal{Y}}$. Then the function g, defined by $g(p) = f_l(p)$ if $p \in l$ and $g(p) = f(p)$ if $p \in \overline{M - \cup \mathcal{Y}}$, is an autohomeomorphism of M.*

Lemma 18. *If M, P and N are subset of $\mathbb{R}^{2n+1}$ that are admissible n-manifolds with boundary, $h : M \to P$ is a p.l.-homeomorphism, $X_1 \in (M, N)^*$, and $X_2 \in (P, N)^*$, then there is a homeomorphism $f : X_1 \to X_2$ such that $f|\partial M = h|\partial M$.*

Proof. Let each of $\mathcal{K}$ and $\mathcal{L}$ be a good set of p.l.-n-cells in M and P respectively. For each $k \in \mathcal{K}$, let $g_k : F(N) \to F(N)_k$ so that $g_k(\partial F(N)) = \partial k$ and $\{g_k|\partial F(N) : k \in \mathcal{K}\}$ is a dense orientation of $\mathcal{K}$. For each $l \in \mathcal{L}$, let $g_l : F(N) \to F(N)_l$ so that $g_l(\partial F(N)) = \partial l$ and $\{g_l|\partial F(N) : l \in \mathcal{L}\}$ is a dense orientation of $\mathcal{L}$. Let $X_1 = (M - S(\mathcal{K})) \cup (\cup\{f(N)_k : k \in \mathcal{K}\})$, and $X_2 = (P - S(\mathcal{L})) \cup (\cup\{f(N)_l : l \in \mathcal{L}\})$. Then, note by Lemma 13, that there exists a one-to-one and onto function $\mu : \mathcal{K} \to \mathcal{L}$ and a homeomorphism $g : M - S(\mathcal{K}) \to P - S(\mathcal{L})$ such that $g|\partial M = h|\partial M$, and $g|\partial k = g_{\mu(k)} \circ g_k^{-1}|\partial k$ for all $k \in \mathcal{K}$. Hence, if we let $f(x) = (g_l \circ g_k^{-1})(x)$ for $x \in F(N)_k$ and $k \in \mathcal{K}$, and $f(x) = g(x)$ for all $x \in M - S(\mathcal{K})$, then we get the required homeomorphism.

Lemma 19. *If P is a subpolyhedron of K_m, which is an n-manifold, then $\mathcal{B}_m(P) \cap C(M, N) \in (P, N)^*$. In particular $C(M, N) \in (K_m, N)^*$.*

Proof. It follows from the construction of $C(M, N)$ that $K_m \cap K_{m+1} = K_m - S(\mathcal{L}_i)$, where $\mathcal{L}_i$ is a finite set of disjoint p.l.-n-cells in K_m for each m. Let $\mathcal{L} = \cup\{\mathcal{L}_i : i \in \mathbb{Z}^+\}$, and note that $\mathcal{L}$ is a good set of p.l.-n-cells in K_m. Let $\mathcal{L}^P = \{z \in \mathcal{L} : z \subset P\}$, and note that $\mathcal{L}^P$ is a good set of p.l.-n-cells in P. Hence, $\mathcal{B}_m(P) \cap C(M, N) = (P - S(\mathcal{L}^P)) \cup (\cup\{F(N)_l : l \in \mathcal{L}^P\})$, where $F(N)_l = \mathcal{B}_{m+i}(l) \cap C(M, N)$ with $l \in \mathcal{L}_i$. Therefore, $\mathcal{B}_m(P) \cap C(M, N) \in (P, N)^*$.

Lemma 20. *For every admissible n-manifold M with boundary, if $X \in (M, N)^*$, then $X \in (M \# T, N)^*$, where T is an admissible copy of N, and $M \# T$ is either of the connected sums of M and T in $\mathbb{R}^{2n+1}$.*

Proof. Let $\mathcal{K}$ be a good set of p.l.-n-cells in M, and for each $k \in \mathcal{K}$ let $g_k : F(N) \to F(N)_k$ so that $g_k | \partial F(N) : \partial F(N) \to \partial k$, and so that $\{ g_k | \partial F(N) : k \in \mathcal{K} \}$ is a dense orientation of $\mathcal{K}$, and let $X = (M - (\mathcal{L})) \cup (\cup \{ F(N)_l : l \in \mathcal{L} \})$. Pick an $\alpha \in \mathcal{K}$ and note that by Lemma 18 that $F(N)_\alpha \in (T - \mathrm{Int}\,(\alpha), N)^*$ where, T is a copy of N that was used in the first level of construction of F_α. It should be noted that since two copies of N were attached in two ways, either orientation may be picked. Hence, $F(N)_\alpha = (T - S(\mathcal{Y})) \cup (\cup \{ F_y : y \in \mathcal{Y} \})$, where $\mathcal{Y}$ is a good set of p.l.-n-cells in $T - \mathrm{Int}\,(\alpha)$ and the sets F_y are as in the definition of $(T - \mathrm{Int}\,(\alpha), N)^*$. Letting $\mathcal{T} = \mathcal{Y} \cup (\mathcal{K} - \{ \alpha \})$, we have that $\mathcal{T}$ is a good set of p.l.-n-cells in $M \# T$ and that $X = ((M \# T) - S(\mathcal{T})) \cup (\cup \{ F_\beta : \beta \in \mathcal{T} \})$. Therefore, $X \in (M \# T, N)^*$.

Now, we prove Theorem 4 stated in Section 4:

Proof. Let $p \in C(M, N)$, and let U be a neighborhood of p in $C(M, N)$. Note that there exist a positive integer m and a vertex v of K_m such that, if W is the star of v in K_m, then $V = \mathcal{B}_m(W) \cap C(M, N) - \partial W$ is an open subset of U that contains p. Let $q \in V$. Let $D_0, D_1, D_2, D_3, \ldots$ be a sequence of open neighborhoods of p in $C(M, N)$ such that $v = D_0$, $p = \cap \{ D_i : i \in \mathbb{Z}^+ \cup \{ 0 \} \}$, and so that the closure of D_i is a subset of D_r for all $r < i$. And, let $S_0, S_1, S_2, S_3, \ldots$ be a sequence of open neighborhoods of q in $C(M, N)$ such that $V = S_0$, $q = \cap \{ S_i : i \in \mathbb{Z}^+ \cup \{ 0 \} \}$, and so that the closure of S_i is a subset of S_r for all $r < i$. Moreover, pick each D_i so that there exist a positive integer $d(i)$ and a vertex $v_{d(i)}$ for $K_{d(i)}$ such that, if $W_{d(i)}$ is the star of $v_{d(i)}$ in $K_{d(i)}$, then $D_i = \mathcal{B}_{d(i)}(W_{d(i)}) \cap C(M, N) - \partial W_{d(i)}$. Also, pick each S_i so that there exists a positive integer $s(i)$ and a vertex $v_{s(i)}$ of $K_{s(i)}$ such that if $L_{s(i)}$ is the star of $v_{s(i)}$ in $K_{s(i)}$, then $S_i = \mathcal{B}_{s(i)}(L_{s(i)}) \cap C(M, N) - \partial K_{s(i)}$.

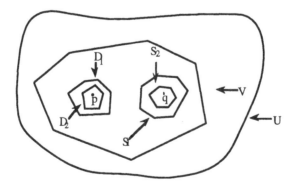

FIGURE 6.   A Picture of p, q, V, U, and the D's and S's.

Note that

$$\mathcal{B}_{d(i)}(W_{d(i)}) \cap (K_{d(i+1)} - \mathrm{Int}\,(W_{d(i+1)})) = (M \# N_1 \# N_2 \ldots \# N_b) - \mathrm{Int}\,(P_1) \cup \mathrm{Int}\,(P_2))$$

and that

$$\mathcal{B}_{s(i)}(L_{s(i)}) \cap (K_{s(i+1)} - \text{Int}\,(L_{s(i+1)})) = (M \# N_1 \# N_2 \ldots \# N_c) - (\text{Int}\,(P_3) \cup \text{Int}\,(P_4))$$

for some integers b and c, where (P_1, P_2), and (P_3, P_4) are two pairs of closed disjoint p.l.-n-cells, and N_1, $N_2, \ldots$ are copies of N. Hence, by Lemma 19,

$$\bar{D}_i - D_{i+1} \in ((M \# N_1 \ldots \# N_b) - (\text{Int}\,(P_1) \cup \text{Int}\,(P_2)), N)^*$$

and

$$\bar{S}_i - S_{i+1} \in ((M \# N_1 \ldots \# N_c) - (\text{Int}\,(P_3) \cup \text{Int}\,(P_4)), N)^*.$$

Therefore, by Lemma 20, $\bar{D}_i - D_{i+1}$ and $\bar{S}_i - S_{i+1}$ are elements of $(M - (\,\text{Int}\,(P_5) \cup \text{Int}\,(P_6)), N)^*$ where P_5 and P_6 are disjoint closed p.l.-n-cells. Hence, by Lemma 18, there is a homeomorphism $h_i : \bar{D}_i - D_{i+1} \to \bar{S}_i - S_{i+1}$ such that $h_i|\partial D_i = \partial S_i$ and $h_i|\partial D_{i+1} = h_{i+1}|\partial D_{i+1}$. Let $h(x) = x$ if $x \in C(M, N) - V$, $h(x) = h_i(x)$ if $x \in D_i - D_{i+1}$ for i a nonegative integer, and let $h(p) = q$. Note that h is a homeomorphism that maps p to q and it is the identity on the complement of V, and therefore on the complement of U. Hence, $C(M, N)$ is representable.

References

[1] F. D. Ancel and L. C. Siebenmann, *The construction of homogeneous homology manifolds*, Abstracts Amer. Math. Soc. 6 (1985), 92.

[2] R.D. Anderson, *A characterization of the universal curve and a proof of its homogeneity*, Ann. of Math. 67 (1958), 313–324.

[3] R.B. Bennett, *Countable dense homogeneous spaces*, Fund. Math. 74 (1972), 189–194.

[4] B. Fitzpatrick, Zhou Hao Xuan, *Countable dense homogeneity and the Baire property*, Topology Appl. 43 (1992), 1–14.

[5] J.F.P. Hudson, Piecewise Linear Topology, Benjamin, New York 1969.

[6] W. Jakobsche, *The Bing-Borsuk conjecture is stronger than the Poincaré conjecture*, Fund. Math. 106 (1980), 127–134.

[7] W. Jakobsche, *Homogeneous cohomology manifolds which are inverse limits*, Fund. Math. 137 (1991), 81–95.

[8] K. Kuperberg, W. Kuperberg, and W.R.R. Transue, *On the 2-homogeneity of Cartesian products*, Fund. Math. 110 (1980), 131–134.

[9] D.V. Meyer, *A decomposition of E^3 into points and a null family of tame 3-cells in E^3*, Ann. of Math. 78 (1963), 600–604.

[10] C.P. Rourke, B.J. Sanderson, Introduction to Piecewise-Linear Topology, Springer-Verlag 1982.

[11] G.S. Ungar, *On all kinds of homogeneous spaces*, Trans. Amer. Math. Soc. 212 (1975), 393–400.

[12] G.S. Ungar, *Countable dense homogeneity and n-homogeneity*, Fund. Math. 99 (1978), 155–160.

[13] R.F. Williams, *A useful functor and three famous examples in topology*, Trans. Amer. Math. Soc. 106 (1963), 319-329.

Part III

The Houston Problem Book

A LIST OF PROBLEMS KNOWN
AS HOUSTON PROBLEM BOOK

HOWARD COOK   Department of Mathematics, University of Houston, Houston, Texas 77204, USA

W. T. INGRAM   Department of Mathematics, University of Missouri - Rolla, Rolla, Missouri 65401, USA

ANDREW LELEK   Department of Mathematics, University of Houston, Houston, Texas 77204, USA

INTRODUCTION.

There is some evidence that the term "Houston Problem Book" is becoming a buzz-word, so we decided to explain. First of all, nobody can be better equipped with buzz-words than a college administrator. The third of us, namely A.L., remembers talking to one of the deans. The dean was clearly interested in topology. During the conversation, the activities of mathematicians living in Texas were also mentioned. Obviously, there has been something about topology in Texas that made it distinguishable. Maybe it was the personality and legend of R. L. Moore, or also, maybe, rather unusual connections with schools abroad, among them, and perhaps in particular, with the mathematical school in Poland. When A.L. joined the faculty at Houston in the Fall of 1970, we conceived an idea of a seminar which would be modeled after Knaster's seminar in Wroclaw, which, in turn, was modeled after Hilbert's seminar in Göttingen. Our idea materialized in 1971. The seminar was named "Cook-Lelek Topology Seminar" because W.T.I. was still a junior faculty member. The memorandum announcing the first meeting read as follows.

TO: Faculty and Students, Department of Mathematics, University of Houston, and all other persons who might be interested in a seminar work.
FROM: Dr. Howard Cook and Dr. Andrew Lelek.
DATE: August 17, 1971.

You are invited to attend the first meeting of the

<u>Cook-Lelek Topology Seminar</u>

to be held on Wednesday, September 1, 1971, from 2:00 p.m. through 4:00 p.m. in Room 302, Arnold Hall. The program of 1-st meeting:

A. Lelek, Opening address.

B. B. Epps, Jr., Results and problems of E. D. Tymchatyn concerning hereditarily locally connected continua.

For the Fall Semester of the academic year 1971-1972, this seminar is scheduled to meet each Wednesday at time and place as given above. The students can take credit from the seminar as either MTH 630S or MTH 830S; students who wish to arrange for credit for participation in the seminar should see one of Professors Cook, Ingram or Lelek. In dealing with the seminar the following rules will be adopted:

(1) The seminar holds its meetings on the permanent basis, once a week, during the normal class period of the Fall and Spring Semesters.

(2) The subject of the seminar is that of the research work and of the interest of the organizers and the participants, and covers a wide selection of themes belonging to metric topology, as well as some topics from general, geometric, and algebraic topologies which have a suitable metrico-set-theoretical flavour.

(3) The seminar is devoted primarily to studying and discussing new ideas and results which are yet unpublished or published recently.

(4) The new results obtained by the participants of the seminar have the priority over any other results in scheduling the talks for the seminar meetings.

(5) The new results obtained by the students have the priority over those of the faculty.

(6) The seminar has its own secretary who takes care of the seminar's notebook and makes notes during the meetings.

(7) The note-book is supposed to circulate, during the meetings, among the participants who are encouraged to skim it, check, read the notes, and make their comments both in the note-book and aloud.

(8) The participants are encouraged to interrupt any speaker, ask questions, and raise problems.

(9) The seminar puts special emphasis on discussing unsolved problems, both old and new, which appear in its area of interest.

(10) The organizers invite each topologist who visits the city to come and participate in the seminar, and to give a talk if he wishes to do it.

There had been many visitors, some spending part of their sabbaticals in Houston. The names include: Marcy Barge, Ralph Bean, Harold Bell, David Bellamy, Morton Brown, C.E. Burgess, Calvin Chang, W. Debski, Edwin Duda, J. B. Fugate, Samuel Gitler, E. E. Grace, C. L. Hagopian, J. H. V. Hunt, S. Iliadis, F. Burton Jones, H. Kato, J. Krasinkiewicz, Wayne Lewis, W. Mahavier, E. J. Miller, Jr., P. Minc, Lee Mohler, Sam B. Nadler, Jr., Michel Smith, R. E. Smithson, E. D. Tymchatyn, Paul Weiss, and others.

The note-book had become an important part of the seminar's work. The note-book which was kept in the seminar was a pre-bound book of numbered, but otherwise blank, pages. A record of each meeting was kept in this note-book. The record included the names of those in attendance, the name of the speaker, the mathematics which was presented and the questions and problems which arose. Both W.T.I. and A.L. had served as secretaries and note-book keepers throughout the years, as well as W. Kuperberg, J. Grispolakis and Lex Oversteegen. The seminar had continued in this format even when H.C. left for Australia in 1972 for two years, and had gone to Wayne State in Detroit with A.L. for the period 1974-1976. It terminated in 1989 when W.T.I. moved permanently to Missouri. This made the duration of the seminar exactly 18 years, so then achieving maturity.

In 1979, there was a conference organized by R. D. Mauldin at North Texas, devoted to another connection with Poland, namely to the so-called Scottish Book, a famous list of problems held by Polish mathematicians at the Scottish Cafe in Lvov, Poland, in the 1930's. Preparing for Mauldin's conference, H.C. thought that we might be able to compete there using a list of open problems taken from the notes of our seminar. This was the origin of Houston Problem Book. That first edition, made in a computerized form, went to Denton and was distributed among the participants of the conference. It was continued with subsequent editions as the seminar was still in progress, and comments, some solutions, and new problems were added through 1989. In those years, it was available to everybody by means of the electronic mail. We now decided to update it as completely as we can, and to publish it in a printed form herein.

University of Houston Mathematics Problem Book

Problem 1. *Suppose X is a continuum which cannot be embedded in any continuum Y such that Y is the union of a countable family of arcs. Does X contain a connected subset that is not path-connected? (E. D. Tymchatyn, 09/15/71)*

Comment: The answer is no. (E. D. Tymchatyn, 09/30/81)

Problem 2. *Suppose X is a continuum such that for each positive number ϵ there are at most finitely many pairwise disjoint connected sets in X of diameter greater than ϵ. Suppose, if Y is any continuous, monotonic, Hausdorff image of X, then Y can be embedded in a continuum Z which is the union of a countable family of arcs. Is every connected subset of X path-connected? (E. D. Tymchatyn, 09/15/71)*

Problem 3. *If X is a continuum such that every subcontinuum C of X contains a point that locally separates C in X, is X regular? (E. D. Tymchatyn, 09/15/71)*

Definition. A point p of a locally compact separable metric space L is a **local separating point** of L provided there exists an open set U of L containing p and two points x and y of the component containing p of U such that $U \setminus \{p\}$ is the sum of two mutually separated point sets, one containing x and the other containing y.

Definition. A continuum is **regular** if each two of its points can be separated by a finite point set.

Comment: The answer is no. (E. D. Tymchatyn, 03/19/79)

Problem 4. *If X is a Suslinean curve, does there exist a countable set A in X such that A intersects every nondegenerate subcontinuum of X? (A. Lelek, 09/22/71)*

Definition. A continuum is **Suslinean** if it does not contain uncountably many mutually exclusive nondegenerate subcontinua.

Comment: No. (P. Minc, 06/15/79)

Problem 5. *Can every hereditarily decomposable chainable continuum be embedded in the plane in such a way that each end point is accessible from its complement? (H. Cook, 09/29/71)*

Definition. A **simple chain** is a finite sequence $L_1, L_2, \cdots, L_n$ of open sets such that L_i intersects L_j if and only if $|i - j| \leq 1$. The terms of the sequence $L_1, L_2, \cdots, L_n$ are called the **links** of the chain. An ϵ-**chain** is a chain each of whose links has diameter less than ϵ. A continuum M is **chainable** if, for each positive number ϵ, M can be covered by an ϵ-chain.

Comment: No. (P. Minc and W. R. R. Transue, 02/27/90)

Problem 6. *Is it true that a chainable continuum can be embedded in the plane in such a way that every point is accessible from the complement if and only if it is Suslinean? (H. Cook, 09/29/71)*

Comment: That such a chainable continuum must be Suslinean follows from a result in *Bull. Polish Acad. Sci.* **9** (1960), 271–276. (A. Lelek, 09/29/71)

Comment: Yes. (P. Minc and W. R. R. Transue, 02/27/90)

Problem 7. *Is it true that the rim-type of any real (or half-ray) curve is at most 3? (A. Lelek, 09/29/71)*

Definition. A continuum is a **real curve** if it is a continuous 1-1 image of the real line and a **half-ray curve** if it is a continuous 1-1 image of the nonnegative reals.

Definition. A continuum is **rational** if each two of its points can be separated by a countable point set.

Definition. Suppose that X is a rational curve and a is the least ordinal for which there exists a basis $\mathcal{B}$ for X such that the a-th derivative of the boundary of each element of $\mathcal{B}$ is empty. Then a is the **rim-type** of X.

Comment: A positive solution follows from Nadler's work. (W. Kuperberg and P. Minc, 03/28/79)

Problem 8. *Does there exist a rational curve which topologically contains all real (or half-ray) curves? (A. Lelek, 09/29/71)*

Comment: Yes. (W. Kuperberg and P. Minc, 03/28/79)

Problem 9. *Does there exist a real curve which can be mapped onto each (weakly chainable) real curve? If not, does there exist a rational curve which can be mapped onto each (weakly chainable) real curve? (A. Lelek, 09/29/71)*

Definition. A continuum is **weakly chainable** if it is a continuous image of a chainable continuum.

Comment: Delete the parentheses, weakly chainable is needed. (W. Kuperberg and P. Minc, 03/28/79)

Problem 10. *Suppose D is a dendroid, S is a compact, finitely generated commutative semigroup of monotone surjections on D which has a fixed end point. Does S have another fixed point? (L. E. Ward, 10/20/71)*

Comment: Yes, even for λ-dendroids. (W. J. Gray and S. Williams, *Bull. Polish Acad. Sci.* **27** (1979), 599–604)

Problem 11. *Is each strongly Hurewicz space an absolute F_σ?*
(A. Lelek, 11/03/71)

Comment: No. (E. K. van Douwen, 05/06/79)

Comment: The answer still seems to be unknown for the metric case. (A. Lelek, (08/31/94)

Definition. A regular space is a **Hurewicz space** if and only if, for each sequence of open coverings $Y_1, Y_2, \ldots$, there exist finite subcollections $\mathcal{A}_i$ of Y_i such that $\mathcal{A}_1 \cup \mathcal{A}_2 \cup \cdots$ is a covering. A regular space X is **strongly Hurewicz** if and only if, for each sequence of open coverings $Y_1, Y_2, \cdots$ of X, there exist finite subcollections $\mathcal{A}_i$ of Y_i such that X is the union over i of the intersection over j of $\mathcal{A}_{i+j}^*$.

Notation. If G is a collection of sets, then G^* denotes the sum of all the sets in G.

Problem 12. *Is each product of two Hurewicz metric spaces a Hurewicz space? (A. Lelek, 11/03/71)*

Comment: No. (J. M. O'Farrell, 03/23/84)

Problem 13. *Is the product of two strongly Hurewicz metric spaces a strongly Hurewicz space? (H. Cook, 11/03/71)*

Problem 14. *Is every Lindelöf totally paracompact space Hurewicz? (D. W. Curtis, 11/10/71)*

Definition. A space X is **totally paracompact** if every basis for X has a locally finite subcollection covering X.

Problem 15. *Suppose f is an open mapping of a compact metric space X to the Sierpiński curve. Do there exist arbitrarily small closed neighborhoods U of x in X for which y is in Int $f(U)$ and $f|U$ is confluent? (A. Lelek, 11/24/71)*

Definition. The **Sierpiński curve** is a locally connected planar curve with infinitely many complementary domains whose boundaries are mutually exclusive simple closed curves.

Definition. A mapping f of a compact space X onto a compact space Y is **confluent** if, for each continuum C in Y, each component of $f^{-1}(C)$ is mapped onto C by f.

Comment: No. (T. Maćkowiak and E. D. Tymchatyn, 01/25/81)

Problem 16. *Suppose f is a locally confluent and light mapping of a compact metric space X to the Sierpiński curve. Do there exist arbitrarily small closed neighborhoods U of x in X for which y is in Int $f(U)$ and $f|U$ is confluent? (A. Lelek, 11/24/71)*

Definition. A mapping f from a space X onto a space Y is **locally confluent** if, for each point y of Y, there is an open set U containing y such that $f^{-1}(\overline{U})$ is confluent.

Definition. A mapping is **light** if each point inverse is totally disconnected.

Comment: Yes. (T. Maćkowiak and E. D. Tymchatyn, 01/25/81)

Problem 17. *Is there a space X on which the identity is strongly homotopically stable but there is an open mapping onto X which is not strongly homotopically stable? (H. Cook, 11/24/71)*

Definition. Suppose f is a mapping of a metric space X into a metric space Y and a is a point of X. Then f is

    (1) **unstable** at a provided, for each $\epsilon > 0$, there exists a mapping g from X into Y such that $d(f(x), g(x)) < \epsilon$ for x in X and $f(a)$ in $g(X)$;

    (2) **homotopically unstable** at a provided, for each $\epsilon > 0$, there exists a homotopy h of $X \times [0,1]$ into Y such that if x is in X and t is in $I = [0,1]$, then

$$h(x,0) = f(x), \qquad d\big(h(x,0), h(x,t)\big) < \epsilon, \quad \text{and} \qquad f(a) \neq h(x,1);$$

    (3) **strongly homotopically stable** at a provided for each homotopy h of $X \times I$ into Y, where $h(x,0) = f(x)$, we have $h(a,t) = f(a)$ for t in I.

Problem 18. *Is it true that each open mapping of a continuum X onto the Sierpiński curve (or the Menger curve) is strongly homotopically stable at each point of X? (A. Lelek, 11/24/71)*

Comment: No. (T. Maćkowiak and E. D. Tymchatyn, 01/25/81)

Problem 19. *Does each strictly non-mutually aposyndetic continuum with no weak cut point contain uncountably many mutually exclusive triods? (H. Cook, 12/08/71)*

Definition. A continuum M is **mutually aposyndetic** if, for each two points A and B of M, there exist mutually exclusive subcontinua H and K of M containing A and B, respectively, in their interiors. A continuum is **strictly non-mutually aposyndetic** if each two of its subcontinua with interiors intersect.

Definition. A point x of a continuum M is a **weak cut point** of M if there are two points p and q of M such that every subcontinuum of M that contains both p and q also contains x.

Comment: E. E. Grace has shown that every planar strictly non-mutually aposyndetic continuum has a weak cut point.

Problem 20. *Suppose f is an open continuous mapping of a continuum X onto Y. Does there exist a space X^* such that X is a subset of X^*, X^* is a locally connected continuum, and there is an extension f^* of f from X^* to Y^* such that f^* is open and $f^*(X)$ does not intersect $f^*(X^* \setminus X)$? Can X^* be the Hilbert cube? If X is a curve, can X^* be the Menger universal curve? (A. Lelek, 12/08/71)*

Comment: The answer to the first and the last question is yes. (T. Maćkowiak and E. D. Tymchatyn, 01/25/81)

Problem 21. *If X is a planar continuum which has only a finite number of complementary domains and X has property A then is X connected im kleinen? (A. Lelek, 01/26/72)*

Definition. A connected space X is **aposyndetic** at H with respect to K if there is a closed connected subset of X with H in its interior and not intersecting K, and X is **aposyndetic** if it is aposyndetic at each point with respect to every other point.

Definition. A space is **connected im kleinen** if it is aposyndetic at each point with respect to each closed point set not containing that point.

Definition. Suppose M is a connected metric space and x and y are two points of M. If, for each connected and closed subset of K of $M - \{x\}$ containing y, M is aposyndetic at x with respect to K, then M is said to have **property** A at x with respect to y. A continuum has **property** A if it has property A at each point with respect to every other point.

Comment: Yes, quite general solution. (T. Maćkowiak and E. D. Tymchatyn, 09/30/81)

Problem 22. *Does every totally non-semi-locally connected continuum contain a dense G_δ set of weak cut points? (E. E. Grace, 02/02/72)*

Definition. A continuum M is **semi-locally connected** at the point p if, for every domain D containing p, there exits a domain E lying in D and containing p such that $M \setminus E$ has only a finite number of components. A continuum is **totally non-semi-locally connected** if it is not semi-locally connected at any point.

Problem 23. *Does every totally non-semi-locally connected bicompact Hausdorff continuum contain a weak cut point? (E. E. Grace, 02/02/72)*

Problem 24. *Is it true that if X is a simple tree and $w(X) > \epsilon > 0$, then there is a simple triod T in X such that $w(T) > \epsilon$? (A. Lelek, 02/09/72)*

Definition. For any compact metric space X, the **width** $w(X)$ of X is the l.u.b. of the set of all real numbers a which satisfy the following condition: for each $\epsilon > 0$, there exists a finite open cover $\mathcal{C}$ of X such that mesh($\mathcal{C}$) $< \epsilon$ and for each chain $\mathcal{C}'$ which is a subcollection of $\mathcal{C}$ there is a member A of $\mathcal{C}$ such that $d(A, \mathcal{C}'^*) \geq a$.

Comment: No. (F. O. McDonald, 06/01/74)

Problem 25. *Suppose X is a tree-like continuum, $w(X) > 0$, and f is a continuous mapping from X to a chainable continuum Y. Is it true that there exist, for each a such that $0 < a < w(X)$, two points x and y such that $d(x, y) = a$ and $f(x) = f(y)$? (A. Lelek, 02/09/72)*

Definition. If X is a metric space, a mapping f from X to a space Y is an ϵ-**map** if, for each point y of Y, diam$\left(f^{-1}(y)\right) \leq \epsilon$. If $\mathcal{C}$ is a collection of continua, a continuum M is $\mathcal{C}$-**like** if, for every positive number ϵ, there exists an ϵ-map of M onto an element of $\mathcal{C}$. In particular, a continuum is **tree-like** if, for some collection $\mathcal{C}$ of trees, M is $\mathcal{C}$-like.

Problem 26. *Let M be a locally compact metric space and a be a point of M. It is known that a is homotopically unstable if and only if there exists a homotopy $f(\cdot, t)$ of M to M such that $f(\cdot, 0)$ is the identity and a is in $f(M, t)$ for each $t > 0$. Equivalently, a is homotopically unstable if and only if there exists a retraction r of $M \times [0, 1]$ to $M \times \{0\}$ such that $r^{-1}((a, 0)) = \{(a, 0)\}$. Is local compactness necessary in this theorem? (W. Kuperberg, 09/13/72) (For definitions see Problem 17.)*

Problem 27. *Suppose M is an n-dimensional compact metric space and A is the set of all unstable points of M. Does there exist a homotopy $f(\cdot, t)$ of M to M such that $f(\cdot, 0)$ is the identity and $f(M, t)$ and A do not intersect for each $t > 0$? (W. Kuperberg, 09/13/72)*

Comment: No. (W. Kuperberg and P. Minc, 03/28/79)

Problem 28. *Prove that the $\sin(1/x)$ curve is not pseudo-contractible. (W. Kuperberg, 09/20/72)*

Comment: Done. (W. Debski, 05/08/90)

Definition. Let C be a continuum and a and b be two points of C. Suppose f and g are two mappings from a space X into a space Y. Then f and g are (C, a, b)-**homotopic** provided there exists a mapping F of $X \times C$ into Y such that $F(x, a) = f(x)$ and $F(x, b) = g(x)$, for each x in X. We say that f is **pseudo-homotopic** to g provided there exist C, a, and b such that f and g are (C, a, b)-homotopic. A space is **pseudo-contractible** provided the identity mapping is pseudo-homotopic to a constant.

Problem 29. *Does there exist a 1-dimensional continuum which is pseudo-contractible but is not contractible? (W. Kuperberg, 09/20/72)*

Problem 30. *Suppose X and Y are continua (or even polyhedra), x is in X, y is in Y, and there exist neighborhoods U and V of x and y in X and Y, respectively, such that a homeomorphism of U onto V transforms x onto y. Is it true that if x is pseudo-unstable in X, then y is pseudo-unstable in Y? (W. Kuperberg, 09/20/72)*

Definition. A point of M is C_a-**unstable** (where a is in C, C a continuum) provided there exists a mapping F of $M \times C$ into M such that

(1) $F(x, a) = x$ for x in M, and
(2) $F(x, c) \neq y$ for x in M, c in C, and $a \neq c$.

We say that y is **pseudo-unstable** provided there exists a continuum C such that y is C_a-unstable.

Problem 31. *Is it true that the pseudo-arc is not pseudo-contractible? (W. Kuperberg, 09/20/72)*

Problem 32. *Does there exist a mapping f of the circle or the plane onto itself such that f^n has a fixed point for each $n \geq 2$, but f does not have a fixed point? (W. Kuperberg, 10/11/72)*

Comment: Yes; L. Block, Trans. Amer. Math. Soc. **260** (1980), 553–562, and, independently, M. Cook, oral communication, 1988. (W. T. Ingram, 06/30/89)

Problem 33. *Suppose P is a subset of the positive integers with the property that, if k is in P, then $n \cdot k$ is in P for each positive integer n. Does there exist a locally connected continuum X and a mapping f of X onto X such that f^n has a fixed point if and only if n is in P? (W. Kuperberg, 10/11/72)*

Comment: Yes. (M. Cook, 02/20/89)

Problem 34. *Is it true that if f is a mapping of a tree-like continuum into itself, then there exits an n such that f^n has a fixed point? (W. T. Ingram, 10/11/72)*

Comment: No. (P. Minc, 06/15/91)

Problem 35. *Suppose f is a continuous mapping of a continuum X onto a continuum Y, $Y = H \cup K$ is a decomposition of Y into subcontinua H and K, $f|f^{-1}(H)$ and $f|f^{-1}(K)$ are confluent, and $H \cap K$ is a continuum which does not cut Y and is an end continuum of both H and K. Is f confluent? (W. T. Ingram, 10/11/72)*

Problem 36. *Suppose f is a weakly confluent and locally confluent mapping of a continuum X onto a tree-like continuum Y. Does there exist a subcontinuum L of X such that $f(L) = Y$ and $f|L$ is confluent? (D. R. Read, 10/25/72)*

Comment: No. (T. Maćkowiak and E. D. Tymchatyn, 01/25/81)

Problem 37. *Suppose $(X, \leq)$ is a finite partially ordered set. By taking $\overline{\{y\}} = \{x \mid x \leq y\}$ one gets a T_0 topology on X; $\overline{A}$ is the union of the closures of the points of A, and then a mapping f of X into X is continuous if and only if f is order preserving. A retraction r of X onto Y is* **relating** *provided that x in $X \setminus Y$ implies x and $r(x)$ are related. Does there exist a unique minimal relating retract of X? (L. E. Ward, 11/29/72)*

Problem 38. *Suppose X in the previous problem is $A \cup B$, where A and B are closed sets and A, B, and $A \cap B$ have the fixed point property. Does X have the fixed point property? (L. E. Ward, 11/29/72)*

Problem 39. *Suppose X is a finite partially ordered set with the fixed point property. Is X unicoherent? (L. E. Ward, 11/29/72)*

Problem 40. *Does there exist a nonseparating (locally compact) subset of the plane which is neither compact nor open in the plane such that if f is a continuous and reversible function of X onto a subset of the plane, then f is a homeomorphism? (A. Lelek, 02/07/73)*

Problem 41. *Suppose that the plane is the union of two disjoint sets A and B neither of which contains a Cantor set. Is each continuous and 1-1 function of A into the plane a homeomorphism? (A. Lelek, 02/07/73)*

Comment: No. (E. K. van Douwen, 07/14/80)

Problem 42. *Suppose X is a compact metric space and S is the set of all stable points of X. Suppose $\dim X$ is finite and a is an infinite cycle in S. Is it true that if $a \sim 0$ in X, then $a \sim 0$ in S? (T. Ganea, 03/07/73)*

Comment: Yes, solved by Namioka. (W. Kuperberg and P. Minc, 03/28/79)

Problem 43. *Suppose X is a curve which is contractible, or pseudo-contractible, or just pseudo-deformable to a proper subset. Does there exist in X a point which is unstable (or pseudo-unstable, respectively)? (W. Kuperberg, 03/07/73)*

Problem 44. *Suppose X is a metric space; does 'superstable' in X imply 'stable'? (W. Kuperberg, 03/07/73)*

Definition. Given a point p of a space X, p is **superstable** if it is not pseudo-unstable.

Problem 45. *Is it true that if Y is an arcwise connected continuum such that each continuous mapping of a continuum onto Y is weakly confluent, then Y is an arc? (A. Lelek, 03/21/73)*

Comment: Yes, since Y is irreducible, unicoherent, and not a triod. (J. Grispolakis, 03/19/79)

Problem 46. *Suppose Y is a compact metric space such that the Borsuk modified fundamental group of Y is 0 and f is a local homeomorphism of X onto Y. Must f be a homeomorphism? (W. Kuperberg, 04/11/73)*

Problem 47. *Do atomic mappings preserve semi-aposyndesis (or mutual aposyndesis) of continua? (W. T. Ingram, 04/18/73)*

Definition. If f is a mapping of a continuum X onto a continuum Y and, for each subcontinuum K of X such that $f(K)$ is nondegenerate $K = f^{-1}(f(K))$, then f is **atomic**. Atomic maps are monotone.

Comment: Yes, such mappings must be either homeomorphisms or constant mappings. (T. Maćkowiak, 09/30/81)

Problem 48. *Is each unicoherent and mutually aposyndetic continuum locally connected? (L. Gibson, 04/18/73)*

Comment: No, but the answer is unknown for such continua which are 1-dimensional. (T. Maćkowiak, 09/30/81)

Problem 49. *Suppose f is an open mapping of a continuum X onto Y and y is a branch-point of Y. Is there a branch-point x of X such that $f(x) = y$? (V. Parr, 04/25/73)*

Definition. A **branch point** of a continuum is the vertex of a simple triod lying in that continuum.

Comment: No, but the answer is yes if f is light. (T. Maćkowiak, 09/30/81)

Problem 50. *Suppose f is a weakly confluent mapping of a continuum X onto Y and y is a branch-point of Y. Is there a branch-point x of X such that $f(x) = y$? (V. Parr, 04/25/73)*

Comment: No. (T. Maćkowiak, 09/30/81)

Problem 51. *Is it true that each polyhedron is the weakly confluent image of a polyhedron whose fundamental group is trivial?*

Comment: False, T2. (J. Grispolakis, 03/19/79)

Problem 52. *Are strongly regular curves inverse limits of connected graphs with monotone simplicial bonding maps? (B. B. Epps, 09/19/73)*

Comment: No. (E. D. Tymchatyn, 10/27/74)

Problem 53. *If X is the inverse limit of connected graphs with monotone simplicial retractions as bonding maps, is X the weakly confluent image of a dendrite? (B. B. Epps, 09/19/73)*

Comment: Yes. (J. Grispolakis and E. D. Tymchatyn, 03/19/79)

Problem 54. *Is it true that X is the inverse limit of connected graphs with monotone simplicial retractions as bonding maps if and only if X is locally connected and each cyclic element of X is a graph? (B. B. Epps, 09/19/73)*

Definition. If M is a semi-locally connected continuum, a **cyclic element** of M is either a cut point of M, an end point of M, or a nondegenerate subset of M which is maximal with respect to being a connected subset of M without cut points.

Problem 55. *Does each continuous mapping of a continuum onto a chainable continuum have property F? (A. Lelek, 09/26/73)*

Definition. A mapping f of a continuum X onto a continuum Y has **property F** provided that there exists a point x in X such that if K is a subcontinuum of Y containing $f(x)$ and C is the component of $f^{-1}(K)$ containing x, then $f(C) = K$.

Comment: No. (J. B. Fugate, 06/16/79)

Problem 56. *Does each (weakly confluent mapping of a continuum onto an irreducible continuum have property F? (A. Lelek, 09/26/73)*

Comment: No. (J. B. Fugate, 06/16/79)

Problem 57. *Is it true that each uniformly pathwise connected continuum is uniformly arcwise connected? (W. Kuperberg, 09/26/73)*

Definition. Suppose X is a metric space. A **path** is a continuous mapping of $[0, 1]$ into X. A collection P of paths in X is **uniform** provided, for each $\epsilon > 0$, there exists a positive integer n such that, for each p in P, there are points

$$0 = t_0 < t_1 < t_2 < \cdots < t_n = 1$$

such that diam $([t_{i-1}, t_i]) \le \epsilon$ for $i = 1, \cdots, n$. We say X is **uniformly pathwise connected** provided that, for each two points a and b of X, there exists a uniform collection P of paths in X joining a and b such that the union of their images is X. We say that X is **uniformly arcwise connected** provided that, for each two points a and b of X, there exists a uniform collection P of homeomorphisms of $[0, 1]$ into X joining a and b such that the union of their images is the union of all arcs in X joining a and b.

Problem 58. *Suppose f is a continuous mapping of a chainable continuum X onto a nonchainable continuum Y. Does there exist a subcontinuum X' of X mapped onto Y under f such that $f|X'$ is not weakly confluent? (A. Lelek, 10/03/73)*

Comment: No. (T. Maćkowiak and E. D. Tymchatyn, 01/25/81)

Problem 59. *Does a continuum with zero surjective span have zero span?*
(A. Lelek, 11/07/73)

Definition. Let f be a mapping from a (connected if necessary) space X to a metric space Y and let S be the least upper bound of numbers a such that there exists a connected set C in $X \times X$ both of whose projections are the same point set and, for each point (x, y) in C, $d(f(x), f(y)) \geq a$. Then S is the **span** of f. If we require that both projections of C be all of X, then S is the **surjective span** of f. If we require only that one projection of C be a subset of the other, then S is the **semi-span** of f; and if we require that one projection of C be all of X, then S is the **surjective semi-span** of f. If X is a (connected) metric space, then the **span**, **surjective span**, **semi-span**, and **surjective semi-span** of X are the span, surjective span, semi-span, and surjective semi-span, respectively, of the identity map of X onto itself.

Notation. If Z is a (connected) metric space or a mapping from a (connected) metric space to a metric space, then

  (1) the span of Z is denoted by $\sigma(Z)$,
  (2) the surjective span of Z by $\sigma^*(Z)$,
  (3) the semi-span of Z by $\sigma_0(Z)$, and
  (4) the surjective semi-span of Z by $\sigma_0^*(Z)$.

Problem 60. *Is it true that if $Y = G_1 \cup G_2$ where G_1 and G_2 are open and $f | f^{-1}(\overline{G_i})$ is an MO-mapping for $i = 1, 2$, then f is an MO-mapping? (A. Lelek, 11/21/73)*

Definition. An **MO-mapping** is a composition gf where g is monotone and f is open.

Comment: Spaces are compact metric. (A. Lelek, 11/21/73)

Problem 61. *Suppose $\mathcal{E}$ is the class of all finite one point unions of circles. Is it true that if X and Y are shape-irreducible continua possessing the same shape and there exist curves C and C' in $\mathcal{E}$ such that X is C-like and Y is C'-like, then there exists a C'' in $\mathcal{E}$ such that both X and Y are C''-like? (A. Lelek, 12/05/73)*

Definition. A continuum X is said to be **shape-irreducible** if and only if no proper subcontinuum of X is of the same shape as X.

Comment: Trivially yes, but there even exists a shape-irreducible figure-eight-like continuum with the shape of a circle that is not circle-like. (J. Segal and S. Spiez, 05/07/86)

Problem 62. *Are all shapes of polyhedra stable? (W. Kuperberg, 04/03/74)*

Problem 63. *Given a polyhedron P. Does there exist a polyhedron Q of the same shape as P and shape stable? (W. Kuperberg, 04/03/74)*

Problem 64. *Is each uniformly path-connected continuum g-contractible?*
(D. Bellamy, 04/03/74)

Definition. A space X is g-contractible provided there exists a mapping f of X onto X such that f is homotopic to a constant mapping.

Problem 65. *Is it true that, for each uniformly path-connected continuum X, there exists a compact metric space Y and two mapping f, from X onto the cone over Y, and g, from the cone over Y onto X? (D. Bellamy, 04/03/74)*

Comment: No. (J. Prajs, 03/01/89)

Problem 66. *If C is a connected Borel subset of a finitely Suslinean continuum X, is C arcwise connected? (E. D. Tymchatyn, 09/26/74)*

Definition. A continuum X is **finitely Suslinean** if, for each positive number ϵ, X does not contain infinitely many mutually exclusive subcontinua of diameter greater than ϵ.

Definition. The family $\mathcal{F}$ of **Borel sets** of a space S is the smallest family satisfying the conditions:

   (1) Every closed set belongs to $\mathcal{F}$;
   (2) If X is in $\mathcal{F}$, then $S \setminus X$ is in $\mathcal{F}$; and
   (3) The countable intersection of elements of $\mathcal{F}$ belongs to $\mathcal{F}$.

Comment: Yes, for subsets of regular curves. (D. H. Fremlin, 04/12/90)

Problem 67. *Suppose X is a continuum such that, if C is a subcontinuum of X then the set of all local separating points of C is not the union of countably many closed disjoint proper subsets of C. Then, is every connected subset of X arcwise connected? (E. D. Tymchatyn, 09/26/74)*

Problem 68. *If for each subcontinuum C of the continuum X, the set of all local separating points of C is connected, then is every connected subset of X arcwise connected? (E. D. Tymchatyn, 10/03/74)*

Problem 69. *Do hereditarily indecomposable tree-like continua have the fixed point property? (B. Knaster, 11/21/74)*

Problem 70. *Do uniquely λ-connected (or uniquely δ-connected) continua have the fixed point property for homeomorphisms? (A. Lelek, 11/21/74)*

Definition. A continuum is δ-**connected** if each two of its points can be connected by an hereditarily decomposable irreducible subcontinuum. A continuum is λ-**connected** if each two of its points p and q can be joined by an irreducible continuum of type λ, i.e. an irreducible continuum from p to q which has arcatomic subsets, no one of which has interior.

Problem 71. *Suppose that X is a continuum such that, if $\epsilon > 0$, $C_1, C_2, \cdots$ are mutually separated connected sets of diameter greater than ϵ, and $C_1 \cup C_2 \cup \cdots$ is connected, then there exist mutually exclusive arcs $A_1, A_2, \cdots$ such that, for some subsequence $D_1, D_2, \cdots$ of $C_1, C_2, \cdots$ and each i, A_i is a subset of D_i, and $\mathrm{diam}(A_i) > \epsilon$. Is X locally connected? (E. D. Tymchatyn, 01/27/75)*

Problem 72. *Suppose X is a continuum such that each σ-connected subset of X is a semi-continuum. Is X locally connected? (J. Grispolakis, 01/27/75)*

Definition. A **semi-continuum** is a continuumwise connected point set.

Problem 73. *Suppose X is a continuum such that each σ-connected F_σ subset of X is a semi-continuum. Is X semi-aposyndetic? (J. Grispolakis, 01/27/75)*

Definition. A space X is **semi-aposyndetic** if, for each two of its points, X is aposyndetic at one of them with respect to the other.

Problem 74. *Let f be a perfect pseudo-confluent mapping of an hereditarily normal, hereditarily locally connected and hereditarily σ-connected space onto a complete metric space Y. Is Y hereditarily σ-connected? (J. Grispolakis, 02/03/75)*

Definition. A mapping from X to Y is **pseudo-confluent** if every irreducible continuum in Y is the image of a continuum in X.

Comment: The answer is yes if X is also a $(Q = C)$-space; i.e., if the quasi-components of any subset of X are its components. (J. Grispolakis, 02/03/75)

Problem 75. *Suppose X is a separable metric space which possesses an open basis $\mathcal{B}$ such that, the set $X \setminus G$ is the union of a collection of closed open subsets of X. Is X embeddable in an hereditarily locally connected space? (E. D. Tymchatyn, 03/03/75)*

Comment: Yes. (L. G. Oversteegen and E. D. Tymchatyn, 08/18/92)

Problem 76. *Suppose X is a separable metric space such that, if A and B are connected subsets of X, then $A \cap B$ is connected. Is it true that if C is a set of non-cut points of X, then $X \setminus C$ is connected? (L. E. Ward, 03/03/75)*

Problem 77. *Is it true that a continuum X is regular if and only if every infinite sequence of mutually disjoint connected subsets of X is a null sequence? (E. D. Tymchatyn, 03/03/75)*

Problem 78. *Is it true that, for each decomposition of a finitely Suslinean continuum into countably many disjoint connected sets, at least one of them must be rim-compact? (T. Nishiura, 03/03/75)*

Definition. A space is said to be **locally peripherally compact**, or **rim compact**, at p if every open set containing p contains an open set containing p and having a compact boundary.

Comment: No. (E. D. Tymchatyn, 09/30/81)

Problem 79. *Is it true that no biconnected set with a dispersion point can be embedded into a rational continuum? (J. Grispolakis, 03/03/75)*

Definition. A point p of a nondegenerate connected space X is an **explosion point**, or **dispersion point**, of X provided that $X \setminus \{p\}$ is totally disconnected.

Definition. A space is **biconnected** if it is not the sum of two mutually exclusive nondegenerate connected point sets.

Problem 80. *Is it true that a separable metric space is embeddable into a rational continuum if and only if it possesses an open basis whose elements have scattered boundaries? (E. D. Tymchatyn, 03/03/75)*

Definition. A point set X is **scattered** if every subset Y of X has a point that is not a limit point of Y.

Comment: Yes. (E. D. Tymchatyn, 09/30/81)

Problem 81. *Is it true that each continuum of span zero is chainable? (H. Cook and A. Lelek, 05/15/75)*

Problem 82. *Is it true that each continuum of surjective semi-span zero is arc-like? (A. Lelek, 05/15/75)*

Problem 83. *Suppose X is a connected metric space.*

   (1) *Is the span of X less than or equal to twice the surjective span of X?*
   (2) *Is the semi-span of X less than or equal to twice the surjective semi-span of X?*
   (3) *Is the surjective semi-span of X less than or equal to twice the surjective span of X?*
   (4) *If T is a simple triod, is the surjective span of T equal to the surjective semi-span of T?*

(A. Lelek, 05/15/75)

Problem 84. *Is the confluent image of an arc-like continuum arc-like? (A. Lelek, 10/21/75)*

Problem 85. *If f is a confluent mapping of an acyclic (or tree-like or arc-like) continuum onto a continuum Y, is $f \times f$ confluent? (A. Lelek, 10/21/75)*

Problem 86. *Do confluent maps of continua preserve span zero? (H. Cook and A. Lelek, 10/21/75)*

Problem 87. *Is every regularly submetrizable Moore space completely regular? (H. Cook, 10/13/76)*

Definition. If (X, T_1) is a topological space and there exists a subcollection T_2 of T_1 such that (X, T_2) is metrizable, then (X, T_1) is **submetrizable**. If, in addition, for each point p and open set O in T_1 containing p, there exists an open set D in T_1 containing p whose closure with respect to T_2 is a subset of O, then (X, T_1) is **regularly submetrizable**.

Problem 88. *Is every homogeneous continuum bihomogeneous? (B. Fitzpatrick, 10/27/76)*

Definition. Suppose X is a space such that, for each two points a and b, there is a homeomorphism of X onto itself such that $h(a) = b$ [and $h(b) = a$]. Then X is **homogeneous** [respectively, **bihomogeneous**].

Comment: No. (K. Kuperberg, 02/02/88)

Problem 89. *Does there exist a noncombinatorial triangulation of the 4-sphere? (C. E. Burgess, 10/12/77)*

Problem 90. *Is there an hereditarily equivalent continuum other than an arc or a pseudo-arc? (H. Cook, 11/02/77)*

Definition. A continuum is **hereditarily equivalent** if it is homeomorphic to each of its nondegenerate subcontinua.

Problem 91. *Do hereditarily equivalent continua have span zero? (H. Cook, 11/02/77)*

Problem 92. *If M is a continuum with positive span such that each of its proper subcontinua has span zero, does every nondegenerate monotone continuous image of M have positive span? (H. Cook, 11/02/77)*

Comment: No. (J. F. Davis and W. T. Ingram, 04/30/86; Fund. Math. **131** (1988), 13–24)

Problem 93. *Does there exist a homogeneous tree-like continuum of positive span? (W. T. Ingram, 02/08/78)*

Comment: Not in the plane. (L. G. Oversteegen and E. D. Tymchatyn, 07/23/80)

Problem 94. *If M is a plane continuum with no weak cut point, is M λ-connected? (C. L. Hagopian, 04/12/78)*

Comment: Yes, but there exists a counterexample in the 3-dimensional Euclidean space. (C. L. Hagopian, 04/01/79)

Problem 95. *Is the countable product of λ-connected continua λ-connected? (C. L. Hagopian, 04/12/78)*

Comment: Yes, but it is unknown if the product of any two continua is λ-connected. (C. L. Hagopian, 04/07/86)

Problem 96. *Is the image under a local homeomorphism of a δ-connected continuum also δ-connected? (C. L. Hagopian, 04/12/78)*

Problem 97. *Can it be proven, without extraordinary logical assumptions, that the plane is not the sum of fewer than c mutually exclusive continua? (H. Cook, 04/19/78)*

Comment: The plane is not the sum of fewer than c mutually exclusive continua each of which is either Suslinean or locally connected. (H. Cook, 04/19/78)

Problem 98. *Is it true that if (X, T_1) is normal, (X, T_2) is compact, T_2 is a subcollection of T_1, $\text{ind}(X, T_1) = 0$, and $\text{ind}(X, T_2) > 0$, then (X, T_1) fails to be a Hurewicz space? (A. Lelek, 09/13/78)*

Problem 99. *Is the Sorgenfrey line totally paracompact? (A. Lelek, 09/13/78)*

Comment: No. (J. M. O'Farrell, 09/01/80)

Problem 100. *Suppose f is a light open mapping from a continuum M onto a continuum N, B is a smooth dendroid lying in N, and x is a point of $f^{-1}(B)$. Does there exist a smooth dendroid A in M such that x is in A and $f|A$ is a homeomorphism? (J. B. Fugate, 10/18/78)*

Definition. A dendroid X is **smooth** if there is a point p in X such that, if $x_1, x_2, \cdots$ is a sequence of points of X converging to the point x, then the superior limit of the arcs $[p, x_n]$ is the arc $[p, x]$.

Comment: Example 2 of *Colloq. Math.* **38** (1978), 193–196, gives a negative solution. (T. Maćkowiak, 01/25/81)

Problem 101. *Suppose M is a continuum and f is a monotone mapping of M onto $[0, 1]$ such that, if D is a closed proper subset of M which is mapped onto $[0, 1]$ by f, then $f|D$ is not monotone. Is M irreducible? (L. Mohler and L. E. Ward, 10/18/78)*

Problem 102. *Is there a non-locally connected continuum M such that there exists a retraction of 2^M to $C(M)$? If M is the cone over a compact set with only one limit point, is $C(M)$ a retract of 2^M? (J. B. Fugate, 10/18/78)*

Problem 103. *Suppose M is an hereditarily unicoherent and hereditarily decomposable continuum and f is a continuous mapping M into 2^M. Is there a point x of M such that x is in $f(x)$? (J. B. Fugate, 10/18/78)*

Problem 104. *Is the open image of a circle-like continuum always circle-like or arc-like? (J. B. Fugate, 10/18/78)*

Problem 105. *Suppose M is an atriodic 1-dimensional continuum and G is an upper semi-continuous collection of continua filling up M such that M/G and every element of G are chainable. Is M chainable? (H. Cook and J. B. Fugate, 10/18/78)*

Comment: It follows from a result of Sher that, even if M contains a triod, if M/G and every element of G are tree-like, then M is tree-like. (H. Cook, 10/18/78)

Comment: Yes if the requirement that M is 1-dimensional is replaced by M is strongly unicoherent and M/G is hereditarily decomposable. (W. Dwayne Collins, 03/15/82)

Definition. A continuum X is **strongly unicoherent** provided X is unicoherent and each proper subcontinuum with interior is unicoherent.

Comment: Yes if M is hereditarily indecomposable and both M/G and every element of G are pseudo-arcs. (W. Lewis, 05/10/82)

Comment: No. (J. F. Davis and W. T. Ingram, 04/30/86; *Fund. Math.* **131** (1988), 13–24)

Problem 106. *If M is tree-like and every proper subcontinuum of M is chainable, is M almost chainable? (J. B. Fugate, 10/18/78)*

Definition. A continuum M is **almost chainable** if, for each positive number ϵ, there exists an open cover D of M with mesh ϵ and a chain C of elements of D and an end link L of C such that no element of $D \setminus C$ intersects any link of C other than L and every point of M is at a distance less than ϵ from some link of C.

Problem 107. *Suppose $M_1, M_2, \cdots$ is a sequence of mutually disjoint continua in the plane converging to the continuum M homeomorphically. Is M circle-like or chainable? (J. B. Fugate, 10/18/78)*

Definition. The statement that the sequence $M_1, M_2, \cdots$ **converges homeomorphically to the continuum** M means there exists a sequence $h_1, h_2, \cdots$ of homeomorphisms such that, for each positive integer i, h_i is a homeomorphism from M_i onto M and for each positive number ϵ there exists a positive integer N such that if $j > N$ then, for all x, $\text{dist}(h_j(x), x) < \epsilon$.

Problem 108. *If M is a uniquely arcwise connected continuum, does each light open mapping of M onto itself have a fixed point? (J. B. Fugate and B. McLean, 10/18/78)*

Comment: No. (L. G. Oversteegen, 01/22/80)

Problem 109. *Do pointwise periodic homeomorphisms on tree-like continua have a fixed point? (J. B. Fugate and B. McLean, 10/18/78)*

Problem 110. *Do disk-like continua have the fixed point property for periodic homeomorphisms? (J. B. Fugate and B. McLean, 10/18/78)*

Problem 111. *If M is a tree-like continuum with the fixed point property, does $M \times [0, 1]$ have the fixed point property? (J. B. Fugate, 10/18/78)*

Problem 112. *If M is a contractible continuum, do periodic homeomorphisms on M have fixed points? (J. B. Fugate, 10/18/78)*

Problem 113. *Suppose M is a continuum which is not the sum of a countable monotonic collection of proper subcontinua. Is M irreducible about some finite set? (J. B. Fugate, 10/18/78)*

Problem 114. *Is it true that, if X is a compact metric space, P is a 1-dimensional connected polyhedron with a geodesic metric and f is an essential mapping of X into P, then the span of f is greater than or equal to the span of P? (A. Lelek, 12/06/78)*

Comment: No, but a similar and quite strong result in this direction has been recently established by H. Kato, A. Koyama, and E. D. Tymchatyn. (A. Lelek, 05/15/89)

Problem 115. *Does every atriodic 2-dimensional continuum contain a 2-dimensional indecomposable continuum?*
(H. Cook, 02/19/79)

Problem 116. *If S is a compact, uniquely divisible, topological semigroup with 0, 1, and no other idempotents, must S have non-zero cancellation? (D. R. Brown, 03/02/79)*

Problem 117. *Suppose that M is a tree-like continuum such that, for each two points A and B of M, the diagonal in M × M intersects every continuum containing both (A, B) and (B, A). Does M have span zero? (H. Cook, 03/02/79)*

Problem 118. *Suppose H and K are two continua with span zero whose intersection is connected and whose sum is atriodic. Does their sum have span zero? (E. Duda, 03/02/79)*

Problem 119. *Suppose Y is a nondegenerate locally connected continuum and each cyclic element of Y is a completely regular continuum. Is it true that there exists a continuum X in the plane and a monotone open map of X onto Y such that $f^{-1}(y)$ is a (nondegenerate) decomposable continuum for each y in Y? (J. Krasinkiewicz, 03/07/79)*

Definition. A continuum is **completely regular** if each nondegenerate subcontinuum has a nonempty interior.

Comment: The answer is yes if Y itself is a completely regular continuum in the plane. (T. Maćkowiak and E. D. Tymchatyn, 01/25/81)

Problem 120. *Suppose M is the 'canonical' Knaster indecomposable continuum (obtainable by identifying each point of the dyadic solenoid with its inverse) and h is a homeomorphism of M onto itself. Does h leave two points fixed? (W. S. Mahavier, 03/14/79)*

Problem 121. *Does every hereditarily decomposable continuum contain an irreducible continuum with a composant whose complement is degenerate? (W. S. Mahavier, 03/14/79)*

Comment: Yes. (L. G. Oversteegen and E. D. Tymchatyn, 09/30/79)

Problem 122. *Suppose n is a positive integer and M is a continuum such that, for every positive number ϵ, there exists a weakly confluent ϵ-map of M onto an n-cell. Does M have the fixed point property? (H. Cook, 03/15/79)*

Comment: Yes for $n = 2$. (S. B. Nadler, Fund. Math. **110** (1980), pp. 231-232)

Problem 123. *Does there exist a widely connected complete metric space? (H. Cook, 03/16/79)*

Definition. A nondegenerate connected space X is **widely connected** if each nondegenerate connected subset of X is dense in X.

Problem 124. *Suppose M is the 'canonical' Knaster indecomposable continuum. Does M have a nonseparating closed set that intersects every composant of M? (H. Cook, 03/16/79)*

Comment: Yes. (W. Debski, 05/08/90)

Problem 125. *Does there exist a universal hereditarily indecomposable continuum? (H. Cook, 03/16/79)*

Comment: Yes. (T. Maćkowiak and P. Minc, 01/24/83)

Problem 126. *If M is an hereditarily indecomposable continuum containing a pseudo-arc P, is P a retract of M? (B. Knaster, 03/16/79)*

Comment: J. L. Cornette has shown that each subcontinuum the pseudo-arc is a retract of it. (H. Cook, 03/16/79)

Comment: If M is an hereditarily indecomposable continuum, K is a subcontinuum of M, and f is a mapping of K into a pseudo-arc P, then f can be extended to a mapping of M into P. (D. Bellamy, 05/05/79)

Problem 127. *Does a mapping of the plane into itself with bounded orbits have a fixed point? (H. Cook, 03/23/79)*

Comment: For mappings of the 3-dimensional Euclidean space, the answer is no. (K. Kuperberg and Coke Reed, *Fund. Math.*, *Vol.* **114** (1981), p. 229)

Problem 128. *Given $n > 0$, is there a continuous of the $(2n + 1)$-sphere into itself such that the orbit of each point is dense? (S. Fajtlowicz and D. Mauldin, 03/23/79)*

Problem 129. *Does there exist a chainable continuum M in the plane such that, if K is a chainable continuum in the plane, there exists a homeomorphism h of the plane onto itself that takes K into M? (R. H. Bing, 03/23/79)*

Problem 130. *Is it true that, if X is a continuum, f is a mapping of X into Hilbert space, and f has span zero, then f is almost factorable through $[0,1]$? (H. Cook, 03/23/79)*

Definition. Suppose X is a continuum, Y is a metric space, $\mathcal{C}$ is a collection of continua, and f is a mapping of X into Y. If ϵ is a positive number, f is ϵ-**almost factorable** through $\mathcal{C}$ if there exists a mapping k of X onto an element T of $\mathcal{C}$ and a mapping p of T into Y such that, for each point x of X, $d(f(x), pk(x)) \le \epsilon$; and f is **almost factorable** through $\mathcal{C}$ if it is ϵ-almost factorable through $\mathcal{C}$ for every positive number ϵ. (M. Marsh, 03/23/79)

Problem 131. *Suppose M is a continuum such that, if G is an uncountable collection of nondegenerate subcontinua of M, then some two elements of G have a nondegenerate continuum in their intersection. Does M contain a countable point set that intersects every nondegenerate subcontinuum of M? (H. Cook, 03/26/79)*

Problem 132. *Is there an atriodic tree-like continuum which cannot be embedded in the plane? (W. T. Ingram, 03/27/79)*

Comment: Yes. (L. G. Oversteegen and E. D. Tymchatyn, 09/30/81)

Problem 133. *If M is an atriodic tree-like continuum in the plane, does there exist an uncountable collection of mutually exclusive continua in the plane each member of which is homeomorphic to M? (R. H. Bing, 03/27/79)*

Comment: No. (L. G. Oversteegen and E. D. Tymchatyn, 09/30/81)

Problem 134. *Is there an atriodic tree-like continuum M with positive span which has the property that there exists an uncountable collection G of mutually exclusive continua in the plane such that each member of G is homeomorphic to M? (W. T. Ingram, 03/27/79)*

Problem 135. *Suppose M is an hereditarily indecomposable simple triod-like continuum such that every proper subcontinuum of M is a pseudo-arc. Can M be embedded in the plane? (C. E. Burgess, 03/27/79)*

Problem 136. *Given a set X of n points on the plane, a line is* **ordinary** *if it contains exactly two points of X. A point is* **ordinary** *if it is on two ordinary lines. Is it true that, if not all points of X are on one line, then X contains an ordinary point? (S. Fajtlowicz, 03/28/79)*

Comment: No, for $n = 6$. (David Jones, 04/16/79)

Comment: What is the answer if n is odd? (S. Fajtlowicz, 04/16/79)

Problem 137. *Is there a monotonely refinable map from a regular curve of finite order onto a topologically different regular curve of finite order? (E. E. Grace, 04/11/79)*

Definition. A map r from a compact metric space X onto a compact metric space Y is **(monotonely) refinable** if, for each positive number ϵ, there is a (monotone) ϵ-map f from X onto Y such that, for each x in X, $d(f(x), r(x)) < \epsilon$.

Problem 138. *Does there exist, for every $k \leq m \leq \omega$, a space X such that:*
   (1) *X^n is paracompact if and only if $n < m$, and*
   (2) *X^n is Lindelöf if and only if $n < k$?*
(T. C. Przymusinski, 04/11/79)

Problem 139. *Does there exist a Lindelöf space X and a complete separable metric space M such that the product space $X \times M$ is not Lindelöf or (equivalently) normal? (T. C. Przymusinski, 4/11/79)*

Problem 140. *Is a para-Lindelöf space paracompact? (W. G. Fleissner and G. M. Reed, 04/11/79)*

Problem 141. *Is a collectionwise normal space with a σ-locally countable base metrizable? (W. G. Fleissner and G. M. Reed, 04/11/79)*

Problem 142. *Is a sL-cwH, metacompact space para-Lindelöf? (W. G. Fleissner and G. M. Reed, 04/11/79)*

Definition. A space X is sL-cwH (strongly Lindelöf collectionwise Hausdorff) if every discrete collection of closed Lindelöf sets can be separated by a disjoint family of open sets.

Problem 143. *Does a para-Lindelöf space with a base of countable order have a σ-locally countable base? (W. G. Fleissner and G. M. Reed, 04/11/79)*

Problem 144. *Is there an honest (i.e. provable from ZFC) example of a space of cardinality ω_1 with a σ-locally countable base which is not perfect? not metacompact? (W. G. Fleissner and G. M. Reed, 04/11/79)*

Definition. A space is said to be **perfect** if every closed set is a G_δ.

Problem 145. *If X has a σ-disjoint base B and a σ-locally countable base B', must X have a base B'' which is simultaneously σ-disjoint and σ-locally countable? (W. G. Fleissner and G. M. Reed, 04/11/79)*

Problem 146. *A finite collection of congruences $m \equiv a_i \pmod{n_i}$, $n_1 < n_2 < \cdots < n_k$, is called a **covering** if every integer satisfies at least one of the congruences. Let c be an arbitrary constant. Is there always a covering satisfying $n_1 > c$? (P. Erdős, 04/18/79)*

Comment: $1,000.00 for proof or disproof. Choi has, in *Math. of Computation*, a system with $n_1 = 20$. (P. Erdős, 04/18/79)

Problem 147. *Let there be given n points in the plane, no $n - k$ on a line. Join every two of the points. Prove that they determine at least $a \cdot k \cdot n$ distinct lines where a is an absolute constant independent of k and n. (P. Erdős, 04/18/79)*

Comment: $100.00 for proof or disproof. (P. Erdős, 04/18/79)

Comment: Proven. (J. Beck, *Combinatorica* **3** (1983), 281–293)

Problem 148. *Let $G(n)$ be a graph of n vertices. A Theorem of Goodman-Posa and myself states that the edges of our graph $G(n)$ can be covered by at most $n^2/4$ cliques, (i.e. maximal complete subgraphs). In fact the cliques can be assumed to be edge-disjoint and it suffices to use edges and triangles. Assume now that every edge of $G(n)$ is contained in a triangle. It seems that very much fewer than $n^2/4$ cliques will suffice to cover all edges. Determine the value of the least $g(n)$ so that the edges of such a graph can be covered by $g(n)$ cliques. (If too hard then try to determine $\lim g(n)/n^2$.) (P. Erdős, 04/18/79)*

Comment: The number of covering cliques is essentially the same; consider a complete tripartite graph in which two parts are equal and the third one consists of one element. (S. Fajtlowicz, 05/01/79)

Problem 149. *Is it true that the subset E of the complex plane is a type-A [type-B] convergence set if and only if $V_A(E)$ [$V_B(E)$] is bounded? (F. A. Roach, 04/18/79)*

Definition. A continued fraction is called a **type-A [type-B]** continued fraction provided every partial denominator [partial numerator] is unity. If E is a subset of the complex plane, $V_A(E)$ [$V_B(E)$] denotes the set of all values of finite type-A [type-B] continued fractions with elements from E. The set E is called a **type-A [type-B] convergence set** provided that every type-A [type-B] continued fraction with elements from E converges.

Comment: If 'complex plane' is replaced by 'real line', the resulting statements are true.

Problem 150. *Let $\mathcal{F}$ be a class of mappings such that all homeomorphism are in $\mathcal{F}$ and the composition of any two functions in $\mathcal{F}$ is also in $\mathcal{F}$. If X is homogeneous with respect to $\mathcal{F}$, is there a continuum which is $\mathcal{F}$-equivalent to X and which is homogeneous? (H. Cook, 04/25/79)*

Definition. A continuum X is **homogeneous** with respect to $\mathcal{F}$ if, for each two points a and b of X, there is a mapping f in $\mathcal{F}$ of X onto X such that $f(a) = b$.

Definition. If X and Y are continua, we say that Y is $\mathcal{F}$-**equivalent** to X provided there is a mapping in $\mathcal{F}$ from X onto Y and a mapping in $\mathcal{F}$ from Y onto X.

Comment: The answer is no for the class $\mathcal{F}$ consisting of all homeomorphisms and all mappings whose range is not homogeneous. (David Jones, 05/09/79)

Comment: What is the answer if $\mathcal{F}$ is the class of all mappings? All monotone mappings? All open mappings? All finite to one mappings? All confluent mappings? (H. Cook, 5/11/79)

Comment: No, if $\mathcal{F}$ is the class of all confluent mappings. (H. Kato, 08/09/84)

Comment: No, if $\mathcal{F}$ is the class of all mappings; take X to be a harmonic fan as in the paper of P. Krupski in *Houston J. Math.* **5** (1979), 345–356. (J. Prajs, 03/01/89)

Problem 151. *Suppose, in the previous problem, X is a homogeneous continuum, Y is a continuum which is $\mathcal{F}$-equivalent to X, and ϵ is a positive number. Does there exist a positive number δ such that if a and b are two points of Y such that the distance from a to b is less than δ, then there is a mapping f in $\mathcal{F}$ from Y onto Y such that $f(a) = b$ and no point of Y is moved a distance more than ϵ? (H. Cook, 04/25/79)*

Comment: If the answer to problem 150 is 'yes', then the answer to this problem is 'no'. (H. Cook, 04/25/79)

Problem 152. *Let M_n be the lattice consisting of 0, 1, and n mutually complementary elements and P_m—the lattice of all partitions of an m-element set. Let f be a function (1-1 function) of M_n into P_m. What is the probability that f is a homomorphism (an automorphism)? (S. Fajtlowicz, 05/20/79)*

Problem 153. *Suppose M is a continuum and f is a mapping of M onto itself such that, for each point x, M is irreducible from x to f(x). Is there an essential mapping of M onto a circle? (H. Cook, 06/06/79)*

Problem 154. *Is the space of automorphisms of the pseudo-arc connected? (J. Krasinkiewicz, 11/14/79)*

Problem 155. *Are planar dendroids (arcwise connected continua) weakly chainable? (J. Krasinkiewicz, 11/14/79)*

Problem 156. *Let X be a nondegenerate homogeneous hereditarily decomposable continuum. Is it true that X is homeomorphic to a circle? (J. Krasinkiewicz, 11/14/79)*

Problem 157. *Does there exist a finite (countable) to one mapping from a hereditarily indecomposable continuum onto a hereditarily decomposable continuum? (J. Krasinkiewicz, 11/14/79)*

Problem 158. *Let X be a nondegenerate continuum such that there exists a continuous decomposition of the plane into elements homeomorphic to X. Must X be the pseudo-arc? (J. Krasinkiewicz, 11/14/79)*

Problem 159. *Does there exist a dendroid D such that the set E of end points is closed and each point of $D \backslash E$ is a ramification point? (J. Krasinkiewicz, 11/14/79)*

Comment: Yes. (J. Nikiel, 01/01/84)

Problem 160. *Let X be a continuum such that the cone over X embeds in Euclidean 3-space. Does X embed in the 2-sphere? (J. Krasinkiewicz, 11/14/79)*

Problem 161. *If the set function T is continuous for the Hausdorff continuum S, is true that S is T-additive? (D. Bellamy, 02/18/80)*

Comment: A bushel of *Extra Fancy Stayman Winesap* apples for a solution. (D. Bellamy, 02/21/80)

Definition. Let S be a compact Hausdorff space. The set function T is defined by: For every subset A of S,

$$T(A) = S - \{ p \in S : p \text{ has a continuum neighborhood } W \text{ which misses } S \}.$$

Definition. S is T-additive if and only if, for all pairs of closed sets A and B in S, $T(A + B) = T(A) + T(B)$.

Problem 162. *If T is continuous for the (Hausdorff) continuum S, is it true that the collection* $\{T(p) \mid p \in S\}$ *is a continuous decomposition of S such that the quotient space is locally connected? (D. Bellamy, 02/18/80)*

Definition. T is **continuous for** S means that T is a continuous function from the hyperspace of closed subsets of S to itself.

Problem 163. *If T is continuous for S and there is a point p in S such that T(p) has nonempty interior, is S indecomposable? (D. Bellamy, 02/18/80)*

Problem 164. *If X and Y are indecomposable continua, is T idempotent on X × Y? Even for only the closed sets in X × Y? (D. Bellamy, 02/18/80)*

Definition. T is **idempotent** on a space S if $T(T(A)) = T(A)$ for all subsets A of S.

Problem 165. *If X and Y are indecomposable continua, $S = X \times Y$, and W is a subcontinuum of S with nonempty interior, is $T(W) = S$? (F. B. Jones, 02/18/80)*

Problem 166. *If X is a compact metric continuum, p is in X, $Y = X - \{p\}$, and $\beta(Y) - Y$ is an indecomposable continuum, must X be locally connected (connected im kleinen) at p? Is it true that $X = M \cup K$ where M is compact, p is not in M, and K is irreducible from some point q to p? (D. Bellamy, 02/18/80)*

Comment: $\beta(X)$ is the Stone-Čech compactification of X.

Problem 167. *Suppose M is a noncompact n-manifold for some $n > 1$ and M has no two-point compactification. Is $\beta(M) - M$ necessarily an aposyndetic continuum? (D. Bellamy, 02/18/80)*

Comment: This is true if M is Euclidean n-space.

Problem 168. *Does there exist a thin space with infinitely many points? (P. H. Doyle, 02/18/80)*

Definition. A topological space S is **thin** if and only if for each two homeomorphic subset A and B of S, there is a homeomorphism h of S onto itself such that $h(A) = B$.

Comment: The only known examples of thin spaces are finite spaces which are the product of a discrete space with an indiscrete space (i.e. a space S whose only open sets are S and the empty set).

Problem 169. *Is every aposyndetic, homogeneous, one-dimensional continuum locally connected? (J. T. Rogers, 07/22/80)*

Comment: No. (J. T. Rogers, 04/26/82)

Problem 170. *Do there exist even integers i and j such that, for every odd integer k and sequence $A_1, A_2, A_3, \cdots$ such that $A_1 = k$ and, for each n, $A_{n+1} - A_n$ is either i or j, some term of that sequence is prime? (J. T. Lloyd, 11/14/80)*

Comment: If the prime k-tuple conjecture of Hardy and Littlewood is true, then A_n is a prime infinitely often for any i and j. In case $i = 2$ and $j = 4$, infinitely many prime twins suffice. (P. Erdős, 11/24/81)

Problem 171. *Is every weakly chainable atriodic tree-like continuum chainable? (Lee Mohler, 04/16/81)*

Problem 172. *Does there exist a continuum M such that no monotone continuous image of M contains a chainable continuum? (H. Cook, 05/13/81)*

Problem 173. *Do there exist, in the plane, two simple closed curves J and C such that C in in the bounded complementary domain of J and the span of C is greater than the span of J? (H. Cook, 05/15/81)*

Problem 174. *In a topological space X, each set A has its derived set,*

$$A^d = \left\{\, x \in X \mid x \in \overline{A - \{x\}} \,\right\}.$$

Does there exist a T_0-space X and a subset A of X such that the second derived set $(A^d)^d$ is not closed in X? (A. Lelek, 05/15/81)

Problem 175. *In how many ways can one put k dominoes on an $n \times n$ chessboard so that each covers exactly two squares? (S. Fajtlowicz, 05/15/81)*

Comment: The answer is unknown for $k < 3$. (S. Fajtlowicz, 03/12/86)

Problem 176. *Suppose $X_1, X_2, X_3, \cdots$ is a sequence of positive numbers which converges to zero. Is there a set of positive measure which does not contain a set similar to $\{X_1, X_2, X_3, \cdots\}$? (P. Erdős, 11/24/81)*

Problem 177. *The equation $n! = m!\, k!$ $(m > k)$ is solved for $n = 10$, $m = 7$, $k = 6$, and also (trivially) if $n = k!$, $m = k! - 1$. Are there any other solutions? (D. Levine, 07/13/82)*

Problem 178. *Does there exist, for each t in $(\frac{1}{2}, 1)$, a simple triod $X(t)$ on the plane such that, with the natural metric, we have*

$$\frac{\sigma^*\big(X(t)\big)}{\sigma\big(X(t)\big)} = t?$$

(A. Lelek, 09/27/82)

Comment: $\sigma^*\left(X\left(t\right)\right)$ and $\sigma\left(X\left(t\right)\right)$ denote, respectively, the surjective span and the span of $X(t)$.

Problem 179. *Is there a tree-like continuum X on the plane such that X is weakly chainable but a subcontinuum of X is not? (P. Minc, 01/24/83)*

Problem 180. *Suppose X is an arcwise connected continuum with a free arc and such that, for each point p of X, there exists a homeomorphism $h: X \to X$ with $h(x) = x$ if and only if $x = p$. Is X a simple closed curve? (H. Cook, 10/31/83)*

Comment: No. (H. Gladdines, 06/15/94)

Problem 181. *Suppose X is an arcwise connected continuum with a free arc and with the fixed set property for monotone onto maps. Is X a simple closed curve? (Y. Ohsuda, 10/31/83)*

Definition. A topological space X is said to have the **fixed set property** for a certain class C of maps of X onto itself provided there exists, for each non-empty closed set A in X, a map f in C such that $f(x) = x$ if and only if x is in A.

Problem 182. *Is it true that if T and T' are trees with T' contained in T, then the surjective span of T is greater than or equal to one-half the width of T'? What about continua T and T' with T' contained in T? (A. Lelek, 10/12/84)*

Problem 183. *Let R be a space having a topological property (P). Is there an R-monolithic (locally R-monolithic) space with property (P), where (P) is one of the following properties: countable, second countable, Moore? (S. Iliadis, 02/08/85)*

Definition. Let R be a topological space. A space X is called R-**monolithic** if every mapping from X into R is constant.

Problem 184. *Does an open mapping always preserve span zero of continua? (E. Duda, 02/22/85)*

Comment: Yes. (K. Kawamura, 03/16/87)

Problem 185. *Is the product of two unicoherent continua always unicoherent? (E. Duda, 02/22/85)*

Comment: No. (A. García-Máynez and A. Illanes, 06/15/89)

Problem 186. *Characterize mappings $f: X \twoheadrightarrow Y$ such that if H is a proper subcontinuum of Y, then there is a proper subcontinuum K of X such that $f(K)$ contains H. (E. Duda, 02/22/85)*

Problem 187. *Does the class of approximable mappings coincide with that of the confluent ones? (W. Debski, 03/22/85)*

Definition. Suppose P is an inverse limit of an inverse limit sequence, where each factor space is $[0, 1]$ and each bonding map is open. A mapping $f: P \to [0, 1]$ is called **approximable** provided, for each i, there is an open map $f_i: I_i \to [0, 1]$ such that the composite $f_i p_i$ (where p_i denotes the projection of P onto I_i) converges uniformly to f. A mapping $f: P \to Q$ between two such inverse limits P and Q is called **approximable** provided, for each j, the composite $q_j f$ is approximable (where q_j denotes the projection of Q onto I_j).

Problem 188. *Is each continuum of span zero continuously ray extendable?* *(W. T. Ingram, 10/14/85)*

Definition. A continuum M is **continuously ray extendable** provided, for each mapping $f \colon X \to M$ of a continuum X onto M and for each ray L such that $L \cup M$ is a compactification of L with remainder M, there exists a ray (closed half-line) R, with $R \cup X$ a compactification of R and X its remainder, such that f can be extended continuously to a mapping of $R \cup X$ onto $L \cup M$. (C. Wayne Proctor, 10/14/85)

Problem 189. *Does every strongly infinite-dimensional absolute G_δ separable metric space contain a G_δ subspace which is totally disconnected and hereditarily strongly infinite-dimensional?* *(A. Lelek, 11/04/85)*

Definition. A space is **hereditarily strongly infinite-dimensional** provided every non-empty subset is 0-dimensional or strongly infinite-dimensional.

Problem 190. *Is it true that if A is a subset of the Euclidean n-space, $\mathbb{R}^n$, and $\dim A \leq n - 2$, then every two points in $\mathbb{R}^n \setminus A$ can be joined by a 1-dimensional continuum contained in $\mathbb{R}^n \setminus A$?* *(J. Krasinkiewicz, 05/16/86)*

Problem 191. *Is it true that if A is a subset of $\mathbb{R}^n$, then there exists a subset B of $\mathbb{R}^n$ which contains A and is such that $\dim B = \dim A$ and $\dim(\mathbb{R}^n \setminus B) = n - \dim B - 1$?* *(J. Krasinkiewicz, 05/16/86)*

Comment: An affirmative solution of problem 191 would imply an affirmative solution of problem 190. (J. Krasinkiewicz, 05/16/86)

Problem 192. *Does there exist a hereditarily infinite-dimensional metric continuum with trivial shape?* *(J. Krasinkiewicz, 06/12/86)*

Problem 193. *Is it true that if X is a homogeneous finite-dimensional metric non-degenerate continuum with trivial shape, then X is 1-dimensional?* *(J. Krasinkiewicz, 06/12/86)*

Problem 194. *Let X be a homogeneous curve such that the rank, r, of the first Čech cohomology group of X with integer coefficients is finite. Is $r \leq 1$?* *(J. Krasinkiewicz, 06/12/86)*

Problem 195. *Let $X_1, X_2, \cdots$ be an inverse sequence of polyhedra with bonding maps $P_n \colon X_{n+1} \to X_n$ such that the inverse limit is a hereditarily indecomposable continuum. Let F_n be a continuous mapping of the 2-sphere into X_n such that F_n is homotopic to the composite $P_n[F_{n+1}]$ for $n = 1, 2, \ldots$. Is F_1 homotopic to a constant mapping?* *(J. Krasinkiewicz, 06/12/86)*

Problem 196. *Does there exist a piece-wise linear mapping f of a tree onto itself such that $\sigma(f^2) > 0$ and $\lim_{n \to \infty} \sigma(f^n) = 0$? (W. T. Ingram, 11/10/86)*

Notation. $\sigma(f^n)$ denotes the span of f^n

Comment: Yes. In fact, for each positive integer n there is a piece-wise linear mapping of a triod onto itself so that $\sigma(f^i) > 0$ for each $i \leq n$ and $\sigma(f^i) = 0$ for $i > n$. (S. W. Young, 11/02/89)

Problem 197. *Does there exist a separable metric space X whose dimension is greater than 1 and such that X satisfies condition (R_1)? (L. G. Oversteegen, 11/13/87)*

Definition. A topological space X satisfies **condition (R_1)** provided there exists a basis $\mathcal{B}$ of closed sets in X such that if U is in $\mathcal{B}$ and x is any point of $X \setminus U$, then X is not connected between $\{x\}$ and U.

Definition. A collection $\mathcal{B}$ of closed subsets of a topological space X is a **basis of closed sets** in X provided there exists, for each point p of X and each neighborhood V of p in X, a set U in $\mathcal{B}$ such that U is contained in V and p belongs to the interior of U in X.

Comment: No. (L. G. Oversteegen and E. D. Tymchatyn, 08/18/92)

Problem 198. *Is the property of being weakly chainable a Whitney property? (H. Kato, 11/14/88)*

Definition. A topological property P is called a **Whitney property** if whenever a continuum X a continuum X has property P, so does every point-inverse under a Whitney map from $C(X)$ to $\mathbb{R}$, where $C(X)$ is the hyperspace of subcontinua of X.

Problem 199. *If X is a tree-like continuum (or a dendroid), does $C(X)$ have the fixed point property? (H. Kato, 11/14/88)*

Comment: In the case of arc-like continua, this has been proven by J. Segal. (H. Kato, 11/14/88)

Problem 200. *Is it true that there do not exist expansive homeomorphisms on any Peano curve (that is, a locally connected 1-dimensional continuum, in particular, the Menger universal curve)? (H. Kato, 11/21/88)*

Definition. If X is a compact metric space, a homeomorphism f of X onto itself is called **expansive** provided there exists a number $c > 0$ such that, for any two different points x and y of X, there is an integer n (possibly negative) with $d(f^n(x), f^n(y)) > c$.

Comment: Yes. (K. Kawamura, H. M. Tuncali and E. D. Tymchatyn, 05/15/93)

Problem 201. *If C is a simple closed curve, is the span of C equal to the semi-span of C? (A. Lelek, 01/16/89)*

Comment: An affirmative answer to part (4) of problem 83 would establish an analogous result for simple triods. (A. Lelek, 01/16/89)

Problem 202. *Is each separable metric, hereditarily locally connected space of dimension less than or equal to 1? (E. D. Tymchatyn, 02/27/89)*

Problem 203. *Can the space of homeomorphisms of the Menger universal curve be embedded as the set of end-points of an $\mathbb{R}$-tree? (E. D. Tymchatyn, 02/27/89)*

Definition. An $\mathbb{R}$-**tree** is a separable metric, uniquely arcwise connected, locally arcwise connected space.

Comment: Yes. (K. Kawamura, L. G. Oversteegen and E. D. Tymchatyn, 10/15/93)

Problem 204. *For separable metric spaces, does condition (R_1) imply condition (R_2)? (L. G. Oversteegen and E. D. Tymchatyn, 04/14/89)*

Definition. A topological space X satisfies **condition** (R_2) provided there exists a basis $\mathcal{B}$ of closed sets in X such that if U and W are disjoint sets from $\mathcal{B}$, then X is not connected between U and W.

Comment: If the answer to problem 204 is yes, then the answer to problem 197 is no. (E. D. Tymchatyn, 04/14/89)

Comment: Yes. (L. G. Oversteegen and E. D. Tymchatyn, 08/18/92)

Problem 205. *Suppose X is a separable metric space and there exists a countable subset C of X such that $X \setminus C$ satisfies condition (R_2). Is $\dim X \leq 1$? (E. D. Tymchatyn, 04/14/89)*

Comment: An affirmative solution of problem 205 would imply an affirmative solution of problem 202. (E. D. Tymchatyn, 04/14/89)

Problem 206. *Are the sets of all end-points and of all branch-points (in the classical sense) of each dendroid Borel? (Analytic?) (A. Lelek and J. Nikiel, 04/21/89)*

Comment: The fact that the set of all end-points of each planar dendroid is $G_{\delta\sigma\delta}$ has been established in *Fund. Math.* **49** (1961), pp. 301–319. (A. Lelek, 04/21/89)

Comment: No, neither set need be Borel. However, the set of end-points is coanalytic and the set of branch-points is analytic. (J. Nikiel and E. D. Tymchatyn, 02/06/90)

Problem 207. *Can each n-dimensional, where $n > 1$, locally connected (metric) continuum be represented as the inverse limit of a sequence of n-dimensional polyhedra with monotone surjections as bonding maps? (J. Nikiel, 04/21/89)*

Comment: If $n = 1$, the answer is no. (J. Nikiel, 04/21/89)

Problem 208. *Suppose X is the inverse limit of a sequence of Hausdorff continua such that each of them is a continuous image of a Hausdorff arc and the bonding maps are monotone surjections. Is X also a continuous image of a Hausdorff arc? (J. Nikiel, 04/21/89)*

Comment: Yes. (J. Nikiel, H. M. Tuncali and E. D. Tymchatyn, 06/29/91)

Problem 209. *Let X be a Hausdorff space such that X is a continuous image of an orderable, compact Hausdorff space. Is X supercompact? (Regular supercompact?) (J. Nikiel, 04/21/89)*

Definition. A Hausdorff space X is **supercompact** if it admits a subbasis $\mathcal{S}$ (for open sets) such that each subcollection of $\mathcal{S}$ which is a covering of X contains a two-member subcollection which is a covering of X. The space X is **regular supercompact** provided, in addition, $\mathcal{S}$ has the property that if U is the union of a finite number of members of $\mathcal{S}$, then U is the interior of the closure of U in X.

Comment: It is known that if X is assumed, in addition, to be zero-dimensional or metrizable, then X is regular supercompact. A less general question was posed in 1977 by J. van Mill; he asked if each rim-finite continuum is supercompact. (J. Nikiel, 04/21/89)

Comment: Yes. (W. Bula, J. Nikiel, H. M. Tuncali and E. D. Tymchatyn, 11/15/90)

Problem 210. *Suppose X is a monotonically normal, compact space.*

1. *Is X a continuous image of an orderable, compact Hausdorff space?*
2. *Is X supercompact?*
3. *Is X regular supercompact?*
4. *What are the answers to the first three questions if, in addition, X is assumed to be zero-dimensional?*

(J. Nikiel, 04/21/89)

Comment: It is known that continuous images of orderable, compact Hausdorff spaces are monotonically normal. Moreover, P. Nyikos and S. Purisch proved in 1987 that monotonically normal, scattered, compact spaces are continuous images of well-ordered, compact Hausdorff spaces. Part (1) is related to the following problem raised in 1973 by S. Purisch: Is each monotonically normal, separable, zero-dimensional, compact space always orderable? (J. Nikiel, 04/21/89)

Problem 211. *Is each monotonically normal, compact space a continuous image of a monotonically normal, zero-dimensional, compact space? (J. Nikiel, 04/21/89)*

Problem 212. *Suppose X is a monotonically normal, separable, zero-dimensional, compact space. Is it true that if C is a collection of mutually disjoint closed subsets of X such that C is a null-family, then the collection of sets belonging to C which have at least 3 elements is countable? (J. Nikiel, 04/21/89)*

Definition. A collection $\mathcal{C}$ of subsets of a topological space X is a **null family** provided, for each open cover $\mathcal{U}$ of X, the collection of sets belonging to $\mathcal{C}$ which are not contained in any set from $\mathcal{U}$ is finite.

Problem 213. *Let X be a dendron. Does there exist a hereditarily indecomposable Hausdorff continuum Y such that X can be embedded in the hyperspace C(Y) of subcontinua of Y? (J. Nikiel, 04/21/89)*

Definition. A Hausdorff continuum X is called a **dendron** provided, for every two distinct points x and y of X, there exists a point z of X such that x and y belong to distinct components of $X \setminus \{z\}$.

Comment: If X is a metrizable dendron, then X can be embedded in $C(Y)$ for each hereditarily indecomposable metric continuum Y. (J. Nikiel, 04/21/89)

Problem 214. *Does there exist a universal totally regular continuum? (J. Nikiel, 04/21/89)*

Definition. A continuum X is called **totally regular** provided X is metrizable and there exists, for each countable subset C of X, an open basis $\mathcal{B}$ in X such that the boundary of each set from $\mathcal{B}$ is finite and contained in $X \setminus C$.

Comment: Yes. (J. Buskirk, 10/11/91)

Problem 215. *Does there exist a continuous function f defined on the closed unit interval with values in a metric space such that f is at most n-to-one, for some positive integer n, and f is not finitely linear? (A. Lelek, 05/15/89)*

Definition. A function $f\colon I \to Y$, where I is a closed interval of the real line, is called **finitely linear** provided there exists a positive integer m such that I can be decomposed, for each $\epsilon > 0$, into a finite number of closed subintervals $I_1, I_2, \cdots, I_k$ each of length less than ϵ and with the property that the set $f(I_i)$ meets at most m of the sets $f(I_1), f(I_2), \cdots, f(I_k)$ for $i = 1, 2, \cdots, k$.

Definition. The usefulness of finitely linear functions has been shown in *Fund. Math.* **55** (1964), pp. 199–214. (A. Lelek, 05/15/89)

Problem 216. *Does there exist a monotone retraction $r\colon D \to C$ of the disk D onto its boundary C (that is, $r^{-1}(y)$ connected and $r(y) = y$ for each point y in C, and r not necessarily continuous) such that all but a finite number of points of C are values of continuity of r? (A. Lelek, 05/15/89)*

Definition. A point y in Y is called a **value of continuity** of a function $f\colon X \to Y$ provided $\lim_{n\to\infty} y_n = y$ implies that $\limsup f^{-1}(y_n)$ is contained in $f^{-1}(y)$ for any sequence $y_1, y_2, \cdots$ of points of Y.

Problem 217. *Does there exist, for every integer $n > 2$, a finite-dimensional separable metric space X and its finite-dimensional metric compactification cX such that if dX is a metric compactification of X and cX follows dX, then $\dim dX \geq n + \dim X$? (A. Lelek, 05/15/89)*

Definition. If cX and dX are compactifications of a topological space X, we say that cX **follows** dX provided there exists a continuous mapping $f\colon cX \to dX$ which commutes with the embeddings c and d, that is, $f \circ c = d$.

Comment: The answer is yes for $n = 2$. (A. Lelek, 05/15/89)

Problem 218. *Let G be the 3-dimensional unit cube in the Euclidean 3-space $\mathbb{R}^3$, and let A be a countable union of planes contained in $\mathbb{R}^3$ (not necessarily parallel). Suppose f is a continuous mapping of G into a metric space Y such that if $f^{-1}(y)$ is non-degenerate, for a point y in Y, then $f^{-1}(y)$ is contained in A. Is $\dim f(G) \geq 3$? (A. Lelek, 05/14/89)*

Comment: An affirmative answer to problem 218 implies an affirmative answer to problem 217 for $n = 3$. The relationship between these problems and some other similar compactification problems was described in the proceedings of a symposium: *Contributions to Extension Theory of Topological Structures*, Berlin 1969, pp. 147–148. The importance of compactifications is not undermined by the fact that they have been only sporadically discussed at the topology seminar of whose 18 years this book is record, more or less. (A. Lelek, 05/15/89)

E-mail address: ingram@ umr.edu "Tom Ingram"

E-mail address: asl@ math.uh.edu "Andrew Lelek"

INDEX

Printed and bound by CPI Group (UK) Ltd, Croydon, CR0 4YY

22/10/2024

01777637-0016